CW00500212

Engineering the Revolution

About the Author

Ken Alder is professor of history and the Milton H. Wilson Professor in the Humanities at Northwestern University. He is also the author of the books *The Measure of All Things: The Seven-Year Odyssey and Hidden Error That Transformed the World* (Free Press, 2002), which has been translated into twelve languages, and *The Lie Detectors: The History of an American Obsession* (Free Press, 2007; Bison Books, 2009).

Engineering the Revolution

ARMS AND ENLIGHTENMENT IN FRANCE, 1763 – 1815

Ken Alder

THE UNIVERSITY OF CHICAGO PRESS

CHICAGO & LONDON

The University of Chicago Press, Chicago 60637
The University of Chicago Press, Ltd., London
© 1997 by Ken Alder
All rights reserved.
First published in 1997 by Princeton University Press,
Princeton, New Jersey
University of Chicago Press paperback edition 2010

Printed in the United States of America

19 18 17 16 15 14 13 12 11 10   1 2 3 4 5

ISBN: 978-0-226-01264-3 (paper)
ISBN: 0-226-01264-6 (paper)

Library of Congress Cataloging-in-Publication Data

Alder, Ken.
Engineering the Revolution : arms and Enlightenment in France, 1763–1815 / Ken Alder. — University of Chicago Press paperback ed.
p. cm.
Originally published: Princeton, N.J. : Princeton University Press, 1997.
Includes bibliographical references and index.
ISBN-13: 978-0-226-01264-3 (alk. paper)
ISBN-10: 0-226-01264-6 (alk. paper)
1. France—History—Revolution, 1789–1799—Influence. 2. France—Politics and government—1789–1815. 3. Artillery—Technological innovations—France—History—18th century. 4. Military engineers—Political activity—France—History—18th century. 5. Enlightenment—France—Influence. 6. Technology and civilization—Political aspects—France. 7. France—History, Military—1789–1815. I. Title.
DC151.A58 2010
944.04—dc22
2009043759

⊗ The paper used in this publication meets the minimum requirements of the American National Standard for Information Sciences—Permanence of Paper for Printed Library Materials, ANSI Z39.48-1992.

For Bronwyn

CONTENTS

ILLUSTRATIONS

MAPS

FIGURES

PREFACE

A screaming comes across the sky.
—Thomas Pyncheon, *Gravity's Rainbow*

ON THE RUN from the Revolutionary police, Condorcet's mind raced toward a better future. Hidden away in a Parisian garret, the last of the philosophes imagined humankind borne inexorably toward perfection. It was the year II of the French Revolution (1794 by the old calendar), and Condorcet looked beyond the violence around him—men and women throwing off centuries of servitude—to gauge the deep currents of progress. The trajectory of knowledge was ever upward. As communications expanded, the threat of reversal receded. In short order, superstition would be driven from every corner of the globe. With knowledge more widely shared, people of all sorts—women, blacks, the propertyless—were becoming capable of self-government. Soon they too, like the French and Americans, would demand the rights and duties of citizenship. Hand in hand with this vision of intellectual and political progress went the expectation of greater and more widely shared material comfort. Scientific knowledge was at last being applied to practical affairs. Everywhere, new machines multiplied wealth, new technologies improved communications, and new medicines prolonged life—all spurring further social, economic, and political equality. Condorcet held that every useful art served human betterment. Even the invention of gunpowder, which had precipitated "an unexpected revolution in the art of war," had actually made battles less murderous, warriors less ferocious, and military expeditions more costly. The perfection of firearms meant that wealthy, well-organized nations no longer needed to fear the blind courage of barbarous peoples. At the same time, the gun had cut down the advantage of the mounted nobility over the common soldier, leveling this last obstacle to freedom and equality. Henceforth, the field of battle would be dominated by prosperous nations of free citizens, until such time as war, the greatest scourge of humanity, would vanish.[1]

Condorcet's utopian vision still compels admiration, both among those who have partially reaped his promise of relative equality and comfort (behind fortified frontiers), as well as among those who still wait for that promise to be realized (some inside the frontiers, and many millions more outside). But two centuries of bitter history oblige us to acknowledge what Condorcet ignored: the dark underside of knowledge-making, technological achievement, and national sovereignty. In this book I examine how, at the end of the eighteenth century, the nation-state was reformulated so as to expand greatly its capacity to wage war. More specifically, I lay out the Enlightenment program of the French military engineers and describe their labors to bring that program to fruition. These engineers sought to design gunpowder weapons of improved destructiveness, produce them with unprecedented precision, and deploy them to devastating effect. Their efforts oblige us to confront a rather different relationship between technology and politics: not simply because these guns were put to destructive *ends*, but because the *means* of creating these guns and the *meanings* invested in them depended on a distinct form of the "technological life." By the phrase, "technological life," I mean a coherent social and ideological world which gives purpose and meaning to a set of material objects. The engineers' technological life revolved around the management of large systems of workers, soldiers, and weapons; it presupposed new forms of technological knowledge and innovation; and it was energized by a radical ideology that justified social hierarchy by reference to national service. This book argues that this engineering technological life was as constitutive of the modern era—and of the French Revolution, in particular—as Condorcet's vision. After all, dissimilar as they are, there is much these two visions share: in general, a utopian aspiration to regenerate society, and in particular, a meritocratic and instrumentalist ethos hostile to the collectivist privileges which governed life in the ancien régime. Both imagined a new polity founded upon these principles; both are part of the *contested* legacy of the French Revolution. It is this contest which I propose to investigate in this book. Our modern technological life is a political creation, and we can read in its artifacts the history of the struggles and negotiations which gave it birth.

I realize that it may seem perverse to probe the great transformations of the French Enlightenment and Revolution through the medium of an artifact like the gun. My intention is not to suggest that gunpowder weapons—either field artillery or handheld muskets—were the principal technological achievement of this period. On the contrary, despite great efforts these smoothbore guns saw relatively minor alterations between the late seventeenth and mid-nineteenth century. Nor is it my intention to suggest that this period can be understood solely through its war-making activities—although the age was racked by continuous warfare, and war-making ranked as the main activity of the state. Rather, I have chosen the gun because I believe it can serve as a revealing site to examine the relationship between the material and political revolutions of the late eighteenth century.

Technology is the material embodiment of human labor and human thought. It is also the substantiation of the choices of particular people in particular times and places. For better or worse, we live in a world that is designed. The question is: how constrained are we by these technologies, and what new possibilities do they create? Over a decade ago, Langdon Winner posed the provocative question: "Do artifacts have politics?" The time has come to pose the next question: What *kind* of politics do artifacts have? How do they acquire these political qualities, and how are they manifested? On one side we hear that artifacts are neutral tools, liberating us to achieve our human potential. On the other, we hear that they set the terms for all social and political relations, radically circumscribing human freedom. Between these two positions, there is a vast ground upon which the dialectics of Enlightenment play. We are born into a world that is designed, yet we redesign our world as best we can.[2]

That weapons can acquire political significance will hardly surprise twentieth-century Americans. Guns, like all artifacts, participate in a material culture which imbues objects with meaning. But guns cannot be understood solely as cultural signifiers. As Mao and the other masters of police and armies have all too well understood, power flows out of the barrel of a gun. Guns *kill.* For this reason, the management of guns has always preoccupied those who hold (or desire) power, and they have expended considerable effort to keep those weapons pointing in the "right" direction. From the end of the Seven Years' War in 1763 to the end of the Napoleonic wars in 1815, France was the site of continual conflict over the proper design of guns, their proper mode of production, and their proper military deployment. As an artifact whose design was amenable (in principle, anyway) to scientific analysis, the gun was tied to transformations in the social organization of technological knowledge. As an artifact on which the state expended growing quantities of capital and labor, the gun was the site of attempts to alter radically the means of production. And as an artifact whose effectiveness depended on tactical arrangements (themselves caught up in strategies for offensive or defensive war), the gun was deployed in ways which implied a particular relationship between officers and soldiers, as well as a new definition of the nation-state. In this deeper, structural sense, the political history of the late eighteenth century was inscribed in its weaponry. To the extent that our own era can plausibly be understood by reference to the V2 and the ICBM, so can the era of the French Revolution be understood by reference to the gun.

Approaching the era of the French Revolution through its material culture is meant to offer a new perspective on the prevailing interpretations of that period. In particular, the attempt to write something like a total history of an artifact will mean confronting directly the current idealist trend in the historiography of the late eighteenth century. Two decades of revisionist history have deemphasized social causes in the explanation of the French Revolution in favor of a purely political narrative. In doing so, these historians have largely obscured the relationship between the political struggles of the eighteenth century and the material conditions under which most French men and women lived. These revisionists have much to teach us about the slippery uses of language in the Revolutionary setting. But rather than limit political activity to the realm of symbols and representations, as many revisionists have done, this book seeks to *expand* our understanding of politics to include contests over the terms of the material life. Artifacts are material vessels which funnel enormous social forces, but they are vessels shaped by human hands, and as such, are the negotiated outcome of diverse interests.

I hope this book will also perform a more general mediating service. One of its main ambitions is to show how the history of science and technology can help bridge the gap between our current cultural-linguistic histories and the older social-materialist histories. This does not mean showing that technology drives history; nor that social interests condition the development of new technology. Both approaches elide the difficult question of why particular technologies have assumed particular forms in particular times and places. Taking the history of technology seriously has implications, not just for "general" historians, but for historians of technology as well. It means applying to technology the same historicist approach with which historians have long plied their craft. Until we understand the development of science and technology from the point of view of its human creators—and take seriously the claims of their opponents—we will never come to grips with our own technological life. Nor will we understand how things might be different.

ACKNOWLEDGMENTS

LIKE MANY a state, both monarchical and democratic, I am awash in debt, and in vain do I imagine that a book that goes out under my name offers sufficient repayment. Yet it will have to do. Earlier drafts of the text have been read by Tessie Liu, Joel Mokyr, Ted Porter, Jessica Riskin, Merritt Roe Smith, Yves Cohen, and several anonymous reviewers. I would also like to thank the following people for valuable criticism and reactions to my work: Carolyn Cooper, Dario Gaggio, Robert Gordon, Sharon Helsel, Patrice Higonnet, Sarah Maza, Ayval Ramati, and Mary Terrall. Although these readers have inevitably sought to pull me in different directions, my book is certainly far better for their advice. I thank them all for the care and attention they bestowed. I have also benefited from the many comments elicited by this material when I presented it at academic conferences and seminars. Not least, I wish to thank my colleagues in the History Department at Northwestern University and in the Program for the History and Philosophy of Science, especially Arthur Fine, David Joravsky, David Hull, Tom Ryckman, and Stuart Strickland; they have provided me with an intellectual home. Finally, I would like to signal my gratitude to Everett Mendelsohn and Merritt Roe Smith for helping me first get this project off the ground.

In France and in America I received assistance from the staffs of many libraries and research facilities. In America: the libraries of Northwestern University, Harvard University, the Newberry Foundation, and the University of Chicago. In the Paris region: the Service Historique de l'Armée de Terre at Vincennes, the Bibliothèque Nationale, the Archives Nationales, the Archives de l'Académie des Sciences, the Archives de l'Ecole des Ponts et Chaussées, the Bibliothèque du Musée de l'Armée, and the Conservatoire National des Arts et Métiers. And elsewhere in France: the Bibliothèque Municipale de Saint-Etienne, the Archives Municipales de Saint-Etienne, the Archives Départementales de la Loire, the Archives Départementales du Rhône, and the Bibliothèque Municipale de Roanne. I am also extremely grateful for having received financial support from several sources. As a graduate student: a National Science Foundation Graduate Fellowship, a Mellon Graduate Fellowship in the Humanities, and a Whiting Fellowship in the Humanities. And as a faculty member: a fellowship from the Northwestern University Center for the Humanities, and a Northwestern University Summer Research Grant.

Personal debts, on the other hand, are incalculable. Friends and family may be far afield, but I have never stopped caring. To my mother and father—a teacher of French and a physicist, respectively—this history of French science cannot be anything less than the conjoining of their loves in me. My daughter, Madeleine, another such conjoining, contributed too, if only by delightful distraction. And as for my wife, Bronwyn Rae, she has brought her own wry tolerance to this effort, without letting it overly distract her from her far more dramatic task. Although I dedicate this book to her, I know full well that it is inadequate recompense.

Engineering the Revolution

A REVOLUTION OF ENGINEERS?

THE SUMMER after his arrival in Paris as American ambassador, Thomas Jefferson paid a visit to the Château de Vincennes on the eastern reaches of the city. The date was 8 July 1785, and over thirty-five years had passed since Jean-Jacques Rousseau traveled there to visit Denis Diderot—briefly imprisoned for his *Letters on the Blind*. There, in the courtyard of the Château, the two philosophes had debated the essay question of the Dijon Academy: "Has the progress of science and the arts corrupted or refined civilization?"[1] Thirty-five years later the American ambassador might have asked himself the same question. The year of Jefferson's visit the dungeon had ceased to function as a royal prison, and its last inmate, the marquis de Sade, had been transferred to the Bastille to make room for two very different occupants. By royal dispensation, a portion of the castle keep had been leased to a baker. The other occupant, whom Jefferson had come to visit, was an inventor and military gunsmith named Honoré Blanc. Under the aegis of the French artillery service, Blanc had established in the dungeon an experimental workshop. In the last decade of the ancien régime, the corps of artillery engineers, led by Jean-Baptiste Vaquette de Gribeauval, had intensified their efforts to reorganize the production of muskets in the kingdom's armories. Under their patronage, Blanc had developed a revolutionary method for manufacturing the musket's flintlock mechanism. Jefferson watched in astonishment as Blanc sorted into bins the pieces of fifty locks—tumblers, lock plates, frizzens, pans, cocks, sears, bridals, screws, and springs—and then selecting from the bins, assembled a number of working gunlocks.[2]

Jefferson wrote home lauding this new method of manufacturing gun parts "interchangeably." He hoped the United States Congress and the State of Virginia would employ the technique in their new armories. Exact replacement parts would simplify the repair of muskets, and the method might save money. Having "abridged the labor," Blanc expected to reduce the cost of the musket by two *livres*, or nearly 10 percent. Eventually, he planned to render every part of the firearm interchangeable—literally lock, stock, and barrel. The French government was planning to found a large manufacture along these principles; in two or three years the muskets would be produced in quantity.[3]

Interchangeable parts manufacturing is one of the key elements of modern mass production. Yet it better expresses an ideal than describes the methods of achieving that ideal. Interchangeability is a sign that the parts of an artifact have been made so precisely that they can be assembled without a final "fitting." To reach this goal, producers usually define tasks operationally so that they can be executed under a strict analytical division of labor, with the precise tooling and gauging of all parts. Integrating these various technological practices means taking a "systems" approach to production, in which both tools and human beings are consciously arranged into a purposeful whole. This generally removes know-how about production from the hands of "makers" and concentrates that knowledge in the office of "planner-managers" who have an overall view of the organization. This substitution of mechanical controls for immediate human judgment is necessarily an institutional achievement.

In its humble way, Blanc's methods fit this model. Blanc relied on steel dies to drop-forge pieces with precision, filing jigs and hollow milling machines to shape them accurately, and an elaborate set of gauges to verify that the pieces fit within the requisite tolerances. He had also refined the tempering process so that the lock pieces remained interchangeable even after case hardening. To preserve the accuracy of his dies and gauges over time, he had constructed a rigorous set of standard steel models that served as a fixed reference. Although many of his procedures still relied on hand tools and human muscle, execution (in principle, anyway) depended on mechanical guides. His method devalued certain artisanal skills in place of new ones. Each worker was now obliged to produce pieces to the requisite tolerance so that they would fit into the final assembly. In this sense, the production process itself acted as an intrinsic check on the good conduct and workmanship of the artisan. And the "fit" of the artifact was a measure of the rigor with which the social order was policed.

On the eve of his departure in 1789, Jefferson shipped a half dozen of the gunlocks back to New York.[4] However, the matter did not rest there. He extolled the virtues of interchangeability on his return to the United States, and eventually, the idea came to the attention of Eli Whitney. From this point onward, the history of the "American system of manufactures" is a well-known tale. Robert Woodbury has told of Whitney's checkered struggle to master the new technology. In the coming decades, the army's Ordnance Department sponsored the technique at its armories at Harpers Ferry and Springfield. As Merritt Roe Smith has emphasized, their success was due to the contributions of able mechanical inventors, but it was also the achievement of military engineers inspired by the rationalist mentality of the eighteenth-century French artillerists.[5] By the 1840s, interchangeable parts manufacturing had been adopted in the private arms industry of the Connecticut Valley, whence it came to the attention of British officials at the Crystal Palace Exhibition of 1851. In recognition of its apparent origin, the English dubbed it the "American system" and imported the machine tools wholesale for their new armory at Enfield. The transfer of this technology from the New World to the Old World has been seen (then and since) as a sign of America's technological "coming of age" and of the emergence of that nation's distinctive style of industrialization. As David Hounshell has shown, by the end of the nineteenth century, this American system had culminated in that quintessential modern vision of the technological life, the assembly-line production of Henry Ford.[6]

But this brief sketch of the filiation of a technological ideal is *not* meant to equate Honoré Blanc's machine shop in the Vincennes dungeon with Henry Ford's Highland Park plant. To achieve mass production, Ford combined a factory setting, a relentless division of labor, and cradle-to-grave mechanization. In his scheme, interchangeability of parts meant that low-skill wage laborers might assemble the final product. Coupled with the unimpeded flow-through of the assembly line, he thereby generated a prodigious output of Model T's at low unit costs. So that not only were Ford's *methods* of mass production different, so too was the sort of labor he used, the market he cultivated, and (above all) the ends he pursued. Under Fordism, interchangeability was driven by the hunger for a return on capital. This emblematic form of mass production did not materialize, however, until the early twentieth century. This book is written in the belief that the end result should not be mistaken for the cause. Any particular form of production must solve the problems of its own time. The use of terms such as "mass production" to describe eighteenth-century manufacturing would obscure many aspects of those early developments; hence, I will refer in this text to "interchangeable parts manufacturing" or "the uniformity system."[7]

After all, even this well-rehearsed history supposes a strange disjuncture. Why had the French state, which had initiated the program of interchangeable production at the end of the eighteenth century, repudiated it in the early nineteenth century? What had happened in the meantime to Honoré Blanc and the grandiose plans of 1785? How can we explain the origins of this technique in Enlightenment France and the demise which seems to have followed? Although many scholars have noted the importance of the topic to the history of industrialization, no one has yet made more than a cursory reference to it.[8] Yet this strange "technological amnesia" should pique our interest, not dampen it. This French trajectory of success and failure invites us to rethink the early history of "mass production." Such a reversal challenges our sense that there is a self-evident direction of technological development, a logic by which production must necessarily become ever more "rational." As Charles Sabel and Jonathan Zeitlin have pointed out, by assuming that production tended inevitably toward the Fordist ideal, economic historians (both Classical and Marxist) have obscured the history of alternative forms of industrialization. Jettisoning those teleological accounts has enabled Sabel and Zeitlin to uncover the flourishing nineteenth-century history of flexible specialization, the use of adaptable machinery by smaller firms who seek to respond quickly to market changes.[9] Less noticed has been the effect of these teleological accounts on histories of the origins of "mass production." As this book will show, the effort to make things identical was part of a larger Enlightenment project to replace the corporate order with a more innovative technological régime. This was not, however, a proto-Fordist scheme driven by capitalist entrepreneurs seeking profit. It was an engineering project driven by state bureaucrats following their own operational logic.

To be sure, this engineering revolution was a *failed* industrial revolution. Yet even failures can be consequential and illuminating. Consequential, because even failed techniques may come to find successful application; as we know, the engineering approach to technology, once it was subordinated to capitalist organization in the late nineteenth century, was to have a transformative effect on industrial practice. And illuminating, because understanding the (temporary) failure of the program of the eighteenth-century engineers will help us explain the pattern of industrialization as it did emerge in the nineteenth century. In particular, it will help us understand the role of the French Revolution in reconfiguring the relationship between society and the machine.

That Thomas Jefferson, implacable foe of factory tyranny, played Prometheus to this new technology, underscores a revealing irony. Recent scholarship has painted a nuanced portrait of Jefferson and his ambivalence about technology: a philosophe fascinated by machinery and repelled by industrialization. Jefferson's commitment to a republic of autonomous citizens existed uneasily alongside his lifelong pursuit of labor-saving devices. Jefferson did not share Condorcet's optimism. He worried that radical new forms of technological organization—the factory system, most prominently—were incompatible with a democratic polity. The historian John Kasson has shown how this tension between republicanism and technology was central to the unfolding identity of the new United States. An analogous tension was also central to the unfolding identity of the French Republic.[10]

The French Revolution, Industrialization, and the State

One of our most familiar historical commonplaces has it that in the late eighteenth century, the English mounted their Industrial Revolution, and the French their political and social Revolution—and from the conjunction of these two revolutions, the modern world took shape. In fact, historians have generally sidestepped the issue of exactly how these two great transformations linked up in their own time. Among the illustrious exceptions is E. P. Thompson, whose classic study explains why the far-reaching changes in the English economy between 1790 and 1830 did *not*—despite the living example of France—result in a political revolution, however much they recast English society.[11] Meanwhile, twenty years of scholarship on late-eighteenth-century France has largely obscured the connection between the French Revolution and changes in the material basis of society there.

For it was roughly twenty years ago that historians realized that they would have to reconcile France's profound social and political Revolution with something they had known all along: that that nation had somehow missed the boat on the Industrial Revolution then unfolding in Britain. The revisionist school of Revolutionary historiography sprang in large part from this effort. A Marxist school of interpretation had dominated French academic life for the previous half-century under Albert Mathiez, Georges Lefebvre, and Albert Soboul. These post-Jacobin historians had assumed that the Revolution had been made by an ascendant bourgeoisie. But when Anglo-American historians such as Alfred Cobban and George Taylor were unable to find evidence of significant change in socioeconomic structures—or of capitalist activity of any sort—in the late ancien régime, they began to wonder in whose name the Revolution had been proclaimed. For François Furet, the solution has been to return to Tocqueville's insight that the French Revolution was born primarily of a crisis of the state, albeit a crisis that served ultimately to consolidate the position of the bourgeoisie.[12]

Locating the rupture in the political realm has drawn the attention of many scholars to the study of rhetoric and myth-making, to the ways in which the new sovereignty was constituted "discursively." This revisionist movement has had many beneficial results. It has refocused scrutiny on the categories of citizenship and virtue which the historical actors considered central to their endeavor. It has reemphasized the contingency of their struggles for power. And it has freed Revolutionary historiography from the pat correspondences of class and party that long dominated analysis. The result has been a valuable reminder to historians not to mistake Revolutionary rhetoric for social reality, and (simultaneously) an invitation to take the representations of the Revolutionaries seriously on their own terms. Certainly, the Revolution did not spring unadorned from changes in the mode of production, as the materialist hypothesis once proclaimed. But that is not to say that the use and meaning of technology was not bitterly contested in the late eighteenth century, nor that material interests had no bearing on political action.[13]

To be sure, not all the revisionists have advocated a complete detachment of politics from social conditions. Alfred Cobban, who initiated the critique of the Marxist interpretation, long held out hope for a social analysis of the causes of the Revolution, which he located in the frustrations of the ascendant office-holding notability. His work has been amplified by Colin Lucas. And Lynn Hunt has always insisted that the political culture spawned by the Revolution had a basis in the contending groups of the late eighteenth century. Recently, several younger scholars have further broadened the purview of politics to show how certain groups—civil servants, professionals, academicians—appropriated this Revolutionary political culture to advance their social interests. They have thereby begun to take up that (neglected) aspect of Furet's Tocquevillean critique which focused on the regeneration of state structures during the French Revolution. Yet these analyses have all sidestepped the central objection to the Marxist interpretation identified by George Taylor; that is, how could the French Revolution have been fueled by the bourgeoisie if the ancien régime lacked anything like an industrial revolution? This book is meant to suggest that the problem with the materialist thesis was not its (plausible) assumption that the political world is entwined with socioeconomic relations in the realm of production, nor its (plausible) intuition that vast social consequences should have social causes of comparable magnitude. This book suggests that the problem with the materialist thesis was its narrow characterization of French industrialization, a conception shared by both the Marxists *and* early revisionists. Freed from a teleological model in which the factory and capitalist mass production represent the sole version of "modern" production, we can see how conflict over material culture set the terms for political conflict during the late eighteenth century. Recently, Colin Jones has begun to hold out hope for a revival of the social interpretation, now founded on the vitality of commercial life and the professional classes in the late eighteenth century.[14]

Clues for this interpretation already abound in the literature. In the past two decades economic historians have also touted their own revisionist line, shifting our assessment of the eighteenth-century French economy. The picture that emerges is complex. France possessed a smattering of large-scale manufactures, an extensive range of private ateliers run on the guild system (what in France is called the corporate system), as well as a variegated sector engaged in "domestic production" (ranging from artisans engaging in luxury production to outworkers dependent on merchants). The growing consensus is that French production was more innovative and dynamic than previously recognized. But while economic historians now suggest that the French economy was launched on the road to growth by the early nineteenth century, they have done so by reemphasizing the peculiarly "nonindustrial" character of French industrialization.[15]

A transforming economy that was not industrial; a vast social revolution undertaken to revamp the state; a class structure without a basis in economic relations: how can these elements be reconciled? A number of historians, most notably William H. Sewell, Jr., Michael Sonenscher, William Reddy, and Gail Bossenga, have begun to wrestle productively with this conundrum about the relationship of work and politics in the Revolutionary period. Despite differences among them, they have tended to locate the continuities and disjunctures in the *meanings* attached to production; that is, in the realm of culture. In *Work and Revolution*, William Sewell set out to explain the persistence of "anachronistic" language among the revolutionaries of 1848, and uncovered the mechanisms by which the corporate idiom of the ancien régime informed the mentality of working people long after the official abolition of the guilds in 1791. Michael Sonenscher situated the work practices of ancien régime artisans in their legal and cultural codes, thereby recovering a mentality that evoked natural rights and privilege as the badge of autonomy. William Reddy in *The Rise of Market Culture* emphasized how the theory of political economy served as the organizing metaphor through which elite analysts examined French industry, even as workers responded to the advent of capitalism in ways which defied market "logic." And Gail Bossenga has identified the persistence of a corporatist ideology among state officials well into the nineteenth century, and their collaboration with merchants interested in cornering a monopoly on trade.[16]

These writers have all concluded that no industrial revolution took place in France between 1750 and 1830, when England was experiencing her boom. Twenty years of scholarship have placed the artisan, not the factory worker, at the center of early French labor history. This book lies squarely within this general framework; indeed, it adds considerable evidence to the widely observed resilience of artisanal modes of production. It also takes to heart the methodological lesson of this scholarship: that the nineteenth-century vocabulary of political economy and class antagonism cannot be imposed on the eighteenth-century workplace without doing disservice to the shifting boundaries between merchant and artisan, manufacturer and outworker, master and journeyman. Certainly, social relations of production were stratified in the eighteenth century, but as Tessie Liu has recently reminded us, the historian cannot (teleologically) assume the triumph of a particular type of class relation any more than he or she can assume the triumph of a particular form of production.[17] This does not mean surrendering all hope of describing the patterns by which the productive life was transformed—nor of addressing the perennial question of *why* France did not industrialize as England did. But it does mean paying close attention to ways in which the ideology of the ancien régime—as refracted through its legal language—shaped the efforts of artisans and merchants to secure their livelihoods. It invites us to examine anew the ways in which the political aspirations of the Revolutionary period gave temporary rein to labor movements which otherwise defied long-standing relations in the workplace. And it reminds us again of the central role played by the French state as the guarantor of the productive order.

Indeed, students of French industrialization have always pointed to the distinct role of the state in the economy there. Under absolutism, the monarch legitimated all activities through the grant of privileges that were particularist, and hence, in principle, revocable. The charters that validated the trade corporations were entitlements of this sort, as were the tax exemptions of the aristocrat and the restrictions on who could hold public office. In return, the monarch assured the public order, including the supply of bread in times of dearth and the regulation of economic exchange. The central achievement of the Revolution was (supposedly) to have swept this particularist world aside and to have established a new state which guaranteed equality before a set of public laws, on the one hand, and an absolute right to property and the freedom to engage in production and commerce, on the other. Yet the above scholars have recently emphasized that the post-Revolutionary state continued to play a central role in regulating commerce. But understanding the exact terms of that statist accommodation with private capitalism—so central to the definition of modern France—will mean being far more precise about the nature of the state's involvement in the economy both before and after the Revolution, an involvement that deliberately eschewed direct intervention in the realm of production and turned instead on the regulation of exchange and controls over public standards.

The military engineers of the artillery corps served as the one of the principal mediators between the state and private capital throughout the eighteenth and nineteenth centuries. And they were among the main architects of this new relationship. These men had their own distinct vision of how the nation's technological life should be organized. Their engineering vision was not tied to the capitalist's desire to generate private wealth. Nor were they committed to Condorcet's vision of social equality and material comfort leading to political emancipation. Least of all were they sympathetic to the collectivist associations by which ancien régime artisans organized their affairs. Instead, these engineers paid homage to a "higher" form of state service, and sought the greater security of the nation through the waging of more efficient war. Rather than gauge their success in the emerging consumer marketplace or in the march of political liberation, they measured their progress by bureaucratic markers of merit. These engineers pioneered new forms of technological knowledge and founded new institutional structures to give them scope. In large measure it was men like these engineers who defined the state's involvement with the productive order at the end of the eighteenth century, and they were, of course, crucial to the state's ability to project its national interest abroad. In both these capacities, the military engineers were at the center of the French state's Enlightenment reforms, its Revolutionary upheaval, and its post-Revolutionary reconstruction.

Revolutionary Engineers?

The eighteenth-century artillery engineers operated at the nexus of coercion and capital that gave war its terrifying violence, and the centralized nation-state its predominance over other forms of social organization. As the emissaries of the state in the realm of private capital, they supervised the kingdom's armaments production, directing the artisans and merchants of these armories to deliver specified instruments of destruction (e.g., cannon, muskets, swords) at a specified quality, quantity, and price. As the emissaries of the state in the realm of military coercion, they commanded soldiers and cannon in battle so as to deliver the explosive power of this capital investment at a precise time and place. The degree to which these two roles could be fused was a matter of some contention, however. Throughout the ancien régime and Revolutionary period, the artillerists struggled to transpose the model of military coercion onto their role as the managers of capital, just as they tried to introduce the logic of efficiency into their military endeavors. The extent to which they succeeded in combining the two can serve as a measure of the evolving role of the state in the material life of the nation.

As a subspecies of *Homo faber*, engineers are engaged in one of the most quintessential human activities. Though only one of several groups involved in the design, production, and deployment of technology, engineers have come to play a dominant role in the pattern of Western industrialization. Indeed, since the late nineteenth century (the period known as the "Second Industrial Revolution"), their activities have come to account for much of what is superficially distinctive about Western society: its methods of machine production, its public infrastructure, its consumer cornucopia, and its ability to wage war. Yet very little scholarly attention has been paid to engineers.

In large measure this is because engineers themselves *insist* on their own lack of importance. After all, in the world of twentieth-century corporate capitalism it is *not* the engineers who hold power. At least, not engineers *qua* engineers. In America, at the turn of the century, astute social critics, such as Thorstein Veblen, argued that engineers *should* be in charge, that they alone had the knowledge and disinterestedness to manage society. For a time, there was even something like a cult of engineers in the United States. Engineers were the innovators who conceived the instruments of technological progress. They were the ones who provided the generals with the ability to wage war and the industrialists with abundant profits. Yet the engineers themselves were largely impervious to the hatreds of nationalist rivalry or the greed of capitalism. American engineers, however, never became the sort of revolutionary vanguard Veblen expected. Nor did they become the reformist leaders some of the Progressives once hoped. Nowadays, the phrase "revolutionary engineers" sounds distinctly incongruous. This, of course, poses the interesting question of *why* engineers have remained loyal to their corporate and military paymasters. David Noble and Edwin Layton have shown how the engineering profession was deliberately shaped by American corporate capitalism to become its hired technical help. Which is not, of course, to deny that engineering was also a route to social advancement for many individuals.[18]

This, of course, is an American phenomenon. In France, as anyone who knows that country will realize, engineers are *not* subservient. On the contrary, engineers there sit at the very pinnacle of the social hierarchy. For two centuries, the graduates of the Grandes Ecoles (the Ecole Polytechnique and its specialist Ecoles d'Application) have filled the highest posts in the state. Yet in France, too, the profession of engineering was a deliberate creation. French engineering education has been carefully nurtured by the state for over 250 years to provide it with a loyal cadre of technicians—with this important proviso, that in France, during the time of the French Revolution, the engineers came, in some sense, to constitute the state. Yet even in France, it was in the nature of their technocratic rule to remain invisible and to deny that their activities involved political judgment. It is precisely this cloak of invisibility which I wish to pull aside.

Rather than seeking "deep truths" about nature, engineers *design*. They practice the "science of the artificial." Where science is supposedly directed toward what *is*, engineering is directed toward what *ought* to be. This means that engineering is a purposeful, future-oriented activity, one which takes cognizance of present circumstances only insofar as they can be shaped to achieve desired results. And who decides what results are desired? The engineers? Their paymasters? Or those who actually perform the labor which brings those results into being? Such are the questions which make engineering a worthy topic of study. They remind us that engineering is a contentious art, that engineering is always social engineering, and that designing an artifact is in some sense a political act.[19]

This is not always immediately apparent. The cloak that makes engineering invisible is the claim that their activities are somehow outside the domain of analysis, either because technology is "applied science" or because technology is "derived society." Both of these views deny that the development of technology has its own history by portraying technologists as mere conduits for external forces. "Applied science" does this by implying that technology is a direct translation of the available stock of knowledge about the natural world, and hence, that the history of technology can be studied by tracking the trail of true scientific knowledge. "Derived society" does this by implying that technology is the direct translation of social interests, and hence that the history of technology can be studied by following the evolving demands of the relevant "marketplace" for technology. Both these claims are, of course, trivially true. Clearly, natural process and social interests set the outer limits for what kinds of technologies are possible. This does not, however, answer the important *historical* question of how particular technologies have come to acquire their particular qualities at particular times and places. Artifacts—as M. Norton Wise has argued—are meeting grounds for diverse and conflicting interests, as well as instruments which mediate between the social and epistemological worlds. Explaining their design, therefore, means coming to terms with the values of technologists, and their social and epistemological presuppositions.[20]

This is easier said than done. The view that technology is an unproblematic instrument, unworthy of closer analysis, is often referred to as the "black box" concept of technology. This phrase implies that an artifact serves a purely functional purpose: what goes in determines exactly what comes out. Black boxes are supposed to be foolproof; they therefore resist an outsider's attempt to decipher the decisions that lie behind the arrangement of their inner mechanisms. Engineers work very hard to transform artifacts into black boxes, but that very effort suggests that something more interesting is going on. Shaping the material world is a laborious process. Physical matter is lumpy and recalcitrant. Even apparently well-made artifacts often prove fragile in actual use. Connections come undone, parts wear out, things break. And artifacts must often operate in diverse and unpredictable circumstances. To guide them in this process of design, therefore, artifact-makers have for centuries accumulated sets of rules and recipes. In their most codified form, these rules have come to be known as engineering science.

One might imagine, therefore, that on its own terms, this engineering science comprises a secure repository of knowledge upon which to build artifacts, a set of technological "facts" beyond reproach. Unfortunately, the generalizability of any engineering rule is always in tension with the particularity of the artifact and the diversity of local circumstances. Unlike the abstractions of an idealized physics, engineers work with a material world that is "thick." Bulk matters. To change the size or shape of a physical object is to transform its performance in ways that are difficult to predict. One cannot simply build a model bridge, expand it uniformly one hundred times, and expect the span to hold across the Menai Straits. This is what engineers call the problem of scale. As we will see, problems of these sort are particularly acute for the flight of a cannonball. Seventeenth-century savants such as Galileo and Newton touted ballistics as one of the central triumphs of their new rational mechanics. Yet these natural philosophers did not fully appreciate that the flight of a cannonball through the atmosphere depended on the characteristics of the cannonball itself. It was the insight of the eighteenth-century engineers, by contrast, to consider the behavior of firearms and their projectiles as "thick things." In this text the term "thick things" alludes to the inability of mathematical theory to fully encompass the behavior of material objects, and consequentially to the possibility of multiple interpretations of those objects. This is what François Blondel, the foremost ballistic theorist of the late seventeenth century, called "the resistance and obstinacy of matter." Overcoming this resistance, Blondel noted, would be possible only through a new amalgam of theoretical and practical knowledge.

> In a word, one cannot say that the sole knowledge of principles is sufficient to bring the [mechanical] arts to perfection. One must apply them, and this application always reveals the resistance and obstinacy (*la résistance et opiniâtre*) of matter. A thousand obstacles arise, obstacles which one can discover and overcome only with the help of practice and experience.[21]

But such a novel approach to technical knowledge was also to have its own pitfalls, as we will see. Having sacrificed any claim to know the absolute truth about how these objects perform, engineers open themselves up to legitimate disputes about whether any particular design is in fact the best able to achieve its purported goal. That is, even when an artifact does not fail in absolute terms, it is always open to some kind of reproach. And when such reproaches are leveled by rival groups of engineers—as was the case in eighteenth-century gun design—the black box of technology is opened for the inspection of the historian.[22]

Working material is hard enough; working it into forms that serve human needs is a daunting challenge. It requires effort, patience, ingenuity, and organization. Human interests are diverse: whose demands should be satisfied, and how? Nor can engineers simply respond passively to these demands. They must take an active role in deciding which social interests to consider and how to meet their needs. After all, much of what makes an artifact successful is the extent to which its design already takes into account the diverse circumstances in which it will be used. Engineers try to shape both the content *and* the context of their innovations. Largely this means expanding the black box to include more and more of its "external" environment; that is, building a technological system.

Thomas P. Hughes, the foremost analyst of the systems metaphor, has argued that technological systems—such as electric power grids—develop as system-builders solve "critical problems" which *in their view* would otherwise impede the coherence and growth of their system. Critical problems may be solved with physical devices, such as transformers and repeaters to overcome the problems of transporting electricity long distances, or they may be solved with new institutions, such as financial arrangements to allow utilities to share power. In my reading of Hughes's metaphor, then, technological systems are something more than merely the historian's unit of analysis, yet they are not "natural" entities or autonomous agents with intentions of their own. Rather, they are the deliberate creations of particular human agents who seek to give their world a meaningful structure. As John Law has noted, these system-builders accomplish this feat by bringing potentially disruptive elements (social, technical, and natural) under their domain. As an example of the diverse elements that comprise such systems, he cites the compass, caravel vessel, Atlantic winds, and lure of spices which enabled the Portuguese explorers to round the Horn of Africa. The diversity of these elements has led Law to call the process of linking them together "heterogeneous engineering." Vast amounts of social and cultural work must go into building and maintaining such durable, yet malleable socio-technological systems.[23]

The important thing to keep in mind about heterogeneous engineering is that it is a purposeful activity. At a minimum, this means that technology cannot be understood without taking into account the interests of system-builders. Moreover, because even the most seemingly coherent system will necessarily encompass diverse elements, there will be tension among them. And this is where a systems approach needs to be supplemented by a consideration of contradictions of the social and ideological context within which system-builders operate.

As we will see, this form of systems thinking—replete with its contradictions—was characteristic of the Enlightenment engineers who operated under the aegis of the absolutist state. Trained in the first schools in Europe devoted to scientific instruction, Gribeauval and his cadre of artillerists deployed new forms of knowledge to design more potent cannon and muskets. Yet this achievement was not merely "technical." It also involved transforming the ways in which those guns would be used in the battle and the social role of the men who served them. As we will see, it also led them to take a radical new approach to the production of armaments. In this way, the Gribeauvalists hoped to forge men, weapons, and tactics into a purposeful "machine" that would prove the most potent military force in Europe. Indeed, their ambition was realized when their mantle was taken up by their successors among the Revolutionary engineers—Lazare Carnot, Claude-Antoine Prieur de la Côte-d'Or, and their fellow savants—when these men found themselves in charge of the state in the years 1793–95. And in the end, their methods and ideologies enabled one of their disciples, the artillerist Napoléon Bonaparte, to make himself Emperor of all the French, and much of the Continent besides.

After all, in the broadest sense, engineering is a revolutionary enterprise, perhaps the quintessential revolutionary activity. In principle, engineering operates on a simple, but radical assumption: that the present is nothing more than the raw material from which to construct a better future. In this process, no existing arrangement is to be considered sacrosanct, everything is to be examined in the light of present aspirations, and all practices refashioned according to the dictates of reason. The immediate purview of engineering is technological design; but as we have seen, engineers thereby seek to shape an entire techno-social world. In this sense, engineering denies history. It has no sympathy with the conflict, compromise, and happenstance that brought the world to its present state. It collapses *is* into *ought* to build (or promise to build) a world more consonant with human desires, as it understands them. To be sure, engineering is also a practical, everyday activity that seeks to satisfy immediate human needs. And to carry out its mission, engineering almost inevitably depends on those very powers which most wish to preserve the status quo. For all its utopian aspirations, then, engineering is enmeshed in its historical circumstances, and its grand ambitions are harnessed to particular (and petty) interests. As we will see in this book, it has been as often turned to destructive ends as those commonly thought of as constructive. None of which is meant to deny that engineering expresses a noble human ambition to transcend the dreary limits of the past.

From this perspective, we can see why the French Revolution brought engineering (and engineers) to the fore. The French Revolution can itself be understood as a vast engineering project. Famously described as a utopian enterprise, it was an attempt to sever a nation's ties to its past and remake society according to the dictates of nature and reason. In this process, corrupt institutions were to be swept aside, human relations refashioned, and the contract between the people and the sovereignty rewritten. In short, society was to be consciously *designed*. To be sure, the French Revolution was also a practical affair, filled with ordinary men and women protesting shortages of bread, bureaucrats carrying out the routines of administration, legislators quarreling over the fine print. Everywhere the historian looks, he or she sees self-interest, pettiness, personal foibles. And in the end, as Tocqueville so brilliantly understood, the Revolutionaries were obliged to make their new world out of the pieces of the old, charting a future that differed surprisingly little from the past. Yet even these mundane acts and thwarted ambitions were suffused with an animating spirit, savage in its destructive intensity, yet inspiring in its hope of transcending history.

Politics and Technology

Like other material objects, weapons were invested with new meanings during the course of this passionate Revolutionary effort. Mona Ozouf, Maurice Agulhon, and Lynn Hunt were the first to show how the cockade, the liberty tree, Marianne, and all the paraphernalia of the festival processions acquired new political significance during the struggles of the Revolutionary decade. The emphasis of these historians has been on the ways in which these icons, some of them adapted from folk culture, mapped out a novel political culture, one which resonated throughout the nineteenth century.[24] As we will see, weaponry also acquired this sort of significance in the Revolutionary period. No artifact stands outside the hermeneutic circle—neither in times of revolution, nor at any other historical moment. Societies continually generate meanings about the artifacts they use.

More specifically, historians have recently rediscovered the vitality of eighteenth-century material culture and its cornucopia of consumer goods. Daniel Roche has turned to clothing, Cissie Fairchilds to "populuxe" goods, and Colin Jones to medical nostrums, each in an attempt to underscore the cultural dimension of commercialization. In the growing diversity of private possessions and in the cultivation of new tastes, these historians have found evidence for a stratified set of urban identities. The relationship between changing fashions and the social order is read in these accounts as a kind of political script.[25] Pistols, muskets, and other weapons were of course consumer products of this sort, and they too were subject to changes in taste and fashion, and marked the social status of their owners. This book, however, will primarily focus on the meaning of material artifacts for the operation of the state, rather than on this emerging private sphere of consumption. Doing so will make the implicit question about the relationship between technology and politics all the more acute.

After all, the example of guns also raises some difficult questions about the relationship between material culture and deep patterns of economic, social, and epistemological change. That is because guns—unlike more purely iconographic objects—also must prove themselves in practice. A cockade may not be revolutionary unless it is red, and breeches may be reactionary if they are worn in 1794, but the color red is not inherently revolutionary, and breeches are not inherently reactionary, apart from what human memory assigns them. Whatever symbolic significance the Revolutionaries attached to the gun, they had also, presumably, to take cognizance of the performance of guns in battle.

I say "presumably" because even here we must not underestimate the importance of human consciousness in shaping material outcomes. After all, in war, morale matters. The meanings attached to weaponry will not only affect the ways in which generals plan battles, but the sort of skills with which artillerists aim their cannon, and the enthusiasm with which common soldiers take up their muskets. Still, there is undoubtedly some sense in which guns (in tandem with a vast military organization) must prove themselves on the field of battle. What then do I mean when I say that guns have political qualities?

I realize that to invoke the category of politics to describe the inanimate world of artifacts may seem troubling to some. Nothing in this book should be taken to imply that there is no correspondence between the design of machinery and its performance, as if technology were *merely* politics. This book is profoundly committed to the proposition that technology *works*, that cannonballs hit their targets, and that bullets kill and maim soldiers. Just beyond the pages of this book lie the corpses of over a million young men killed in the Revolutionary and Napoleonic wars. To deny the potency of modern weaponry would be to mock their deaths. Yet the science of ballistics was (and is) inherently imprecise and must serve many different purposes; hence, even technological knowledge will often be a matter of dispute. For while it is true that one cannon design may well be operationally superior to another, at least according to some set of well-defined criteria, no single design will satisfy all criteria equally. Moreover, the criteria themselves are subject to dispute. The important historical question, then, is how and why certain criteria come to be seen as paramount.

Nor am I arguing that the social values embedded in technologies persist regardless of the changing context. The artillery cannon of the ancien régime expressed the king's absolutist rule, yet they were turned against the Bastille. As long as the values embedded in any technology are themselves contradictory, the meaning and purpose of any technology will be subject to a process of continual reinvention. The important historical question to answer is when and why that happens.

Perhaps I would do well at this point to be clear about what I mean by politics. In this book I will invoke four interrelated definitions of politics. On the most basic level, politics is the process by which a group of people come together in the recognition that they have common interests, and act accordingly. This is politics with a small "p," and it can operate through the maneuverings of patronage, by means of bureaucratic intrigue, in the unrecorded struggles of the workplace, as well as in the broader light of public debate. This last is *modern* politics with a capital "P," the explicit, public articulation of social interests which is usually said to begin with the French Revolution. At that time the language of left and right took shape, and political parties emerged which claimed to speak in the name of well-defined social interests. Even after Thermidor, however, the match of party and the articulation of interest emerged only fitfully, with political rhetoric as much an attempt to forge solidarities as a reflection of "authentic" interests. Thirdly, on the most structural level, politics is that set of half-examined assumptions—presumably useful to their holders—to which these social groups refer as a way of giving meaning and purpose to their activities. This level of politics is commonly referred to as "ideology," and its operation is most visible in such slogans as "privilege" or "meritocracy" or "patriotism." Finally, by the word politics, I also mean to invoke the possibility (the naive hope?) that were a different alignment of interests to prevail, the social life would also be transformed. To say then that an artifact is political is to imply one or all of the following: (1) it represents the outcome of a particular alignment of the interests of various groups, (2) it becomes the subject of debate in public life, (3) its physical attributes in some sense reflect the ways in which the social life is conceived, and (4) its physical attributes would somehow be different were a different alignment of interests to predominate.

To summarize then: Ours is a world in which human interests are myriad and circumstances vary. In such a world, technology is the physical embodiment of that form of knowledge which we call power. This knowledge is not the outcome of private ratiocination or access to some abstract truth, but of a public engagement with recalcitrant "thick" objects. Shaping these obdurate objects so that they can be brought to bear on social problems is a daunting challenge, and one which admits of many possible solutions. As a result, there will always be, within any society, disagreements about how this should best be done and who should reap the immediate advantages. Engineering is a social art, and a contentious one. Consequently it has a history. And artifacts are the physical embodiment of that set of (temporary) treaties which the contending parties strike as they pursue their interests as they see them.

The central implication of such a view is that things might be arranged differently. And, of course, they often are. Noel Perrin, for instance, has provided us with a history of a country which "gave up the gun." Japan in the late sixteenth century was a gun-infested land. The first musket had been introduced by European adventurers in 1543. By 1575 as many as ten thousand Japanese soldiers wielded matchlocks at the battle of Nagashino. These firearms, moreover, were manufactured locally, and with novel adaptations for night battle and battle in the rain. Mass armies of musketeers were sent to invade Korea in the coming decades, and soon were fighting to consolidate control over the home islands. Yet in the early seventeenth century, the newly unified state obliged its army to renounce firearms and set about exorcising all knowledge of gun-making. For the next two centuries, guns played a negligible role in Japanese life. Such an extreme case is a stark reminder that even the most "rational" technology is a political choice.[26]

Writing about a subject closer to our own time, Donald MacKenzie has shown how the ballistic missile is likewise the outcome of a political process. Just as there are alternative conceptions of nuclear war-making (each with its institutional devotees), so too are there alternative designs of missile guidance systems. Where policy hacks once bowed their heads before "inevitable" improvements in the accuracy of these missiles, MacKenzie demonstrates that their doubled and redoubled accuracy is actually the result of bureaucratic infighting. The provocative inference is that the invention of atomic weaponry is in some sense open to reversal, albeit by a radical recantation of our social priorities. Yet such things do happen.[27]

My aim is more broadly historical, but I take my cue from these liberating examples. In this book I hold fast to the historian's responsibility to see the past with a double vision. By this I mean that he or she must endeavor to understand the rationale of an era on its own terms, keeping in sight the unrealized futures which the different people of the past labored to bring forth, while simultaneously explaining how the world we inherited came (seemingly inevitably) into being.[28] In the pages that follow, I bring this double vision to engineering rationality, the invention of uniform production, and the organization of mass war. These developments are usually explained as the outcome of an inevitable process of "modernization"; indeed, they are often considered constitutive of our modern era. What is engineering rationality if not modern rationality? What is uniform production if not the optimal way to organize capital and markets? What is mass war if not the inevitable logic of the populist nation-state? Yet in each instance there were (and are) alternatives. The making of modernity proved to be a protracted struggle, born—like the artifacts that are its emblems—out of process of conflict and negotiation.

The Organization of the Book

One final word on the structure of this book. The text is divided into three parts. Parts One and Two cover the same chronology, and carry the reader from the mid-eighteenth century across the threshold of the Revolution (1763–93). Both emphasize the continuity of the Revolution with the developments of the late eighteenth century. Part One is concerned with the artillery engineers in their role as the converters of capital into coercion. There, I examine the conflicts over the proper design of guns and their battlefield deployment in the context of state formation and engineering rationality. I lay out the logic of the particular "marketplace" which the artillery engineers sought to serve and shape, and describe their struggle to give France a new social and political structure. Part Two is concerned with the engineers in their role as the converters of coercion into capital. There, I examine the engineers' attempt to transpose this battlefield logic onto the workplace. Out of this attempt—and the resistance it engendered—came new "objective" methods of production, culminating in the method of interchangeable parts manufacturing. I describe the conflicts over the production of guns as part of a process which defined the limits of the state's involvement in capitalist production, and hence the relationship of the state to its citizen-producers. Both Parts One and Two shed light on the process of "reform" and "reaction" which propelled France from the Enlightenment to the Revolution. Both emphasize how Enlightenment reforms in military organization and military production were exploited by the Revolutionaries to dramatic effect. At the same time, both highlight the resistance these changes aroused in elite military officers and provincial producers, a resistance which fueled the passionate struggle of the Revolutionary period.

Part Three combines the themes of the first two parts of the book, and carries the story forward from the Revolution through the Napoleonic reaction (1794–1815). Here I stress the novel elements introduced by the Revolutionary experience, and how these played themselves out in the early nineteenth century. I examine how the military engineers, from their new position at the head of the Revolutionary state, set about redesigning, reproducing, and redeploying the gun. By showing how their program was contested, I directly confront the claim that the technological life is apolitical. The French Revolution, I argue, was also a technocratic revolution. By this I do not simply mean that the French Revolution gave scope to a new régime of engineer-experts, but also that these experts masked the nature of their rule through a "technocratic pose." This pose was meant to make the hierarchical structure of the post-Thermidorean order appear inevitable and natural. Paradoxically, it also signaled a partial retreat of the state from the world of production, and hence the emergence of a new social alliance to govern post-Revolutionary France. We can read the terms of this modern political settlement in its material artifacts.

PART ONE

Engineering Design: Capital into Coercion, 1763–1793

The act of killing and destroying a man, continued my father, raising his voice—and turning to my uncle Toby—you see, is glorious—and the weapons by which we do it are honourable—We march with them upon our shoulders—We strut with them by our sides—We gild them—We carve them—We in-lay them—We enrich them—Nay, if it be but a *scoundrel* cannon, we cast an ornament upon the breech of it.

—Laurence Sterne,
The Life and Opinions of Tristam Shandy

Chapter One

THE LAST ARGUMENT OF THE KING

GRIBEAUVAL and his artillery engineers came to power in the wake of the Seven Years' War when humiliating French defeats prompted a bitter debate about field artillery: a dispute about cannon design, which thinly cloaked a struggle over military strategy, which itself thinly cloaked a battle for influence and power within the state. Certainly, this quarrel over field maneuvers attracted a much wider audience than such technical and morbid subjects usually warrant. It was the "Star Wars" dispute of its day, a public debate over the offensive and defensive capabilities of the nation and the effectiveness of high-tech gadgetry. In the end, the salons and academies of Enlightenment Paris were enlisted in a debate over what sort of ballistics could best redeem French national power.

Though both parties in this battle of artillery systems claimed to be vindicated by the most prestigious arbiter of the day—scientific experiment—in the end, the triumph of Gribeauval's "Moderns" turned less on narrow technical grounds than on their coherent vision of an aggressive new war of movement, and on the support these military engineers consequently received from a reform-minded faction of the army high command. For the artillery bureaucracy itself, the main outcome of Gribeauval's victory was to consolidate the service's hold over the world's largest proto-military-industrial complex (map 1.1). The result? From 1763 to 1789—with a dramatic hiatus in 1772–74—they refashioned the artillery service into a new kind of military machine that, in conjunction with the Revolutionary mass armies, spearheaded the destruction of the old Europe under the command of their disciple, Napoléon Bonaparte.

Historians of this episode have invariably been drawn to the bitter polemic that accompanied the transition from the old Vallière system of artillery to the new Gribeauval system. Yet almost without exception, they have recounted the Gribeauvalist triumph as a self-evident outcome, a chapter in the tale of the inexorable modernization of France. In doing so, they have followed the tone set by the young Louis-Napoléon in the multivolume history of artillery he wrote while waiting to fill his uncle's shoes: "The history of artillery," he said, "is the history of the progress of science, and consequently, of civilization."[1] The story of scientific progress, it would seem, is so compelling a narrative that it can even be measured by a body count. And the engine of progress,

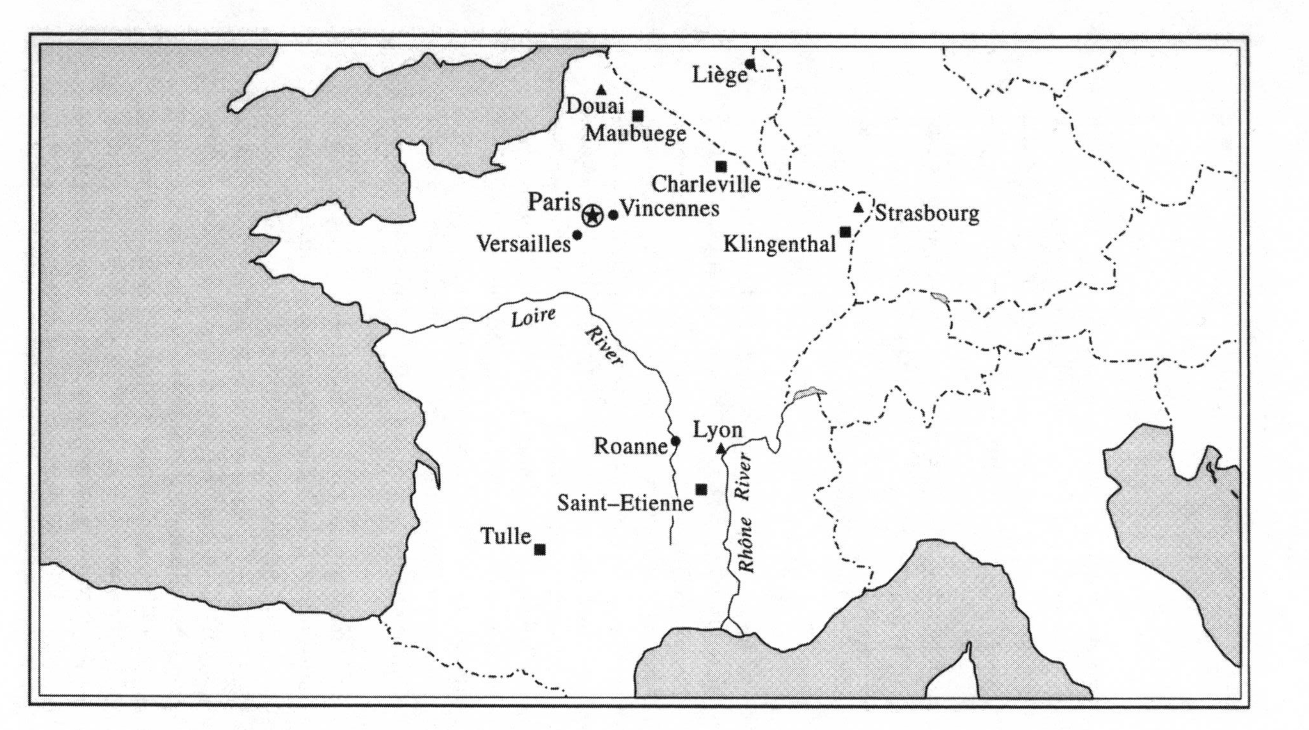

Map 1.1. The Major Armories of France, 1763–1815. This map shows the principal sites of the French military-industrial complex. It indicates the three major cannon foundries run by the artillery at Strasbourg, Douai, and Lyon. It also indicates the five French armories for small arms: the three musket manufactures at Saint-Etienne, Maubeuge, and Charleville, plus the manufacture for sabers and bayonets at Klingenthal and the naval manufacture at Tulle. During the Revolution and Empire a number of new armories were established, not all shown here. Among them was Blanc's factory for interchangeable gunlocks in the town of Roanne. Liège, a thriving armory town for many centuries, served the French military during the occupation. Map by Chris Brest.

here as in so many other areas of Western history, is said to be the knowledge which tells us how the world "really" works and those instruments which can be built with that knowledge. Told about guns, this story begins with Galileo discovering the principles of parabolic flight, and then takes a short trip along the trajectory of time to the ICBM. Adoption of the new artillery weapons, it could be argued, required new forms of central battlefield coordination, hastening the staff and line organization of the army. This change permeated military values: courage was demoted, and skill in calculation highlighted. Rather than an elite of blood, capable officers were now recruited from the "middling ranks of society." The ancien régime breaks down, everyone is a citizen, and civilization spreads around the world—according to Condorcet, because the courageous, barbarous peoples can now be defeated by those who are wealthy. Better ballistics for better artillery for a better world.[2]

The appeal of this sort of narrative follows from its explanatory efficiency. It juxtaposes a narrow definition of science (rational mechanics) and a narrow definition of technology (the gun) with a big temporal and geographic frame (the history of the modern West). What should put us on the alert against this sort of explanation is that it is one of the central arguments developed by the interested parties themselves. Indeed, the reading of history as the unfolding of a technological determinism was one of the main armatures of the group which took the name "Moderns." Under closer investigation this determinism argument breaks down. The Gribeauvalists did not passively supply artillery know-how to their patrons. Like all effective creators of new technology, they actively defined the *context* as well as the *content* of their innovation. In doing so, they spearheaded a larger transformation in the nature of French sovereignty.

In this regard it is telling that the Gribeauvalists' new methods of uniform production precluded the ornamentation which had once adorned royal cannon (figs. 1.1 and 1.2). On the big guns which played such a prominent role in the wars of the ancien régime, the Bourbons had inscribed the motto, *Ultima Ratio Regis*, "the last argument of the king." For many of the traditional artillerists, the Gribeauvalists' erasure of the king's motto symbolized the radical rupture of the new system. Implicitly, these new motto-less cannon posed the question: *whose* argument were they? Speaking for the Gribeauvalists, Tronson Du Coudray refused to "mourn" the passing of these "costly ornaments." The design of the new artillery, he said, "allows no thought for the superfluous."[3] The question was: in the searching light of this leveling instrumentalism, what else might be considered superfluous? This chapter examines the changing nature of state power in the eighteenth century as refracted through the contemporary debate over the artillery. In doing so, it highlights those contradictions within the ancien régime's logic of power that ultimately threatened to tear the state apart.

Fig. 1.1. Artillery Cannon, Vallière System. This siege cannon was typical of the large caliber Vallière pieces. Note the elaborate embossing work. The motto engraved near the muzzle declares: "Ultima Ratio Regis." From the Musée de l'Armée (N134).

State Power and the Monopolization of Violence

The royal states of early modern Europe only gradually established dominion over feudal rivals, privileged cities, and peasant producers. Throughout this fitful and uneven process, the centralized national state triumphed largely to the extent it could monopolize violence. How and where this occurred depended on two factors: first, the state's position within the system of states; and second, the particular mix of coercion and capital the sovereign employed domestically as he pursued his interests and those of the ruling elite. The absolutist state functioned as a vast machine for waging war on behalf of an aristocratic class for whom the spoils of battle meant gratifying gains in economic surplus and prestige. Rather than acting with "absolute" authority, however, the scale and diversity of war-making activities meant that states relied on intermediate powers: local magnates, private financiers, merchant dealers, and artisanal producers—all "capitalists" of one sort or another, ranging up and down the social ranks. Indeed, the largest and most successful states of Western Europe (France and England) increasingly used a system of extracting revenue to *purchase* the coercive power that guaranteed a further supply of domestic revenue, and hence the ability to wage more war. In return, these capitalists received protection for their commerce, an expanding military market for their goods and services, and social and financial privileges. Charles

Fig. 1.2. Artillery Cannon and Carriage, Gribeauval System. This field cannon is a 12-pounder from the Gribeauval system. It was forged in the year II and served in the campaigns of 1794. There is only light etching on the surface of the cannon. The elevating screw (*vis de pointage*) is visible between the spokes of the wheel. The parts of the gun carriage would have been interchangeable. From the Musée de l'Armée (N182).

Tilly has likened this rondo of coercion, extraction, and protection to organized crime.[4]

Increasingly in this period, the actual business of battle was turned over to men trained for a variety of specialized tasks and coordinated by a central power. This occurred because, quite simply, such organization paid—not just in victories on the field, but in profits and social standing for domestic interests allied with the sovereign. Two innovations in particular gave the advantage to those centralized powers which organized for war during times of peace. By the early seventeenth century, Maurice of Nassau and Gustavus Adolphus of Sweden had shown that the repetitive drilling of rank-and-file soldiers transformed them into a victorious battlefield instrument. Ranged in long lines and firing their smoothbore muskets in volleys, these troops constituted the first effective response to the massed squares of "Swiss" pikemen who had once dominated European battle. The drill enabled commanders to make effective soldiers out of urban riffraff and impoverished peasants' sons without disturbing those productive members of society whose taxes maintained armies in the field. This regular drilling also demanded greater self-discipline from officers, men almost always from the noble classes, whose duties and values became increasingly distinct from those of the men they commanded. The larger armies made possible by drill also increased the need for military equipment, providing burgeoning profits for military contractors. The second cluster of innovations centered on fortress warfare. Increasingly, the northern European states built walled defenses—the *trace italienne*—which placed a premium on mobilizing vast sums of capital, technical expertise, and manpower. These defenses called for a corresponding investment in the cannon to breach those walls, and in the capital, expertise, and manpower needed to bring these guns to bear at a specified time and place. The changes surrounding these twin tactical, technical, and socioeconomic innovations comprise what historians call the "Military Revolution," now seen as central to defining and sustaining the absolutist state.[5]

Once France adopted this formula in the mid-seventeenth century, her monarch was able to pursue his dynastic ambitions across Western Europe. Louis XIV and his lieutenants—Colbert, Vauban, Le Tellier, and Louvois—dramatically increased the size of the French armies and imposed ministerial authority over aristocratic officers. The adoption of a military uniform (though the pattern still varied from regiment to regiment) neatly symbolized the new routine of military life. Similarly, the War Office standardized the supply of weaponry by designating a sole buyer on its behalf. At the same time, these ministers also cultivated a proficient corps of military engineers, and gave them resources to build fortresses and armaments to fight the siege wars which characterized the period. And last but not least, they transformed the army into a standing peacetime force largely insulated from the civilian population. This standing army not only guarded against foreign enemies, but assured internal peace, and hence the revenue to fund further war.[6]

The spread of this "bureaucratization of violence," however, produced continual frictions within the ancien régime. At the beginning of the seventeenth century, the peak wartime army barely equaled 60,000 men. Richelieu increased this to 125,000 men, largely relying on the purse strings of commanders. By the end of the century, the Sun King fielded at least 340,000 men, plus a peacetime contingent of 150,000. Remarkably, this expansion had been funded out of his own coffers.[7] But while combating the coalitions ranged against him in the later years of his rule, Louis XIV strained France's ability to raise, supply, finance, and coordinate these troops. The effort put pressure on the constitution of the absolutist state, and on the system of privileges which defined social status in the ancien régime. The nobility, which provided virtually all the members of the officer class, had historically claimed to merit these positions by their family tradition of personal service, as recognized by the king. In the eyes of some reformers, however, effective battlefield coordination required subordinate but knowledgeable officers, merits not necessarily promoted within the vast machinery of patronage and purchase that was the key to advancement in the ancien régime. As a result the rationalization of the military was decidedly incomplete in this period.

Commanders still retained a substantial proprietary interest in their units, advancement still depended upon birth and courtly connections, and war aims were still shaped by dynastic interests. It is important to stress the contradictions of a system which waged war for the benefit of the noble elite and was staffed by them, but which worked against their particular interests on behalf of a centralizing power. This tension—which the ancien régime proved unable to sustain—was built directly into its instruments of war.

The Vallière System: "Absolutist Rationality"

As the state's emissaries to the nation's privately held armaments works, and as the operators of the cannon that spearheaded the king's army, the artillery engineers were central to the state's ability to convert capital into coercion, and vice versa. In this respect, the reform of the French artillery in the early modern period lay in the main line of a two-hundred-year effort by the northern European states to bring military force to new levels of potency.

The Vallière cannon, introduced in 1732 (fig. 1.1), belonged to a carefully conceived *system* of weapons, social organization, and tactics, which operated within this framework, and which suffered from its contradictions as well. The Gribeauvalists were later to accuse the Vallièrists of "irrationality," of being bound to meaningless traditions. But as Joseph-Florent de Vallière, *fils*, noted, his father had not settled on his system of artillery "arbitrarily or by mere conjecture," but on the basis of his twenty-eight years of experience fighting the wars of Louis XIV.[8]

Indeed, while the Vallièrists and Gribeauvalists themselves divided their ranks into "Ancients" and "Moderns," they shared professional attitudes which the line army was only slowly beginning to accommodate. Within the military, the artillery represented a case apart: atypical in the extent to which professional values permeated the corps, though quite typical in their mistrust of the civilian constitution of France. Known, along with the royal fortification engineers (Corps du Génie), as an *arme savante*, the artillery service differed from the main body of infantry and cavalry not only in its tactical role, but in its social composition and mentality.

Cannon had been essential to warfare and French national power since the resurrection of the monarchy from the Hundred Years' War and the Italian campaigns of the Renaissance. These big guns could obliterate or defend a fortress, and decimate or protect massed troops. They enabled political organizations with sufficient financial and bureaucratic resources (the national state) to prepare this explosive force in advance and deliver it with considerable precision. The effectiveness of these cannon, however, depended on various factors: maneuverability, throw-weight, rate of fire, and accuracy. As no single gun could maximize all these parameters, artillery design involved difficult tradeoffs. Prior to the Vallièrists, the solution was specialization. Even though the advantages of a uniform caliber size had been apparent since the time of François I, the varieties of cannon proliferated throughout the seventeenth century.[9] To cope with the diversity of situations encountered in war, artillerists needed distinct classes of siege guns and field guns. Cannon also varied in the thickness of their cast, and hence in their maneuverability and the strength of the blast they could withstand. Furthermore, each French foundry—under its separate regional administration—cast pieces of sufficiently different caliber that their shots were not interchangeable. Finally, these tubes were mounted individually on carriages whose dimensions depended on local methods of production and local needs, so that, for instance, the axle length matched the requirements of local roads.[10]

In this regard, Vallière's program of standardization represented a radical reform. Appointed director-general of the artillery shortly after the death of Louis XIV in 1715, Jean-Florent de Vallière, *père*, set out to bring order to the French arsenals. In 1732, he instituted his famous system of standardized artillery pieces, specifying the five classes of cannon—4-, 8-, 12-, 16-, and 24-pounders—whose dimensions and weights were not "under any pretext" to deviate from the official guidelines as inscribed in the official master drawings.[11] Like his absolutist sovereign, Vallière sought to reign over a highly centralized and stable domain. The Vallière system combined a simplified administrative structure with a conscious appraisal of the wars of the period. The same artillery pieces were now to be used for coastal and fortress defenses *and* for sieges and campaigns in the field. To enable them to fit into prepared emplacements, all the cannons had been given the same length. And because they had to batter down the walls of fortresses from perimeter trenches, they had to be capable of propelling the heaviest caliber shot a considerable distance, and hence possess sufficient thickness to withstand a tremendous charge.

Vallière also reorganized the service of these weapons. At the end of Louis XIV's reign, two distinct bodies claimed jurisdiction over the artillery: the Corps Royal and the Royal Artillerie. The former operated under the aegis of a Grand Master, usually a prince of the blood who answered directly to the king. In the first half of the seventeenth century, the Grand Master had engaged gunnery experts—often of bourgeois origin—for the duration of a campaign. In the late seventeenth century, however, the Secretary for War, Louvois, began recruiting semipermanent artillerists and subjecting them to military discipline. These technically proficient and specialized officers oversaw the manufacture and maintenance of all war matériel. Furthermore, they organized the tremendous artillery trains, such as the 1747 expedition to Flanders, which included 150 cannon, 397 wagons, and 2,965 draft horses, plus ancillary equipment. Once in battle, they ranged their pieces and directed their fire. These officers' responsibilities were parceled out with bureaucratic nicety: the *ingénieurs* drew up plans for batteries; the *conducteurs* moved the pieces into place; the *connétables* distributed powder and balls; and the *commissaires* had jurisdiction over the gun in action, et cetera. Generally speaking, these officers were managers of capital rather than masters of coercion; they collected a fee for their services, and they did not command troops.[12]

Authority over troops in the field was the prerogative of the Royal Artillerie, officers of infantrymen assigned to protect the artillery, who eventually came to assist in the firing of cannon. These were the masters of coercion, rather than managers of capital. They identified closely with the line army, and almost all came from the nobility. Throughout the eighteenth century, these officers and those of the Corps Royal argued over pay, rates of promotion, and their respective spheres of authority, despite attempts at amalgamation. Even after the resignation of the last Grand Master in 1755, there is evidence that the distinction between "technicians" and "warriors" persisted.[13]

The artillerists experimented with various solutions to this conundrum—central to the triumph of the nation-state—of how to marry technical competence (expert knowledge) and martial values (loyalty). The temporary introduction of venal offices in 1703 should be seen in this light. This practice proved unsatisfactory, however, and in 1720 the Regent authorized permanent artillery schools in the five garrison towns.[14] These schools provided a formal and uniform education for all officers, thereby ending the variation which "clouded the mind and disheartened officers as they pass from one garrison school to another."[15] I will discuss the schools and the "art of artillery" at greater length in the next two chapters; here it is enough to note that this uniform course of studies involved heavy doses of "mixed" mathematics, technical drawing, metallurgy, as well as practical instruction in firing cannon. This was the sort of "science" these men claimed as a resource. In this period, ballistics theory—the representation of a cannonball's flight—was nominally derived from the mechanics of moving bodies as laid out by Galileo. Experienced men, however, knew perfectly well that the trajectory of a cannonball could not be predicted by this sort of theory, and on the battlefield they continued to ply their trade as a craft, a skilled "art" which resembled the kind of rule-based knowledge used by artisans.

Siege War and the Science of Limited Destruction

What gave meaning to this arrangement of weapons, personnel, and art was its central role in the attack and defense of heavy fortresses. Siege battle dominated warfare between 1680 and 1747, a period in which war itself was almost continuous.[16] Sébastien Le Prestre de Vauban, Louis XIV's illustrious military engineer, had bestowed on France a perimeter of double fortifications to defend the nation, so that the kingdom itself came to be seen as a fortress with outlying "works" to give it defense in depth. Behind these "natural" frontiers—fashioned out of stone and at a cost of a million soldiers' lives—presided a king whose glory haunted the eighteenth century.[17] But the legacy of these state-builders was ambiguous. In the latter part of his reign, Louis XIV had used Vauban's barrier as a springboard to project his dynastic ambitions across the continent, only to be hurled back in costly and humiliating defeat. Vauban, who disapproved of his monarch's expansionist aims, had nonetheless contributed more to the art of taking fortresses than to the art of defending them. He introduced the zigzag trench attack that captured Maastricht in 1673. And at Ath in 1696, he first used ricochet fire to sweep the ramparts of defenders. The imitation of his methods by enemy powers meant that his "impregnable" double row of fortresses had been broken in several places by the end of the Sun King's reign. So despite the supreme confidence which eighteenth-century French elites placed in their membership in a powerful state, the relative capabilities of offensive and defensive war-making—and of siege warfare in particular—constituted an unspoken instability in the definition of the nation. In the second half of the eighteenth century, intelligent contemporaries such as the artillerist-littérateur Choderlos de Laclos began to realize this.[18]

The "science" of siege warfare was an attempt to control the uncertainties of warfare. Vauban's fame rested on his three systems of fortresses built along geometric principles. These were not simplistic Cartesian abstractions, however. Vauban's cardinal rule was that fortresses must take advantage of natural terrain. A similar mix of art and science can be found in his influential method for reducing the *conduct* of a siege to a scenario whose permutations could be predicted by both parties. In his famous *Traité des sièges*, Vauban listed the twelve stages of a siege, from investment to capitulation, along with a handy forty-eight-day schedule. For every action a town might take (countermines, sorties, etc.), he laid out the appropriate response of the attackers. And he spelled out the duties of each defender, from the commissaires in charge of individual cannon to the governor—"avoid tensions (*émotions*) between the soldiers and civilians." But unless a town was relieved by a friendly army, the outcome was not in question; no town could hold out beyond a set period.[19]

To bring his art to this level of certainty, Vauban had partially brought the artillery under his authority and coordinated their bombardments with his schedule. A siege was largely an affair of the artillery. Their guns cleared the ramparts, destroyed enemy stores, breached the walls. Vallière's massive artillery was designed for just these roles. They were not expected to move or fire rapidly. They had to pack a wallop and survive a long campaign. So long as the battles were fought in Flanders (as during most of this period), river transport could bring the cannon into position. Nevertheless, a siege took tremendous organizational skill. For their efforts, the artillery received generous payment. The king accorded the Grand Master a flat fee for each cannon brought to the trenches and for each day the guns saw action; out of this he paid the soldiers and officers. And should the town fire in its defense, the artillery was rewarded upon capitulation with all town's bells, window grates, rain pipes, and metal kitchen utensils (or their equivalent in specie).[20]

These "scientific" procedures and customary practices standardized the long-standing rituals that bounded siege warfare. The uncontrolled slaughter of the Thirty Years' War was still vivid. Now the dangerous "leveling" firepower which could fell a commander as easily as a foot soldier would be directed in predictable fashion. The pageantry of the initial parlay, the formal rules of surrender, and the honors accorded the vanquished according to their status and conduct, all guided the participants through situations with enormous potential for uncontrolled violence. Behind such rules lay the savagery of bombardment and the carnage of a final assault. A citadel that refused to surrender once its walls were breached was fair game for looting and rape. This was in no one's interest, certainly not that of the local bourgeoisie or the defending garrison. Nor was a ravaged town valuable to the conquerors as a future base of operations or bargaining chip. And officers did not like to lose control of their troops during the chaos following an assault.[21] In time, the prescriptions of the theoreticians became the standard against which participants measured themselves. In 1705, Louis XIV pleaded with the governors of his fortresses to put up more than a symbolic defense after Vauban's forty-eight-day period had elapsed.[22]

In practice, the length and casualties of sieges still varied. But Vauban brought the average duration down from approximately seventy-six days to around forty days, while reducing their variation. He did this by increasing the size of the besieging army without thereby increasing the number of casualties.[23] His famous admonition, "Burn your powder and spare your blood!" expressed his confidence in science over courage, and method over fury. In short, the "science" of siege warfare minimized the destruction and uncertainty entailed in the transfer of valuable capital, and spared the lives of highly trained men.[24] Sieges bounded the ravages of war. Moreover, Vauban's method could be taught. It was a learned and honorable art.

In the eyes of contemporaries, siege warfare unfolded like a classical drama. The dazzling theater of power acted out at Versailles was but a mirror of that vaster theater of power playing out on the French perimeter. As Michel Parent has noted, a siege preserved the unity of time, place, and action in the manner of the great tragedies of Racine (who attended one of Vauban's "stagings"). This dramatization was itself an important constituent of state power, both within France and abroad. Writing to Racine of the calamity that would ensue were France to lose even a single fortress along the frontier, Vauban said, "Are they so uninformed in the Council of State to not know that states maintain themselves more by reputation than by force?"[25] In this theater, as at Versailles, the king was the principal *metteur en scène*. Of the forty-eight sieges led by Vauban, Louis XIV personally commanded the troops at twenty. Siege war proceeded with all the deliberation of a court review. From the heights overlooking the recalcitrant town, the monarch directed his expert technicians, men of the bourgeoisie and lower nobility engaged in a quasi-mechanical trade. Their guns, which bore his name, represented his body in its warlike aspect. And their power was a direct expression of the *Ratio* of his rule. That way, while his great aristocratic rivals waited on the sidelines or back at Versailles, all the glory redounded onto him.[26]

Field battle, on the other hand, was unpredictable, ungovernable, and uncomfortable. Wise generals avoided it as much as possible. The particular agony for the artillery was getting their three-ton weapons into position. The volleys of musket balls made infantry assaults perilous, and the bayonet (after 1700) protected troops from cavalry charges. No one's heart was really in it. Eighteenth-century field battle was a futile and inconclusive maneuvering for position, in which armies spent most of their time "dodging" one another.[27]

In sum, siege warfare was a display of power that brought the logistical capacities of the absolutist state to bear on points of resistance. Aerial bombardment undercut urban privilege and aristocratic virtues. At the same time, a siege was a deliberately circumscribed affair, bounded by rituals and codified by technicians. Its "science"—a patina of geometry superimposed on a wealth of craft knowledge—limited destruction. To that extent, Vallière's artillerists, even as they rained down death and demolition, could assert their allegiance to a measured civilization. Similarly, the dynastic ambitions of a Louis XIV were always tempered by a readiness to negotiate, especially if turmoil threatened his kingdom's economic and social base.[28]

In this context, the security of the state depended on the equilibrium of offensive and defensive capabilities (as must be the case where states are embroiled in a system of competing powers). Taken as a whole, Vauban's massive building effort had raised the cost of an enemy incursion into France. But in individual engagements, his art gave the advantage to the offensive forces. Hence, early eighteenth-century reformers sought to restore the balance to the defense, and innovations favoring offensive war were covered with opprobrium. Symptomatic of this was the debate over explosive mines between Vallière, *père*, and Bernard Forest de Bélidor, an engineering professor in the La Fère artillery school. Vallière insisted that theoretical science, because it marked the upper limits of destruction that could be achieved with buried explosives, favored the use of countermines, and hence the defense of fortresses. Whereas the practical methods of Bélidor, he argued, by promising ever larger destruction, favored the mines of the attackers, and hence destabilized the status quo.[29] As we will see, Vallière also argued that ballistics theory placed limits on what was practically achievable—to the advantage of the defense. His science was a science of containment, a Cartesian geometrization that circumscribed the possible.

Guillaume Le Blond, the chief artillery theoretician of the mid-eighteenth century, noted the imbalance that lay on the side of offensive weapons, and succinctly laid down this warning to his monarch.

> For the good of humanity, as well as their own advantage, sovereigns ought to unanimously refuse all inventions whose object is to render our offensive arms still more pernicious and dangerous, and offer prizes to those who invent defensive arms.[30]

Le Blond was an Encyclopédiste, math tutor to the royal children, and a partisan of Vallière's cause. He noted that the wealth of kings depended on the number and well-being of their subjects. Given the rapid transfer of military knowledge among the European nations, no side could gain a permanent advantage by innovation. This eminently civilized logic explains why the Vallièrists took such a dogmatic tone when confronted with the Gribeauvalist reforms—going so far as to ban "all innovations."[31]

The Vallière system of cannon was a fully integrated *system* and any change threatened its larger coherent purpose. For its Cartesian administrators, standardization and science offered a bulwark against chaos within their corps and the destabilization of the régime they served. Vallière's corps had every reason to consider itself in the forefront of professionalism. Its officers were trained in the most advanced technical schools in Europe, experienced in their craft, and increasingly assimilated to their noble military calling. The corps' massive and standardized cannon—emblazoned with the king's motto and representing him in bodily form—corresponded to a seventeenth-century conception of power. Centralized under the aegis of a princely Grand Master, rationalized through its uniform matériel, and permeated with a Cartesian "esprit de système," the Vallière system was a mirror image of the absolutist regime it defended.

When Vallière, *père*, died in 1759 at the age of ninety-three, the Seven Years' War had been underway for over three years. The terms of the equation had changed. But the artillery reformers of the Enlightenment who then argued that France should forsake the bounded destruction of siege craft for the unbounded violence of offensive war were not just tinkering with the self-declared mission of the artillery service, they were overturning one of the core definitions of the kingdom.

The Seven Years' Fiasco

The Seven Years' War was indeed to prove the "king's last argument." During its disastrous course the artillery came in for a good deal of the blame. Critics made much of its deplorable state in comparison with its three rivals to the east: Prussia, Austria, and Russia. With the war still in progress, the French leadership tried to modify the organization, matériel, and tactics of the artillery. In 1755 the War Office pushed aside the current Grand Master. The Royal Artillerie and the Corps Royal were finally fused, with the fortification engineers temporarily added as well. In theory, the new corps operated under Joseph-Florent Vallière, *fils*, who succeeded his father. Day-to-day control, however, lay in the hands of seven inspector-generals, who for the first time answered directly to the minister of war. At the same time, a central school for officer-cadets was created at La Fère to train fifty recruits a year. More generally, there was an attempt to increase the number of artillerists in response to the large number of guns fielded by the Prussians.[32]

The matériel of the artillery had also proved inadequate. French 4-pounders weighed twice as much as their Prussian equivalents, and the Prussians had more of them. The French found themselves continually outmaneuvered and outgunned. These were not the fields of Flanders, laced with canals.[33] At Rossbach (1761), Frederick II won a stunning victory over a French force twice the size of his army, in part by repositioning his artillery to great effect. With horses in short supply, the French could not bring their cannon to bear as they wished, and on retreat had to abandon a number of their pieces.[34] Between campaigns a frustrated Marshal Broglie took his Vallièrist cannon to Strasbourg to have them rebored to accept larger projectiles; he also had foundry master Jean Maritz produce new cannon that were shorter and less thickly cast. These irregular light pieces were manufactured without the active participation and oversight of the artillery service. The artillerists warned that they would not stand up to lengthy service and noted ominously that they might burst on use.[35] They also objected to the distribution of these improvised pieces to infantry formations in imitation of the Prussians. The artillery was losing its monopoly over the production and management of cannon, the service's entire *raison d'être*. The rationally conceived Vallière system was being undermined. At the end of the war, the artillery pieces in the possession of the French army of the Lower Rhine included a hodgepodge of different calibers.[36] In this sense, Gribeauval's reforms were motivated by a desire to *preserve* the rational structure of Vallière's system, and to ensure the artillery retained its role as the sole supplier and manager of war matériel.

Jean-Baptiste Vaquette de Gribeauval (fig. 1.3) owed his meteoric rise to the fact that he emerged as one of the few French heroes of the Seven Years' War—that, and his ability to acquire influential patrons. He was born to a recently ennobled family in 1715.[37] As a youth he showed mathematical talent, and was thereby set on one of the routes to position open to a young man of modest fortune. His father, a lawyer and mayor of Amiens, placed him in the garrison artillery school at La Fère at the age of seventeen. There, he came under the influence of Bernard Forest de Bélidor, the mathematics professor whose empirical approach to ballistics and other engineering sciences attracted powerful admirers and enemies. Gribeauval, like his mentor, was an

Fig. 1.3. Jean-Baptiste Vaquette de Gribeauval. This portrait by Louis Charmontelle is said to be of Gribeauval (1715-89), who led the artillery service after 1763 and became first inspector-general of the artillery in 1775. The original is held in the Musée Condé à Chantilly. Photo by Lauros-Giraudon.

innovator. Tests of his prototype siege carriage brought him to the attention of then Minister of War Argenson.[38] In the summer of 1755, with France and Prussia still nominal allies, the French War Office learned that Frederick was planning to adopt a new light cannon. Sent as the French observer for the trial maneuvers of the piece, Gribeauval met the Prussian monarch and returned to Paris with a set of plans. From these, the French foundries quickly produced a working model, which Gribeauval demonstrated before the king. In his official report on his mission, however, he expressed an ambivalence about the Prussian matériel; he admired its maneuverability, but it lacked sufficient power.[39] This was to be a tradeoff he would often confront in the years ahead.

Within a year, France and Prussia were at war, and Gribeauval was again sent on a mission to Central Europe, this time under the command of the comte de Broglie. Both Broglie brothers later became Gribeauval's advocates in their secret correspondence with Louis XV. By 1757, Gribeauval had gone into the service of the Hapsburgs, now allies of the French. The Austrians had assembled a first-rate artillery matériel since their defeat in the War of Austrian Succession, but they lacked technical officers. There Gribeauval enjoyed a rank echelons above his status in France. In Vienna, he reported to the French Ambassador, the duc de Choiseul, who became the patron of his reforms after the war. His star rose to international prominence during the siege of Schweidnitz (1761–62) where he directed the defense, holding a large force of Prussians at bay for months until Frederick himself came in person to launch the final assault. The Frenchman's fame was sealed when Frederick, the acknowledged military genius of his day, toasted his heroic conduct and pronounced him one of Europe's finest engineers. When Choiseul was recalled to Versailles as virtual prime minister and initiated his military reforms, he lured Gribeauval back. With the prodding of the duc Du Châtelet, the new French ambassador in Vienna (and another future patron), Gribeauval accepted the directorship of a refashioned French artillery.[40]

Already in 1762–63, the bureau chief Dubois of the War Office had requested that Gribeauval draw up a comparison of the artillery systems of France and Austria. The War Office had been determined to reorganize the artillery since gaining control of the service. With the stocks of cannon depleted, they saw their chance. Their comprehensive proposal for rationalization became the template for what historians have called the Gribeauval reforms. It insisted on uniformity of matériel and professionalism for officers. It also envisaged an increase in the quantity of cannon, a reduction in their weight, and a thorough reappraisal of how they were to be deployed in the field.[41] Gribeauval's report began with a wary appreciation of the obstacles facing reform.

> Our [Austrian] artillery here has a great effect in battle because of its large numbers; it has advantages over that of France, as does the French over it. An enlightened man without passion who understood the [relevant] details and had suffi-

> cient credit to cut straight to the truth, would find in these two artilleries the means to compose a single one which would win almost every battle in the field. But ignorance, vanity, and jealousy always intervene: it is the devil's work and cannot be changed as easily as a suit of clothes, it costs too much and one runs a great danger if one is not sure of success.[42]

Even the clothes were to prove difficult to change.

The Rise and Fall—and Resurrection—of the Gribeauval System

The quarrel between the Reds and the Blues—the Vallièrists wore red doublets and the Gribeauvalists wore blue—offers a particularly telling instance of the tensions generated by the reform of the French military after the Seven Years' War. Recognizing that their arm of the military would have to be radically altered if it wished to participate in the wars of the future, Gribeauval's cadre of artillery officers repudiated Vallière's giant cannon, which they termed "paralytic." They denounced siege warfare and placed themselves at the forefront of the new tactical thinking that emphasized mobility in the field. They thereby initiated a bitter polemic within the artillery service, termed by its combatants an "internal war."[43] The controversy was further exacerbated by dramatic shifts in government policy. After Choiseul fell from grace in 1770, Gribeauval was vulnerable. When ministerial portfolios changed hands in 1772, the Vallièrists were returned to power. Within the year, production of the new ordnance ceased, methods of serving cannon reverted to the old system, and a quarter of the officer corps were furloughed (mostly young lieutenants). Gribeauval's partisans were either discharged or sent into the colonial service. Reports on officers were prepared, "so His Majesty can know how much confidence to place in each one." Gribeauval himself retired to his estate near Amiens. Then, following another change of ministers in 1774, the Gribeauvalists were reinstated.[44]

What was state policy in the central question of its war matériel? How did the Gribeauvalists regain hold of the artillery? Why did the Moderns triumph? The answer usually given is that the Gribeauvalists provided a demonstrably superior artillery. This, however, was by no means clear to contemporaries. Both sides conducted quasi-public tests of their system: the Gribeauvalist at Strasbourg in 1763 and the Vallièrists at Douai in 1771. These tests were witnessed and corroborated by savants, military officials, and powerful patrons. But as we will see in chapter 3, the results were inconclusive, and the artillerists continued to flood the capital with arguments and counterarguments. The topic was debated in the academies, salons, and the press. Yet this second trial—this time by "public opinion"—was also inconclusive.[45] If anything, the Vallièrists had the support of the greater luminaries, among them Voltaire, Condorcet, and especially Buffon. Le Blond, the royal censor on military topics, wrote pro-Vallièrist articles on the artillery for the supplementary volume of the *Encyclopédie*. And the Academy of Sciences served as a bastion of support for the Vallières (both father and son were members).[46] Of course, the Gribeauvalists did not lack allies in "enlightened" circles, but they did not always enjoy the support of those individuals who twentieth-century historians class among the "modern thinkers" of the age.

There have been several explanations for this controversy. Pierre Chalmin has concluded that the clash pitted two generations: the Vallièrists were mostly older men; the Gribeauvalists tended to be younger. Certainly, the Gribeauvalists had the support of the junior officers they recruited between 1765–1772.[47] But advocates of light, mobile cannon had been trying to make their case since the days of Louis XIV, as Chalmin himself is at pains to point out. One must still explain *why* the Gribeauval system suddenly attracted the support of the young. The two camps cannot be divided on the basis of their social background either. They cannot be segregated into nobles and commoners. If anything, the "modernizing" Gribeauvalists came disproportionately from the nobility. No doubt personal enmities played a role in the conflict. When the position of Grand Master was abolished without indemnity, some members of the Corps Royal—Jacques Antoine Baratier de Saint-Auban in particular—lost a source of revenue.[48] And Vallière, *fils*, was piqued by Gribeauval's meteoric rise to his father's position, which he had plainly expected would be his. Still, none of this explains the larger grouping of their adherents, nor the appeal of the Moderns outside of the artillery.

The crucial turning point was the convocation of a blue-ribbon panel of those field marshals who had led the French armies during the Seven Years' War (Richelieu, Soubise, Contades, and Broglie). In 1774 they reviewed the data and elicited comments from Vallière and Gribeauval. The salons snickered that the field marshals had been asked to witness "physics demonstrations well suited to increase the difficulty of their decision." But physics had nothing to do with it. Ballistics tests might be difficult to interpret but the field marshals had no trouble interpreting the failure of the Seven Years' War. They noted that those powers against which France was likely to go to war had all adopted light artillery pieces which "followed troops through all the evolutions of battle." France, they argued, could not fail to match their abilities. This was precisely what the Gribeauvalists claimed their artillery was able to do.[49]

The "hardware" at the core of Gribeauval's reforms was a new field cannon (fig. 1.2). It was smaller and lighter than the old siege pieces, and manufactured using the boring machine of Jean Maritz. The Maritz family of Swiss foundry masters dominated French cannon production at mid-century, operating the Strasbourg, Douai, and Lyon foundries. Previously, each cannon had been cast in an individual mold with an earthen core forming its hollow, and finished on a giant vertical reaming machine (fig. 1.4). Now, the piece was cast solid and bored out on Maritz's horizontal machine (fig. 1.5). By effectively halving the windage (the distance between the inner bore of the can-

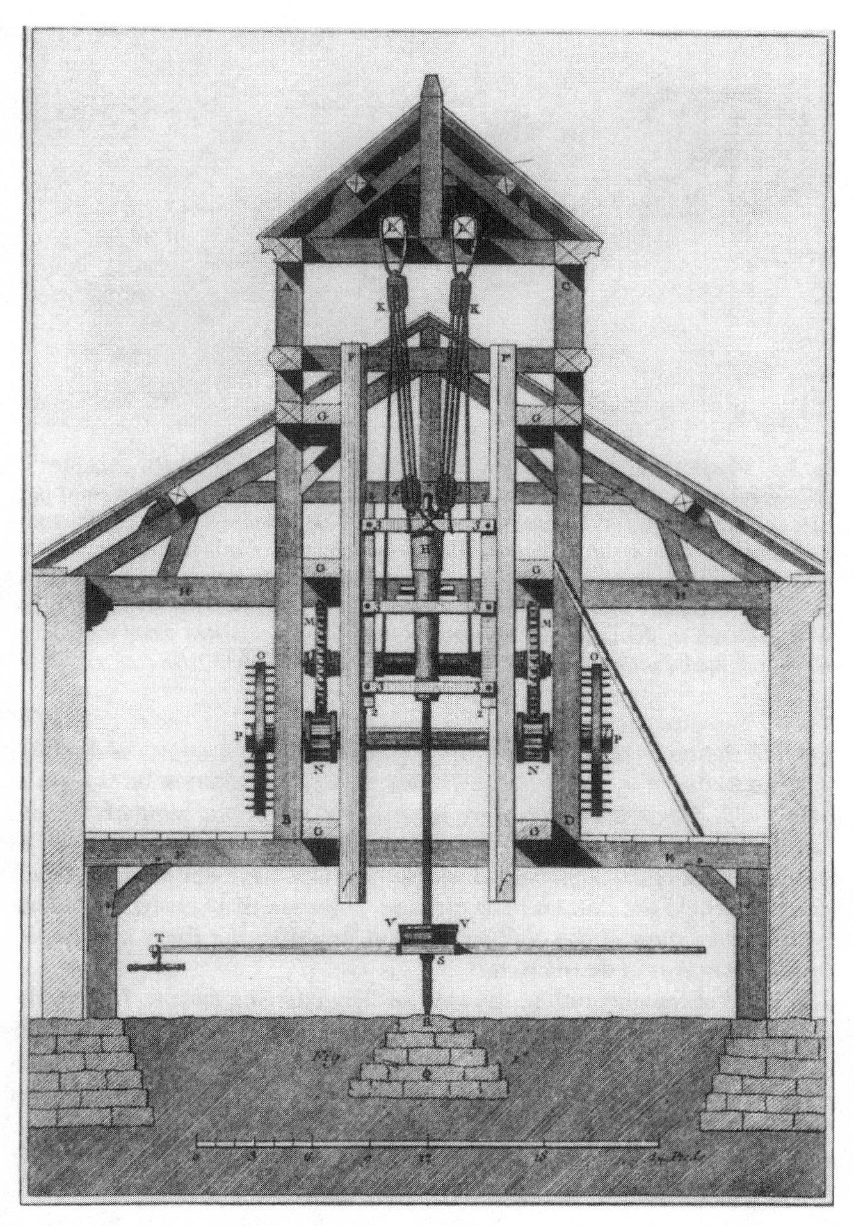

Fig. 1.4. Vertical Cannon-Reaming Machine. This vertical reaming machine was used in the early eighteenth century. The cannon was cast hollow and its core reamed out by lowering the cannon onto a turning bit. From Diderot, ed., "Fonte des canons," *Encyclopédie*, vol. 22, plate 17.

Fig. 1.5. Maritz's Horizontal Cannon-Boring Machine (1/6 scale model). Adopted by Gribeauval in the 1760s, the machine allowed cannon to be cast solid, permitting a more accurate placing of the bore. The cannon (a Gribeauvalist 12-pounder, pictured on the far left) was rotated by horses further to the left, while the bit was advanced from the right. Rotating the cannon increased the chance that the bore would be centric to the axis of the cannon; it also enabled the exterior of the barrel to be turned by the lathe, pictured in the front left. The model indicates the massive stone foundation which stabilized the machine. From the Musée de l'Armée (A13016).

non and the projectile), the new method increased the accuracy of fire from one shot to the next. And because a series of cannon could now be cast in the same mold, different cannon were more likely to perform similarly. It also meant that the cannon could be simultaneously turned on a giant lathe to finish their exterior. This helped accurately place the twin supports (trunnions) that held the cannon in its carriage. Together, these changes ruled out the ornamentation of the Vallière cannon, transforming them into impersonal instruments of destruction.[50]

In place of ornamentation, the new artillery offered a *purpose*. In fact, the new cannon had less range and accuracy than the pieces they replaced. As the Gribeauvalists themselves acknowledged, all these advances in the precision of production were more than canceled out by the fact that the new cannon were shorter than their Vallièrist counterparts and could not withstand as great an explosive charge. Donald MacKenzie has recently made clear that neither accuracy, precision, nor any other objective, is a "natural" or self-evident technological goal.[51] These objectives are in dispute just as much as the means to achieve them. Yet it was here that the Gribeauvalists had the decisive advantage over the Vallièrists for they alone were able to articulate a role for firepower in the wide-ranging debates over military tactics then current in France.

Gribeauval's cannon was only one element in an elaborate techno-social system intended to increase the *mobility* of firepower. The length of the new field pieces varied in proportion to their caliber. The net result was that they weighed roughly 40 percent less than the old cannon. Coupled with extensive changes in the design of the artillery carriages, this made the cannon far more maneuverable.[52] As the Gribeauvalists conceived it, precision was but a stepping-stone to mobility in the field. This larger purpose is evident in a variety of ancillary innovations introduced by the Gribeauvalists. Leather straps allowed the cannon to be repositioned in battle by manpower alone, and a new harness hitched horses in tandem for more efficient transport. Other innovations served to increase the *rate* of fire, a factor more important on the fast-moving field than in siege warfare. An elevation screw (*vis de pointage*) saved gunners the trouble of re-aiming the pieces after each recoil. Prepackaged case shot of mixed powder and balls could be fired with great rapidity, with murderous effect at short ranges.[53]

None of these individual elements was revolutionary in itself. Each was well known and made a marginal improvement at best. Frézeau de la Frézelière had proposed lightening the artillery in the days of Louis XIV, a suggestion Bélidor had reiterated in the 1730s. The *vis de pointage* had been known since early in the century. The use of case shot had been well known in the seventeenth century. Even the Maritz boring machine represented a refinement of a machine he had advocated since the 1740s, and which the French navy had adopted in its independent foundries in the 1750s. As Gribeauval admitted, many of these innovations (like the carriage settings for trunnions to position cannon during transport) had been borrowed from the Austrian or Prussian artillery. What made the Gribeauvalists radical was their willingness to overturn venerable social arrangements to achieve their desired results.[54]

For instance, the reforms transformed the way cannon were served, spelling out the duties of each man who helped fire them. The organization table for the old Royal Artillerie still reflected its origins as an infantry unit. Now artillery units would be staffed around the number and type of cannon needed to accomplish the desired result. All contingencies were planned for; all movements prescribed. Up to that time, the artillery had requisitioned infantry troops to assist with different cannon on different days. Now Gribeauval insisted that the same men and officers serve the same piece for the duration of a campaign. In peacetime they were to train regularly.[55] In a sense, this merely brought cannoneers to the same level of drill that had been the hallmark of the infantry since the seventeenth century. As we will see in the next two chapters, this "rationalization" of the physical work of firing cannon was matched by a corresponding transformation in the "managerial" work of directing their operation.

What gave meaning to these new arrangements was a tactical doctrine which emphasized massive firepower as the basis for offensive warfare. This new doctrine was spelled out most clearly in Jean Du Teil's text of 1778, which military historians have called the blueprint for Napoleonic artillery tactics. The Du Teil brothers were closely associated with the early career of Bonaparte. Jean served as a superior officer in the young lieutenant's first regiment, and Joseph commanded the artillery school at Auxonne where he took Napoléon under his wing. Jean also interceded in 1793 to give Napoléon his big break at Toulon. In Du Teil's doctrine, the central lesson of the Seven Years' War was that to perform effective service in field battle, the artillery needed to be able to keep pace with the troops on the march. Du Teil believed that Gribeauval's artillery made it possible to bring to bear the maximum possible firepower on "the key points which must decide the victory." For this reason, ammunition should be preserved for the "decisive moment," rather than wasted on pointless artillery duels or infantry-bolstering "booms." Instead, in coordination with ground troops, the artillery should close in on the enemy. Battles unfolded in time. And every decision—such as the decision to switch from round shot to case shot—was a matter of "judging the moment."

> It is by means of this science of movement, the speed and intelligence with which the artillery chooses its position, that it gains advantages over the enemy's artillery, when it continually trains its fire on the decisive points and keeps pace with the troops.[56]

This increase in the mobility of cannon was accompanied by an increase in firepower. The new doctrine called for 200 pieces per division, in place of Vallière's 150, plus a disproportionate increase in the number of large caliber pieces, multiplying the "throw-weight" by some 275 percent overall.[57] But this multiplication of cannon threatened to undermine the mobility of the artillery in general. Individually, the Gribeauvalist guns might be more mobile, but *en masse* they constituted a logistical nightmare. The new divisional artillery would need 440 carriages instead of 230, and 2,840 horses instead of 1,720. Advocates of mobile warfare—like Jacques-Antoine-Hippolyte de Guibert—worried that this superabundance of cannon would strangle offensive movement and force the nation to sit back behind its heavily armed coasts and frontiers.[58] Much as he admired Guibert, Du Teil would have none of this. As he pointed out, none of the states with a numerous artillery showed any sign of returning to the past. No power could afford to buck the "progress of reason." "[C]an one compare those times of ignorance and darkness with the progress made by the arts and sciences today?" To oppose the numerous artillery of foreign armies with a small artillery would risk losing all. In these days military knowledge circulated freely and European states had to contend with opponents possessing equal skill. France, he implied, should use its size and wealth to outmatch its enemies on the field, gun for gun. In fact, the ratio of artillery cannon to infantry troops doubled between the wars of Louis XIV and those of Frederick the Great, and stayed at that level throughout the nineteenth century.[59]

These innovations paid off handsomely for the artillery personnel and their clients in the foundries. Instead of retrenching at the end of the Seven Years' War, the corps remained at its swollen wartime level—twice the personnel than at the start of hostilities. The artillery now included 1,065 officer-engineers and 7,409 permanent soldiers.[60] By 1770, the French army had added 1,200 new field cannon, 1,300 carriages, and 2,000 caissons. Though each individual new cannon cost less than the old (because it used less metal), their greater number increased the flow of moneys heading for the foundry masters. Between 1754 (the last year of peace) and 1770, the annual budget of the artillery had nearly doubled: from 2.0 million *livres* to 3.5 million *livres*. A report drawn up in 1771 claimed that in addition to the extra 14.2 million *livres* already spent on the Gribeauval matériel, future spending would include 9.7 million *livres* for new siege cannon, 10.4 million *livres* for munitions, and 8.5 million *livres* for new muskets for the army. This total additional investment of 43 million *livres* in new armaments represented more than a third the typical annual budget of the entire royal army. By the end of the ancien régime, the artillery budget had doubled yet again, reaching 8 million *livres* in the 1780s. Meanwhile the budget for the peacetime army had remained constant at best. Altogether this signaled a subtle but noticeable shift in state spending away from the personnel of the line regiments to the artillery service and to the manufacturers under their supervision.[61]

To be sure, these artillery expenditures paled before the staggering cost of waging war. The military budget during the Seven Years' War ranged between 150 to 200 million *livres* a year. Against this monumental drain on the king's purse, the Gribeauvalists could persuasively argue that to the extent that the artillery decisively influenced the outcome of a battle, the cost was worth it. The result was a spending binge that amounted to an arms race of the eighteenth century. If, as Claude Sturgill has argued, declining military budgets in the eighteenth century actually correspond to a "retreat from loyalty," then the allegiance of the reform-minded artillery was certainly being bought and paid for.[62]

Louis XV's death in 1774 prompted yet another succession of ministers. The second of these, Saint-Germain, was committed to professionalizing the army and reducing the "show" regiments that harbored ineffectual court nobles. On the other hand, the new minister was also committed to reducing the military budget. Would the gyrations of court politics again plunge the artillery into crisis? In his *Mémoires* of 1779, Saint-Germain admitted he was no expert in the subject of artillery. Disavowing any personal responsibility, he confirmed the Gribeauval system, saying that "to re-establish order [in the artillery] it was necessary to decide for one party or the other, and so I gave my preference to the side with the plurality of votes (*suffrages*)."[63] In 1776

Vallière, *fils*, died, and Gribeauval was appointed to the new position of first inspector-general of the artillery, a post he held until his death in 1789. His artillery system served with only minor modifications throughout the Revolutionary and Napoleonic wars, and was not replaced until 1827.

So in the end, the battle of the artillery systems turned not simply on their relative technical merits, but on (1) the unwillingness of those in power to sustain discord in an important government service, and (2) the support of powerful reformers in the upper echelons of the French military, including Choiseul, Broglie, Guibert, and Du Châtelet. When these reformers compared France's army with its most recent adversary, Prussia, they were not happy. But their program to meet this challenge—of which the artillery reform was but a part—dangerously heightened tensions within the ancien régime military.

The Contradictions of Military Reform in the Enlightenment

The polemic which divided the ancien régime artillery was just one manifestation of the larger contradictions of French state-building. Historians have often pointed to the military reforms of the late ancien régime as laying the groundwork for the triumphs of the Revolutionary and Napoleonic armies.[64] This might be called their "positive" contribution to the Revolution. But the reforms also generated many of the "negative" frictions which tore the ancien régime apart from within.

The intellectual current ran strongly against militarism at mid-century. A large proportion of male elites served as officers, yet the martial virtues which Louis XIV had arrogated to himself had been deliberately excluded from court society. Furthermore, the wars of his latter years had been accounted a disaster. Making a virtue of this fiasco, Montesquieu declared the Sun King's failure a "good thing," not just for other nations, but for the French and Bourbons to boot. A monarchy, he argued, ought to stay within its natural limits, and not fight except in self-defense. The philosophes, as a rule, opposed militarism. To stand on the side of reason was to stand for general prosperity, on the side of commerce, and against senseless violence. War, if and when it had to be fought, ought never disturb those productive citizens whose labor was the engine of wealth and progress. The military man was not an honored figure in polite society.[65]

After 1763, a new tone crept into public discussion about warfare. Humiliation at the hands of the English hurt, and Prussia's rapid rise to great-power status was unprecedented in European memory. Frederick the Great's radical new style of battle tactics stunned Europe, and his militarist constitution provoked near universal admiration. Enlightened despotism, which Frederick soon seemed to incarnate, appeared to succeed by subordinating private interests to the service of the state. Louis XIV had famously identified the state with his dynastic ambitions—as in his apocryphal call: "L'état, c'est moi." But

Frederick proudly called himself the first servant of the state. His battlefield successes seemed to flow directly from the subordination of his officers and men, a discipline which enabled them to achieve troop movements of devastating rapidity. Frederick had devised his radical new tactics because his small kingdom lacked the financial and human resources to sustain a lengthy war. The French reformers now concluded that even a great nation could not sustain a series of indecisive campaigns.[66]

According to these reformers, the French effort in the Seven Years' War had been at once arrogant and apathetic. The ostentatious luxury of the superior officers suddenly offered a painful contrast with Spartan Prussian discipline. The officer corps, full of court nobles angling for promotion, many unprepared for the technical exigencies of war, was so riven with personal animosities that drawing up a battle order was a process of near infinite diplomacy. There was no permanent general staff, and disputes between military and civilian authorities sometimes stopped the army in its tracks. To a degree, this was characteristic of all previous French war efforts and those of her allies and enemies as well, but defeat suddenly made the army seem an unwieldy mess.[67]

So when a reform-minded faction under Choiseul's Ministry (1761–70) set out to reorganize the army, they focused on the officer corps. This continued a Europe-wide process of transforming entrepreneurial commanders, who raised and "owned" their troops, into salaried officers subordinated to a chain of command.[68] Under Choiseul and Saint-Germain the crown took sole charge of recruiting soldiers and began to phase out the venality of military office, two practices by which colonels and captains still claimed proprietary rights to their units. Increasingly too, commanders were asked to familiarize themselves with the details of field maneuvers and drill troops regularly. The reformers' aim was to sideline those court nobles and parvenus who took army posts to embellish their status, and instead reward and retain the poorer nobles for whom the army was a career. Seniority, the primary mechanism by which the army had traditionally promoted such officers, was matched to grades of increasing pensions. Ceremonial prestige positions were reduced. This reformulation of the rights and duties of unit commanders was meant to permit new tactical arrangements, such as the combat division, by which the new general staff could direct self-sufficient units and so pursue field warfare with new speed and flexibility.[69]

A number of young theorists prophesied that this new mobile, aggressive warfare would redeem French national honor. The most famous was the dashing young Jacques-Antoine-Hippolyte de Guibert, whose *Essai général de tactique* of 1772 won him the acclaim of the Parisian salons, Voltaire, and even Frederick II. More than anyone else, Guibert rehabilitated the military hero in polite society in the last decades of the ancien régime. Guibert dedicated his treatise to "*Ma Patrie*" thereby reviving that archaic word. And his fame and high connections soon propelled him into a position to effect considerable reform, first as collaborator of Minister of War Saint-Germain (1775–77), and then as the head of the Conseil de Guerre (1788). There, his advocacy of tactical and organizational reforms made him a highly controversial figure.

From the beginning, Guibert recognized that a radical new type of warfare required a corresponding transformation in French military and political institutions. He had withering contempt for the conduct of war under Europe's monarchs—all except Frederick. They ordered war and then refused to fund it. The debt they piled up impoverished the state. Guibert railed against France's over-reliance on fortresses—they wasted treasure and bred complacency. For the same reason he opposed a massive artillery. Under these conditions, campaigns were inconclusive, draining money and demoralizing the troops. What France needed was a citizen army. A regenerated citizenry called to arms and impassioned by their cause would put the monarch's feeble armies to flight.[70]

Guibert professed to hate war. A citizens' army would only fight to defend their homes. But there is no disguising his revanchist program. He boasted that his popular army, once committed to war, would move with unparalleled speed to deliver a crushing blow. Machiavelli, Montesquieu, Diderot, Rousseau, and many others had also advocated citizen armies.[71] Unlike them, Guibert was a professional soldier, whose father had served as chief of staff to Marshal Broglie and was himself rising rapidly in the ranks. Guibert knew one could not just fantasize about a revival of Roman virtues; one had to consider real rivals abroad, the interests of commanders at home, and the way contemporary campaigns were financed, administered, and fought. Even Frederick had agreed that a citizen army would be invincible—without having the least intention of introducing one. There was as little likelihood of such an army under the Bourbons. But if one did not have a citizen army, one would have to invent new kinds of soldiers. Tipping the balance toward offensive war did not just mean inventing new armaments. Guibert proposed peacetime training camps, rigorous instruction for officers and men, and the concentration of central command in a general staff. Subordination was the next best thing to virtue. With it, and with properly conceived science of tactics, one might be able to win the rapid and decisive battles that would make warfare pay.

Inevitably, the reform of institutions so central to the ancien régime provoked considerable resistance. Periods of innovation were punctuated with reversion to former policies. The divisional structure, for instance, first conceived by Broglie in the 1760s and formally proposed by Guibert and Saint-Germain in 1776, was abandoned because regimental commanders objected to surrendering control to a general staff. Similarly, Choiseul's decision to allow officers to fill vacancies as they opened, regardless of regiment, subverted the ability of commanders to reward their own clients. Throughout the last three decades of the ancien régime, the army was subject to a welter of regulations and countermanding laws. This pattern of reform and reversal infuriated all parties, and everyone blamed the confusion on the intrigues of the court.[72]

The underlying problem, however, was that the nobility remained the principal prop of the absolutist state, and the nobility was far from monolithic. Even those whose claim to privilege rested on their military service differed dramatically in their fortune and expectations. By the eighteenth century, promotion in the army essentially operated on a two-track system. Wealthy nobles from the great houses rose with extreme rapidity to the top commands on the basis of family connections and venal purchase. Meanwhile, the poorer provincial nobility rose slowly by seniority through the lower ranks, rarely breaking into a top command. As Jay Michael Smith has convincingly shown, these two groups can be understood as operating within an increasingly contradictory conception of "merit." The court nobility's service to the king was rooted in "birth"; for them, a family history of fidelity and heroic actions "merited" the king's personal recognition and the reward of high office. Alongside this was another kind of "merit"—more familiar to us today—which flowed from the individual officer's innate qualities, as judged by institutional norms. As a practical matter, the rapid increase in the size of his army meant that the king could not personally recognize the accomplishments of each officer. Hence, the administrators to whom the king had delegated the authority to make these promotions increasingly turned to objective measures of worthiness: the number of battle actions, the number of wounds, and especially, seniority. This preserved the king's metaphorical role as regulator of all rank and precedence in his kingdom, while filling the institutions of the monarchy with loyal and competent officers. Minister of War Argenson put it this way at mid-century: "Were everyone the child of his achievements and merits, then justice would be done and the state would be better served." And in the 1770s, Minister Saint-Germain went so far as to condemn the social distinctions among nobles as injurious to the effectiveness of the royal army. The king, he noted, was often personally urged to promote favorites who worked against the letter of his own law. The court nobility "had everything . . . , without meriting it," and the provincial nobility, "had nothing . . . , even when they merit [better]." The result was to erode the proper performance of military duties. "Many have pretensions," he noted, "but few put themselves to the trouble of justifying it by their services and talents."[73]

These reform-minded administrators certainly believed that they were serving the sovereign's interest. But in the long run the new institutional norms had two corrosive effects on the system of privilege which defined social status under absolutist monarchy: they threatened to set aside the "accident" of birth, and they subtly detached state service from personal loyalty to the king. As some military men recognized, this had profound implications for the nature of French sovereignty. Guibert himself noted this tension in 1775:

> So long as one is employed in the service of the state, one owes her all one's abilities without exception. One must show for the public good—if such a phrase

> is permitted in a nation which is not republican—a profound veneration as if it were sacred, and one must put it in advance of one's own interest, and serve it regardless of the circumstances or persons involved.[74]

Old forms of personal service—with proximity to the king being the paramount measure of worth—were no longer deemed sufficient to mark out officers capable of doing their job effectively. In return, the nature of loyalty was itself being reformulated in terms of public service. This did not necessarily mean that the nobility's days were finished. For over a century, the nobility had been in transformation, increasingly turning to formal education to resolve the tension between its claim to rule arising from heredity, on the one hand, and that arising from ability, on the other. But the definition of merit is notoriously vexed. What constitutes good officership? Bravery, as exemplified by bold action on the field? Experience, as counted by years in service? Some ineffable quality of leadership, as recognized by discerning superiors (or by admiring subordinates)? Expert ability, as measured by examination in one of the "sciences" of war? All might be desirable qualities which the army leadership wished to promote. The problem was finding an institutional arrangement that could confer legitimacy on any particular definition of merit—and this proved controversial even among the reformers.[75]

The most famous attempt to solve this conundrum was the notorious Ségur Law of 1781 which restricted access to the officer corps to candidates with four generations of nobility. Denounced during the Revolution as the work of exclusionary aristocrats, the law, as David Bien has shown, was actually intended to create a dedicated, hard-working officer corps, willing to master the expert knowledge that made for effective battlefield action. Reformers like the prominent lieutenant-generals of the Comité Militaire (1781–83) thereby hoped to exclude from the army the status-hungry sons of those "new men" who had purchased ennobling civil offices and lacked the family tradition that inculcated martial virtues among the young.[76]

These conflicting pressures also came to bear on Guibert's infamous Conseil de Guerre. The Conseil was established in 1787 by the Brienne Ministry, which was hoping to forestall a declaration of bankruptcy by the crown. As the largest item in the king's expenditures, the military budget would have to be brought down. Guibert attempted to do this by giving the French army a more "military" constitution. His plan was to make the two-track system of promotion explicit. Henceforth, court nobles ("five or six hundred families") who burdened the state with their sumptuous offices and who refused to subordinate themselves to central command would be relegated to largely ceremonial ranks. Meanwhile, the workhorses of the officer corps (poor nobles from military families) would climb the traditional table of ranks. These officers would compete for promotion, and their advancement would depend on a review of the written record of their service, as judged by a committee of impartial military superiors.[77] But even reformers who shared Guibert's goal, like the generals of the Comité Militaire, had repudiated this sort of invidious distinction among nobles. The nobility was, at bottom, a legal title. Better to stick with the present system, they argued, which operated "as if an invisible hand classed each according to his birth, talents, conduct, and position in the world and at court." As the Comité put it, that way "all is simple and uniform . . . , and officers are all of one *corps*."[78] The proposal was never adopted.

It is also telling that Guibert's Conseil proved unwilling to reduce the budget for the artillery. As we have seen, Guibert himself was not sympathetic to expansion of the artillery. But he announced that he could not reduce the pay of officers who had risen by hard work and seniority rather than favoritism. And a cutback on technical officers would never "get the votes" of the service. Furthermore, he reminded everyone of the danger of laying off officers who could go over to enemy powers (as some artillerists had been preparing to do during Gribeauval's fall from power). This favoritism shown the technical branches of the army greatly irritated senior officers in the general army, who were being asked to accept cuts. Infantry commanders sneered that the artillery should no longer be known as an *arme savante*, but an *arme de luxe*. They complained about the preponderant voice of the artillery and fortification engineers on the Conseil, where Gribeauval, though he was now very ill, had a prominent seat.[79]

The bitter resistance to these reforms dangerously increased disaffection within the army on the eve of the Estates General. As historians have long recognized, the advocates of aristocratic privilege used the run-up to the Estates to undermine royal authority.[80] When Guibert presented himself as a candidate for election in 1789, the nobility of Berry stamped their feet shouting "No, to Guibert! (*point de Guibert!*)."[81] Guibert was defeated by the members of his own *état*. A sometime democrat, he had once fantasized that a conqueror would arise to destroy the privileges and vestiges of the past, and having set the world to rights, graciously retire to private life. Historians have seen in this a prophesy of Napoléon, but I suspect that Guibert had someone else in mind for the role. In 1790, the last year of his life, he wrote: "I was the precursor of many of the opinions which today are the foundations of liberty, and which I propagated in a time when it took courage . . . to do so."[82] The military visionary died at the age of forty-seven, and it was left to others to carry out his program.

Military Enlightenment and Political Revolution

War-making was intrinsic to absolutism. But the "Military Revolution" that multiplied the coercive powers of the centralized state between the sixteenth and eighteenth century also accentuated its social contradictions. The origin of this Military Revolution has been the subject of some controversy, but scholars have all agreed that its terminus coincides with the rise of national warfare during the French Revolution. Why was the monarchist state unable to sustain its machinery of violence?

Historians have always stressed the financial pressures that precipitated the French Revolution, and have linked this burden of debt to military expenditures, especially to those incurred in the Seven Years' War and American War.[83] In the tradition of Paul Kennedy's *The Rise and Fall of the Great Powers*, French military modernization in the eighteenth century can be seen as an inspired response to the rise of Prussian power on the Continent and British colonial competition overseas, which the financial apparatus of France could not sustain—until refurbished by the Revolution. Other historians have looked to features common to all the European states, such as demographic and market pressures, to explain the failure of the monarchy. This is the argument in William McNeill's *Pursuit of Power*: that the European monarchies initiated the process of bureaucratization to conduct their wars on a vast scale—including better planning, broader patterns of recruitment, and greater concentration of capital—but were unable to sustain these levels efficiently without mass participation and larger markets. The *Rex* simply did not have enough *Ratio*. Both these views (which are quite compatible) have the virtue of decoupling military history from the usual saga of battles and "forward-looking" ministers, but their failings are similar too. As David Kaiser has pointed out, nothing obliged the French to carry out this particular program of military reform. Indeed, as Bailey Stone has recently shown, taking on these reforms obliged the French state to grapple with divisive domestic questions, among them the sharing of the tax burden, the sale of civil offices, and the restructuring of the army. To understand *why* France embarked on this course we have to understand the political struggle *within* France and the ideological milieu within which it played itself out.[84]

The historian Guglielmo Ferrero has argued that, in the end, the ancien régime was "demolished by Guibert and his disciples, more than by Voltaire, Rousseau and their school." By which Ferrero means: more by the warfare Guibert prophesied, than by the intellectual critique of the philosophes. Fair enough. The seeds of the modern state were sown by the military reformers (and the civilian bureaucrats), not by Diderot, Helvétius, Condorcet, and their intellectual circle. But Ferrero's statement needs amending. Voltaire and Rousseau were never of the same "school" with regard to warfare and such closely related matters as the progress of the arts and sciences. Whereas Guibert *was* a member in good standing of the Enlightenment club: an admirer of Voltaire, the lover of Julie Lespinasse, a (mediocre) playwright, and a man who believed that he had transformed the venerable military art into something like a science. On the one hand, his impact was a "positive" one. Although he never predicted the mass armies of the Revolution, Guibert was the first to provide a practical program for national war, including the means of making effective use of self-disciplined officers and enthusiastic popular participation. As we will see, his tactics (properly modified) played a key role in the victories of the French Revolutionaries. On the other, his impact was "negative." The reforms which he and his confederates hoped would vault France ahead of its military competition also spurred a violent reaction from those committed to the defense of aristocratic privilege at home. The ensuing controversy added greatly to the paralysis of the ancien régime state in the 1780s.[85]

Under the circumstances, that paralysis was to prove disastrous. Theda Skocpol's structural approach to the causes of revolution has highlighted the dangers to old régime states when the officer class proves incapable of holding its own in international military competition. Drawing largely on the work of Samuel Scott, she has noted the deep disaffection of the French officer corps on the eve of the Revolution there. As a practical matter, this may well have been the case, but we must still consider how this disaffection played itself out in ideological terms. In a response to Skocpol, William Sewell has reaffirmed the importance of ideology as a structural cause of revolution, alongside those influences more often thought of as social. In the French case, he notes the growing tension between the ancien régime's foundation in corporate privilege and its growing embrace of Enlightenment universal reason. He concedes that this ideological "split personality" need not have brought down the state; such contradictions can be sustained over many decades. But the French monarchy was vulnerable once these contradictions became coupled to social grievances, particularly among those groups who provided the financial and military backbone of the state.[86]

This places the question of the relationship between the Enlightenment and the Revolution in a rather different light. Scholars seeking to understand that relationship have faced two difficulties. First, they have had to explain the thirty-year gap between the high tide of the philosophes and the outburst of the Revolution. Second, they have needed to explain how "scribblers" could inspire a vast social and political transformation, itself unleashing twenty years of reaction and carnage of the sort they deplored. To fill this lacuna, Robert Darnton has posited a successor-generation of underground "Grub Street" critics who violently denounced the institutions of the ancien régime in terms their Enlightenment forebears would never have countenanced. And recently other historians—Sarah Maza in particular—have broadened this thesis, pointing to an emerging public sphere in which a critical appraisal of absolutist justice became coupled with a critique of ancien régime social mores. There is much to commend this thesis. But the monarchy was not simply pushed over from "outside"; the ancien régime state fell apart from within. Or perhaps one ought to say that in the late ancien régime notions of "inside" and "outside" were themselves being reshuffled.[87]

Here Tocqueville long ago showed the way. He pointed out how the administrators of absolutism found unwitting allies among those thinkers who elevated Reason to the position of "sole despot of the universe." So that when the philosophes trained their fire on privilege, they simply handed the authorities a new rationale for further centralization—even while holding those same authorities to an impossible standard. Equal taxes, standard weights and measures, and routine public administration were all long desired by the monarchy—and long opposed by the bulk of the nobility whose wealth and authority depended on landed property and provincial autonomy. The same was true for the uniform standards by which meritorious military officers were henceforth to be judged for promotion in state service. Though Tocqueville concentrates exclusively on the centralization of civil administration, his analysis applies brilliantly to the ancien régime "military-industrial complex."[88]

In the chapters which follow, I will try to give some specific content to this Tocquevillean thesis, to show how his somewhat abstract theory of "modernization" was the vehicle for the rise of a particular group of engineers whose mission was to extend the military strength of the state through the application of a new form of practical reason. The (mis)use of reason for these purposes had already been satirized within the Enlightenment, of course. At the beginning of the Seven Years' War, Voltaire had petitioned Marshal Richelieu with his own design for a novel attack chariot "capable of killing 10,000 men." He continued:

> I know very well that it is not up to me to meddle with the most convenient means of killing men. I confess to feeling ridiculous; but if a monk with coal, sulfur, and saltpeter could change the art of war in this wicked world, why can't a scribbler of papers like me render some little unknown service?[89]

Such were the grotesque multiplications reason made imaginable. Similarly, when Voltaire heard of the popularity of Guibert's *Essai générale sur les tactiques*, he could not resist composing some verses in "praise" of the work.

> J'étais lundi passé chez mon libraire Caille,
> Qui, dans son magasin, n'a souvent rien qui vaille.
> «J'ai, dit-il, par bonheur, un ouvrage nouveau,
> Nécessaire aux humains, et sage autant que beau.
> C'est à l'étudier qu'il faut que l'on s'applique;
> Il fait seul nos destins: prenez, c'est la *Tactique*.»
> . . . .
> Mes amis! c'était l'art d'égorger son prochain.[90]

> I passed last Monday by my bookseller Caille,
> Who has, in his shop, seldom anything first-rate.
> "I have, by chance," he said, "a work that is new,
> Necessary for mankind, as wise as it is beautiful.
> We must all apply ourselves to its study,
> Guibert's *Tactics* will shape our destiny."
> . . . .
> My friends! It is the art of doing in our neighbor.

The poem then proceeds in this ironic vein to praise the Prussian monarch for being "a more capable murderer than Gustavus [Adolphus]. . . ." And reasoning that the young Guibert will one day surpass Frederick, the poet wonders when the rising crescendo of destruction will end. But the author of *Candide* was himself no ingénu. The poem closes by giving Guibert the last say. War, the young man is made to respond, is a necessary evil for those who wish to defend themselves—and he proceeds to recite a list of French victories over perfidious Albion. In the end the old poet finds himself admitting that "war is the first of all the arts."[91]

Military reform is one way the Enlightenment program was brought to bear on matters central to the maintenance of the state—and to the identity of the ruling elite. David Bien, for one, has seen grounds for an Enlightenment "broad enough to embrace generals and their exclusionary policies, incipient counterrevolution as well as revolution."[92] It is just this sort of Military Enlightenment that Guibert, Gribeauval, Choiseul, and their cohort represent. But to see in detail how this synthesis of absolutism and expertise was achieved, we will need to examine how the Enlightenment's new forms of knowledge were transformed into practical instruments of power. In other words, we will have to turn to a corps of men the state formed for that very purpose, a corps of military engineers who sought to align a new cognitive program (which they claimed would be useful in beating France's international competition) with a social ideology of meritocratic striving.

Chapter Two

A SOCIAL EPISTEMOLOGY OF ENLIGHTENMENT ENGINEERING

IT IS HARDLY sacrilegious nowadays to suggest that knowledge is produced socially. Historians have at last rid themselves of that old-time faith that science and technical know-how have an immaculate conception in method, logic, or unmediated experience. This apostasy has become particularly necessary now that historians have begun to look beyond the air-less, friction-less, point-particle abstractions of an ideal physics, and confront knowledge-claims about the bulky, worn world of thick things. This is *not* to say that technical knowledge is mere whimsy, a relativist free-for-all in which anything goes. The challenge is to explain *how* this knowledge—which seems to translate, however imperfectly, across time and space—is socially produced, and *how* it comes to be invested in physical things. Understanding this process means coming to terms with the social life of knowledge-makers: their values, their ambitions, their loyalties. In their now-classic *Leviathan and the Air-Pump*, Shapin and Schaffer describe the instauration of the London Royal Society as the creation of a "social space" for the practice of the new natural philosophy, a space in which particular social values were inculcated and invested in physical objects. These social values (and these objects) were political in that they reflected the alliances of Restoration England, and offered a model for the larger polity.[1] Similarly, the particular ways of knowing developed by French engineers in the eighteenth century did not just guide proper design, they also guided proper social conduct. The character of this social epistemology was profoundly shaped by the conditions under which the engineers labored. The profession of engineering emerged under the auspices of the absolutist state. Yet the ethos of modern engineering was also at odds with the constitution of the ancien régime, and pointed the way toward the new régime to follow.

An eighteenth-century French engineer was invariably a military engineer. The term "*ingénieur*" (as in Corps du Génie) does not derive from the word "genius" (*génie*), I am sorry to report, but from the Latin *Ingenium*, meaning "engine of war"—with an emphasis on the contrived and artificial nature of that machinery. When Renaissance humanists first took notice of the occupation, the term referred to experts in the new gunpowder weaponry. In 1694, the Académie Française defined an engineer as "one who invents, sketches, and conducts the works and instruments for the attack and defense of fortifications." In 1755, Diderot's *Encyclopédie* still referred exclusively to military engineers, now emphasizing the many mathematical and technical subjects they needed to master.[2] As we have seen, the siege wars of Louis XIV had overwhelmed the supply of such qualified men, and in the eighteenth century the French state turned to new forms of training. The artillery was the first to establish formal engineering schools (1720), followed by the Ponts et Chaussées (1747) and the Corps du Génie (1748). While apprenticeship training of engineers persisted in the French civilian sector through the nineteenth century, and in Anglo-American lands until the twentieth, this type of school training today defines professional engineering around the world.[3]

The artillery schools of ancien régime France were the first institutions in Europe where students received a scientific education. Why did the French state move to train engineers in this fashion? And how did "school culture" shape their activities? This chapter offers an account of the origins of modern engineering that points toward its interlocking epistemological and social bases. It is the story of how a group of "new men" were molded into a (relatively) cohesive social class—and how they developed new ways of knowing to do so. In the process, it attempts to understand the making of expertise, not simply as an institutional arrangement, nor even as an epistemological pose, but as a code for living. The Gribeauvalists sought to ground their form of know-how in a new kind of "technological life." At the center of this new life was a paradoxical allegiance to corporate service and to individual preferment. The artillerists sought to give the lie to Pascal's quip that "Men are not in the habit of inculcating merit, but only in compensating it where they find it ready-made." To this end, they subjected young men to a rigorous education. Their purpose was to create a *self*-disciplined individual, a being whose institutional habitat has been described by Max Weber, whose anatomy has been dissected by Michel Foucault, and whose highest ideal was service. The modern professions inhabit that peculiar panopticon called meritocracy.[4]

This remains, after all, one of the most widely circulated accounts of the radical legacy of the Enlightenment: that its notion of a "career open to talents" was given free reign by the Revolution. The same Napoléon who famously alluded to a marshal's baton in every soldier's pack was himself a professional artillerist. Of course, engineers were not alone in traveling this path. Lawyers, academicians, civil servants, and other proto-professionals likewise confronted an analogous tension between their corporate privileges and the tug of individual preferment. And a variety of social spaces evolved—the academy, the salon, the Masonic lodges, the "public sphere"—where society could in some measure be organized around these new principles. This conflicted history of merit, however, is only now beginning to be written.[5] For engineers, its resolution was a new cognitive program.

Engineering Values and Social Hierarchy

What then is engineering knowledge? How does it differ from craft knowledge and scientific knowledge? Edwin Layton has suggested that science and engineering are locked in a "mirror-image" relationship, whereby the two communities exchange instruments and techniques, even while they invert their relative value. Scientists, for instance, are rewarded for generating public knowledge, and engineers for making proprietary artifacts. According to Layton, the result is that engineers are tied to the interests of their patron (typically the state or a large corporation), while scientists tend to be relatively independent politically.[6]

An important correction to this model has been made by Eda Kranakis, who has shown that not all engineering communities share the same values. For instance, nineteenth-century engineers in France and America ranked mores quite differently and designed bridges differently as a result. American engineers employed by private capital placed the highest premium on experimental tests that required practical skills. Their goal was to achieve cost savings for their employers; hence, they tested cables until they failed and built bridges to minimum specifications. By contrast, the dominant, state-led segment of French engineers placed the highest value on mathematical models that honed their theoretical skills. Their goal was to build public goods of lasting value and (ultimately) achieve membership in the Academy of Sciences; hence, they built bridges based on "scientific" principles and with considerable overcapacity. The dichotomy, of course, was never this clear cut. As Antoine Picon points out, French engineers in the Ponts et Chaussées acquired much of their training on the job, and their construction projects relied on a specifically engineering art—although many did aspire to theoretical mastery. And as Theodore Porter points out, French state engineers also integrated economic considerations into their design decisions—although in ways which highlight their distinctive approach to calculating the public good.[7]

But the most famous characteristic of engineering in post-Revolutionary France was its social and professional stratification. This pattern has been widely imitated.[8] French engineers were ranked in a professional hierarchy reinforced by a well-defined hierarchy of engineering schools, which in turn recruited from a corresponding hierarchy of social classes. At the top of the pyramid stood the Polytechnicians who went on to attend the engineering Grandes Ecoles, themselves graded in an ascending ladder of prestige (Artillerie, Génie, Géographes, Ponts et Chaussées, Mines). Drawn largely from sons of the upper reaches of the civil service and the rentier class, the Polytechniciens were trained in the most abstract mathematics and sciences, and bound for state service at the highest level (often in the military). Below them in status stood the graduates of engineering schools such as the Ecole

Centrale, where the sons of mid-level functionaries and the industrial bourgeoisie learned an extensive range of sciences and mathematics, as well as practical skills in technical drawing and industrial processes. They were generally employed in private industry, or if in state service, as the minions of the Polytechnicians. Below these, stood the industrial foremen, trained in schools such as the Ecoles des Arts et Métiers, where the sons of artisans received basic instruction in mathematics and technical drawing, and extensive shop-floor experience in a specific trade.[9] This hierarchical structure was typical of many of the French professions in the post-Revolutionary period. It operated under the umbrella of the state, and at the same time, came to constitute the structure of the state. Gone were the legal distinctions of the ancien régime. In place of the Ségur Law, which had limited access to the military by birth, were steep barriers in financial and cultural capital: tuition fees, the *baccalauréat*, and a set of rigorous mathematical tests. These locked out most students while allowing for the "meritorious" aspirant. In return, the state apparatus was staffed with a cadre of loyal experts.[10]

It is my contention that these interlocking hierarchies—accepted by contemporaries as the "natural" landscape of nineteenth-century France—were themselves constructed historically. The schools which validated and reinforced these strata are usually said to have originated in the Napoleonic university reforms. Recently, Roger Hahn has alluded to the important precedent set by the engineering schools of the ancien régime, and the artillery schools in particular. Hahn seeks to show that the French savants inhabited a depoliticized administrative space. My claim is that this post-Revolutionary space was a deliberate creation—and hence political. The connections forged between professional status, social strata, institutional structures, epistemology, and technological practice took shape during political conflicts which antedate the French Revolution and which acquired a further stamp during that period. Demonstrating how this came about will mean showing that this structure was the active creation of the interested parties. Portraying it as a contingent process will mean in some measure showing that there were alternative ways of organizing the profession, and hence the design, production, and deployment of artifacts. After all, from the point of view of the ancien régime, there is much that this structure takes for granted. How did these educational institutions obtain their legitimacy? Why did certain forms of knowledge—such as the mathematical sciences—come to serve as the marker of elite knowledge? And how did particular social strata come to be associated with particular strata in the hierarchy of knowledge when such social strata could hardly be said to have existed in 1750? The answers to such questions are historical in nature, and can be found by examining the engineering institutions of the ancien régime.[11]

Theory and Practice in the Enlightenment

French engineering was born of a marriage of Enlightenment rationality and the (military) needs of the state. The artillery service is a telling offspring of this union. From its position as a second-class and vaguely dishonorable branch of the military, the artillery developed an ethos of self-discipline and hierarchy that would prove emblematic of the new social order. Unlike the officers of the line army, artillery officers could not command by dint of their privileged birth alone. For them, command depended on a claim of mastery over the material world. The push to create a new class of trained professionals begun by Vallière, *père*, in the 1720s was intensified by Gribeauval in the 1760s. Drawing on those social strata that historians have come to associate with the "new men," these artillery engineers carved out a place for themselves in ancien régime society by a set of artful positionings. On the one hand, they positioned themselves as officer-experts between military patrons and cannoneer subordinates. Here their role was to translate the state's investment in capital-intensive weaponry into coercive power on the battlefield. And on the other hand, as technical experts, they interposed themselves between the state and private arms merchants and artisanal weapons-makers.[12] Their role here was to translate the coercive power of the state into capital goods of enormous destructive power.

Central to both these mediating roles was the "middle epistemology" the engineers staked out, combining theory and practice in the pursuit of technological novelty. Engineers sought to blend the universal knowledge of the savant with the particularistic knowledge of the skilled craftworker. Their program was ambitious. Faced with the Gordian knot's problem of how to translate knowledge into action, the Enlightenment engineers seized a sword-thrust solution. Instead of studying the hidden mechanisms of nature, they turned directly to the investigation of human-built machinery. Theirs was a science of artifice, the systematic investigation of human constructs. Their method—deceptively straightforward—was to describe quantitatively the relationships among measurable quantities, and then to use these descriptions to seek a region of optimal gain (as they defined it). Like the Enlightenment call for an empiricist natural philosophy, to which it was near allied, this engineering epistemology renounced any claim to a Cartesian understanding of hidden causes. In return, a vast array of social and political problems was expected to yield up easy solutions. In its utopian excesses, this engineering rationality dreamed of sweeping away all customary practices and remaking the world entirely in the light of present exigencies. As Du Coudray, the Gribeauvalists' chief spokesman, put it, "In those arts which have to do with power—such as the artillery—one cannot be bound by the past, but only by the present."[13]

This radical view that all existing arrangements could be subject to criticism found license in a new kind of empirical knowledge. The Enlightenment looked to Francis Bacon as the preeminent advocate for the view that "Theoria" and "Praxis" be combined to remake the world by conscious design. According to Bacon, each approach on its own was insufficient: without experiment, theory degenerated into sterile dogma; without principles, praxis remained mindless collecting. Bacon, moreover, associated each deficient epistemology with a deficient social type. He condemned the "reasoner" (the scholastic), who like the spider, spun "webs out of his own substance." This type of savant was subject to the idols of the tribe, den, marketplace, and theater—physiological, psychological, linguistic, and metaphysical limits, which deceived the rational investigator and distorted his findings. The failings of the savant, then, were the failings of the isolated mind in commerce with false schools.[14] What is less well known is that Bacon, champion of the mechanical arts, also criticized the "empyrick" (the artisan) who, like the ant, merely "collected and used." The rare mechanic who pursued original investigations invariably picked the fruits of knowledge too soon in the hope of seizing an immediate profit. In other words, the failing of the artisan was the result of competitive striving. Against both types, Bacon held up the individual who, like the bee, combined the two approaches, transforming what he found through innate powers of his own. Only such a creature could effect "a true and lawful marriage between the empirical and the rational faculty, the unkind and ill-starred divorce and separation of which has thrown into confusion all the affairs of the human family."[15]

Bacon's goal, famously, was the "relief of man's estate." In his utopian essay, *The New Atlantis*, this meant a collaborative program of technological experimentation carried out in the secretive House of Salomon. Bacon proposed violent intervention in nature to provide knowledge that was both more certain and more useful. He seemed to have in mind the Mathematical Projectors, men who sought empirical rules to guide precision instruments toward practical ends. The compass, the quadrant, and the firearm—the instruments of Renaissance empire—were the tools they sought to master. Bacon prophesied the instauration of just this sort of investigator and just these mechanical arts.[16] And at the core of the House of Salomon, itself the very "eye of the Kingdom," was an "Inginary," a vast experimental workshop to produce destructive weaponry, designed to "[surpass] muskets or any engine that you have; and . . . multiply them more easily . . . , and to make them stronger and more violent than yours are; exceeding your greatest cannons and basilisks."[17]

The French philosophes made Bacon's thought one of the central planks of their epistemological (and political) critique—without acknowledging the destructive undercurrents of his vision. They too called for a reformulation of knowledge-making and its institutions. In the role of the scholastic spiders, they cast the Cartesians, with their sterile *a priori* speculations and penchant for spinning cosmological "systems" at the drop of a thought. And in the role of the empyrick ants, they cast the artisanal corporations, with their secrecy, venal collusion, and adherence to blind routines. In his "Discours préliminaire," Jean Le Rond d'Alembert attacked the Cartesians for their hubris that the mind could infer the hidden workings of nature without recourse to experiment. Theory in isolation was impotent: "to be a savant in some art it is enough to know its theory, [but] to master it one must join practice to theory." At the same time, he attacked those crude practitioners who experimented without principles and therefore failed to see that a set of facts revealed different aspects of the same subject. Diderot and d'Alembert made this two-pronged attack one of the central strategies of the *Encyclopédie*. In between, they promoted a rule-governed *pratique* that was general, flexible, and workable.[18]

The argument for a theoretically guided technology was most fully developed in Diderot's famous article on "Art" (meaning here the mechanical arts). There he made a plea for the dignity of the artisan, and for the mutual aid that the savant and craftworker should offer one another. Theoretical training was counterproductive unless combined with a practical knowledge of basic physical properties. And he warned against expecting rapid progress in the mechanical arts from those smug, self-aggrandizing savants who sought to base all knowledge on "principles." Too many inventions failed because machines designed on paper would not function in the real world. In the same breath, however, Diderot showed his appreciation of the methodological and organizing power of theoretical science by proposing an academy of the mechanical arts. He deplored the chaotic terminology of the trades and called for a "Logician" to invent a "grammar of the arts." Only a general and open discussion—conducted by means of rigorous analysis—would generate new technical knowledge. Artisans needed to forsake their secret ways if they wished to throw off their slavish acquiescence to routine. Against the corporations which restrained producers and stifled free exchange, Diderot called for an innovative applied science.[19]

This was the epistemological and social opening into which engineers inserted themselves. Technological texts had long claimed to marry theory and practice. Driven by a formidable exploitation of Central European mines, Renaissance metallurgical texts had brought "secret" craft knowledge into the open. Writers such as Georgius Agricola sought to translate local practices into a uniform technical vocabulary and general rules. The goal was to give mine managers a general mastery of the specialist trades so they might "explain the method of construction to others." These texts thereby carved out a role for experts intermediary to capitalist patrons and artisanal miners.[20] To define their mediating role, military engineers further developed this middle way of knowing.

Proper Knowledge *Is* Proper Behavior

Eighteenth-century artillery officers were expected to master skills not traditionally associated with the military virtues. Such skills were not to be acquired on the battlefield, parade ground, or in barracks. Not even the virtuous home of a noble military father—much as it might predispose a young man to an illustrious career—could do the trick. In the seventeenth century, the art of artillery had been passed on through a practical apprenticeship.[21] But according to Enlightenment military leaders, certain essential skills could only be acquired in the classroom through mathematical and scientific study. Though never as famous as the Ecole du Génie at Mézières, the contemporary artillery schools were its equal in providing the most up-to-date technical education; and they turned out far more engineers. Nearly one thousand artillery officers served at any one time between 1763 and 1789, as opposed to three hundred with the Génie. The state considered this training of great importance; by the end of the ancien régime, education expenses came to one-eighth of the artillery's total budget.[22]

After 1720, school training included both practice and theory. Weather permitting, officers spent three days a week in the *école de pratique*, which included maneuvers, siege-craft, and gunnery. Other times, officers attended *école de théorie*, taught by a mathematics professor. Instruction included the science of fortification and sieges, plus various branches of mathematics, such as algebra, geometry, trigonometry, planimetry, and mechanics. Stereotomy (the use of geometrical drawings to plan constructions) was taught in a special drawing studio run by a full-time drawing master. All officers of captain's rank and below attended class, as well as those NCOs who might profit, and young aspirants. Every six months the professor of mathematics tested students in the presence of superior officers.[23]

The artillery service justified this dual track of practical training and theoretical knowledge as central to the identity of the engineer. On its own each mode of understanding was insufficient, and each associated with a defective social type.

> There are officers who devote themselves entirely to mechanical details; others regard such details as beneath their notice. Both types are deficient. The latter must be made to realize that [knowledge of] mechanics is an absolute necessity; the officer should know the language of the workman so as to make the workman understand, and on occasion to instruct him. On the other hand, those absorbed only in mechanical details must know that a knowledge of them alone, without a wider view, does not raise them above the level of a cannon founder, a powder-maker or a workman. . . . The artillery officer who knows his profession must not be ignorant of details, but he must know them, as an architect must know more than a mere stone mason.[24]

The challenge was to combine the two roles in a unified course of study. In an influential series of engineering texts published in the 1720s–1730s and used throughout the century, artillery professor Bélidor steered young engineers through this social epistemology. On one side lay the danger of relying solely on experience, like the artisan. Those who did tended to see things in the same way, and hence slavishly followed tradition. It might be true, Bélidor admitted, that only practical experience brought one in contact with novelty, but without theory to discipline one's attention, one would not appreciate its newness. In this regard, he noted how artisans criticized anyone who proposed some new way of doing things. The degrading aspect of manual labor was less implicated here, than the collusion of the artisanal corporations as they pursued their collective, venal interests.[25]

In a more positive vein, engineers, unlike artisans, were to strive for innovation. And rather than collude, they would vie in meritocratic competition. This competition would mark out their achievements for individual preferment—an identity consonant with their dignity as notables—even as it molded them to fit into narrow functional roles. The engineer was to vie in a meritocratic competition whereby the more he excelled, the more he would come to fulfill the service's corporate mission. The edict founding the new schools in 1720 stipulated:

> Those who have ambition (and all should have it) will not content themselves with what they have seen and heard in school, but will study at home and in private lessons. And by deep thought and application, they will surpass the instruction given. The progress of their studies will every day encourage them, every day they will acquire new insights (*lumières*), and they will thereby achieve the highest merit of their profession, which must be the unique goal of every officer.[26]

Vallière insisted that rising officers possess ability consummate with their post. His *Ordonnance* of 1729 stipulated that "no subordinate, no matter what seniority he possesses, may be promoted without a demonstration of intelligence and his capacity to perform the functions required of the artillery." This requirement was emphatically underlined by Gribeauval. In the margin of a 1762 report he wrote that "seniority must be reduced to next to nothing, favoritism abolished, and superior talents given every reward and these initiated before the age when the body begins to lose [its force] and the mind ceases to acquire [knowledge]."[27]

At the same time, however, Bélidor insisted that theory always be subordinated to practice; it must never be used for "ostentation." Most savants, he admitted, took little interest in practical problems, and even when they did, piled on endless calculations to no apparent purpose.[28] Artillerists in charge of the artillery schools echoed this concern about the danger that mathematical study posed for their pupils. "Mathematics," one noted, "are all too alluring for those who have passed the first hurdles; so we must see that officers destined to fill these very important functions do not acquire too much of a taste for this science." Another officer contrasted the artillery engineer with the solitary "metaphysician" who studied science in the silence of his *cabinet*. The problem with abstract studies was not simply that savants could not master tangible instruments, but that they were asocial.[29]

Again, on a more positive note, engineers, unlike savants, were trained to work together as a corps to shape the world. This meant acquiring a detailed knowledge of tools and trades, even if they would never actually have to perform any physical manipulations themselves. Indeed, engineers needed to understand the artisans' craft "better" than the artisans in order to command them and keep an eye on the avaricious world of private contractors. As the engineer and architect Amédée-François Frézier wrote in his treatise on technical drawing, "Let us suppose that the contractor (*Entrepreneur*) has supplied good stonecutters, is it not [still] consonant with the dignity of an engineer to be capable of understanding and examining what they do, to order and decide on the best [method of] construction, and to not allow faults . . . and thereby avoid wasted effort?"[30]

In short, the engineer was obliged to eschew his own venal interests and consider only the good of his employer, the state. This was what Bélidor had in mind when he said in 1737: "Whatever class one is born into; one must serve to the best of one's ability; that is what it means to be a good citizen."[31]

The consequences of this program of institutionalized innovation were not always easy for the artillerists themselves to stomach. Vallière, *père*, had Bélidor exiled in 1732 for presuming to lay down new guidelines for the design of cannon based on his empirical methods. As further punishment, his textbook was banned. Thereafter, artillery students reverted to the purely geometric mathematics of the abbé Camus. After 1756, students entering the cadet school at La Fère had to pass a stringent entrance exam based on the first two volumes of Camus's course of mathematics, an exam administered by Camus himself. Students as young as fourteen could enroll, but the average age was closer to eighteen. Roughly half were mathematically able graduates of the Ecole Militaire. To graduate, students had to pass a final exam on volumes three and four, and these too came to be administered by Camus.[32]

Discipline and Pedagogy

As we have seen, an artillery engineer received a broad and practical technical training. But to outside eyes, his mathematical education, more than anything else, distinguished the school engineer from the craftsworker—or the English apprenticeship-trained mechanic. Skeptics (both then and since) have implied that for those engaged in practical tasks, mathematics served as mere ornament, a learned flourish to confer status on its practitioners. David Bien has argued that the elementary mathematics at the army's Ecole Militaire was intended to discipline its young cadets, rather than provide them with useful knowledge. He shows that Enlightenment military leaders saw the rigor of mathematics as a tool to instill the martial value of subordination. Bien is no doubt right in asserting that the theorems of their school days were of little practical use to officers in their subsequent career in the line army. Can the same be said of the far heavier dose of mathematics served up to officers in such technical services as the artillery?[33]

In *Discipline and Punish* Michel Foucault writes about those institutions of social control founded during the classical period to mold the bodies and minds of "inmates." Foucault cites the Gobelins tapestry school as an example of how the pedagogues of the late ancien régime regimented the unruly children of artisans with regular drawing lessons. In the seventeenth century the Gobelins Manufacture still apprenticed young boys to a master craftsman who taught them practical skills. But in the middle of the eighteenth century a school was founded that offered formal course work under an academic drawing master. Tasks were assigned at regular intervals, students signed their work, and the supervisors recorded the students' activities in a log book.

> The Gobelins school is only one example of an important phenomenon: the development, in the classical period, of a new technique for taking charge of the time of individual bodies and forces; for assuring an accumulation of duration; and for turning to ever-increasing profit or use the movement of passing time.[34]

But a school is not simply an assemblage of ranked desks and methods of coercion, a mere conduit for miscellany; a school is also a curriculum. In their first class, Gobelins students were taught to trace over model pictures; in the second class, they learned to draw "by sight, without tracing"; and in the most advanced section, they added color. From this, Foucault deduces the discipline built into the organization of space (desk position), time (school periods), and supervision (the master's log book). With these instruments of differentiation, the omniscient eye at the center of the panopticon assigned rank and precedence. Living continuously under the scrutiny of this gaze, the students in time would come to internalize the markers of merit, and so become modern self-disciplined individuals.[35] But Foucault's analysis slights the disciplinary power of this particular *subject*. As a result, he misses the bonds forged between knowledge and social function in these institutions. Both elite artillery students and artisanal pupils sat in ranked desks and filed in and out of the classroom at prescribed hours. However, the two groups studied different—but related—subjects. Schools do not simply provide a discipline *quelconque*; they discipline students for particular *roles*.

Certainly the artillerists cared a great deal about the forms of authority. They maintained strict order in the classroom: attendance was taken and tardiness noted; students had to wear appropriate dress and maintain the strictest silence. As professors were typically non-nobles, senior artillery officers stood by to enforce discipline with house detention and even imprisonment. This far-reaching power to survey and punish students gave them wide latitude to shape the pupils under their tutelage.[36]

This regimentation was meant to provide a standard education for aspirants across space and time. In the context of the ancien régime, students did not come from the same social stratum, as we will see. And upon graduation they were rotated through garrisons all over the kingdom. This problem worsened after the central La Fère cadet school was permanently shut down in 1772. Thereafter, only a uniform curriculum could provide the officers of a numerous and fractious corps with a "common language" to discuss their work. According to Napoléon's teacher at Auxonne, this was the reason artillery students passed through a rigorously defined sequence of drawing exercises.[37] Even small deviations from the curriculum were frowned on. Gribeauval received reports on the professors too, and chastised faculty who taught their own particular methods of solving problems.[38] The intention was to impose a uniformity of habit and thought, instilling a solidarity that was the technicians' equivalent of esprit de corps.

While any well-defined subject might have provided this common language, mathematics was particularly well suited for this role. Rhetoric had been the mainstay of secular education since the Renaissance, but the subject was deemed suitable for courtiers, not professional officers. Rhetoric initiated students into the oily art of eloquence, by which they might redescribe virtues and vices, and vice versa. For the natural philosophers of the seventeenth century such equivocation was inimical to the search for the truth.[39] The structure of mathematics, by contrast, not only provided students with a body of common knowledge, it also impressed on students the virtues of uniformity and precision. These were the virtues that made agreement matter.

In what is by now a familiar paradox, this uniform education also offered the artillery leadership a ready scale for ranking pupils. Math and drawing ability were tested regularly, with elaborate checks to prevent cheating. The first inspector-general himself looked over the reports and incapable students were flunked. Recalling his arrival as a pupil, one artillerist evoked the almost comical competition among the first entering class.[40] Gauging achievement against a uniform curriculum was held to foster emulation and zeal. Increasingly, promotion—in peacetime anyway—came through academic prowess. In the first half of the century, examiners had noted whether cadets pursued their studies with "distinction," "did their best," or "did little." They also commented on students' physical health and morals: Captain Missols "brings exactitude to service, but [is of] a complacent character, not capable of command"; and Captain Langlis is "irreproachable, but too concerned with personal affairs." For instance, the examiner Camus passed Freard du Castel "with indulgence" because even though his mathematical abilities were pitiful, his commanders had lauded his exemplary conduct. With Gribeauval's assumption of power, evaluations were increasingly tied to objective performance on examinations. A 1761 law expressly prevented the director of the school from influencing the examiner to pass weaker students. After 1763, the examiner no longer consulted the school's commander before assigning grades. In the late 1760s, the new examiner Bézout began grouping graduates according to mathematical ability. Revealingly, a tie between two young men for first rank in the class of 1768—Tardy, a noble-born officer, and Gassendi, a *roturier*—was broken in favor of Gassendi on the basis of "birth" because he was a distant descendant of the famous natural philosopher. By the 1780s, the examiner Pierre Simon Laplace simply ranked the graduating class numerically.[41]

So that while a standardized education dissolved old social distinctions and bound young artillery engineers together, it simultaneously made them subject to new instruments of differentiation, and hence, social control. As an impersonal marker of merit, based on criteria about which examiners were unlikely to disagree, mathematics was a tool of radical democratization. As a marker of merit which tracked the translation of innate ability into painstakingly inculcated knowledge, however, mathematics reproduced an aristocratic norm which could not be acquired by just anyone. But whatever kind of a barrier mathematics was, it was not a covert stand-in for a financial one. Cadets were paid to study, and housed at government expense. And the entry-level geometry could be acquired in any decent collège. Several in Metz prepared students for the exam.[42]

But mathematics did not simply regulate the internal affairs of the artillery engineers, it also mediated their relationship with the outside world—both with other social groups and with the material objects they were expected to master. This particular body of shared knowledge—and the discipline necessary to acquire it—distinguished the artillerists, as a body, from general army officers. The artillery service felt that the study of mathematics insulated its young officers from the corrupting world of dilettantism they associated with the study of belles-lettres in favor at court. One officer noted that artillery officers needed "a singular aptitude for study" and "an angelic patience for our artillery trains, to which we are shackled." As a result, he said, "we are strangers, even to our fellow soldiers, even to our commanders." This pride in the cold and fastidious details of their craft was a continual theme in the artillery officers' complaints to the War Office, and distanced them from what military reformers had come to see as the decadent and unprofessional court at Versailles.[43]

Finally, this standardized education equipped artillery officers with an authority they could wield over their subordinates in military actions and the civilian world of work. Uniformity in the execution of orders was seen as a prerequisite for military effectiveness on the battlefield. As noted in the previous chapter, there was a prevailing disgust among army reformers with the way junior officers (usually of high social status) seemed to quibble with the commands of their military superiors. One artillery general justified a uniform education for officers principally as a means of maintaining a clear and effective chain of command: "We must regularize instruction [for officers] and render it uniform so that NCOs and soldiers do not execute a single movement that has not been foreseen and so that it is given in the prescribed form by their commanding officer."[44] In other words, authority in a hierarchical social organization flows from clear and consistent rules for members at *all* levels. For subordinates to be properly disciplined, commanders must learn *self*-discipline, which they acquire by adherence to a common regime of knowledge.

The engineers' refuge in mathematics confirms what Theodore Porter sees as the characteristic response of a "weak" community. As we saw in the last chapter, the artillery constantly had to defend its autonomy against powerful patrons (court nobles of high birth and fortune), and against poachers on its bureaucratic turf (wealthy merchants with court connections). Mathematics spoke across social and spatial distances, binding these men into a corps capable of defending its interests. This was the "public face" of mathematics. But mathematics was not meant to turn officers into automatons. Foucault's analysis of how externally coerced regimentation engenders self-discipline cannot be understood apart from the specific cognitive program which gave engineers their particular authority *and* the flexibility we associate with the professional class. The artillerists knew perfectly well that a math exam did not test all the complex skills and virtues an officer needed to carry out his duties. Even the artillery examiners agreed that memorizing formulae was less important than teaching students to use mathematics to appreciate the relative importance of facts. As we have seen, there was considerable hostility within the corps against mathematics "for its own sake." Students smugly satisfied with their facility with equations were sorely misguided, wrote one artillery officer; such men were "speculators," not "actors." We ought not be surprised to learn that mathematics was a marker of merit in contemporary Cambridge University. But the French engineers were not scholars; they were men of action. Their activities could not be reduced to equations or rigid routines. As members of a quasi-noble profession who inflicted death at a distance, the artillerists also insisted on their autonomy as experts to exercise their discretion. Such mathematics as was taught them was therefore intended to train the *judgment* of young officers and orient their approach to practical problem-solving.[45]

Manners of Mathematics

The issue for engineers, then, was not mathematics *per se*, but what *manner* of mathematics. In a broad critique of artillery education at the end of the Seven Years' War, Guibert complained that even after poring over all four volumes of Camus's official geometry course, young artillery officers were still unable to fire a cannon with accuracy. Not having been commissioned by the artillery, the text made no reference to artillery problems (unlike the old text of Bélidor). Worse, this curriculum at cadet school was mindlessly repeated at the garrison school. Now that Camus had promised additional volumes of his textbook, Guibert quipped, perhaps by volume thirteen he would have made astronomers of them. There was a real danger in submitting cadets (some of them adolescents) to this sterile study; it risked turning them into "metaphysicians" who would "despise the thousand necessary tasks which require less wisdom than *pratique*." In its place Guibert proposed a curriculum suited to their ages and talents, one that would include extensive mechanical drawing, as well as optics, geography, and tactics, plus the basics of algebra, geometry, and Newtonian mechanics. Students, he said, should learn "the theory [strictly] necessary to their profession." Otherwise the schools would soon be empty of pupils, and the few who remained would be "pale under the lash."[46]

Gribeauval agreed with his assessment. He strengthened the practical side of the curriculum. But Gribeauval did not toss out the mathematics professors. Instead he reformed the theoretical portion of the course, dividing the lieutenant-pupils into two groups. The less advanced students concentrated on basic algebra and geometry. And, for the first time, the advanced group studied the infinitesimal calculus. They also learned rational mechanics, hydraulics, the principles of machines, plus some physics and chemistry. The goal was a curriculum that replaced the "pure ostentation" of geometry with "a theory that applied to practice."[47] After heated discussion, a four-volume textbook by Etienne Bézout was adapted for the artillery schools in the early 1770s. Prominent artillerists such as Du Puget and Jean-Louis Lombard helped supply ballistics problems, now couched in the new analytical calculus of the Continental savants. The resulting mix of theoretical discussion and practical problem-sets has since become the hallmark of the modern engineering textbook, and this text was used at the Ecole Polytechnique well into the nineteenth century.[48]

The important thing to keep in mind here is that this engineering science served three interrelated purposes. It offered a method of research (the design of new technology), a guide to practice in the field (the production and operation of technology), and a program of pedagogy (the shaping of technologists). I will treat the role of ballistic science in shaping the design and operation of guns in chapter 3, and the uses of technical drawing in organizing the production of guns in chapter 4. Here I want to focus on engineering science as a way of shaping *technologists*. There were two subjects emphasized by the pedagogical reformers: (1) the new analytic "mixed mathematics" (or what we would today call "applied mathematics"), and (2) the methods of geometrized technical drawing (what today goes by the name of the descriptive geometry). In each case, this cognitive training of engineers—the particular way these subjects merged theory and practice—carried particular moral lessons and positioned students for their mediating role in ancien régime society.[49]

Historians have usually ascribed the Enlightenment's orgy of quantification to a recognition that "Newtonian" mechanics offered a framework to *research* all manner of technological problems.[50] This, after all, was considered to be the chief virtue of the new rational mechanics. In the Renaissance, one military engineer had dismissed mathematics as useless because it failed to account for "those impediments which by nature are always conjoined to the matter that is worked on by the mechanics."[51] By this he meant that the corporeality of thick things could never be treated by an idealizing mathematics. Unlike the pure mathematics of the Renaissance, however, the new mixed mathematics could supposedly handle thick objects.[52]

One may agree that Newton's example prompted eighteenth-century investigators to tackle technological questions, and still note how often success eluded them. Time and again, engineers found that Newtonian mechanics did not lead directly to improved designs of guns or water wheels or boat hulls or other thick objects. This is not to say that mathematics was without value to engineers. If anything, mathematics helped them model technological systems whose behavior could *not* be derived from first principles. For the French engineers who created the science of machines in the years 1765–1830, mathematics served as a form of "descriptionism," a way to quantify how changes in certain measurable parameters affected some other relevant parameter. Mathematics, more often than not, enabled engineers to *evade* real causal explanation.[53]

In a sense, this was entirely in line with Newton's own insistence that scientific claims be empirically verifiable at every level of formulation, and not be simply derived from mechanistic principles. After all, Newton was eulogized in the eighteenth century for his refusal to speculate on the underlying causes of gravity in a Cartesian fashion, famously announcing, *Hypotheses non fingo*. Playing on the cliché that nature is a vast machine whose springs are hidden from our view, d'Alembert argued that the best a savant could hope to do was "find among the most evident parts . . . a few which are moved by the same spring." In this spirit, engineering science was simply a more thoroughgoing Newtonianism that absolved itself of claims about "why" the world acted the way it did. At the same time, the limits of mere description were already evident. d'Alembert derided the growing "spirit of calculation," and disapproved of the recent tendency to let "the desire to be able to make calculations . . . guide the choice of principles." The dilemma here was that as the objects of study grew more complex (more like phenomena in the world outside the laboratory), the difficulties of the mathematics grew by leaps and bounds. Experiments on a small scale might be tractable, but they never transcended their particularity. The difficulty was to find experiments that revealed true principles and yet kept calculations manageable.[54]

Operating in this middle ground of calculation and experimentation gave rise to that method of engineering research which Walter Vincenti has termed "variation of the parameters."[55] In the broadest sense, this method allows researchers to isolate (and so assess) a given element of a system by holding everything else constant. This, of course, assumes that some parameters can be isolated and others held constant, and that the investigator knows which are which. A great deal of practical skill goes into generating this data, either by studying artifacts *in situ* or by means of a scale model. Considerable mathematical ingenuity can go into finding a workable formulation. Above all, a vast institutional effort is necessary to stabilize the conditions under which the object and its operator perform.

It was to this end that Gribeauval insisted that artillerists be taught the method of mathematical analysis, particularly in its algebraic form. There are several reasons for this turn. First, there is preeminent *status* of analysis among those savants perched atop the Enlightenment pyramid of learning. The Continental physicists of the eighteenth century (the Bernoullis, d'Alembert, Euler, Lagrange, Laplace) created a new "foundational" rational mechanics. The relationship between these analysts' academic clout and the state engineers was symbiotic. Gribeauval appointed Laplace to the post of artillery examiner, his first important step up the ladder of state patronage. And Laplace, in his turn, used his position to launch his career as a scientific patron in his own right.[56]

Second, analysis was a *linguistic* mode of mathematics in a period that took language as constitutive of thought. For Condillac, the identification was entire: to write an equation was to think rationally. Guibert and others felt this new form of mathematics was easier for students to grasp and retain. Leonard Euler, as we will see, translated Benjamin Robins's Newtonian geometric derivation of the flight of a cannonball into the language of analysis because it was "simpler, clearer, and of greater utility." The education of hundreds of young engineers played an important role in what one historian has called the "degeometrization of analysis" in this period.[57]

Third, analysis was ideal for the representation of *variation*. It was a language that was open-ended and dynamic. This made it an ideal tool for engineers practicing the method of the variation of parameters. The Gribeauvalists admitted that geometric synthesis had the advantage of presenting new subjects to pupils in a "palpable manner" (*manière sensible*), but analysis was superior for those "in the situation of needing to operate [mathematically]" because "it quickened and facilitated the discovery of the unknown," and "eased all problems of calculation." For the engineer Lazare Carnot, the method did more than clarify thought, it acted *upon* thought, leading "infallibly to new truths." For Condillac, this analytical process of decomposition and recomposition was inherently generative. Analysis—as Antoine Picon has recently stressed—was a language of technical innovators.[58]

And fourth, analysis was ideally suited to tackling the problems of *optimization* at the heart of engineering science. Already in the 1730s, Bélidor was warning that geometry might teach one how to raise solid buildings, but one would needlessly waste materials until one learned the language of algebra and rational mechanics. With the calculus, equations could now be subjected to "realistic" constraints and examined for points of maxima and minima. This meant that designs could (in theory, anyway) be examined for moments of greatest strength or weakness. After 1770 the fourth volume of Bézout's textbook taught these techniques to artillery students, with examples that purported to find the maximum range of a gun, its greatest muzzle velocity, et cetera. Ultimately, analysis would enable engineers to factor that all-important variable—capital—into their equations and formulate solutions in terms of least cost or maximum utility.[59]

Thus, the analytic mixed mathematics did more than distinguish the artillery engineer from the rude cannoneers he commanded on the battlefield, or from the artisans he directed in the manufactures. It was more than the sign of a practical theory. Analysis associated the engineer with research, innovation, and a dynamic mode of thought. The geometric methods of the seventeenth century had expressed the limits of theory, and hence the limits of destruction. Analysis was accessible, open-ended, and explosive. And at the same time, it tied the creator of technological novelty to his duty to husband the capital of his royal employer.

Such geometry as was taught to engineers in the eighteenth century was itself increasingly tied to analysis. In this period, engineers learned to apply the methods of mathematical analysis to technical drafting. Indeed, skill in mathematical drawing became so integral to an engineers' education that without it one hardly qualified as an expert anymore. In the 1730s, Amédée-François Frézier, an architect and military engineer, developed a careful amalgam of theory and practice under the sign of projective geometry (a mathematical version of the plans, facades, and elevations long used in construction design). This was followed by Gaspard Monge's even more general descriptive geometry, taught from the 1760s onward at the Ecole du Génie at Mézières. The artillery school curriculum also included a full year of prescribed drawing assignments coordinated with mathematical theory. Students were particularly enjoined to draw the artillery equipment "with a great deal of exactitude and care, so that they are rendered in the most exact proportions."[60]

These forms of technical drafting solved important social and procedural dilemmas for engineers. It gave them a common language, practiced their eye, and taught them the value of precision. It also gave them a way to master details of construction without acquiring the manual skills of the trades or involving themselves in demeaning aspects of manufactures. Not only did drawing define the engineer's role in production, the modalities of this form of drawing had important implications for the organization of work and the design of artifacts. These new, more exact forms of technical drawing permitted an increasingly sharp separation between the patron-sponsor of an artifact, its conceiver-planner, and its executor-maker. It thereby positioned engineers in a hierarchy of knowledge that stretched from military patrons to subordinate laborers.[61]

This explains, for instance, why engineers of this period drew shadows in their illustrations. Strictly speaking, shadows provide no information not already given in the projective views; on rational grounds they are unnecessary. Nevertheless, engineering officers in the ancien régime were taught to calculate shadows, since the mastery of this technique was deemed "necessary to discipline and perfect drawing." But shadows offered more than an interesting exercise in geometric construction; they also "rendered representations more distinct." As engineers recognized, it was often easier to draw an artifact in projective views than to reconstruct it mentally from the multiple drawings. By adding shadows and tints, engineer-writers absorbed some of the difficulty of representation so that worker-readers and patron-readers might more easily interpret their drawings, thereby preserving the correspondence between the hierarchy of expert knowledge and the social hierarchy.[62]

In particular, by freeing engineers to design and manipulate artifacts on paper, technical drawing enabled them to shift control over production from the artisan's atelier to the drafting studio. Engineers who mastered this idiom therefore spoke the language of command. School commanders explicitly ordered artillery students to hold onto all the memoranda and drawings of their school days, so that they "would not be embarrassed" when ordering workers to carry out wood or metal constructions in the arsenals or manufactures. This geometrization of artifacts was to inspire (and enable) them to set exacting specifications for their subordinates.[63]

Student engineers had a chance to draw these artifacts and see these exacting specifications implemented before their eyes in the "arsenals of construction" attached to each garrison. In these state-owned manufactures soldiers of artisanal background built the army's artillery carriages. Under the Gribeauvalists, these carriages were made with the method of interchangeable parts manufacturing. Officers were required to visit there, and learn "all the principal dimensions of the pieces of all the artillery carriages, the prescribed caliber of each type of smelted [cannon], the weight of the principal munitions, and finally, the common price of wood, iron, and other raw materials in the different provinces. . . ." The more technically minded officers were given command over these battalions of worker-soldiers, and managed these factories; the best were later appointed inspectors in the kingdom's privately held cannon foundries and musket manufactures.[64]

As supervisors of France's largest collection of metalworking factories, student-officers needed to know metallurgy as well. Gribeauval charged the math professor with giving lessons on the physics and chemistry of metals. In weekly seminars the captains kept abreast of the most advanced techno-scientific developments of the day. They read papers by leading savants (Lavoisier, Monge, Vandermonde, Berthollet) and reports by fellow officers on the processes used in the cannon foundries and the manufacture of muskets. In the early 1780s a proposal circulated to establish a chemistry lab, but only during the Revolution was one actually built. Whether the practice of metallurgy made significant strides in France during this period is doubtful, but real progress may be less relevant than Gribeauval's public commitment to defining and making use of state-of-the-art science in the training of his officers.[65]

To enhance this impression, the artillery also sponsored innovative technological research outside the school grounds. Gribeauval sponsored Cugnot's showy steam-engine chariot, which in a 1770 test before a royal audience at Meudon, pulled a massive 45-caliber cannon at five kilometers per hour.[66] This sort of presentation technology made the artillery's commitment to innovation abundantly clear to patrons. Gribeauval also regularly consulted Gabriel-Jean Jars, the most noted French metallurgist of the time and maintained close ties to Ignace de Wendel, the son of a prominent forge master, who had been sent to train as an artillery officer and so acquire technical skills and political connections. Wendel later became owner of the musket-making Manufacture of Charleville and established the famous foundry at Le Creusot in the late 1780s with Gribeauval's assistance. This sort of revolving door of personnel and technical know-how was characteristic of the proto-military-industrial-research complex in ancien régime France.[67]

In short, the artillerists' education highlighted the corps' status as expert managers of battlefield firepower and industrial production. This education served primarily a *social* function, one located in the sphere of values. That social function, however, depended on the *content* of that curriculum. Their training in the various mixed mathematics provided the artillery engineers with a social profile, reconciling the service's obligation to master technical details with its pretensions to elite (aristocratic) status. That reconciliation was given shape by a particular social structure.

Making Meritocracy

Gribeauval and his cadre moved the service in what may appear contrary directions. On the one hand, they sought to militarize the artillery, which meant bringing its officer corps in line with the aristocratic composition of the general army. On the other, they insisted that merit be the main criteria for advancement, even though this meant drawing heavily on those classes reviled by the line army. How they found an institutional framework to reconcile these two aims is the story of how a hierarchy of knowledge was matched to the hierarchy of men.

Unlike the line army, the artillery recruited officers exclusively from the *petit noblesse*, the recent *anobli*, and well-to-do commoners (*roturiers*). Perhaps the most noteworthy feature of the service was the almost complete absence of court nobles: 4 percent of artillerists as opposed to 14 percent in the line army had been presented at court or had knightly antecedents.[68] In the line army, these *Grands* were a source of considerable friction because of their near monopoly on the highest ranks. As artillery commands could not be purchased, the main route to rapid advancement in the ancien régime army was closed here. Court nobles seem to have been further dissuaded by the rigorous mathematical training, the socially demeaning association with manufactures, and the inglorious (and poorly rewarded) assignment of inflicting death from a distance. Du Coudray, one of those rare artillerists with court connections (he had been military tutor to the comte d'Artois, the king's brother), had particular insight into the anti-artillery sentiments of the upper nobility.

> [T]he position of artillery and engineering officer in France requires too much work and application for [wealthy court nobles] to make a long stay of it, and find a place for themselves. [Artillery] officers are composed of those whose birth places them among the ordinary nobility, or those not born to that class, [but] who have a relative ease or education which places them above ordinary citizens.[69]

Register lists of entering lieutenants, kept with some thoroughness after 1763, enable us to characterize those who did enter the service. In the final decades of the ancien régime, some 14 percent of new artillerists had a non-noble background, roughly three times as many as in the rest of the army. This, however, almost certainly represents a *reduction* in the number of non-noble officers since the beginning of the century. One indication of this is the striking number of *roturiers* reaching high ranks, who had therefore begun their careers in the first half of the century. Over one-third of artillery generals promoted in the early 1780s were non-noble, whereas the number of non-noble generals in the regular army never exceeded 5 percent.[70]

Alongside this appreciable number of commoners, the artillery recruited the bulk of its officers from noble families attached to the state's financial and juridical apparatus. Between 1763 and 1781, nearly 7 percent of those listed by the scribe had a father employed on the civilian side of military procurements, and 39 percent had a father who worked in a judicial, financial, or municipal office.[71] Gribeauval was typical of the men who came from this stratum of society. These are the main groups against which the notorious Ségur Law was directed, yet even after 1781, they accounted for 34 percent of new artillery lieutenants. In the years between the Seven Years' War and the Revolution, the artillery recruited roughly half its officers from among the sons of the "new men"—that is, the stratum of society which Bergeron and Chaussinand-Nogaret have identified as the backbone of the nineteenth-century notability.[72] And although we have no data on their personal finances, great wealth seems to have been rare. When the artillery closed its regimental school in Grenoble, local notables complained that this would deprive them of one of the few routes open to young men of talent without private fortune. Those young men who did get in, therefore, were winners in that all-important game of eighteenth-century France; what historians such as Colin Lucas have identified as the manic hunt for office in the late ancien régime.[73]

The final large grouping of artillerists came from families with a history of artillery service. The Aboville, Le Pelletier, Du Teil, and Sénarmont clans each supplied several generations of recruits. In the last three decades of the ancien régime, 23 percent of entering lieutenants had an immediate relative in the artillery (a father, a brother, or an uncle); and of these, half had a father in the service.[74] On its face, this intensive "auto-recruitment" might seem incompatible with the service's meritocratic ethic. Why give preferment to insiders? The limits of this seeming contradiction were reached on the eve of Revolution, when one proposal circulated that would have closed the ranks of the service to all but those with an immediate relative in the artillery. This proposal was rejected as "shocking and . . . contrary to the essence of a *corps à talents*," but no one ever attacked the analogous pattern of recruiting almost exclusively among those of noble descent. No doubt this pattern was justified by the widespread sentiment, legitimated in a dim way by the sensationalist psychology of the period, that one's upbringing determined one's abilities, particularly those appropriate to military service. In this way, the artillery service reconciled the two long-standing faces of aristocratic self-identity—its sense of kinship and its ideal of service—under the Enlightenment sign of talent.[75]

Ironically, once they had entered the service, most officers had similar expectations; in other words, an excruciatingly slow rise through the ranks. An anonymous "Reflections" asserted that a regular army officer managed in eight years to reach the same grade that took an artillery officer twenty to twenty-five years. In peacetime there was virtually no movement. The famous artillerist-littérateur, Choderlos de Laclos, was typical of this plodding rise through the ranks. Despite superlative reports from his superiors, the cadet at the La Fère school in 1760 made first captain only in 1780. He was still a captain when he quit in 1788, after twenty-eight years in the service. Those entering in the early 1770s found themselves at the back of a long line.[76] At such desultory rates, "emulation" was hardly encouraged. What was the use in being meritorious in such a situation?

This is the context in which we must understand the most seemingly radical innovation of the Gribeauvalist party: their method of promoting officers on the basis of election. When a vacancy appeared in one of the upper levels of the corps (colonel, lieutenant-colonel, *chef de brigade*, and major) the members of that rank elected three nominees from the rank below "exclusively on the basis of merit"; the final choice was then made by the minister of war on the recommendation of the first inspector-general. For the rank of captain and below, Gribeauval instituted a hybrid system of seniority and election, with three-fifths of first captains elected and two-fifths advanced by seniority, and one-third of lieutenants elected and two-thirds advanced by seniority.[77] In certain respects this resembles the hybrid system of promotion used by the Revolutionary armies of the year II (1793–94) to balance seniority and talent, democracy and subordination. Historians of the Revolution have universally ascribed these innovations to the new ideals of 1789; in fact, certain elements can be found in the ancien régime artillery. But rather than representing a *flattening* of the military hierarchy, this innovation reinforced it. This was not a bottom-up democratic system in which soldiers elected their leaders—even the Revolutionaries only allowed the practice in limited circumstances, and then temporarily—but a professional structure in which officers selected worthy subordinates to join them. Indeed, the artillery corps was to prove as hostile to the populist leveling of expertise as it was to the interference of patronage.[78]

Certainly, elections reduced the play of patronage and the discretion of the first inspector-general. This mollified those in the higher ranks at a time when Gribeauval's reforms had ruffled feathers in the service.[79] But primarily it insulated the artillery from powerful outside interests, particularly from those without sufficient knowledge to discriminate among the meritorious. The result was that the consensus of the artillery service actually mattered. Indeed, Saint-Germain recognized as much when he conceded that Gribeauval had the "votes" of his corps. Merit in the professions is a judgment made by other experts. This is not exclusively a feature of "modern" organizations. Insider judgments also enabled the artisanal corporations to preserve standards and so control access to their trade. In this sense, the artillery engineers acted within the corporate framework that still underlies the modern professions.[80]

The Gribeauvalists extended this hierarchy of knowledge both up and down the "chain of service"—tying "each grade of officer to their talents and their accomplishments."[81] This can be seen in their establishment of two controversial new ranks. The first was the *garçon-major*, an officer with the rank of third lieutenant, filled by worthy NCOs. While occasional non-noble "soldiers of fortune" had served as officers in all branches of the military—albeit more prominently in the artillery (15%) than the infantry (10%)[82]—the fact that the artillery had now *institutionalized* their presence, and with pay and rank equivalent to that of noble-born officers, shocked contemporaries. The Vallièrists objected bitterly to admitting into the officer corps these men "without birth, education, talent, knowledge of mathematics, and often without principles of moral and [good] conduct." Pecuniary need would force them into acts unbecoming an officer; favoritism would guide their promotion. And worst of all, they would believe that they had risen on the basis of merit! As Saint-Auban put it, "A sergeant will always be more docile and obedient than this new officer, who, believing himself promoted for his merits, will have [all the more] pretensions." In short, these usurpers threatened to disrupt the social order that guaranteed effective command. Upon resuming power in 1772, the Vallièrists abolished the position.[83]

The Gribeauvalists justified this innovation on several grounds: first, it inspired NCOs with zeal for the service; second, it introduced more battlefield experience into the lower officer ranks; and third, it freed the higher ranks from the tedium of discipline and allowed them to devote their time to the theory and practice of their profession. In other words, this reform represented a further division of the tasks surrounding the service of cannon in battle. It signaled a shift from a hierarchy of birth to a more carefully graded chain of command, based on function and education. Prospects for further advancement for *garçon-majors*, moreover, remained limited, and a formidable social barrier still lay between them and those officers who had entered through the cadet school.[84] The Gribeauvalists promoted only those NCOs who demonstrated "irreproachable conduct, acknowledged wisdom, were well-instructed and versed in the particulars of their company, able to teach soldiers, and above all, had never practiced a vile trade; they [were] also to propose no married men without making certain that His Majesty's pay was enough to keep them from being distracted from the exactitude they owe the service by recourse to means foreign to their state."[85] Still, the Gribeauvalists could not resist chiding the Vallièrists for their caste snobbery. Had not they argued that access to the officer corps should be open to all estates, and that merit alone should guide promotion? Du Coudray disingenuously needled Saint-Auban (himself ennobled only after years of military service), that "of all people, he should know there is no such thing as class in the military."[86]

Matching the table of ranks to a hierarchy of knowledge also meant supplying education even to those at the lower ends of the scale. The formal education granted NCOs was another shocking innovation of the Gribeauvalists. But again, this education always operated within strict limits. In the flush of the new reforms, math and drawing classes were extended to NCOs on a regular basis, to stimulate their interest in the artillery during the long peace.[87] A special Ecole des Sergents met three times a week. Once students had made sufficient progress in basic arithmetic, they were given lessons in "practical geometry" to enable them to "translate" the plans of their superiors into instructions for their subordinates. They were also taught skills such as measuring distances with instruments and the forces of simple machines. Those that could not write neatly and correctly were taught to do so, primarily to enable them to take inventory, one of their frequent tasks. A superior officer was always present during class, both to maintain order and to report on those students who showed progress.[88] The best of these might qualify for the rank of *garçon-major*. But the Gribeauvalists were careful to give the sergeants only as much "theoretical principles" as was consonant with their social status and function.

> But we must guard against pushing them too far in this direction, or we risk making them into mathematicians, which would be a great pity because they would then despise their jobs (*functions*) as sergeants. We must therefore limit [their education] to that which is strictly necessary for them.[89]

This same concern with rank and function can be seen at the high end of the hierarchy where the Gribeauvalists created the position of *chef de brigade*. These were elite officers promoted from among the first captains and given command over an autonomous artillery unit that could support an army division through all its maneuvers. From this rank they could overrule the long-standing disputes between the "militarists" of the old Royal Artillerie and the "technicians" of the old Corps Royal. In the person of the *chef de brigade* the peacetime and wartime operations of the artillery were finally merged, achieving the synthesis of capital and coercion that had eluded a century of reformers. Gribeauval also ended the last customary prerogatives of Corps Royal, putting these men on salary. This "rationalization" of military management was part and parcel of the "rationalization" of the soldierly labor of those who physically tended the guns. The hierarchy now stretched from the lowest cannoneer to the first inspector-general. It would ultimately extend to the chief artillerist of them all, Napoléon Bonaparte.[90]

Rank and Revolution

In describing the structure of the engineering corps as a "hierarchy," I have been using a word that French military reformers refashioned for just this purpose. "Hierarchy" has often been employed by social historians as a synonym for social stratification of any kind. But the "orders" of the ancien régime were defined by legal entitlement. This *état* might be acquired by birth or by purchase. But it did not refer to a well-defined ladder in which advancement depended on individual achievement as recognized by one's superiors. Until the 1780s, the term had been reserved exclusively for the rungs of celestial angels, and by extension, for the hieratic order of ecclesiastics. Yet this was how one artillery captain described the ascent of the *garçon-major* in the mid-1780s (in the first such use of *hiérarchie* I can locate):

> In the promotion (*avancement*) of the sergeants and sergeant-majors there exists a sort of hierarchy (*hiérarchie*), a progression which at all time feeds their zeal, and [acts as] a lure to ambition. Before becoming a *garçon-major*, one must pass through all the grades. . . ."[91]

This new secular usage attracted attention only in 1788 when Guibert entitled his law restructuring the army: "Ordinance Regulating the Duties of the Hierarchy of All Military Personnel." And to make clear what he meant by hierarchy, he noted that henceforth, "all must obey the commands of those in the rank above them" because only this would calm the "universal ferment which the lack of rules produces." Applying this term to a secular organization like the army excited considerable amusement from Parisian wits. The term, however, stuck. In 1791 Necker, writing from the safety of London exile, spoke of the *hiérarchie politique* established by the National Assembly. In doing so, he acknowledged that "etymologically it ought to be solely reserved for the realm of the sacred," but he reluctantly agreed to "submit to the extension which usage has recently given it."[92]

J.-G. Lacuée was a career officer in the ancien régime (and onetime aspirant to the artillery) who rose to prominence in that National Assembly. But in 1789 he was still composing his volume of the *Art militaire* for the *Encyclopédie méthodique*. In an article entitled "Cadet," written immediately prior to the Revolution, he attacked the endless mathematics and drawing classes forced down the throats of *cadets-gentilshommes* who wanted to be nothing more than infantry officers. Somewhere between "C" and "E" the Revolution hit home. By the time he got around to writing the article entitled "Exam," Lacuée was openly asking the legislators to "consecrate anew this eternal truth: *Men are born equal in rights*, and this principle of the constitution: *All citizens are equally admissible to public dignities, positions and posts, without any other distinction than those of talent and virtues*." Lacuée opposed training these new *citoyens-officiers* in public schools; the expense could not be borne by the state. On the other hand, military science, he acknowledged, was difficult; its principles and rules could not be acquired by experience alone. So in the place of schools, he advocated a system of examinations, like those used by the artillery and fortification engineers. Apart from exams, he admitted he saw "no other way which was genuinely constitutional . . . [and] could eliminate the arbitrary in the selection of officers." Only an exam would enable the army to select capable men, not simply those with ambition or a "name."[93] The search for some "non-arbitrary" sorting mechanism asserted itself with renewed force with the breakdown of the ancien régime. Lacuée eventually came around to the idea of state-sponsored education. Indeed, as director of the Ecole Polytechnique under Napoléon, he played a critical role in forging the post-Revolutionary links between a hierarchy of knowledge and professional status.

France has long been held up as the nation that first developed a competitive educational apparatus to produce and reproduce a stable elite. This "planned meritocracy," as Harold Perkin has called it, has been said to be the distinctive feature of modern France. The state's monopoly on education—itself the outcome of a nineteenth-century contest over which institutions would define the elite—offered markers of merit by which the various social strata could legitimate their position. This carefully managed hierarchy has proved remarkably enduring; flexible enough to incorporate new groups as they won economic power, and imposing enough to command assent from those not yet let in.[94] Historians have placed the origins of this system in the Napoleonic period. They have thereby begged the question of how the different social strata and types of knowledge came to be connected and ranked. I have suggested that the connections between the social, epistemological, and professional hierarchies were already being forged in the ancien régime. Let me briefly review the argument so far.

In the midst of a corporate society founded on distinctions in "orders," the artillery engineers of the Enlightenment instituted a bureaucratic hierarchy founded on merit and function. I have not assumed that this was a "natural" development. Rather, I have tried to show how the artillery created the social strata from which it drew: a hybrid of the poorer nobility and comfortable commoners. To achieve this crossbreed, they gave institutional form to a middle epistemology appropriate to the dignity of these new men, which would render them fit to serve the state. That middle epistemology was a way of knowing that invited the active manipulation of human instruments to purposeful ends. It partook neither of the savant's speculative study of nature, nor the artisan's immersion in the particularistic details of his craft. The one was useless, and the other unoriginal. Engineers were therefore cautioned against the sociability associated with these deficient epistemologies. An epistemological pose is also a moral code. Engineers were to be neither aloof and doctrinaire like the savant, nor venal and collusive like the artisan. Instead, they were to compete individually for places within a strict hierarchy—though that competition was to make them ever more attached to the corporate mission of the service. And they were to pursue novelty, though again, always in such a way as to serve the state better. The result was an oxymoron: institutionalized innovation.

Meritocracy is said to transcend politics; today it is considered a system in which "the best individual" triumphs. Even within this modern "democratic" notion of merit as the outward reflection of innate talent and effort, constructing a meritocracy presupposes agreement on the standards of judgment about what those talents are. Such agreement must necessarily be mediated by social institutions. When set against the corporate order of the ancien régime, meritocracy emerges as a deliberate construct, a social space in which a particular kind of achievement was rewarded and directed toward particular ends. In the late ancien régime these ends were increasingly detached from personal loyalty to the king. Promoted on the basis of criteria as judged by their immediate superiors, the artillerists were many layers removed from royal largesse. Their loyalty was therefore institutional rather than personal, and directed toward the abstraction of the state rather than the person of the king. Self-disciplined and self-made, the engineers were pioneers in the creation of a new social space which we associate with the modern professions. Some scholars, such as Colin Jones, have taken a leaf from Edmund Burke's accusation against the talented "men of letters" and fingered these professional classes as the instigators of the French Revolution.[95] Are they therefore the missing "revolutionary bourgeoisie" for which historians of the ancien régime have been searching?

Engineers were not the only group seeking to reconcile individual preferment and corporate service. To take just one well-studied example: the legal profession of eighteenth-century France was riven with an analogous conflict. Yet most lawyers—like most engineers—were content with their social status in the ancien régime. Whatever contradictions we may see in their condition, eighteenth-century professionals did not necessarily translate their divided loyalties into revolutionary action. Prominent as some lawyers were to become in the new legislatures and the Napoleonic notability, many others were appalled by the Revolution's assault on their corporate privileges. The lawyers of the ancien régime cannot be fingered as an aggrieved "revolutionary bourgeoisie," any more than the engineers can. Yet many lawyers *were* open to enlightened ideas, including some which placed them in opposition to the established order. Many chafed under the rule of the aristocratic magistracy. And some began to conceive of themselves as self-made individuals, whose innate abilities were the sole avenue to success. As one lawyer put it in 1777, "The lawyer acts alone, without outside help, owing nothing to such fleeting shadows as birth, heredity, credit, venality, or blind faith. All his glory is personal, all of it drawn from his own resources."[96]

For the French military engineers, individual achievement always operated within the framework of their larger corporate mission. They thereby reconciled their individual merits with the traditional rankings of birth through a notion of collective service, which required loyalty to the state. No wonder sentiment in favor of the Revolution ran high among artillery officers, at least in its early patriotic phase. Certainly, as time wore on, a sizable percentage of officers recoiled; the artillerists were distinctly hostile to the populist tenor of republicanism that began to be heard after 1791. Yet in 1791, according to one artillerist's memoirs, the officers in his company were equally divided between revolutionaries and royalists. Very few artillery officers refused to swear allegiance to the Constitution after the disastrous flight of Louis XVI. And whereas only 18 percent of infantry and cavalry officers active in 1789 were still in service in September 1792, the comparable number for the artillery was 42 percent.[97]

Even at the very height of the Terror, when only 5 percent of ancien régime officers remained, 20 pecent of artillery officers were still in French service. And most of the dramatic turnover of personnel took place at the lower ranks. The top ranks of the service almost entirely reflected the training of the ancien régime. They also reflected the mixed social composition of the ancien régime artillery. Gribeauvalist innovations, such as the *garçon-major*, supplied an experienced leadership to replace émigré officers. Many of these soldiers took up the Revolutionary cause with fervor. The continuities of the Revolutionary personnel can be conveniently gauged from the vantage point of 1796, after the Terror had run its course, but before the émigrés had been allowed back. At that point, all nine inspector-generals were products of the pre-Revolutionary schools. Only one, Jean Elbé, had risen from the ranks, and thanks to the Gribeauvalist reforms, he had been promoted to the officer class *before* 1789. Of the seventy-six *chefs de brigade*, 30 percent of the artillerists holding that key rank in 1796 had begun their careers as NCOs in the ancien régime; but more than half of these had risen to officer class prior to 1789. Of the rest (70%), all had trained as officers during the Gribeauvalist reforms, 66 percent were ex-nobles and 34 percent had been born commoners. To summarize: 99 percent of the top four ranks in the French artillery of the early 1790s were occupied by men who had either been officers (46%) or NCOs (53%) in the ancien régime. Of those who had entered the service as officers, 75 percent were noble born. Of those who had risen from the ranks, more than half had reached officer class before the Revolution (table 2.1). The rump artillery of the Revolution was not composed of recent volunteers; nor was it a "bourgeois" artillery. It was composed of aristocratic officers, elite commoners, and their favored NCOs. It was a professional artillery.[98]

As we will see in chapter 8, the reaffirmation of a hierarchical social structure first sketched out during the ancien régime was not a foregone conclusion. Elite monopolies of all sorts were under novel and potent democratic pressures in the years 1793–95. But the new social space engineered by the artillery service did triumph. From one point of view, this pattern is understandable; the new state needed technicians. Even during the Terror, the hostility of the Jacobin War Ministry to "aristocratic" experts fell by the wayside once the Committee of Public Safety had eliminated its populist opponents and seized control of the military. Lazare Carnot and even Saint-Just spared valuable officers of the artillery, protecting *ci-devant* nobles "useful to the Republic." Among these was François-Marie d'Aboville who had commanded the cannon at the critical battle of Valmy. Briefly imprisoned during the Terror, he later succeeded Gribeauval to the post of first inspector-general of the artillery. But the continuities here involved more than personnel; they include Tocquevillean continuities of structure. Indeed, the Revolution saw large portions of the state apparatus organized under just the sort of hierarchy the Gribeauvalists had pioneered.[99]

But why did the artillerists (even those noble-born) accept the new regime? My data suggest that rather than look to social origins to explain the allegiance of the artillerists, we would do well to consider the ideology of the technicians which explicitly *detached* loyalty from class allegiance, and decoupled the military hierarchy from the *ci-devant* social order. Professionalism meant a large investment of time. It meant that one's sense of self-worth was bound up in one's task, rather than in one's "private" status. Professionalism was one of the bonds that kept relatively many artillerists loyal to the French state in the early years of the Revolution. Gribeauval's meritocratic and elective structures prepared the artillery to survive the political changes of the years 1789–95, even as they provided a model of national service which transcended party loyalties.[100] The career of captain Pion des Loches is instructive here.

Born to the *roturier* class in 1770, Pion des Loches was studying to enter the lay priesthood and become a school teacher when he was swept up against his will into the "volunteer" army of 1794. He saw battle on the German front, social turmoil, and political wrangling before setting out for the new Revolu-

TABLE 2.1

The Social Origins of Senior Artillery Officers Serving in 1796

Rank (birth known/total)	Noble birth (% of birth known)	Common birth (% of birth known)	Common birth, Entered as officer (% of commoners)	Common birth, Entered as NCO (% of commoners)	Revolutionary volunteer (% of commoners)	Entered as NCO, Officer pre-1789 (% of NCOs)
Général de division (8/9)	5 (63%)	3 (39%)	2 (66%)	1 (33%)	0 (0%)	1 (100%)
Général de brigade (8/12)	3 (39%)	5 (63%)	3 (60%)	2 (40%)	0 (0%)	1 (50%)
Chef de brigade (61/76)	25 (41%)	36 (59%)	13 (36%)	23 (64%)	0 (0%)	14 (61%)
Chef de batallion (90/114)	23 (26%)	67 (74%)	3 (4%)	62 (96%)	2 (3%)	38 (61%)
Total (167/211)	56 (34%)	111 (66%)	21 (19%)	88 (79%)	2 (2%)	54 (61%)

Source: Data are from A.N. AF III 153 *Etat par ordre d'ancienneté*, nivôse, year V [December 1796]. Birth status has been checked against S.H.A.T. Yb668-Yb670 *Registres*, 1763–1789; and A.N. AF II 293c *Registres*, [1795].

tionary artillery school at Châlons. That year's class was full and he returned to the front, but the next year he took the entrance exam after a draconian program of self-discipline. Twenty-five years old and a veteran of several campaigns, he derived from this mathematics exam a profound sense of his own worth in the midst of the Revolution's social chaos.

> It was at the beginning of 1795 that I became a man. Under my cannoneer's uniform I was preparing a peaceable life, which intrigue and cabals would not trouble; the most frugal diet, the most diligent work, privations without number, all this only redoubled my zeal. . . ."[101]

The artillery offered Pion a refuge from political troubles. "In the regiments," he said, "we knew neither aristocrats nor patriots, our leaders saw in us only good or poor soldiers." Thanks to his efforts, he had made himself a career (*un état*). Later, he would see how inadequately his theoretical training had prepared him for fifteen years of battle across the length and breadth of Europe. But the mathematical instruction—and the self-discipline it had fostered—were essential to his esprit de corps and sense of himself as a worthy servant of the state.[102]

The middle epistemology of engineering structured the corporate ethos toward action in the name of a higher good. Innovation was put in the service of the nation. As the artillery instructor Lombard put it in the preface to his study of ballistics (published posthumously in 1796): "[The artillery officer] is devoted to perfecting his art in order to make himself better able to render the most notable services during war, which make him known for good reason as the most solid support of the state."[103] Engineers were *designed* to serve.

Chapter Three

DESIGN AND DEPLOYMENT

THE PREVIOUS CHAPTER laid out the social and epistemological basis of Enlightenment engineering as the story of formation of a new polity within the body of the ancien régime. Here I continue that story into the heart of the gun. To that end, I examine how the design and operation of eighteenth-century military weaponry reflected the artillery engineers' suppositions, and hence expressed particular political values. My purpose is to show how technological change in the late eighteenth century was an integral part of the transformations which preceded and accompanied the French Revolution. By this I mean more than that experts squabbled over the proper way to use guns. I want to argue that the outcome of these disagreements came to be embedded into the very design of guns and made a difference to the way they were deployed. The qualities of physical artifacts are the (temporary) resolution to social conflicts, themselves part of the broader reconfiguration of the polity.

Making this argument stick will mean confronting two venerable accounts of technological change, each of which denies in its own way that artifacts possess political qualities. First is the claim that technology is the direct translation of scientific knowledge, or the fallacy of "applied science." In such a view, the design of an artifact is dictated by the available stock of knowledge about basic natural processes. Second is the claim that technology is an unmediated response to social demands, or the fallacy of "derived society." In such a view, the only artifacts which survive are those which produce results sufficiently desirable to some sort of sponsor or consumer. Implicit in both accounts is the claim that technology-makers (in the aggregate anyway) are neutral conduits who passively mediate between the epistemological and social world around them. Certainly this is what engineers have often wanted outsiders to believe. And certainly both claims are trivially true. *Of course*, the "facts of nature" and "social interests" constrain the sort of technologies we can expect to find at any time and place. But these accounts of technology merely set the outer limits of the technologically possible. Only in the messy middle will we find the answer to the difficult, but all-important *historical* question: Why are particular technologies designed and used in particular ways at particular times? To answer this question, we must necessarily confront the active role of technology-makers.[1]

One approach to this conundrum has been suggested by Walter Vincenti with his idea of a "design hierarchy." Vincenti argues that design specifications on the broadest scale may be set by external social needs; in his example the overall design of a new fighter plane must meet the U.S. Air Force's appraisal of its tactical (and institutional) needs. But he argues that as one moves down a design hierarchy—to the shape of a wing section, say—engineering knowledge increasingly reflects the "way things really are." Surely, this argument goes, there is a core of "reality" upon which technology is built.[2]

To the extent that gun design is based on years of experience with firearms, codified as engineering know-how, this is plainly right. Nor do I deny that modern guns are superior to antique guns, if by "superior" one means that they out-perform older guns with reference to specific criteria, such as accuracy, or rate of fire, or throw-weight, or even a combination of all these. And I certainly acknowledge that systematic experience about these artifacts has contributed in essential ways to that superiority. That said, one must still confront the question of how this knowledge is produced, validated, and used. In other words, a historian must examine the question of technological knowledge from the point of view of its human creators and practitioners. From this perspective, as Donald MacKenzie has recently made clear, technological "facts" can prove on closer inspection to be highly flexible constructions. Whatever verisimilitude such knowledge may ultimately seem to possess *to us*, it still must be produced by parties who are uncertain as to the outcome and who have a stake in the result. Technical knowledge, moreover, cannot simply explain how the world *is*. One cannot simply announce that the trajectory of a cannonball follows Euler's equation (itself based on a large number of approximations and limiting assumptions) and leave it at that. Designers must make choices. Given the tradeoffs inherent in any technological design, how did artillerists decide what sort of criteria would be important and which configuration would produce the desired result? How were they to judge and validate such technological decisions *in advance*? And finally, knowledge must also guide practitioners in the field. Equations are not a substitute for meaning. Practitioners must still decide how to *act* in the world. Given the many situations encountered in war, how did the artillerists decide which instruments to use and how?

To this end, this chapter takes up two problematic "sciences" of the eighteenth century: the science of ballistics and the science of tactics. These two subjects had long pedigrees, yet Enlightenment practitioners of both used new tools to discover (and justify) new techno-social arrangements. While today we consider ballistic science an established "fact" and dismiss military science as a chimerical enterprise, eighteenth-century savants did not make so sharp a distinction.[3] What was then unsettled about each seemed similar: on one hand, the relative status of controlled experiment and lived experience; and on the other, the relative organizing power of mathematical theory and social codes. Moreover, finding any (temporary) solution to one of these sciences depended on finding a (temporary) resolution to the other. Although gunpowder weapons had influenced warfare since the Renaissance, their role in battle was still contentious in the Enlightenment, and not just on ethical grounds either. For the Gribeauvalists, the science of ballistics was part of their effort to make a case for the *importance* of guns in warfare, and this, in turn, meant defining a type of warfare—a system of tactics—in which those guns (and their managers) played a central role.

This chapter, therefore, pushes in two different (but interrelated) directions. First, it travels deeper in the realm of technological "fact" to show that the design of hardware embodies the suppositions of its creators and users, not the rote application of scientific knowledge. And second, it travels out into the broader world to show how the criteria "external" to the hardware are shaped by the creators of the technology. My procedure, therefore, will be to begin at the "bottom-most" rung of Vincenti's design hierarchy. I will start by laying out my case for how engineers actively mediate between their science and the artifacts they design and use. I will then address their active role in defining the larger social context in which those artifacts are deployed.

The Art of Artillery

The early modern art of artillery—like that of hat-making or gun-making—required the coordination of large numbers of men whose tasks had become increasingly specialized. When speaking of a "gunner" then, we are speaking of an officer-manager who coordinated ten to fifteen soldiers in the aiming and firing of cannon as part of a larger process of delivering explosive power to the battlefield. This gunner operated under the supervision of a military commander, himself embedded in a bureaucracy that reached up to those who nominally directed the battle. So long as gunners and their subordinates were expected to deliver results in the diverse circumstances of warfare (the "market" for their craft), no single set of tools or procedures could produce everything a military commander might want. The best that could be done was to stabilize large portions of artillery practice and hardware, while still providing recipes and tools flexible enough to meet the eventualities of combat. In essence, this meant creating a stable set of conditions on the battlefield, a sort of laboratory-like oasis amidst the chaos and destruction of war. This was necessarily an *institutional* achievement, requiring coordination and discipline among a hierarchy of well-trained men.[4] To understand *how* it was achieved, we must examine how gunners integrated theory and practice. Only by understanding what gunners actually *did* can we appreciate how the artillerists' social life was inscribed onto the weapons themselves.

Early modern firearms were not simply devices for launching a projectile. As one artillery theorist noted, the gun was itself an instrument which guided its own proper use. That is, early modern cannon and muskets served as their own "sights" because gunners used the exterior shape of the gun to aim their weapon. For this reason, the very dimensions of the barrel implied a set of social practices and a theory of projectile motion. These weapons, therefore, were scientific instruments of the type noted by Gaston Bachelard. They were, of course, products of the foundries, forged of bronze and shaped by human labor and industrial machines, but they were also the substantiation of physical theories. That is, guns are thick objects given meaning by reference to procedures, other instruments, and a working theory about how they function.[5]

What was that theory? Previous accounts of ballistic science since the Renaissance have divided its development into three major phases. The first was initiated by Niccolò Tartaglia in the sixteenth century. His formal mathematical treatment was offered as an aid to practical gunners, and as a learned confirmation of their prejudices. Tartaglia represented flight as an Aristotelian "mixed" motion, which included a violent portion (the initial straight line of fire), a curved portion (an intermediate arc, which he took to be circular), and a natural portion (the final vertical descent).[6] The second phase was initiated by Galileo Galilei in the early seventeenth century. His reformulated ballistics was one of the most striking demonstrations of his new mechanical philosophy. Deliberately setting aside the impediment of air resistance, Galileo used the principles of inertia and acceleration to calculate that the trajectory followed a parabolic curve. The triumph of this mechanical philosophy in the hands of Descartes, Huygens, Newton, and the giants of seventeenth-century natural philosophy is well known. Specifically, Torricelli used Galileo's theory to produce tables and instruments for mortar fire, and Newton attempted to provide a theoretical formulation of air resistance.[7] The third phase was initiated by Robins, Euler, and the ballistics engineers of the mid-eighteenth century. Combining new empirical tools and mathematical techniques, they attempted to take air resistance into account. With this new knowledge, they suggested how to improve the design of guns and the practice of gunnery. Told in this manner, this story of ballistics knowledge is a tale of progress, a stream that joins the river of true understanding of the world. Indeed, ballistics has often been cited as the prime proof that scientific knowledge produces useful technologies. Why then did Enlightenment ballistic theorists fail to persuade gunners to give up the three hundred-year-old Tartaglian model—or indeed, take account of any model at all?

For an answer, historians have followed A. Rupert Hall, who since the 1950s has argued that Galilean theory (phase two, above) did little to help early modern gunners hit their targets. For all their assertions of the utility of their endeavor, Galileo, Newton, and the other natural philosophers treated ballistics as a mathematical "gymnasium," and had little real interest in the practical problems faced by artillerists. Not until the nineteenth century, Hall claims, when technological mastery of metallurgy had created stable conditions within the cannon, did scientific theory offer a useful guide to gunners.[8] With this account, Hall hoped to refute Boris Hessen's 1931 accusation that commercial and military interests (shipping, warfare, etc.) had prompted the Scientific Revolution. More generally, he took issue with Robert Merton's 1938 cultural sociology that posited a (religiously inflected) practical life as the motive for the new form of scientific inquiry. Hall's signal contribution was to warn us not to take at face value the word of early modern apologists that scientific knowledge contributed to warfare, nor read twentieth-century "successes" back into the preindustrial past. The problem here is Hall's narrow conception of the "science" involved in firing a cannon, and his unwillingness to examine other forms of knowledge which—however imperfectly—guided activities in the field.[9]

Recently, Brett Steele has taken issue with Hall, plumping for a "ballistics revolution" of the eighteenth century (phase three, above), which generated *usable* knowledge about guns and gunnery. In this regard, Steele has performed an extremely valuable service, reminding us that engineering is not derived from "pure" science. There are, however, two fundamental problems with his approach. First, while Steele properly criticizes Hall's view of science as having an "immaculate conception" in philosophy, he himself offers no alternative account of the social production of technological knowledge. To be sure, Steele argues that the Military Revolution then underway in Europe *motivated* the search for technological knowledge. But this, in itself, is hardly a surprising claim, and tells us nothing about the means by which such knowledge was socially validated or implemented. Second, his analysis assumes that technological knowledge *drives* social and political change; hence, his claim that the ballistics revolution prompted the rise of a rigorous mathematical education for European artillerists, suggested new weapons, altered the nature of combat, and enabled Napoléon to conquer Europe. Such a model of causation crowds off the historical stage a host of more plausible and complex techno-*social* explanations for these momentous events. And Steele goes further. His hero Benjamin Robins is also said to have offered the first successful explanation of why rifling worked, initiated the study of thermodynamics, discovered the sound barrier, modeled the first internal combustion engine, and "anticipated" certain results regarding the aspect ratio in airfoils. This list of anachronistic "firsts" begs important questions. Why did Robins (or Bélidor, or Euler) set out upon this approach to ballistics at just this juncture? And why were their proposals either ignored or bitterly contested by the orthodox artillerists of the day? These are important historical questions because they frame the eighteenth-century origin of engineering science in terms of the interests it served (and subverted) in its own time. Steele's "lag" thinking is characteristic of determinist accounts of the relations of science, technology, and society.[10]

Together, Hall and Steele represent the two paradoxical branches of the received view of science: that knowledge-producers are solitary thinkers, and that they are the driving agents of technological progress. From this position, scientific knowledge becomes the unmoved mover, the engine of all progress (for good or ill). Among the several reasons to doubt this account is that it dovetails too neatly with the claims of the knowledge producers themselves. Hall's view, in particular, now appears to be part of a broader intellectual movement, in the wake of the atomic bomb, to absolve pure science from the taint of war-making.[11] This book takes a different perspective. Without denying that "progress" in technological capacity can occur, I take seriously the choices left to human beings faced with trying circumstances and an unknown future. This means thinking about knowledge in a much broader way. As we saw, the engineering "marriage" of theory and practice followed upon a protracted courtship under the aegis of the state. The thing to recall is that this marriage was never consummated. Just as "calculation" guided the craft skills of artisanal gunners, so did "art" continue to govern the practice of ballistics engineers. Yet a shift in the relative mix of art and science does signal a broader change in the form of the engineers' technological life.

Ready . . . Aim . . . Design!

Ballistics theorists of the Enlightenment exerted themselves to make the triumphant new mechanical philosophy useful to gunners. Theory, they claimed, produced new knowledge, synthesized experience, and normalized practice—all with a view toward generating more predictable results (themselves more theoretically plausible). Consider François Blondel's influential *L'art de jetter les bombes* of 1685. There the man who spoke of the "resistance and obstinacy of matter" laid out a history of artillery as a tale of theory-led progress, culminating in the parabolic curves of Galileo. In Blondel's account, ballistic science was firstly an interplay of theory and experiment in search of more certain knowledge. While an "infinite number" of factors influenced the flight of the ball, Blondel believed these could be measured and considered separately. Air resistance was one of them, but it was not a major factor so long as the ball was round and heavy and did not travel at excessive speed. Secondly, ballistics was a mathematical theory that synthesized experience. Gunners could not achieve best practice without knowledge of first principles, especially when they encountered situations outside their ordinary experience. Theory enabled gunners to learn their trade quickly without acquiring their skill on a case-by-case basis. This was theory as pedagogy. And thirdly, ballistics theory guided the normalization of conditions that made theory workable. To be sure, uncertainty dogged the conditions of warfare (wheels were not level, powder varied in quality, etc.). But theory could suggest how practitioners might best reduce these uncertainties. In sum, Blondel did not expect theory to supplant practice—he admitted that his new range tables were inadequate—but he hoped an experimentally informed theory could guide practice.[12]

Against this vision, practical gunners plied their craft as an "art," the same sort of rule-based tacit knowledge that guided artisans in their workshops. In 1716, the expert artillerist, Deschiens de Ressons, took his case to the lion's den, explaining the failure of Blondel's ideal curves to the Academy of Sciences. There, he outlined twenty-five factors that disturbed the operation of a gun in the field, from the eccentricity of a gun's bore to differences in the weight of cannonballs. Did the academicians know that balls made by the same workers varied by as much as five pounds? He then hinted at some of the practical recipes by which artillerists minimized these uncertainties: how they parked a cannon carriage on a stable platform, how they mixed the day's gunpowder in a large sheet *before* doling it out, et cetera. . . . Even so, the best gunner in the corps, shooting three times in a row, achieved three different results. What good was a precise theory in these circumstances?[13] This view of artillery had the backing of Saint-Rémy's monumental *Mémoires d'artillerie*, which sneered at Blondel's presumption. These "artisanal" experts were *not* saying that gunners shot blindly in the dark. On the contrary, a gunner needed years of practice before he could develop an "eye" good enough to direct the cannon's aim while keeping his subordinates working to the rapid pace of a drum beat. Indeed, Saint-Rémy's massive three-volume text reads like a compendium of how to create a disciplined social structure that will sufficiently stabilize conditions so that technological *judgments* can be made.[14]

So early eighteenth-century artillery proceeded on two levels. At the Academy of Sciences and in their mathematics classes, ballistic theorists, including the Vallièrists, professed the new rational mechanics. Meanwhile, in the field, gunners (including those under the Vallièrists) adjusted their pieces by trial and error, guided by procedures, instruments, and "practical rules" that were an anachronistic, metaphysical mess. Indeed, something like Tartaglia's picture of mixed motion persisted among these artisanal gunners because it provided a ready guide to practice in the field. This was less a matter of the "true" trajectory, than of calculable results. According to the mortar expert, François Malthus, projectiles sometimes followed a mixed motion and sometimes a parabolic motion and sometimes a hyperbolic motion, but he preferred to think of the trajectory as a mixed motion because that way "it is easy to find the rules by which one can achieve the desired goal. . . ."[15]

The Tartaglian model did this by distinguishing among the different circumstances encountered in war. It thereby segmented "the market" for the artillerists' craft, and suggested the appropriate rules of calculation, measuring instruments, and weaponry for each situation. For the short-range shots that had sufficient impact to damage fortress walls (the violent portion of flight), gunners attached a "collar" of wood to the muzzle of their long cannon to create a line of sight parallel to the line of fire—and blasted away.[16] More often, when the battle or siege called for longer-range fire (thereby including

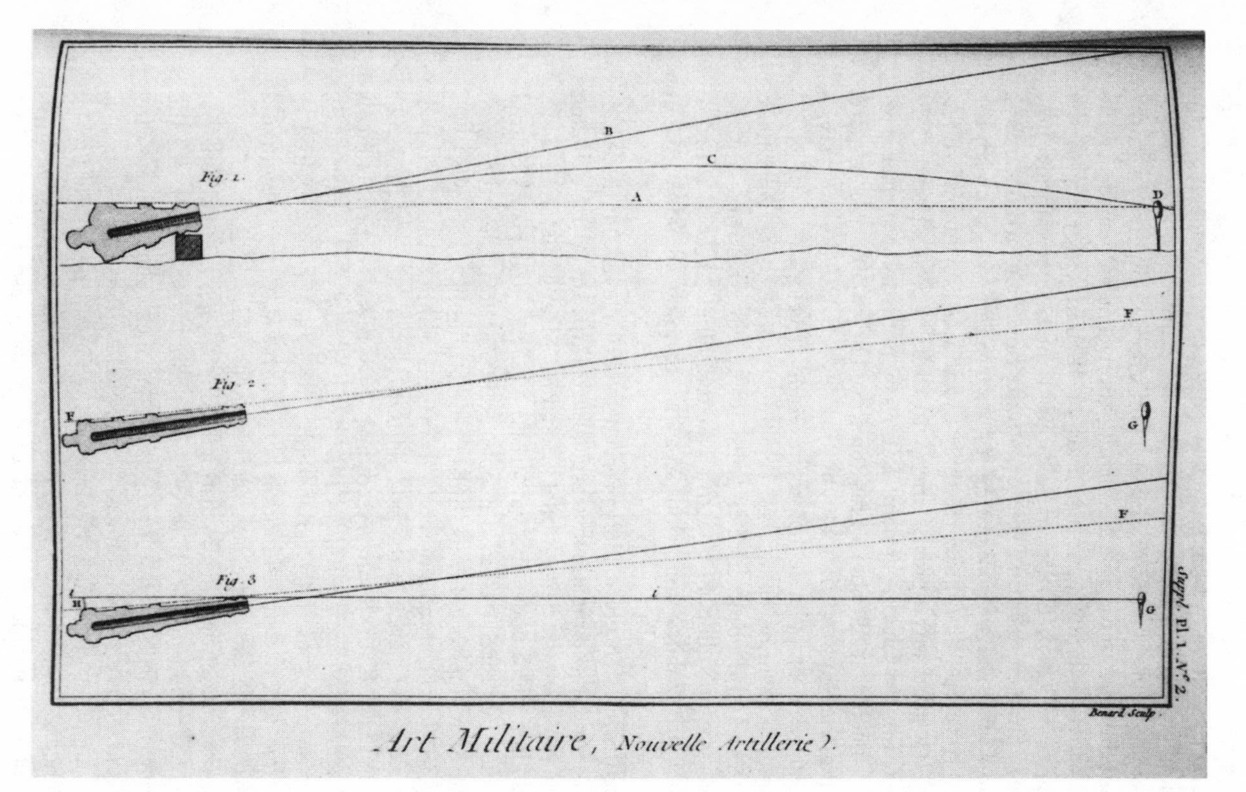

Fig. 3.1. Aiming Cannon Using the Point-Blank Method. Composed by a pro-Vallièrist, these drawings indicate how to aim a cannon by sighting along the external dimensions of the barrel. The top image shows how the old heavy cannon was calibrated to hit the target *D* at point-blank range. The middle image shows how the new Gribeauvalist cannon, set at the same angle would not allow one to sight the same target G. The bottom image shows how the use of a tangent sight at the breech of the cannon (the *hausse H*) would enable one to sight the target along *i*. The illustrator neglected to draw the curving path of the incoming cannonball. [Le Blond], "Art militaire," *Supplément à l'Encyclopédie*, plate 1, no. 2.

some curved portion of flight), gunners used the principle of the *but en blanc* (fig. 3.1). Indeed, the first thing an early modern gunner wanted to know about his cannon was this "point-blank range," that is, the distance at which the line of sight intersected with the downward curve of the shot. That way, having fired a test shot to determine the parameters for that particular gun, an artillerist might estimate the correct elevation by means of a simple ratio of the test range to the desired range (assuming, of course, that everything else was held constant). If the next salvo did not succeed, the method could be used to fine tune the next shot, and so on. Other texts supplied tables for "rebating" these ratios if steeper degrees of elevation were needed. According to the Vallièrists, this constituted the "natural" way to aim a gun.[17] Finally, when the shot had to be lobbed at an elevated angle over a fortress wall, artillerists used different procedures, instruments, and even a different type of gun (typically, a mortar). Here, learned authors incorporated elements of the Galilean parabolic theory into this essentially Aristotelian scheme, creating a welter of algorithms and recipes, sufficient to baffle any but the most knowledgeable expert.[18]

Each of these schemes presupposed ancillary instruments that would normalize the activities of gunners. Tartaglia's quadrant enabled a gunner to determine the cannon's elevation on a twelve-point scale. Revised to read in the conventional 90° scale, this quadrant defined practice with reference to the range tables. In an analogous way, Torricelli's quadrant normalized practice for the parabolic theory by expressing the elevation on a scale that varied as the sine of twice the distance.[19] In fact, even the simplest calculations and instruments were set aside in the heat of battle. Instead, gunners used quoins, triangular wooden wedges, which they jammed under the gun's breech to raise the cannon to a set elevation. Gunners were assisted in this operation by three subordinates who levered the piece with handspikes. The gunner then marked the quoins to keep track of his previous angle of fire. Vallière's 1732 system went so far as to remove the muzzle sight and breech groove from the official French cannon so that gunners would not try for fancy aiming. As Vallière, *fils*, put it: "The *coup d'oeil* of a resolute and well-trained officer, and the skill of a cannoneer are worth more than these ingenious bagatelles."[20]

What was important about such rules and instruments was that they trained the "eye" of young practitioners. The art of artillery was just this sort of acquired skill. In the 1780s, the commander at Auxonne ordered his young officers to train first with instruments, and then without, to accustom them to work rapidly in wartime conditions.[21] This conceit of the gunners' "eye" set the art of artillery within a larger military lore about the *coup d'oeil* indispensable to every great commander. Military authors from Folard to Frederick to Guibert all agreed that a warrior's "eye" for terrain and field position could only be acquired by experience; the study of cartography or geometry were not in themselves sufficient. As one prominent Vallièrist put it, an officer needed this "eye" not just to know where to shoot, but when. This "eye" was the sign of the ineffable and tacit knowledge that lay at the heart of the art of artillery and warfare. It was the elite, cerebral equivalent of the tacit, bodily knowledge that constituted the mystery at the core of the artisanal corporation.[22]

The hardware of the Vallière system substantiated these practices. Vallière standardized the point-blank range by fixing the amount of powder at two-thirds the weight of the ball and by designing all the cannon of his system with roughly the same outer dimensions. Hence, sighting would be the same on each.[23]

From this perspective, the Gribeauvalists stood accused of making the guns more inaccurate, not simply because they *fired* less accurately, but because they were less accurate to *aim*. The shortened barrel increased sighting errors, especially if the gun was not true-bored. More important, the narrowed angle between the line of sight and the line of fire foreshortened the point-blank range (fig. 3.1). The Vallièrists argued that henceforth gunners would be forced to aim below their target at short distances, an "unnatural" way to shoot. At larger distances, where their target lay beyond the point-blank range, they would be forced to increase the angle of fire, and hence shoot with less accuracy (and with less chance to produce deadly ricochets).[24] These objections may appear absurd. Why should a modification in the *outer* dimensions of a gun affect the ball's flight or in itself force the gunner to aim at a higher angle to hit a given target? Such objections, however, make perfect sense from the point of view of the *practice* of artillery, practices built into the very design of the cannon.

What these objections overlooked was that the Gribeauvalists' new cannon had been redesigned to operate with a different set of social practices. First, rather than building guns with fixed *outer* dimensions, the new artillery pieces were built with a constant ratio between the caliber and the *inner* length of the tube. This ratio gestured toward a notion of efficient fire—an assurance that the smaller cannon of the Gribeauvalists would be worthy successors of the bigger, more powerful guns of the Vallièrists. Certainly, this ratio, which set the length of each tube at seventeen times its caliber, should not be equated with some "scientifically optimal" design. Even Gribeauval's partisans admitted the ratio was a politic compromise between the Prussian cannon (fourteen caliber lengths) and the old French cannon (which ranged upwards from twenty-four caliber lengths). It simply signaled a new compromise among competing parameters, most notably the need to shorten the pieces so they would be light enough to maneuver.[25]

Second, the Gribeauvalists partially disassociated the practice of aiming artillery from the exterior dimensions of the gun itself. They did so by means of a retractable brass sight (the *hausse*) attached to the cannon's breech (fig. 3.1). The gunner was to adjust this scaled sight so that it lined up his target with the fixed sight on the muzzle. He could then use the sight's scale to keep track of how changes in the gun's elevation affected its range. Equally important was the large screw (the *vis de pointage*) which the Gribeauvalists incorporated into the limber which enabled the gunner single-handedly to raise or lower the cannon's angle of elevation (figs. 1.2 and 4.2). B. P. Hughes has asserted that these improvements together constitute the most important innovation in smoothbore artillery in its final two centuries. Now the gunner had sole control over the vertical aim of the piece: a rationalization of the management of aiming artillery that went hand in hand with the rationalization of its loading and firing. As Du Coudray noted:

> With the *vis à pointage*, the gunner himself positions his piece, . . . whereas with the quoins, he is obliged to negotiate (*s'entendre*) with the men who assist him; which because it requires agreement, multiplies the uncertainty (*tattonnement*) and convinces the gunner, always impatient to fire, especially with enemy nearby, to be satisfied with an approximation.[26]

To the extent that the soldiers had their fixed routine, the specialist gunner was now free to concentrate on expert decision-making. Yet contrary to the claims of some Vallièrists (and some recent historians), the *hausse* did not require the use of mathematical skills or numerical tables. Du Coudray insisted that "no kind of science," was needed and that the *hausse* could be used by gunners "without even knowledge of what a 'degree' is." To be sure, it greatly facilitated the use of tables, and in later decades came to be associated with primitive ballistic calculations. But its purpose was still to train the eye; now, however, the eye could relate the range of a gun to marks on a uniform scale. Moreover, because the *hausse* was useless for short ranges (as the Vallièrists had noted), it reminded gunners to switch to the new Gribeauvalist case shot when enemy troops drew near, thereby resegmenting the market for their craft.[27]

The Vallièrists practiced gunnery as an art, an elaborate rule-bound skill associated with the corporate structure of ancien régime society. This was a tacit, experiential knowledge, whose rules and practices were closely bound to a unique artifact. Even after being tempered by an elaborate social organization, the operation of such an idiosyncratic object still required great judgment (an "eye") on the part of the user. The Gribeauvalists practiced gunnery as engineering, an elaborate rule-bound skill associated with the structure of modern society. This was an explicit, formal knowledge, whose rules and practices were abstracted from the object itself. However, the operation of this object still required judgment, except that now this judgment was increasingly about mathematics, as we will see.[28]

The thing to remember here is that "art" involved calculation, and "engineering" involved judgment. Tacit knowledge—as Michael Polanyi has pointed out—is central to all manners of technical work. For the practitioner, however, making the transition demanded new skills and devalued old ones.

New forms of knowledge imply a reevaluation of experience. The transition was necessarily painful for those reared in the old methods. On these grounds the French artillerist Tousard explained the "dislike of innovation . . . found in old soldiers. . . ."

> [B]ecause by adopting new weapons, and consequently a new exercise, the old man and expert soldier finds himself in a worse situation than a raw recruit. He not only has a new [trade] to learn, which after a certain age is not an easy matter, but also the old one to forget.[29]

We are now in a position to understand the Vallièrists' vehement resistance to the new artillery system. Gribeauval's methods challenged the Vallièrists' "technological life." His redesign of the cannon devalued skills that older gunners had developed over years. His guns challenged practices that were the sinews of the gunners' social life and the art at the core of their corporate identity. As a result, when the Gribeauvalists announced their new system of artillery in 1763 it precipitated a bitter polemic over which gun was superior.

Test, Countertest

The relationship between experiment and experience lay at the center of the debate between the Vallièrists and Gribeauvalists. Both sides claimed to be vindicated by "scientific" demonstrations, and both appealed to battlefield experience. But the Gribeauvalists challenged the disjuncture—as they saw it—between the Vallièrists' craft-based art and the exigencies of future warfare. In its place they proposed an engaged experimental program. In rebuttal, the Vallièrists attacked the experiments of the Gribeauvalists as inadequate proof of the new cannon's suitability for battle, which they asserted was the only relevant test. Out of this polemic emerged a searching exposé of the ways in which technological knowledge is produced and validated.

The Gribeauvalist program of engineering experimentation was initiated (however imperfectly) by Bernard Forest de Bélidor. For Bélidor, technological design meant a program of "parameter variation" to find experimentally what combination of factors would maximize the relevant output. In the 1720s, Bélidor asked what combination of a gun's caliber, charge of powder, barrel length, and angle of fire would produce the greatest range. The question was not new; only in retrospect did the tests of seventeenth-century investigators appear to lack "reasoning." Even in ancient Greece, engineers had approached the design of catapults similarly. Bélidor, however, tried to link his tests with mathematical rules. In his textbook of 1725 he proposed that gunfounders progressively shorten a gun barrel with a saw, test it at each length, and *calculate* the optimal amount of powder. Later that decade he undertook a series of tests to show just how much the quantity of gunpowder could be lessened without reducing the gun's carry. Bélidor never found a mathematical relation among the relevant variables that matched his experimental data, however. So he compiled tables based on the parabolic trajectories of Galileo. With these tables, gunners, on the basis of an initial trial shot, could look up the correct angle of fire. Bélidor admitted that such a theory-based table needed the "witness of practitioners [to] give it authority." Groups of assembled artillerists gathered at firing ranges, as at Metz in 1739, to witness repeated trials of his formulae and vouch for his findings.[30]

Though his results proved valid only within certain narrow conditions, Bélidor's engineering approach threatened fundamental aspects of the Vallièrist system. In particular, he understood his experiments to imply that the cannon could be shortened without appreciable loss in firepower. This assault on Vallière's concurrent effort to fix the dimensions of artillery pieces provoked an angry response. As we have seen, the Grand Master had Bélidor stripped of his rank and exiled. Yet the engineer continued to attract support from powerful patrons, such as Marshal Belle-Isle and the duc d'Orléans. They obliged Vallière, *père*, to attend some of Bélidor's experiments. In 1756 Bélidor was elected to the Academy of Sciences, and with the Grand Master deposed, reincorporated into the artillery. When Gribeauval came to power in 1763, he adopted his mentor's program.[31]

For all its alluring simplicity, however, this engineering program had numerous pitfalls. In trying to link experiment and theory to guide "real world" application, engineers encountered four generic problems. First, the fallacy of "descriptionism." It proved naive to think one might simply search for mathematical regularities among experimental data. Any formulation ended up requiring allegiance to some physical theory or other—otherwise one could hardly know which parameters were the stable and relevant ones. Bélidor adopted the Galilean vacuum physics and focused on cannon range; Robins would develop his own version of Newtonian fluid dynamics and focus on initial velocity. Either way, some implicit theory was needed. Second, the frustrations of normalization. Engineers struggled to standardize the performance of the test-object and the conditions under which it operated. Yet it proved confoundedly difficult to calibrate the test-object and measuring instruments so that they produced experimental results consistent enough to make variation visible. For guns, inequalities in gunpowder quality and other inputs meant that results varied among themselves as much as between one design and another. Third, the problem of "scale." Engineers found that the corporeality of physical bodies made it difficult to translate experimental results from laboratory-size objects to full-size objects. In real-world processes, bulk matters; one cannot simply assume that the relationship among parameters will vary in simple linear fashion as one goes from musketballs to cannonballs. Thick things do not obey the rules of an idealized physics. And fourth, the quandary of complexity. Any equation that did try to model real-world conditions (or even experimental results), quickly became too complex to guide practice in the field. Engineers found they had to make simplifying assumptions to generate mathematical formulae that were soluble. Were these simplifications plausible? Were they plausible in all circumstances? And how would the operating gunner be sure?[32]

Each of these doubts was raised, first in the 1730s and 1740s, during the polemic between Bélidor and Vallière, *père*, and then, more publicly in the 1760s and 1770s, during the polemic between Gribeauval and Vallière, *fils*. In both periods, all these doubts coalesced into the central accusation against the nascent engineering sciences: the problem of credibility. Orthodox natural philosophers of the Enlightenment worshipped at the church of deism. They could (and often did) invoke God as the guarantor of nature's lawfulness, and trust that in His benevolence He had shaped human minds so that these truths might be apprehended. The engineers had no such comfort. They operated under a rather different "theological" assumption. Praying as they did at the altar of power, their investigations proceeded without divine warrant. At their disposal lay only human-made instruments designed to measure the operation of human artifacts. The sole guarantee that such a study might yield intelligible results was their conviction that human activity was directed toward intelligible ends. And the only convincing proof that they had succeeded was the success of their design in the world outside the testing ground. But that success could never be predicted in advance, since the design was itself an attempt to shape how the world *ought* to be. This meant persuading those in authority (who held the purse strings) that such a world was desirable. But anyone who lobbied for such a world was inevitably interested in the outcome. So how could they be trusted?

For this reason, technological tests placed enormous strain on those social processes which natural philosophers had developed to legitimate their findings. In the seventeenth century, the credibility of experimental results depended in part on the genteel social status of the practitioner. At that time, the laboratory, as part of the gentleman's house, became the locus of experiment, a quasi-public space open to a discerning few. Results could be debated, but only by those with the appropriate credentials for truth telling.[33] In early eighteenth-century France, savants also began to recruit a lay elite audience in their battle against the scholastics in the universities. This cultivation of a public for science also led them to assemble "theaters of proof," demonstrations conducted with an eye toward convincing their audience of their mastery of nature.[34]

Along similar lines, the Gribeauvalists and Vallièrists staged quasi-public demonstrations to prove the superiority of their respective artillery systems. But the use of this sort of "theater of proof" posed real problems for military engineers. First, military technology was a state secret, and the property of the king. Hence, any open discussion of such tests—as would be expected in a scientific controversy—ran the risk of contravening the *raison d'état*. Second, ranked in a hierarchy, military engineers lacked the gentlemanly liberty to dispute as equals. Could a lieutenant really question the interpretation of his colonel? Finally, since the corps transparently had a collective interest in the outcome, how could they convince anyone of the validity of their test? To counteract these doubts about the objectivity of their results, each side in the debate put great effort into assembling the proper sort of witnesses. Their audience included co-opted rivals within the artillery, powerful patrons in the state, and even a carefully managed "public."

Upon introducing his new system of artillery in 1763, Gribeauval staged elaborate tests at the Strasbourg garrison. The foundry master there, Jean Maritz, hoped to have his new capital-intensive boring machine adopted by the government. To give assurance of impartiality, the tests were conducted under the direction of an elderly partisan of the Vallière system. Privately, however, Minister Choiseul assured Gribeauval that "His Majesty intends that you not be bothered by anyone and have all the authority necessary."[35] The tests—which took place over four months and involved continual modification of the matériel—were witnessed by Marshal Contades, the local artillery garrison, and general staff officers such as Rochambeau and Besenval. Every officer kept his own journal of the proceedings, and these were reconciled by common consent in the final report. The tests compared the old and new cannon for their range at various angles of elevation, their endurance, the solidity of their carriages, and the use of case shot. The final report of 1764 admitted the new cannon's maximum range was somewhat diminished, but noted that neither cannon fired accurately at those extreme ranges anyway. More important, the report glowed about the maneuverability and durability of the cannon. Gribeauval boasted that he had given Marshal Contades "a show every day."[36] The Gribeauvalist partisan, Du Coudray, felt the results were conclusive—and that this conclusion was universally shared.

> Putting aside the confidence inspired by any operation carried out by officers who merit so much confidence, I ask if it is thought possible that so many witnesses, several of whom were interested parties through their understandable attachment to their former opinions, who had carefully examined the facts announced, facts by the way, very simple in their nature and repeated many times over, with the obligation to contradict [them if they felt it right], I ask, is it possible that they would have conspired to certify these facts if they were not the most exact truth?[37]

However, this confidence in officers meriting so much confidence was *not* universal. On behalf of the Vallièrists, Saint-Auban accused the tests of being conducted "mysteriously" by the partisans of the new system, who "kept away officers who by their knowledge and rank could have made embarrassing objections."[38] Thus, upon *his* return to power in 1771, Vallière, *fils*, staged his own counterdemonstration. This time, however, the tests were conducted at Douai, where the foundry master, Bérenger, having quarreled with his father-in-law, Maritz, had allied himself to the Vallièrists. These tests pitted the Gribeauvalist cannon against a *modified* Vallièrist artillery bored with the new Maritz machine. The result: Vallière's modified cannon carried significantly further than the new cannon, particularly at high angles of elevation. Also, whereas the Vallière cannon shot fifteen hundred rounds without damage, the new pieces began to fire erratically after only three hundred to four hundred rounds. Finally, the new carriages broke down under the violent kickback of the new cannon. To bolster the credibility of their tests, the Vallièrists conducted these experiments in front of the Régiment de Navarre, witnesses from the "public," and the comte d'Artois (the brother of the king). The fact that the comte d'Artois was only fourteen years old at the time was irrelevant in this context; his royal presence made the event public. So too did the formal publication of their test results, something the Gribeauvalists never agreed to do. The Vallière cannon were declared superior on the basis of scientific evidence and the foundries were ordered to return to the old system.[39]

Who was to be believed? As the Vallièrists pointed out, the first rule of experimental science was to hold conditions identical. Plainly, they noted, this had not been done at Strasbourg. There, Gribeauval's cannon had been bored with Maritz's new machine, and Vallière's cannon had been bored with the old one; at Douai, they had both been bored with the Maritz machine. And when the Gribeauvalists protested that the Vallièrists had no right to make use of the boring machine, the Vallièrists responded they were simply using a neutral technology which was irrelevant to the fundamental purpose of their system. Furthermore, while the Gribeauvalists had used equal amounts of gunpowder in both cannon at Strasbourg, the Vallièrists insisted that the gunpowder should have been proportional to the cannon's weight. Finally, at Strasbourg, Gribeauval had compensated for the shorter range of his new cannon by increasing their elevation by half a degree; at Douai, the two sets of cannon had always been fired at the same angle. Moreover, the Gribeauvalists' half-degree increase in elevation would reduce the deadliness of the ricochet shot. To this, the Gribeauvalists responded that the half-degree difference was trivial; what was outrageous was the use of unrealistically high angles to demonstrate that the Vallièrist cannon had outperformed the Gribeauvalist cannon. And so on. . . . Clearly, the rival artillerists had rigged the test conditions to impress their witness-patrons. Or, as one Vallièrist put it: "to mislead the unsuspecting and inattentive spectator." This often happens when jobs or funding are riding on the performance of a presentation technology.[40]

After closely following the claims of both parties, the military philosophe Guibert threw up his hands. Clearly, the science of ballistics remained inadequate. Artillerists lacked a master equation to guarantee good aim. Understandably, then, they turned to experiments to resolve their questions. But often, he noted, these experiments were conducted in such a way that "the result was known in advance." This occurred because the hierarchy of rank or the interests of "party" could not be set aside. The attachments of military society undermined the impartiality necessary to be a good witness, "either because the authority of the presiding officers carries [the day] and covers up all [contrary] opinions, or because each observer comes with his own prejudice rather than his judgment, and the opinion that he wishes to maintain, rather than the impartiality that waits to see before judging."[41] For this reason, the interpretation of tests on the practice ground always "conformed to the dominant opinion." That a field marshal and the royal presence were needed to validate these results suggest how far the artillerists had to go before they could reproduce the social mechanisms for generating technological "facts."

Does this mean that all tests are charades? As long as an experimental plan is laid out in advance, and the criteria of judgment are out in the open, might not a neutral party adjudicate rationally among various technological schemes? Proposals of just this sort circulated in the 1760s and 1770s. Both sides were to sign a prior agreement on the conditions of the test, and then conduct it under the supervision of a field marshal, members of the Academy of Sciences, and artillerists of all persuasions.[42] Wouldn't this have cleared up the controversy? After all, some cannon do shoot farther and more accurately than others, and experiments can make clear which. Here, at least, is a hard technological "fact." Or is it?

Blowing in the Wind

As part of their defense of their cannon system, the Vallièrists laid out a social critique of technological "facts." Even as he directed the Douai tests, Vallière, *fils*, republished a thirty-year-old memorandum of his father's which dismissed all hope for an experimental science of ballistics. This memorandum pointed out that many factors disturbed the flight of a cannonball: variations in air density, temperature, projectile size, gunpowder, et cetera. Out of any hundred shots no two results would be identical. There might be *some* relation between the amount of charge and the distance traveled by a cannonball, but theory would never describe it for any *individual* shot.[43] Vallière did not therefore conclude that artillerists could not hit targets, nor that theory was entirely useless. But at best theory placed *limits* within which practitioners might operate. This was, of course, literally true: the parabolic theory, by ignoring atmospheric resistance, placed an upper limit on the range of a shot. So that even if a parabola did not resemble the actual flight of the ball, theory directed the "eye" of the gunner.

> Theory enables us to know the possible impediments, and the *pratique* which it illuminates, teaches us to prevent or diminish these to a degree sufficient to deliver, to a very close approximation, all the different operations demanded by war.[44]

The danger of experiment, by contrast, was that it made "facts" seem self-evident. "Everyone thinks they can do [experiments] because few people grasp the difficulties of performing a decisive one, especially in artillery." There were many sources of error in the measuring instruments and those who wielded them. Before one could begin an experiment, one had to have a *plan raisonnée*. Experimentation might appeal to an uninformed public, but its execution needed to be left in the hands of experts. Indeed, Vallière, *fils*, denounced experimentation as a kind of popular despotism:

> "The method of experimentation has an air of authority and dominion (*empire*) which imposes itself on our reason and tyrannizes our belief: 'One cannot and must not contest the facts (*les faits*).' But produced (*faits*) in a particular way are [facts] in themselves illuminating?"[45]

The Vallièrist critique of technological facts, then, turned on their deceptive transparency. The uninitiated might think such facts readily attainable and impressively persuasive. But experts understood how easily (spurious) technological facts could be created, and how ambiguous they proved in practice. Hence, decisions about technological systems ought best be left to those experts and not paraded before an ignorant public. The Vallièrists would never have engaged in such tests had they not been forced out of office by bureaucratic rivals. Obliged to prove the superiority of their cannon to patrons, they too turned to public demonstrations and publication.

Yet there was considerable merit to their argument. Indeed, this assault on the disjuncture between ballistics theory and experiment mirrored the objections posed by the new school of *engineering* ballistics. This school was also founded on a critique of the ambiguities of creating a Galilean experimental ballistics. It claimed, however, to offer a means of superseding those ambiguities, and thereby establish ballistics on a firm empirical footing. Benjamin Robins had launched this new approach thirty years before the controversy over the Gribeauvalist cannon. Robins was an English civilian who in 1742 published a book, *New Principles of Gunnery*, which challenged the British Ordnance Department. The core of Robins's critique was a demonstration that the impact of air resistance meant the *range* of a cannon was the wrong parameter to study, and he suggested that gun designers focus on the initial *velocity* of the cannonball instead. In 1747 Leonard Euler, the century's preeminent mathematical physicist, translated Robins's text into German, commented on it, and further developed his mathematical analysis of the trajectory of a ball moving through a resisting medium. Together, these men put ballistics on a new basis—one with its own ambiguities, however.

These works were familiar to French artillerists. True, it was not until 1783 that artillery professor Jean-Louis Lombard published his French translation of Robins and Euler, announcing with great fanfare that the effect of air resistance on a cannonball was no longer "secret knowledge." But widespread publicity had been given to Robins's theory by the Irish-French academician Patrick d'Arcy in the 1750s. The French physicist and naval engineer Jean-Charles de Borda had published an analysis of the trajectory in 1769. In 1771, an assistant professor at the artillery school of Grenoble published a first French translation of Robins. And in 1774, Condorcet brought Robins's and Euler's work to the attention of Turgot, who recommended a translation in a letter to the king.[46] What is more, both sides in the artillery dispute cited Robins in support of their position. The Vallièrists quoted him to the effect that longer gun barrels fired greater distances. The Gribeauvalists quoted him to the effect that this extra range was irrelevant since guns were only accurate at less than this maximum. Yet as Brett Steele notes, both sides ignored Robins's principal lesson: that muzzle velocity—as measured by a ballistics pendulum—was the only reliable indicator of a cannon's performance.[47]

Robins had noted the discrepancy between a gun's range as determined by Galilean theory and as found by experiment, and had showed that the effect of air resistance on range swamped any change produced by a new gun design.[48] No wonder artillerists were able to support any theory they wished.

> It is not therefore to be wondered at if this vague method of experimentation, which according as the trials were selected might be urged in confirmation of different and even opposite assertions, [nor] that this loose and inconclusive experience hath been urged in support of the most erroneous opinions, and by its authority with those who were ignorant of its fallacy hath greatly tended to the establishment of the many prejudices and groundless persuasions which at present prevail amongst the modern artillerists.[49]

Robins criticized more than the artillerists' experimental setup; he attacked the artillery establishment of his day for their deliberate manipulation of these uncertainties so as to remain the beneficiaries of much "treasure . . . largesses and emoluments for their supposed dexterity and skill therein."[50] By contrast, his own experiments produced more "certain" results because they depended on the initial velocity of the shot, which he showed was a stable parameter because it remained constant so long as the other variables were held constant. His great success was to derive a value for this initial velocity from an analysis of the gunpowder's explosive force (one thousand times the atmospheric pressure) and then show that this matched the value found experimentally with his ballistics pendulum. He thereby connected an analysis of interior ballistics (the gunpowder explosion) with an analysis of exterior ballistics (the shot's flight through the air). From this, he went on to derive and corroborate an equation for the ball's flight, assuming a simplified model of air resistance.[51]

Robins showed that ballistic theorists needed to consider physical principles; without such reasoning, one cannot know *which* parameters are relevant. (The fallacy of descriptionism.) His work, as Steele has noted, operated within Newton's program of research; it combined the new mechanics with experiments that set the coefficients of the equations. But Robins's work also marked a departure, in that it possessed many of the characteristics of engineering knowledge—and many of its pitfalls as well. His ballistics pendulum enabled him to overcome the frustrations of normalization. However, other difficulties remained. First was the problem of scale. Robins had done his experimental work with musketballs, but as he noted, his model was sensitive to the speed, size, and shape of the shot because of the nonlinear relationship between air resistance and the velocity of the ball. Not until 1775 did Charles Hutton carry out some tests at Woolwich using Robins's methods with six-pound balls.[52] Second were the quandaries of complexity. As Leonard Euler pointed out, Robins had made some unjustifiable approximations, for example, assuming the explosion was instantaneous and that no gas leaked around the ball. These ought to have led to a large disagreement with experiment. The fact that they had not, showed that his value of 1,000 was nothing more than an experimentally determined coefficient. "The truth," Euler remarked, "had less to do with [Robins's] research, than the need to square the results of his theory with those of practice." Yet both Euler and Robins soon realized that the ballistics equations could only be solved by introducing a far simpler assumption about the effect of air resistance than Robins had first proposed. And even then the results had to be divided into types of curves and recalculated as numerical tables for use in the field. The question that remained was whether these simplifying assumptions were valid, and whether the results could be made accessible to practical artillerists.[53]

But all these problems paled beside the challenge Robins's method presented for artillery partisans seeking to legitimate their innovations: the problem of credibility. After all, artillerists focused their attention on a gun's range for good reasons. Distance was an unproblematic parameter, measurable with familiar instruments (e.g., eyeball estimates, rulers, sextants). Moreover, distance could be understood (and witnessed) by patrons in dramatic displays. By contrast, expert knowledge of impact theory was needed to interpret a ballistics pendulum, and this excluded patrons (and many artillerists) from corroborating the findings. As a design-oriented program, Robins's approach—which was to become so fruitful in the next century—still had to be produced and validated socially. Finally, and most significantly, range was the key parameter an artillerist needed to know in order to hit his target. As a guide to practice in the field, Robins's methods had to be made meaningful for gunners on the battlefield.[54]

Consider the controversy when the academician Patrick d'Arcy brought Robins's experimental program to France in the 1750s. To challenge the indifferent artillery corps, d'Arcy took his case for "optimal" gun design to the prestigious public journal, the *Mercure de France* (while simultaneously promoting a new lightweight cannon designed by his fellow Irishmen, Moor and Stark). On behalf of the artillerists and their allies in the foundries, Saint-

Auban defended the expert knowledge of the corps. "The most seemingly true (*vrai semblable*) subjects in physics," he intoned, "are often those which lie farthest from the truth." Not that he disapproved of experimentation; the artillery service possessed a wealth of data. But even Robins's pendulum had not eliminated variation in all the relevant parameters.[55] To this, d'Arcy readily agreed. He was pleased to learn the French artillerists performed tests, but why was their data not available? Had they waited between shots to let the barrel cool? Did they use accurate scales? To be credible, the results had to be open for public inspection, like the investigations of the Royal Society. "The authority of M. de Saint-Auban is without doubt of great weight, but he cannot dissemble that in matters of fact one cannot have proof unless one confronts experiment with experiment."[56]

Against this image of ballistics as a public science, Saint-Auban stood firm. He assured readers of the *Mercure* that the corps was open to suggestions, but in the final analysis only *gens du métier* could judge this dispute. At this juncture, d'Arcy gave up. The dilemma, as he acknowledged, was that he could only carry his experimental program so far on his own. Like Robins, he had halted his researches for lack of money.[57] Only the state (i.e., those who controlled the purse strings of the artillery) could afford to perform tests with full-scale cannon. And a cannon-sized ballistics pendulum was expensive. None was built in France until the Restoration. Even the emperor-artillerist Napoléon—familiar with Robins's work—refused to pay eighty thousand francs for one in 1804.[58] In other words, d'Arcy lacked the institutional support to force the corps to reexamine its methods of knowledge production. As we have seen, both sides in the artillery debates of the 1760s and 1770s still validated their cannon designs in ways calculated to appeal to patrons and the public, rather than in ways which savants believed got at the truth.

Even after the Gribeauvalists adopted Robins's perspective, they acknowledged that the true flight of the cannonball mattered less than new "rules of practice." In the 1780s, Professor Lombard drew up new range tables based on Robins's equations, which he normalized for the dimensions of the Gribeauval cannon and its *hausse*. The Revolutionary and Napoleonic artillerists went to war with these tables. Shooting was not, however, a matter of simply plugging in numbers. Lombard admitted that his tables remained inadequate. Unlike astronomical tables, they were based on "hypotheses too often belied by experience." To account for the idiosyncrasies of their particular weapon, gunners still needed to fire a test shot before looking up the right adjustment. Moreover, this adjustment would be uncertain in that gunners were still plugging in the *range* of the test shot. Unfortunately, as Lombard pointed out, the muzzle velocity was unavailable to gunners in the field.[59] Now, rather than possessing a *coup d'oeil* for the battlefield, expert judgment now consisted of what Lombard called a *coup d'oeil* for the tables. Gunners had long insisted on summary tables so that they could assimilate data in an "initial glance." Now they would make expert judgments about mathematics. Even if the new tables did not always make gunners shoot more accurately, Lombard noted, "at least they will have a more exact notion regarding the most important object of his service."[60]

This underlines a point Theodore Porter has recently made about the nature of technical expertise. Rather than being the slaves of numbers, expertise involves the ability (and latitude) to make judgments *about* numbers. We saw in the last chapter how French artillerists used the rigor and universality of mathematics to assert their status in a society where aristocratic codes made their profession less than honorable. To do so, they married their mathematical knowledge to an ethic of noble service. Their embrace of mathematics was not meant to turn them into automatons, however. Indeed, their professional status depended on preserving their *discretion* about how to use mathematics (including the right to privately not use it at all).[61]

Each of the antagonists in this quarrel of the artillery systems owed allegiance to instruments, practices, and knowledge that formed a meaningful whole. The Vallièrists belonged to the corporate world of absolutist France. Dragged reluctantly into the world of public demonstrations, they defended their expert status on the basis of the unrationalizable art at the center of their craft. By contrast, the Gribeauvalists anticipated (in some respects) the modern world of Revolutionary France. This was not simply a matter of removing the king's motto from the breech of their cannon (symbolic as that might be), but of redesigning the physical dimensions of the weapon, reconceiving its mode of use, and refashioning the social life of its operators. However, the design of their cannon did not follow directly from natural knowledge about the "way things are." Indeed, the Gribeauvalists ignored the best scientific knowledge of their day. Rather, their design reflected their technological life, in which expertise was to be refounded on a new kind of engineering knowledge about material objects.

Toward a Science of War

But what of the effectiveness of these cannon in battle? To the extent that the tests at Strasbourg and Douai offered little evidence for the marked superiority of one system over the other—at least with respect to the gun's range and accuracy—this supported the Vallièrists' argument. Science and experiment were beside the point, they said, what counted was the fact that the Vallièrist cannon had proved themselves in battle, the only "theater of proof" that mattered. War alone possessed the variety of circumstances needed to test a cannon properly. No matter how effective the new artillery may have seemed on the practice polygon, only the old system had undergone a "baptism of fire."[62]

Against this argument, the Gribeauvalists offered something more exciting: a vision of redeemed French honor. According to the Gribeauvalists, the demonstration of a rough equivalence in the range and accuracy of the two types of cannon meant that they could be compared on other grounds. In their redesign of French cannon, the Gribeauvalists faced a series of tradeoffs. As I indicated in chapter 1, the Gribeauvalists "traded" the precision offered by the Maritz boring machine for more mobile cannon. The constraints on cannon design were not merely "technical" (integral to the object), but involved the expectations of particular type of warfare (social interests). Infantry commanders, reform-minded patrons, and other nonexperts increasingly believed that France must turn—if only for financial reasons—to a more mobile type of offensive tactics. The Gribeauvalist cannon were designed for just those ends. Yet the Gribeauvalists did not merely respond passively to these expectations (what Vincenti calls the "external" social demands on technologists). The top brass might adopt a mobile tactics, but they need not necessarily include the artillery in their plans. The Gribeauvalists were obliged to take an active role in the controversies over which *type* of mobile tactics France should adopt in the last decades of the ancien régime. This meant making an active case for the importance of gunpowder weaponry in battle, both for cannon (which the artillerists operated) *and* for handheld muskets (which the artillerists supplied to all the rest of the army).

Just as the controversy over gun design had meaning only in the context of disputes over artillery theory and practice, so did the artillery debates themselves turn on the ancient and contentious science of war. In a European states system where competing nations borrowed avidly from recent victors, military leaders continually confronted the question of reform. The French army was marched through a bewildering number of changes in infantry tactics between the 1750s and 1790s, a period in which the soldier's basic weapon—a musket with an attached bayonet—remained relatively constant. In such circumstances, military leaders had an interest in predicting how these untried tactics would affect battles as yet unfought. In the Enlightenment this meant using the methods of "scientific analysis."

The dispute over the "technology" of tactics was the occasion for an outpouring of books by active commanders, retired brigadiers, professors of mathematics, and armchair theoreticians. By one count, the number of such texts quadrupled in the latter half of the century.[63] In these books, the lessons of history were weighed against "experimental proofs," and national martial codes were set against raw calculations of musket balls raining down on a mass of human bodies. Was Saxe's victory at Fontenoy due to his use of light artillery? Was a parade-ground maneuver proof that infantry could effectively deploy from a marching column to a three-man shooting file, and thence to an attack column? Was the French character suited to precision drilling? How many men would a cannonball kill if the troops formed a square . . . , a line . . . , a column? Answers to questions such as these were marshaled in support of the author's contention that a particular drill should be taught to soldiers during the present peace and would guarantee victory in the next war. At stake was the manner of social relations among officers, and between officers and their troops. At stake too were matters of high policy, such as whether France should take an aggressive or defensive posture vis-à-vis her continental rivals. Together, answers to such questions implied a definition of the national (martial) identity, and a program to shape that identity. Were French soldiers intrepid and independent, or disciplined and patient? Could they master the intricacies of German maneuvers? Should they even try? The authors of these competing tactics were "engineers" in the realm of social technology. They were not content to explain the world, they offered a vision of how the world *ought* to be. If war is international politics by other means, then this debate over tactics was domestic politics by other means.

Almost all writers agreed that the technology of tactics, like the technology of guns, must be adaptable to the varied circumstances in which an army might find itself during a campaign. Generals wanted their troops to take advantage of terrain, the enemy's position, and myriad other factors. Indeed, generalship in the eighteenth century consisted in large measure of achieving just these transformations; field battle was a perpetual struggle for position. All participants agreed that this flexibility meant designing a drill book which enabled troops to shift from one formation to another without dissolving into a chaotic crowd that no longer constituted an army. And the majority agreed—following the example of Frederick the Great—that rapidity of movement was the key to offensive war. But here the consensus broke down. The ostensible focus of controversy was the role of gunpowder weaponry in battle, but its resolution turned on the outcome of a political struggle to define just what sort of France (and what sort of Frenchmen) should wield those guns—and in what manner.

The eighteenth-century debates over tactics generally pitted the "Ancients," who advocated massed columns of pike-bearing infantry (the deep order), against the "Moderns," who championed thin lines of musket-firing troops (the thin order). The Ancients were by no means bested in these inky conflicts. Indeed, a majority of the orthodox military thinkers of the Enlightenment denied the value of gunpowder weapons—both muskets and cannon—and not on humanitarian grounds either. They argued that the increasing reliance upon firearms by generals in the field had had a deleterious effect on French martial virtues and hence on French military fortunes. They blamed the guns for the futile French efforts during the Seven Years' War. And together, they pioneered the forms of persuasion that gave the debate its increasingly "scientific" character, including public demonstrations of field maneuvers and mathematical analysis of imaginary battles.

The seventeenth-century general Puységur was among the first to reduce tactics to Cartesian axioms and geometrized battle orders. The theoreticians who followed his lead in the eighteenth century did not, however, accept his dictum that firepower played an important role in battle.[64] Rather than rely on recent battlefield experience (as Puységur did), these authors looked to antiquity for their inspiration. This ought not be dismissed as pedantry; early modern officers found a model for troop discipline and cohesion in the doctrines of Sparta and Rome. Maurice of Nassau had even justified his introduction of the musket drill on Roman precedents.[65] More often, however, partisans of the classical order dismissed such newfangled inventions as gunpowder. The preeminent strategist of the early eighteenth century, Jean-Charles de Folard, went so far as to argue that the catapults of the Romans were superior to modern firearms.[66] Even Marshal Saxe, the most renowned French general of the mid-century, advised against a reliance on firearms in his widely read *Rêveries* (though he wrote this text before the battle of Fontenoy, where he used a light, mobile artillery to great effect).[67]

More than any other "Ancient," Mesnil-Durand belittled modern weaponry. In the contentious period of the Military Enlightenment, he was the most influential (and prolific) advocate of a Folard-like "deep order" in which the infantry would attack in columns with long pikes, masked by only a light screen of gun-bearing troops. Folard's fame rested on his reassertion of the maneuverability of columns of troops. Now Mesnil-Durand offered "geometric proof" that his attacking phalanxes were impervious to enemy fire, and took advantage of the genius of French soldiers for aggressive hand-to-hand combat. Mesnil-Durand denigrated his opponents for imposing foreign (Prussian) tactics on the valorous French. This was not merely an academic exercise. In the decade after the Seven Years' War, Mesnil-Durand gained the ear of Marshal Broglie, France's foremost commander. Broglie had always admired Folard's phalanx, but his official infantry drill of 1760 had promulgated the thin line order. Then, in the late 1770s, Broglie pulled a volte-face and pressed Louis XVI to adopt a new form of pike-led attack column, which Mesnil-Durand had taken to calling the "French Order."[68] How was such a position tenable in eighteenth-century Europe after four hundred years of experience with firearms?

Anachronistic as these classicists may appear, they had a valid point. Musket fire remained miserably inaccurate in this period; as few as 0.1–0.5 percent of shots reached their target. This is why commanders insisted on massed volleys aimed at massed troops; they compensated for inaccuracy with a hail of fire. Rifling the barrel was no solution, because even though it improved the accuracy of fire—as had been well known since the sixteenth century—the absence of an effective breech-loading mechanism meant that rifling dramatically slowed the *rate* of fire. Soldiers on open terrain were better served by smoothbore muskets than rifles and Napoléon actually banned their use in 1805. The Prussians went so far as to design their smoothbore muskets without a proper sight and stock, so that infantrymen would not be tempted to take aim; they were to fire from the hip with mechanical speed. Even so, a common rule of thumb of the eighteenth century suggested that to kill a man with musket fire took roughly seven times his body weight in lead! The same logic held for the larger artillery pieces. That is not to say that muskets and cannon did not produce casualties or have an enormous impact on the way battles were fought. The question was their relative effectiveness and cost.[69]

The function of the artillery, in the eyes of the Ancients, was to breech the walls of fortified towns and create a morale-boosting "boom" for wavering infantrymen. The old artillery was designed for just these ends. But this placed the Vallièrists in an awkward position. They had too great a stake in gunpowder weapons to agree entirely with the Folardists regarding field battle. Yet, as they acknowledged, Modern field tactics would doom the old artillery.[70] That left it to the Gribeauvalists to make the case that, properly concentrated and positioned, cannon and muskets, however inaccurate, had a significant effect on field battle. In other words, the Gribeauvalists hitched their star to the thin-line order and mobile warfare that Guibert had advocated in his *Essai général de tactique*. But the triumph of this particular sort of mobile tactics was by no means assured—especially now that Guibert's and Gribeauval's patron, Broglie, had gone over to the Ancients.

The Killing System

Military historians have told the story of the Moderns' triumph in the final decades of the ancien régime as setting the terms for Revolutionary warfare. The French tactical innovations of 1763–91, a time when France was at peace, are seen as key to the new Revolutionary warfare. In this, they follow the great military historian Jean Colin without always showing his attention to the ambiguities of the Moderns' "victory." Yet even Colin assumes that the Moderns' methods were validated by empirical tests. But once again, these tests turn out to be another name for a forum—a "theater of proof"—in which a social settlement could be reached.[71] The reason for this turn to "experimentation" was the failure of other methods of legitimation.

The historical record was particularly subject to contention. Whereas the Ancients read the history of warfare as a morality play, with the nation (and its army) as hero, the Moderns read history as driven by technological determinism. Firearms had transformed battle. The proof was in the real choices of generals on the battlefield. The artillerist Du Coudray noted that Folard's doctrines might appeal to the sentiments of military *authors*, but that over the past four centuries real commanders in the field had used progressively thinner infantry lines: from thirty men deep during the Renaissance, to eight deep during the wars of Louis XIV, to five at the end of the seventeenth century, until Frederick the Great had arrived at a three-man line in the Seven Years' War. The reason for this was simple. Even if the line had a certain "vulnerability" in the face of an attack column, it compensated for this by maximizing the fire of its own muskets and offering the least target to enemy guns. This fact had caused Saxe to change his mind about firearms in later years.[72] Here, technological determinism was being taken up as an argument in favor of adopting the plan of the masters of technology—just as modernization is the doctrine of the Moderns.

The problem with this reading of the historical record was that no one could positively separate the various factors that decided the outcome of battle. The Ancients could cite recent engagements won by tactics analogous to their own. Even at Fontenoy, where the Gribeauvalists had argued that a light artillery had proved decisive, the battle had actually ended in a successful column charge. The Moderns answered that the column used at Fontenoy belonged to the set of maneuvers associated with thin-line tactics, and were a legitimate part of their drill book. The interpretation of these events was as contentious as the tactics themselves. Moreover, Mesnil-Durand himself doubted the existence of a trend toward a thinner line. Or even if it were true, why was that necessarily for the best? Had not generals protested against it in every age? In any case, he argued, past experience could never offer a basis for deciding if a future innovation would succeed.[73] What was needed was a way to determine the effectiveness of a new tactics without recourse to the messy historical record.

Just as the artillerists had hoped quasi-public demonstrations would convince the relevant community of the superiority of their cannon, the Ancients offered to put their system to a public test. But testing a "technology" like field tactics magnified the problems that had plagued the cannon tests. The challenge here was to "run" the technology in a controlled setting that also allowed an imaginative reconstruction of how it might perform in the more open-ended (and deadly) circumstances of real war. For this, the army needed a vast theater of proof. In 1775 Broglie conducted field demonstrations with four battalions garrisoned in Metz. These reportedly proved the superior maneuverability of the deep order to the thin order. These tests, however, were not sufficient in scale, nor were they widely witnessed, so Broglie pressed for a larger and more public demonstration. The War Office had every incentive to resolve the debate quickly. The polemic threatened to disrupt the daily exercises that formed the principal activity (and discipline) of the peacetime army. Commenting on the uncertainty, Baron Wimpfen sadly noted that he lived in a time without an "active central power, when intrigue and favoritism held sway." It seemed unlikely the king would have an effective army without a "happy revolution" in the organization of the army. Otherwise, France—the most powerful nation in Europe in terms of natural endowments—would fall prey to those nations around her which were governed by better principles.[74]

At Vaussieux, Normandy, in September 1778, the War Office sponsored a vast experiment in military science which has since become known as the field exercise. Jean Colin considers this event the crucial turning point in French tactics, setting the stage for Revolutionary tactics. There, French troops, massed for a possible invasion of Britain, were put through elaborate test maneuvers that pitted the thin (Modern) order against the deep (Ancient) order.[75] These demonstrations were witnessed by much of the French military leadership, including nine generals, eighteen brigadiers, six lieutenant-generals, among them the Guiberts (father and son), Du Châtelet, Besenval, Gribeauval, Lückner, Rochambeau, and Castries. Many of these men were to play a leading role in the American and French Revolutionary wars. Prominent civilians, both French and foreign, and their female escorts also attended. Under the command of Broglie, four divisions (thirty-two battalions) had trained for three weeks in Mesnil-Durand's deep system. The witnesses then watched these troops being put through maneuvers for three days, followed by two weeks in which they "confronted" an equal number of troops marching to the official Modern order (under the command of Rochambeau, Broglie's brother-in-law).[76]

A majority of officers present (including Guibert and Gribeauval) seem to have decided against Mesnil-Durand's system. Already convoluted on paper, his maneuvers dissolved into chaos on the rolling hills of Normandy. His columns could deploy only from a double column formation; this committed his troops to a straight-ahead charge from the beginning. Attacked from the flank, or worse, from the rear, they were unable to turn to meet the enemy in orderly fashion. Furthermore, they were unable to take advantage of the hedges, ridges, and ditches of the region, or detach small parties to survey the terrain ahead. This "discomfiture" of the Ancients mean that time and time again the Moderns carried the applause of the assembly of witnesses.[77]

Was applause the same as victory? Mesnil-Durand had claimed that if 10 percent of the officers present gave his system their "votes" (*voix*), he should be considered to have achieved a plurality, and if 25 percent, then his system should be adopted by acclamation.[78] Even that threshold appears not to have been met, but *one* vote might yet be enough. So long as Broglie sided with Mesnil-Durand the interpretation of the results remained contested. The field marshal's final report to the king admitted that the system had its detractors, but that he was impressed with its maneuverability and offensive capacity. He proposed that Mesnil-Durand's system be adopted for the whole army and boldly asserted that many senior officers thought as he did.[79] How could he justify this reading of events?

First, he simply dismissed any glitches as the result of inadequate preparation. On opening day, Mesnil-Durand objected that "his" troops had not been given enough time to train, that their marching orders were out of sequence, and their execution was imperfect. The problem here was that these modifications had been ordered by Broglie. And second, in a demonstration of this sort, what was to prevent Broglie from simply declaring victory? The problem here was that the one day when the Ancients had led an indisputably "successful" charge was also the day when Broglie had prescribed *in advance* the precise movements and terrain for *both* sets of troops. One might well hesitate to extend this victory to the uncertain conditions of real battle.[80]

The more basic problem, however, was that the entire demonstration had been arranged to test whether troops charging without guns could defeat troops armed with guns—and yet not one soldier was allowed to fire a shot. Guibert always maintained that even if Mesnil-Durand's maneuvers were executed perfectly, his phalanxes would still be massacred by enemy fire.[81] But how could witnesses to a peacetime maneuver draw this conclusion? One pro-Modern officer present was certain that they could.

> The most ignorant of the spectators, the women even, judged that . . . [the attacking column] would have been reduced to powder by a fire from front and flank. . . . [In] the time that would have been necessary for these columns, marching at the most rapid maneuver pace, to reach the range for a bayonet charge, and calculating the ground they had to cover under artillery fire, each gun would easily have been able to get off 50 shots, and there would have resulted a total destruction of the formidable columns, which probably would have sought their safety in flight.[82]

How could such a "virtual" witnessing be definitive? Here, another kind of imaginary witnessing was called into play. To justify his new method of warfare, Mesnil-Durand had turned to another kind of "scientific" demonstration, a form of mathematical modeling. Geometric manipulations had long been part of the military craft. But Mesnil-Durand's tactics involved a dramatic charge into the teeth of the enemy's firepower. Geometry did not capture this dynamic process. So he placed the key variables in a simple time-dependent algebraic equation, and making some assumptions about the speed of his columns, the rate and accuracy of musket and artillery fire, and the damage wrought per ball, concluded that his columns would arrive victorious at their destination. This, he demonstrated, was true even if the troops confronted the Gribeauvalist artillery as described in Du Coudray's recent book.[83] Infuriated, Du Coudray had recalculated the equation with different values. He concluded that of the 500 men in Mesnil-Durand's charging column, 722 [*sic*] would be killed: 38 by cannonballs, 624 by case shot, and 60 by musket fire. This was overkill indeed! How could Mesnil-Durand's column sustain these losses, Du Coudray wanted to know, "without falling into disorder?"[84] This ludicrous overstatement gave Mesnil-Durand a chance to respond with dripping sarcasm. On Du Coudray's own assumptions, he noted, the new cannon would have an even more devastating effect on the *thin* line, killing 1,841 [*sic*] out of 500 men. By *reductio ad absurdum*, then, the assumptions were false.[85]

The point is not that Mesnil-Durand and Du Coudray employed simplistic models or dubious assumptions. What is interesting is that both believed this game-theoretic approach would persuade readers that they had captured some essential aspect of future battle. Du Coudray even boasted that his imaginative reconstruction of the destructiveness of modern weaponry would dissuade attacks, and would therefore reduce the frequency of war.[86] An exasperated Guibert dismissed this pseudo-precision as patent nonsense. "As far as I am concerned," he said, "when one argues, one ought always aim for the truth."[87] The issue, however, was not the truth, but about what could be credibly "witnessed," and hence believed. It was in the realm of the social imagination that the calculations of Du Coudray and the demonstrations at Vaussieux made their impact.[88]

That social imagination necessarily operated within a set of larger assumptions about the nature of French society. The most damaging accusation against the Moderns was that they sought to transform Frenchmen into Prussians. Mesnil-Durand derided the Prussians for making their "soldiers into machines, and their officers into automatons."[89] This was a familiar dichotomy. The "coarse, lethargic" Prussians responded to blows and strict rules that made them virtual "slaves." The French, being "lively, vain, eager to seek glory," would energetically follow where properly led.[90] Hence, the regimentation that Frederick imposed on his troops was inappropriate for the independent-minded French. Even commanders who believed that "discipline and training are the sure and fundamental basis of military life" admitted that the French national character was not suited to a Germanic style of drill. Even officers who despised the common soldier for his low birth and dissolute ways conceded that he retained a certain personal dignity. Witness the outcry when Saint-Germain introduced the punishment of beating with the flat of the blade—called the *discipline allemande*. It still roused bitterness in the *Cahiers* of 1788, both among the third estate *and* the nobility.[91]

For their part, the Moderns never denied that the national character undergirded French martial virtues. Guibert, we should recall, was passionately committed to a regenerated France, and had once advocated a citizen army to that very end. More to the point, he had always asserted—with Montesquieu—that there was a particularly French style of battle. He even agreed that a nation needed a tactics appropriate to "its genius." But Guibert also believed that an army's tactics should compensate for those national characteristics which, in the wrong situations, might be considered a vice. "Even admitting that the French are born for offensive war. . . . [We must] give them a battle order to compensate for their weaknesses." Tactics, for Guibert, was a technology to shape society, not simply to reflect it. His "mixed" tactics were themselves a form of heterogeneous engineering designed to produce a particular kind of soldier.[92]

Over the next decade the consensus set against Mesnil-Durand's system. Even advocates of the deep order, such as Marshal Castries, wrote to the king deploring the inflexibility of Mesnil-Durand's version of the attack column. A maneuverable column, Castries argued, could be achieved by finding a "middle way between the two extremes." This had long been the opinion of Guibert. Mesnil-Durand was discredited, and Broglie, who fell into disfavor at court, was relieved of his command.[93] Ten years later, Guibert's Conseil de Guerre of 1788 announced the famous "mixed order." This compromise operated on the premise that only the thin-line tactics took advantage of the firepower of gunpowder weaponry *and* possessed the requisite flexibility to form attack columns. Although the officers on the Conseil who drew up this new regulation were partisans of the deep order, they worked within the framework of Guibert's tactics. The Moderns' triumph, then, was also a sociopolitical compromise within the army high command, as much as a technical compromise. It also was a compromise about what it meant to be a French soldier.[94]

Pikes vs. Muskets: Tactics, Technology, Republicanism

That is not to say that the story of tactics was finished. On the contrary, the Revolutionaries passionately debated the tactics and weaponry appropriate for a liberated people. In 1789, Broglie was made minister of war. He suppressed the Conseil and abolished the "mixed" order. Its provisions were, however, revived in the regulation of 1791. Shortly thereafter, the levies for soldier volunteers began. Yet even after the Republic declared hostilities in April 1792, the *means* by which France would go to war were no more self-evident than the *ends* she should pursue. The officer corps was abandoning the flag; the government in Paris had trouble settling on a high command. The successive calls for recruits—Danton's 1792 declaration of the *patrie en danger*, and Carnot's 1793 announcement of the *levée en masse*—themselves innovative political acts, also generated enormous new problems. How were officers to be chosen? How would raw recruits operate vis-à-vis the remnants of the old line army? Given the constraints of time and experience, how should the army be trained? And finally, how were the political qualities that the ancien régime had invested in weaponry and tactics to be transformed by the creation of a popular army? All these questions were subject to reexamination as the situation in Paris changed, as the fortunes of war shifted, and as the reconstituted army acquired its own revolutionary traditions. Resolving them depended in large measure on redefining the relationship between the state and its citizenry.

A growing body of scholarship by Bertaud, Scott, Forrest, Wetzler, Lynn, and Blanning has decisively refuted the top-down accounts of the Revolutionary army's battlefield successes. By tying the military life ever more closely to the broader transformation of France, these authors have made the army one of the primary sites for the study of Revolutionary ideology, social reorganization, and modern state-building.[95] Out of this complex matrix, I want to isolate one question: the relationship between the instruments of war and field tactics. This relationship turned in part on the proper characterization of a

French soldier. Not surprisingly, the Revolutionaries offered new characterizations, and consequently novel suggestions for the relationship between tactics and weaponry. As John Lynn was one of the first to emphasize, some lawmakers even advocated that the French go to war armed with pikes, rather than muskets.[96] How was it possible for republican leaders to consider ordering their soldiers into battle with a weapon which had not figured in the French army for one hundred years?

The pike has had a long association with popular insurrection. Sometimes no more than a sharp blade on a long stick, the pike helped peasants assault seigniorial châteaux in 1789. The "holy" pike also figured prominently in the iconography of the urban militants known as the sans-culottes. Topped with the red cap of liberty, it quickly became one of the enduring symbols of the people's determination to defend their new freedoms.[97] This symbolism was underwritten by those political leaders who saw an armed people as an effective counterweight to the reactionary threat of the old army. For instance, in a February 1792 speech on how to "save the Republic," Robespierre opposed the declaration of war—he feared it would lead to the usurpation of the legislative by the executive branch—and called instead for arming the people with pikes.[98] Not everyone agreed a universally armed citizenry was desirable. That same month, the Parisian municipality asked all those not enrolled in the National Guard to register their weapons, and insisted that "vagrants" bearing weapons in public would have them confiscated. When the bourgeois leaders of the Parisian National Guard were forced to accept pikemen that summer, they insisted they march in orderly fashion and learn to "love their brothers."[99] As a weapon associated with overturning established authorities, the pike also appealed to Revolutionary women. Members of the Society of Republican Women carried daggers and pikes. A delegation of armed women appeared before the Legislative Assembly to demand that the legislature confirm their right to bear arms. Pikes were found in the house of the radical leader, Claire Lacombe, upon her arrest.[100] Weaponry in Revolutionary France was loaded with political symbolism, and the pike was an emblem of popular democracy. But is that enough to explain why the pike was seriously contemplated as a weapon of war?

Shortly after the war had begun, a captain Scott took to the floor of the Assembly and proposed arming the troops with pikes. Lawmakers heaped ridicule upon him and called for the opinion of a military expert. Someone shouted for Lazare Carnot, a representative trained as a military engineer (Corps du Génie). To everyone's consternation, he agreed with Scott. Pikes were cheaper and quicker to make, and hence more suitable for rapidly equipping an overnight army. They were also easier for raw recruits to master. Finally, they were more effective in battle than muskets. In support of this last contention, Carnot cited venerable French generals from the reign of Louis XIV and Louis XV, as well as more recent theoreticians of the deep order, such as Folard and Mesnil-Durand. Carnot did not abjure firearms entirely. He believed that a mix of a hundred thousand pikemen and musketmen would be invincible. Especially, he emphasized the suitability of the pike for the French temperament; whereas the Germans excelled at orderly fire, the French were "invincible" with naked steel. This "impetuousness" of the French in combat was given an explicitly political twist. Carnot cited the historic triumphs of democratic Swiss pikemen against aristocratic cavalry.[101] For the same reason, Minister of War Servan, a longtime advocate of a citizen-army, called the pike "the weapon of a free people."[102] Thus, the Revolutionaries transformed the Ancients' claim about the French soldier's dignified bravery into a claim about the qualities of a citizen-soldier.

To be sure, this paean to pike warfare also reflected the current shortage of muskets. The Convention ordered the production of a half-million pikes in August 1792. Recruits actually trained with these weapons, and the artillerist Choderlos de Laclos drilled a company of pikemen on the northern frontier.[103] But even these "practical" considerations were given an explicitly political twist. Collot d'Herbois gave an impassioned plea for the pike as the instrument with which revolutionary enthusiasm would carry the day.

> What delays our production of arms is the dragging routine of manufactures. . . . Isn't it the bayonet, cold steel, that makes the French superior to the slaves of the tyrants? While waiting for the production of guns will you cool off this impulsive and holy enthusiasm that is carrying 300,000 men to our frontiers? Let us arm our soldiers with pikes, and remember the words of a Spartan woman to her son. He said to her, "My sword is a bit short." "My son," replied his republican mother, "you will take one more step forward." We too will take one more step forward and knock down the enemies of liberty all the better.[104]

In the end, however, the great body of French troops did not go to war with pikes, but with muskets and bayonets. Three months before the *levée en masse* of August 1793, the Marais section scrapped plans to supply the army with forty-eight thousand pikes and offered twenty-thousand muskets instead. In September 1793, Carnot, now elevated to a seat on the Committee of Public Safety (thanks in part to his speeches in favor of pikes), suspended all pike production in Paris to devote the resources of the capital to making muskets and bayonets.[105] Why this about-face?

No doubt this reversal can be understood as a recognition of the "realities" of contemporary warfare. But it is not enough to say that an army of pikemen—no matter how inspired—would have been routed by Coalition troops. The reversal also reflects the kind of centralized state the Revolutionaries were able to erect in the intervening year. As we will see in Part Three below, this was a state capable of organizing the "crash" production of firearms for an unprecedented army of seven hundred thousand men. It was also a state which created a military structure to train (however hurriedly) this mass army to use those firearms with a modicum of skill. More generally, this transformation belonged to the larger reestablishment of the state's monopoly on violence, a process whose domestic political face has been called the Terror. This can be seen in the co-optation and suppression of the *armées révolutionnaires*, those quasi-official civilian forces sent by urban radicals to ensure food supplies to the cities, and in the usurpation of crowd violence with the bureaucratic machinery of the Revolutionary tribunal. Even the declaration of the *levée en masse*, far from being the spontaneous "rising" some radicals had demanded, was in fact a relatively measured and coordinated conscription for the army.[106] To repudiate the pike, then, was to repudiate popular insurrection in favor of state-led violence.

Instead of pike charges, partisans of the mass assault now insisted that French victories depended on the ardor of troops charging at close quarters with bayonets. Representative Prieur de la Côte-d'Or, another military engineer (Corps du Genie) brought onto the Committee of Public Safety, called the bayonet "the arm of heroes." The cult of the bayonet became a staple of Revolutionary lore. This was rhetoric that mattered. As John Lynn has pointed out, ascribing victory to Revolutionary élan not only meshed with a particular mythology about the French citizen-soldier, it inspired troops in the field. Morale counts in battle, and here is one site at least where the politics of representation had practical consequences. But even morale is only one factor. Lynn's careful analysis of the military engagements of the Armée du Nord in 1794 shows that battlefield practice was more complex than these exhortations imply. True, the Revolutionary troops concluded many engagements with a columnar bayonet charge, but not all were successful, and linear tactics (and firearms) played important preparatory roles in those that were. Artillery fire, too, was a contributing factor. Lynn thereby confirms the view that the Revolutionaries went to war under the "mixed order," adapting elements of Guibert's system (the maneuver column) to new purposes (an attack column). What commended this mixed order to the Revolutionary tactician was exactly what had commended it to the reformers of the late ancien régime: its mobility and flexibility.[107]

This is not the place to give a full analysis of Revolutionary and Napoleonic warfare. Suffice it to say that under the direction of Carnot, victory came from attacking in superior strength whenever possible. Under the more able generalship of Napoléon, victory came when superior force could be brought to bear on that point which would break the enemy's equilibrium. Both strategies depended on the superior mobility of French troops and cannon, including the new horse artillery. Above all, both depended on the ability of French troops to respond to battlefield contingencies more rapidly than their opponents could. The tactical advantage of this flexibility cannot be separated from its political virtues.[108]

This flexibility was seen as particularly suitable to the new citizen-soldier. No longer expected to obey his officer like a machine, the republican soldier was expected to supply an *internal* discipline, tempering his rights as a free citizen with obedience to martial law. This had always been the sort of soldier which Guibert had believed would be ideal for his style of tactics. The citizen-soldier's self-discipline made him capable of standing in a thin line, charging with a bayonet, or skirmishing in small groups. To be sure, this self-discipline was battened on nationalist fervor and hemmed in by military tribunals. But it also represented a logical extension of the same universalizing meritocracy which had shaped the identity of ancien régime artillerists. The whole nation now was ranked within a "professional" hierarchy.[109]

This flexibility was enacted socially in the famous *amalgam* of 1793–96, which merged the troops of the old line army with the Revolutionary volunteers. The form of this brigading was politically contentious. Moderates wanted to preserve the old army. The Jacobins wanted a unified army, but insisted that the volunteers not be dissolved among the "whites." Only after the *levée en masse* did the process begin in earnest, superintended by Carnot. It produced the first fully autonomous divisions, something the ancien régime had never quite achieved. And it was accompanied by a hybrid system of officer election, which attempted to balance experience, political reliability, and some degree of popular choice—an extension of the methods of the ancien régime artillery.[110]

Finally, this flexibility was consummated technically through the bayonet-musket combination. Of course, the bayonet-musket had been the staple infantry arm of all the European powers for nearly a century. The Revolutionaries neither invented the weapon, nor modified it in any significant way. Rather, the French found a way to exploit this artifact in a new manner, and (as we will see below) to produce it with novel methods and in unprecedented quantities. That they could do so was possible only because the central state—after a period of disarray—was able to complete a transformation already begun under the ancien régime. Tracing that transformation through the evolution of tactics has enabled us to see that even the external context in which guns were used was itself a political creation.

The Cannonade of Modernity

> Right now, I just want to lay siege, use my hollow shells,
> and then let them make a quick and honorable peace
> which will return me to you and my children.[111]
> —General Laclos to Madame Laclos, 7 June 1800

In a classic article on the social dimension of technological change, "Gunfire at Sea," Elting Morison describes the tragicomic resistance of the American navy to an innovative firing system that improved the accuracy of shipboard cannon. At the end of the nineteenth century, an American naval officer,

William S. Sims, learned of a simple gearing mechanism that enabled British gunners to correct for the rolling motion of a ship's deck and thereby dramatically improve the accuracy and rate of their fire. But despite Sims's experimental proof that the new firing system permitted a thirtyfold increase in accuracy, his superiors in Washington refused to credit his reports. At one point, they even sponsored counterexperiments proving that it was beyond human strength to manipulate a cannon with these gears—and seemed not at all troubled by the fact that these tests had been conducted on land where inertia would not favor the maintenance of a level gun. Only after the forcible intervention of President Theodore Roosevelt, a man above the bureaucratic squabbles, did the U.S. Navy adopt the new system.[112]

Morison asks: Why did the navy, an organization whose ostensible goal was to increase the fighting ability of its ships, resist this new technique? His answer is to consider the navy as a society. Those who resisted Sims, he notes, sensed all too well what havoc the new gunfire would wreak on their routines and values. Not only would the system raise to prominence the gunnery officer, formerly of little account, but it would render meaningless the "courage to close," once the mark of a superior crew. Now, naval engagements would play out over vast distances, with ships operating under wholly different tactics. A warship, Morison reminds us, is also a home, and the naval officers were defending their way of life. Sure enough, in the decades following Sims's innovation, naval life was utterly transformed.

This story offers suggestive analogies with the struggle over the French artillery and French tactics in the eighteenth century. Here too, scientific experiments did little to decide the matter. Techniques have meaning only in relation to particular purposes, purposes which are contested because they presuppose different criteria of judgment. In fact, the very bitterness of this struggle over French war matériel alerts us to the fact that much more was at stake than the proper length of a cannon tube or which sort of curve best described the trajectory of a falling body, or even whether there was any point in debating the shape of such a curve at all. So long as these questions were embedded in systems linking men and machines, any significant reform in the art of artillery (or the art of musketry) went hand in hand with new social arrangements and new ideologies to give them meaning. The superiority of Gribeauval's cannon could only become manifest within the context of a wholly different form of warfare, a type of battle not yet fought and largely at odds with the constitution of the ancien régime. And this marked precisely the source of opposition to Gribeauval's system: his insistence on a radical social transformation within the service. This was not simply because the reform gave prominence to new ranks of officers, such as the *chef de brigade* and the *garçon-major*, but because it implied a radical change in the lived experience of artillery officers, the proper measure of their merit, and their relative standing as an elite social group. In the most personal sense, it also asked them to embrace a very different conception of their own life—and death. After all, these military officers knew their destiny was bound up in the service of these cannon in battle. In some very real sense, different trajectories implied different fates. And the same held true for the nation as a whole.

In the "glorious" seventeenth century, Louis XIV and his lieutenants had put on a display of warfare that literally defined the French nation. These battles to mark the frontier of France were primarily siege affairs. Conducted as carefully orchestrated displays of royal authority, their success was central to the maintenance of absolutist authority. It is symbolically fitting, then, that the Gribeauvalists' new manufacturing methods precluded any elaborate ornamentation on the cannon, transforming the brass tubes into nameless instruments of destruction. In the event, the Revolutionary armies made abundant use of Gribeauval's cannon, its artillerists, and their new forms of organization. Tapping the new mobile tactics, the Revolutionary state (and its Napoleonic successor) for twenty years fought a war without limits.

In his letters home to his wife during the Napoleonic Wars, Choderlos de Laclos recounted the anguish of a sixty-nine-year-old general obliged to break camp nearly every night, as he struggled to keep his artillery trains in pace with the Grande Armée. The beds were atrocious, his hemorrhoids tormented him, and just one month after he had congratulated himself for finally attracting a cook who knew how to make the most of army requisitions, he had to decamp for Italy. Field war, he sighed, demanded far more energy than siege war. Laclos had joined the service in the last years of Vallière's administration and had spent the 1780s helping the marquis de Montalembert with his scheme for a "perpendicular" fortification, intended to restore the advantage to the defense. The scheme had failed. But as much as he admired the victories of Napoléon, Laclos always longed for a peaceful fortress France. The Vallièrists had warned that the new artillery would force gunners to scamper across the field of battle. Now they were being asked to scamper across all of Europe. During the Revolutionary Wars, artillery at last took its place as an integral component of field battle.[113]

The great victory at Valmy is the exception that proves the rule. There, the Republic was "saved" by the cannonade of 20 September 1792, in which one-third of all the Gribeauvalist cannon available traded shots with their Prussian opposites. This set-piece battle demonstrated that the French state still had the organizational capacity to deliver explosive force on the battlefield. This was in itself a nontrivial achievement. Philosophes had proclaimed citizen-armies to be the heirs of invincible Roman soldiers. And today we take for granted the compatibility of republicanism with large military operations. But the advancing Prussians clearly expected the bourgeois rabble in Paris to be incapable of mounting a concerted defense. Experienced noble commanders believed that only absolutism had the organizing capacity to direct the complexities of eighteenth-century warfare. Johan Wolfgang Goethe, who was accompanying the Prussians on their jaunt into France, clearly thought so too. They were not counting on the artillery service: a self-organizing, self-disciplining, and self-promoting social body whose loyalty was to the state, rather than to the king. The "humming . . . , gurgling . . . , whistling" sound, which Goethe later told the Prussian troops signaled "the beginning of a new epoch," was the sound of Gribeauval's artillery flying overhead.[114]

On that occasion, the French troops sat silently by. But they too would soon prove their capacity for self-organization and self-discipline. And the combination of the two quickly gave the French the edge in land warfare. That is not to say the artillery immediately achieved the battlefield dominance it would later acquire under Napoléon. Du Teil's ideal of massed and mobile artillery was seldom achieved in the 1790s. In 1793–95, Vauban's hundred-year-old fortresses still contributed to the defense of France. And the "boom" of cannon still served to boost the moral of inexperienced Revolutionary troops. But the use of horse artillery, first introduced by Prussia during the Seven Years' War, and long advocated by Guibert, finally saw action. And over the next twenty years, the artillery frantically pursued victory across the face of Europe.[115]

So complete a rupture in the norms of a social body, Morison argues, proceeds only with the support of a higher authority. Just as Sims and his navy reformers triumphed only with Roosevelt's help, Gribeauval's artillerists rose on the decision of the field marshals' committee and the patronage of reform-minded ministers, such as Choiseul and Saint-Germain. But Morison's invocation of a *deux ex machina* President obscures the larger role that naval power played in America's imperialist agenda in this period (1898–1908). Similarly, the success of the reform-minded officers within the French high command flowed from their ability to articulate a vision of a regenerated army as an instrument of French national power.

Morison suggests that his story has a lesson: that the ability of society to adapt to technical changes depends on its social flexibility. Writing in the optimistic 1960s he believed he saw evidence for this in the American society of his day. One may question whether Americans are so flexible, but certainly the situation was "worse" in the French ancien régime. The social dimension of the Gribeauvalist reforms struck not only at the status of a military caste, but at the social order itself. In the world outside Sims's navy, socio-technological dislocation had become a familiar, if resented, phenomenon. But Gribeauval's program of meritocracy and professionalization was at odds with much of the ancien régime. So too was their program of rational production.

PART TWO

Engineering Production: Coercion into Capital, 1763–1793

Faust (blinded):
"The darkness seems to press about me more and more,
but in my inner being there is radiant light;
I'll hasten the fulfillment of my plans—
only the master's order carries weight.—
Workmen, up from your beds! Up, every man,
and make my bold design reality!
Take up your tools! To work with spade and shovel—
what's been marked off must be completed now!
Prompt effort and strict discipline
will guarantee superb rewards:
to complete a task that's so tremendous,
working as one is worth a thousand hands."
—Goethe, *Faust II (Act V)*

Chapter Four

THE TOOLS OF PRACTICAL REASON

HISTORIANS have long wondered how and why production in Western Europe shifted from the artisanal workshop to the industrial factory. Not the least of the challenges has been to explain the assimilation of "mass production" machinery into the making of textiles, pottery, clocks, guns, and a host of other goods. Various hypotheses have been presented to account for the timing and character of this transition that underpins much of the West's wealth and military power. The approach of economic historians, such as David Landes or Joel Mokyr, is to couple the rise of the factory organization with technological creativity motivated by the heady lure of entrepreneurial profits. In complementary fashion, business historians since Coase, Schumpeter, and Chandler have emphasized the essential role of the entrepreneur as the organizer of production, apportioning tasks and underwriting investment in new machinery. In another tradition, Marx and a host of historians since, have posited a transitional Age of Manufactures, a period before the machine tools of modern industry set the pace and scope of production. In search of a more specific mechanism for this transitional age, Franklin Mendels and his interpreters have proposed the "proto-industrialization" thesis, which suggests how capitalists first gathered outworkers from rural areas under a single factory roof.[1]

Each of these schemes has illuminated different aspects of this great transition; each has its defenders and detractors. Yet all spin some kind of teleological narrative. As recent commentators have noted—William Reddy, Tessie Liu, and Maxine Berg among them—each assumes the success of the phenomenon it seeks to explain: the rise of machine production, the emergence of the entrepreneurial role, or the triumph of capitalists over domestic producers.[2] Up to a point, this form of teleology is understandable and even salutary. So long as their goal is to explain the rise of the factory (and its attendant technologies), historians will seek out those factors which they believe contributed to that development. On some level, teleology focuses the historical attention, and satisfactory historical explanations may well require some degree of over-determination. Without hindsight and presentism, historians would be baffled antiquarians. However, as Sabel and Zeitlin have pointed out in their article on "Historical Alternatives to Mass Production," teleological histories of industrialization have obscured important aspects of that process. Retrieving this buried past has implications because it reminds us that the people of the past made choices, that those choices have affected the range of choices available to us, their successors, and that we nonetheless face alternatives too.[3]

Part Two of this book tells a nonteleological history of a particular form of "mass production," known as interchangeable parts manufacturing, or the "uniformity system." This form of production was first developed under the sponsorship of Gribeauval's artillery engineers and applied to the making of military goods: first to artillery carriages and then to the flintlocks of muskets. While our attention may be drawn to these efforts by the role of interchangeable parts manufacturing in modern (Fordist) mass production, we should not equate the two. Just as there are many ways to organize production, so are there many kinds of mass production. Reading eighteenth-century interchangeable parts manufacturing as leading inevitably to Fordism would misrepresent the historical evidence: the motives of its sponsors, its actual techniques, and its subsequent trajectory. Interchangeable parts manufacturing first emerged as the program of Enlightenment engineers employed by the state.

After all, the French artillerists were more than mere gunners. At the beginning of the eighteenth century the artillery service was designated as the sole intermediary through which the state would acquire all its weaponry: cannon, artillery carriages, munitions, and small arms (muskets, pistols, and sabers). As the managers of a vast assemblage of cannon foundries, musket manufactures, and arsenals of construction, the artillerists supervised the kingdom's largest industrial establishment. Their efforts to transform production in this proto-military-industrial complex were an integral part of their effort to field an effective "killing system" for battle. To this end they sought to discipline the production of artifacts by enforcing a set of rigorous standards. This way they hoped to "normalize" the performance of their weapons, making their operation more regular, reliable, accurate, and deadly. Their goal was to secure capital goods of sufficient quality, quantity, and price to provide the margin of victory.

The engineers sought to accomplish this task by bringing to the workplace the same "system thinking" they had employed on the battlefield. Workplace coercion was a key element in this effort. But artifacts do not spring from the minds of engineers like Athena from the head of Zeus. Translating a design into a single artifact may be a problem in the manipulation of raw materials with the skillful use of tools. But making tens of thousands of similar artifacts is a problem in the politics of production. "To make bold design reality" one must exercise "prompt effort and strict discipline." One must enlist others in the work. The engineers tried to impose their vision of production in the face of considerable resistance from artisanal gunsmiths and arms merchants with very different views of production. This struggle to define the respective spheres of the state and private producers set the stage for the political conflicts of the French Revolution.

Engineering Production

My point of departure, then, is to ask how the uniformity system emerged from an eighteenth-century contest over how the productive life should be organized. In other words, I take the same historicist, actor-centered approach to technology developed in the first part of this book. How did eighteenth-century elites imagine the transition away from artisanal production? Many thinkers of that time were convinced that artisanal production was deficient. In France, those savants associated with the Physiocratic movement launched a concerted attack on the artisanal guilds—called corporations—and hinted at distinct alternatives. These "Economists," as they were called, had an important influence on government policy. One of their adherents, Minister Turgot, banned the corporations in 1776, and although the guilds were revived shortly thereafter, the Revolution abolished them permanently in 1791 (as formal legal entities anyway). The question was: what was to replace them? For all their hostility to the guilds, these savants recognized that the corporations were a coherent world which organized the social life of artisanal producers, as well as daily practices in the workplace. In the absence of the corporations, who would decide how to set up work schedules, and how? What would the rates of compensation be? And what would give coherence to this new world? The answers to such questions had important implications for the distribution of wealth and knowledge in society. Yet these theoreticians of the workplace did not necessarily anticipate the outcome that today leaps to the historian's lips: "the entrepreneur," "the machine," "the market." They *were* adamant, however, that "theory" alone was insufficient to answer them. What they called for was a way to bring theoretical knowledge to bear on practical problems. As we have seen, this amalgam was central to the self-definition of the French engineers. We will see now how the French Enlightenment engineers applied this new form of technical knowledge to the challenge of finding a new approach to production. To that end, they offered new ways of representing the objects of production: novel techniques for drawing artifacts on paper and for defining them with physical tools and machines. And implicit in these new representations was a new conception of work, and a new vision of the social order.[4]

As these innovations were introduced for the production of armaments, it will be useful to recall some of the peculiarities of the military market. Sombart, Weber, and Mumford have all noted how much early industrialization owed to European military competition. Sombart has gone so far as to locate the impetus for industrial capitalism in the militarized state. Thanks to its rondo of revenue-raising, protection, and coercion, the early modern state was a ready source of capital for new technologies. The army also offered a mass market for a relatively cheap and undifferentiated product (uniforms, guns, etc.). Moreover, military life offered a model for the management of complex activities and the regimentation of subordinates. As Weber put it: "[M]ilitary discipline is the ideal model for the modern capitalist factory." And recent commentators, such as David Noble, have condemned military command and control as the precedent for industrial management. This martial influence certainly does not in itself explain the rise of industrial capitalism, as Sombart thought. Nor was the military model of management as easily transferred to the world of commercial relations as Noble implies. Rather, the efforts of the artillery engineers to transform the production of armaments in eighteenth-century France provide a valuable insight into the difficulties of the transition from the putting out system to industrial production. In telling this story we will learn something about the failures of an explicitly engineering approach to production—and the basis upon which the nineteenth-century *capitalist* approach to industrialization was formulated.[5]

The thing to remember about military production in Western Europe is that by and large it has been handled by intermediaries. The military market may have been large and undifferentiated, but it was erratic, rising and falling with the fortunes of war and peace. As a rule, therefore, the Western states of early modern Europe did not own the means of military production. They let merchants and local producers absorb the risks associated with these investments, and in return cloaked these intermediaries in legal privileges and allowed them to reap heady (if intermittent) wartime profits. But if the state did not produce its own military matériel directly, it did send emissaries to make sure that merchants and producers delivered goods for a reasonable price and with some assurance of quality. In France at the end of the ancien régime, these emissaries were the artillery engineers. Over the course of the eighteenth century, these engineers took an increasingly active role in the management of weapons production. By the end of the century they attempted to introduce the production of goods with interchangeable parts. Why did the engineers come to take such an approach to production?

One way to answer such a question is to rattle off those timeless qualities of "rationality"—precision, uniformity, control, efficiency—to which engineers are "innately" committed. This has been the approach of many histories of production, whether they are discussing Taylorist scientific management or numerically controlled machine tools. But this approach only reintroduces teleology through the back door.[6] A central tenet of this book is that rationality is not a set of timeless abstractions; it is a set of social practices which have emerged historically. Rational production is a prime example of this. If we consider how engineers enacted shibboleths such as control or efficiency in practice, we begin to see how a given form of production emerged as a particular solution to a particular set of historical problems.

Take, for instance, the various practical means by which interchangeable parts production is usually said to have been achieved. In Robert Woodbury's article on Eli Whitney, he cites four "tools" as prerequisites for interchangeable parts production: (1) precision machine tools, (2) precision gauging, (3) uniformly accepted standards of measurement, and (4) techniques of mechanical drawings.[7] This formulation has the great advantage of emphasizing the essential role of standards in realizing interchangeable parts production. But Woodbury's checklist of "tools" still reads like a set of preordained criteria. As we will see, the French engineers developed and honed all four of these tools because they lacked the one factor Woodbury takes for granted: they had no direct control over the process of production. With one exception (to be discussed below), military engineers in eighteenth-century France were the supervisors, rather than the owners of capital. Engineers entered the workshop as outsiders, inspectors sent by the state to enforce various standards for acceptable goods. They approached production as an exchange; in return for the king's coin they wanted guns of a particular sort. To aid them in this task, they claimed certain legal powers—including the right to imprison recalcitrant workers—yet these powers were themselves contested. Their managerial role was assumed in the face of considerable opposition from artisans and merchants. As in all such exchanges, there was a strong element of mistrust between the parties.

Sociologists have pointed out that trust is a crucial issue in social relations in the workplace. Mistrust is a structural feature of the interactions of inspectors and workers—indeed of any contractual relationship where the interests of opposing parties are potentially at odds. Stringent demands for quality (especially when backed up by the state's unique powers of enforcement) stir resentment among producers and merchants who may fear that they will bear the costs of those standards. Producers address this problem in various ways, often depending on their position vis-à-vis other producers. Some larger producers (and merchants) will see the standards as a way to lock in their share of the market, keeping competitors at bay. At the same time, all producers (even the largest) will try to subvert those standards whenever they can get away with it. Confronted with these constant attempts to undermine their standards, inspectors will try to make them more difficult to circumvent. Typically, this will involve making them as rule-bound as possible, often to the point of embedding them in physical devices. Their aim is to restrict the discretion enjoyed by the producers. And the producers, for a price, will agree to go along with this, though still trying to cut corners where they can.[8]

In such a scheme, the tools of production emerge not as the willful imposition of rational planners upon irrational makers, but out of a process of conflict and negotiation over the terms of an exchange. And the development of the physical instruments which mediate those exchanges are the *outcome*, not the precondition, of conflict in the workplace—though such seeming resolutions are always occasions for further conflict and renegotiation. The history of these tools, then, charts the changing relations of production in that society.

To the extent that an artifact (here, a gun) is the sum of its standards of production, its qualities reflect the changing nature of those relations. And, consequently, the transformation of objects into objective tokens of exchange—commodities—depends on the relative power of those groups associated with production.[9]

The advantages of such a formulation are several. First, rather than assume that struggle over the terms of production begins with conflict over the use of tools and machines (as do most Luddite-inspired accounts), this approach folds the development of such tools and machines back into the larger history of production. Second, by historicizing the development of those tools and machines, this approach reminds us that they do not define a unique organization for production. Indeed, other ways of organizing production were then current in France, and they too invoked standards as a way to coordinate production. The issue, as we will see, is who *controls* those standards and how they are implemented. And finally, this approach does not prejudge the relative strength of the various parties involved in production. It thereby reminds us that engineers and managers do not always succeed in imposing their vision of production upon artisans or laborers. Obviously, significant power disparities exist between inspector-engineers and worker-producers. Indeed, these asymmetries play an essential role in shaping the development of these rational tools. But relative strength depends on circumstances. In the extreme case of monopoly capitalism (of either the statist or private variety), workers may be wholly without alternatives or resources of their own, and managers may seem to have an unrestricted ability to impose their standards. In other cases, when the inspectors are state engineers and the producers are autonomous artisans, the artisans may at times (under the right market conditions or with the right sorts of social solidarities) dictate terms to the engineers and the central state. The point may well be a general one. Similar patterns of mistrust and conflict obtain wherever managers address themselves to production.[10]

In the rest of this chapter I discuss the development of three of these tools of rational production in the Enlightenment—measures of work, technical drawing, and precision gauges and jigs. In each case, I show that there were competing conceptions of these tools: an engineering vision and an artisanal vision. Each of these conceptions, I argue, implied a different social order, and each had ambiguities which surfaced during the French Revolution. I then turn to the actual implementation of these tools in the workplace, showing how they were deployed during the Gribeauvalists' first introduction of interchangeable parts production.

Measures of Work, Divisions of Labor

One of the most basic issues in production is the measure of work. How work is going to be measured, divided, and rewarded depends necessarily on who is doing the measuring, dividing, and rewarding. The artisanal corporations of the ancien régime had one way of carrying out this operation. In the eighteenth century there emerged an engineering conception of how to measure work.

Everyone by now has heard the tale of how Descartes imagined the human body as a piece of machinery with a tacked-on rational soul. And they have heard its oft-told sequel: how the Enlightenment rationalizers, in their materialist excesses, toyed with automata and their intellectual equivalent, Man-the-Machine. As William Sewell has recently reminded us, however, the corporatist society of ancien régime France also saw the worker with a double eye: as the possessor of a communal "art," and as the suffering body of the laborer. Without the former, a worker was an *homme de peine*, a day laborer outside the legal entitlements that defined communal privileges in the ancien régime. The human-machine imagined by the Enlightenment engineers was a Cartesian analysis of this asocial human being: it was a laboring body—and it was somebody else's.[11]

Engineers had long asked how much labor power this naked body could deliver. As early as Agricola in 1556, "manpower" was expressed as a ratio of "horsepower." By 1700, under the influence of the new mechanical philosophy, manpower was expressed as the height to which a given weight could be raised (units of what physicists today call "work"). And by the end of the eighteenth century there was a vast accumulation of data on prisoners cranking pumps, laborers treading treadmills, porters climbing stairs, and journeymen polishing glass. None of it much reflected or measured anything like the effort of a worker employed at a real task.[12]

Charles-Augustin de Coulomb added a modicum of common sense to this program. Coulomb was a savant and military engineer (Corps du Génie) who spent over a decade on Martinique, supervising the construction of a fort. There, he became the first engineer to consider the effect of fatigue on labor power. He understood that a full day's work is *not* the same as the worker's maximum burst of effort. This insight opened the door to the question of what *was* the worker's maximum effort. Posed as a physiological analysis of the muscles of Man-the-Machine, the solution was impossibly complex. Coulomb, however, took a characteristic engineering approach, and posed the problem in empirical terms as a relationship among measurable parameters. This meant, however, that he faced the same difficulty as the ballistic engineers: Given that engineers seek to define what *ought* to be, rather than what *is*, how can they perform experiments untainted by social norms and expectations? For Coulomb this meant procuring data that did not simply mirror the customary efforts of real workers.[13]

On the assumption that climbing the stairs represented the most efficient work the human body could perform (raising one's own body only), he asked a young laborer to climb a long set of stairs eighteen times. The young man refused. So Coulomb relied on data for sailors climbing a mountain in the Canary Islands. With this benchmark, he then considered the efforts of a local town porter, who assured him he was making his maximum effort. Solving for the efficient maximum between these two cases, Coulomb noted that the porter would be able to move the same amount of wood with slightly less fatigue by taking fifty-three kilos on eighty-eight trips (instead of sixty-eight kilos on sixty-six trips). Coulomb acknowledged that individuals vary in strength, that diet and climate play a role in labor power, and that different tasks place different demands on workers' physical capacity. For that reason, he separately calculated the optimal manpower for the wheelbarrow, the crank, the hoe, and the pulley. But as his empirical data was always derived from the efforts of real workers on the docks and building sites, there was always the risk that all he was measuring was the laborer's notion of "a fair day's work." This was a difficulty that neither he nor his ergonomic successors ever fully solved. Frederick Taylor's time and motion studies—for all their scientific gloss—remained empirical measures of effort and hence extrapolations from existing patterns of work. They therefore required the consent (or co-optation) of at least some workers.[14]

This insight into the ambiguous prehistory of Taylorism reveals the uneasy tension between the Enlightenment engineer's concept of standardized work-effort and artisanal measures of labor in the ancien régime. As Michael Sonenscher has made clear, ancien régime artisans considered themselves to have property in their labor power. In the natural law tradition, labor (like life and liberty) were natural attributes of human beings, which French subjects had surrendered (alienated) to the absolutist sovereign in return for the privilege of organizing their own affairs collectively. Thus, the charter of a trade corporation was a title of an exceptional kind. For the artisan—even one outside the purview of the corporations—the price of this alienation was the wage. This was what distinguished the artisan from the slave or dependent servant. Whether this wage was paid for a day's work, or for the making of a particular article (the *prix de façon*), an artisan retained certain rights in his or her labor (or in the product of his or her labor). This legal "fiction" of ownership, as Sonenscher calls it, had real implications. It allowed journeymen to lay claim to time spent on setup work or to customary rights to the by-products of production. And this in turn structured the division of labor in the workshop, the type of work artisans were willing to do, and the levels (and kind) of compensation they received.[15]

Artisans, of course, possessed more than their labor power. They also possessed skills by virtue of their participation in a collective order. The role of the trade corporations was to safeguard the reproduction of that art. This art derived from a kind of rule-based knowledge, yet it also required judgment and discernment. The making of a commodity—a gun, a hat, a book—involved a series of recipes that could never be fully specified. The tacit and unanalyzable nature of art—what Polanyi has called personal knowledge—made it the central mystery around which the rituals of corporate life were organized. The gun trade was one which prized craft secrecy highly. This meant that instruction could only be acquired by apprenticeship to a master, and this, in turn, enabled practitioners to regulate access to the trade.[16]

Contrast this with the engineering conception of skilled work as expressed by Diderot. The cutler's son was a champion of the mechanical arts. He bemoaned the low social status of manual laborers and sought to express the richness and dynamism of the trades. The plates and text of his *Encyclopédie* pulled aside the veil that hid the processes of production from genteel eyes, serving up a cornucopia of commodities, along with images of the machines and workers who made them. Yet, the same gesture that tore the veil from the production of ordinary things, also exposed the trade secrets that the corporations considered the foundation of their communal life. Diderot denounced the artisanal corporations for stifling innovation. Their secrecy was a prescription for stasis. Ending the restraints on production and beginning a *public* discussion of the arts was the sole remedy for the inefficiencies of current practices. Only such a discussion would subordinate the mechanical arts to science (marry theory and practice) for the greater good of technological progress.[17]

The consequences for most workers were not expected to be pleasant. Diderot's emphasis was on the "sagacity, wisdom and intelligence" of the machines, not the laborers. The hand alone was powerless without the aid of "instruments and rules." To be sure, these ingenious machines were themselves the creation of a tactile intelligence which Diderot greatly admired. But he admitted that for a majority of workers, labor would ultimately become subservient to those instruments and rules. In his famous article, "Art," he made much of the advantages of the division of labor for increasing productivity. This was most famously set out in the pin factory example, which Adam Smith adapted from Diderot, who filched it from the engineer Jean-Rodolphe Perronet's detailed study. Already in the 1740s, Perronet had used time and motion analysis to estimate realistic rates of pin parts production, task by task. He then added the costs for raw materials, included fixed capital costs for machinery, and calculated the profitability of pin-making for various types of pins. To maintain these task rates, Diderot noted, each worker would be obliged to do the same thing for "the rest of his life." Or as d'Alembert put it in his "Discours préliminaire," the worker would be "reduced to mechanical actions."[18]

A question then arises: In the absence of a corporate body to regulate the art of craftworkers, what would hold the productive process together? How would one devise production schedules, divide labor, or control the work standards of these automatic laborers? These were acute questions in the period before those schedules, divisions, and standards were built into special purpose machinery. Diderot's effusions aside, the pace of work in eighteenth-century France was *not* set by automatic machinery. These questions were also a pressing concern before the role of the entrepreneur had been clearly delineated.

The same question that has troubled historians about the transition away from the artisanal atelier also troubled Enlightenment contemporaries. To see how eighteenth-century theorists of work tackled this problem, we must return one last time to Diderot's article on "Art."

The Geometry of the Workshop

For eighteenth-century theoreticians of the workplace, one of the principal tools for organizing the workshop was technical drawing. As we saw in chapter 2, Diderot had proposed a new two-way relationship between "geometry" and "art." He called for a "Logician" to invent a new "grammar of the arts." In the trades, tools with identical functions had different names; other tools had only the most vague and generic names. As a good philosophe, Diderot had a simple solution for this reprehensible state of affairs: each tool that generated a clear and distinct idea should have its own name. This would be possible because even the most complex machine (*machine composée*) was actually an assemblage of a limited number of simple machines. The "Logician's" first task, therefore, would be to develop a quantitative scale to express the various measures of tools (their size, force of action, etc.). His second would be to prepare a morphological analysis of tools. The *Encyclopédie* was offered as a first step in this taxonomic "natural history" of machines. Where art had once organized the social world of production, this public and generally available "science" would now do so. Diderot was not calling for that abhorrent "geometric spirit" so much in disrepute in the Enlightenment. He assured his readers that academic geometry was no substitute for what he called "the geometry of the workshop," and by this he meant the actual rendering of shapes—that is, a program of technical drawing.[19]

It is this practical geometry that explains much about the famous plates of the *Encyclopédie*. Diderot considered the plates an integral part of his project and an essential component of his new grammar of the arts. A verbal description of a machine on its own was inadequate. Yet these illustrations carried a deeper message. William Sewell argues that these representations of work seemed to disassociate production from the existing social order. He contrasts these silent and anonymous artisans, docile at their benches, with seventeenth-century prints showing a community of vibrant artisans, and nineteenth-century prints depicting autonomous, highly individuated craftworkers. This is hardly surprising if one considers the different mission of Diderot's text. The *Encyclopédie* belonged in the tradition of the Renaissance *Theatrum instrumentorum et machinarum*. These compendia sidelined human actors and instead presented views of isolated tools. This was a visual language that conveyed the heroic possibilities of machinery. The plates of the *Encyclopédie* had a similar (though more expansive) goal. (See figs. 1.4, 4.1., and 5.2.) The illustrations begin with an overview, then analytically divide the machine into components. Thousands of machines and tools are displayed in this way,

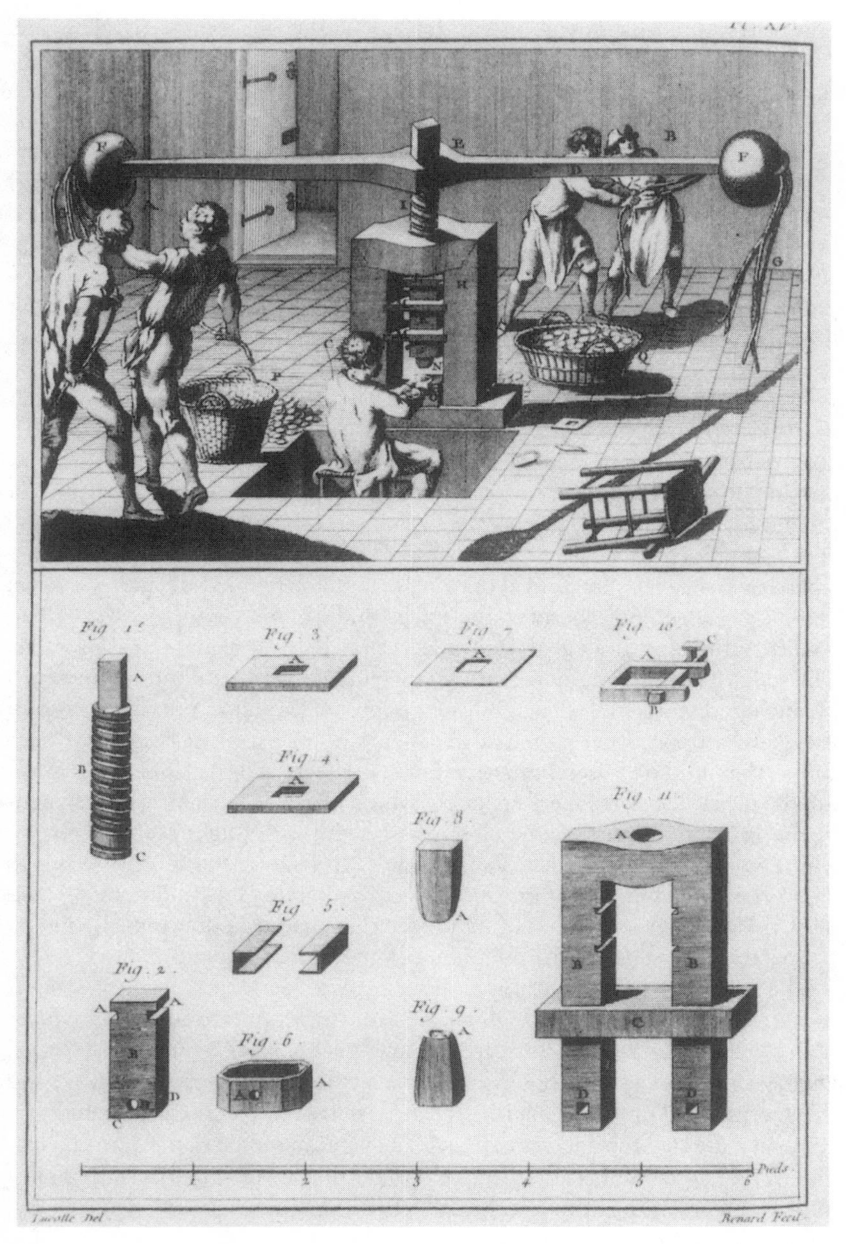

Fig. 4.1. Coining Press, 1750s. This two-meter-high coining press was used in mints in the ancien régime. The operator placed the coin to be stamped in the working space N. He would be sure to get his hands out of the way before the four burly laborers tightened the press home. Note the close-ups and cut-aways of the machine parts. From Diderot, ed., "Monnayage," *Encyclopédie*, vol. 25, plate 15.

drawn from multiple angles and at different scales, in "exploded" views and "cut-aways." Their aim was to highlight the *processes* of production; for instance, sixty-odd plates were devoted to lock-making or silk-weaving alone. As for the workers, Diderot recorded only their "essential movements." The plates, moreover, demonstrated the interconnectedness of technical knowledge. True, these tools and machines were still classified by the trade they served. In the alphabetical order of the text, however, they were catalogued both by trade *and* by function. Moreover, here, as elsewhere in the *Encyclopédie*, the *renvois* dispersed knowledge throughout the volumes. In a metaphorical sense, then, these pictorial representations removed knowledge about production from the particular trades and dispersed it throughout the structure of public knowledge as a whole.[20]

As a practical matter, technical drawing also promised to organize the workshop on both procedural and social levels. First, by distinguishing between the conception of an artifact and its execution, technical drawing suggested how one might redistribute tasks within the workshop and provide a standard for production. And second, by creating a common language for both artisans and elite technologists, it bound and ranked the members of the productive order. These twin aspects of technical drawing as an analytic method and as a social marker appealed enormously to the *Encyclopédistes* and their engineering epigones. Achieving these goals depended critically on a pedagogical program. The sensationalist epistemology that undergirded the Enlightenment posited a pliable mind, and most of the philosophes placed an almost unlimited faith in the power of education to recast the mentality of students. Ever since Descartes, the mind had been understood to see objects with its "inner eye," leaving the physical eye as a mere apparatus which might deceive. Proper judgment in the Enlightenment became closely associated with "right seeing." That is why, speaking of this subject, Diderot said, "It is never too soon to begin to adjust the human mind by furnishing it with examples of the most obvious and most rigorously exact reasoning," and recommended that drawing—including perspective drawing—be the one subject taught to all students.[21]

The mid-eighteenth century in France saw a vogue for technical schools centered on a geometric curriculum. These institutions ranged from the more than twenty part-time apprenticeship drawing schools for artisans to advanced engineering schools like those run by the artillery, the Corps du Génie, and the Corps des Ponts et Chaussées.[22] Across the Revolutionary divide and across the divide of social status, drawing education served as the core curriculum in French technical education. Consider how surprising this is in light of all that one might learn in a technical school: the properties of raw materials, the actions and forces of instruments, not to mention manual skills. In and of itself technical drawing accomplishes nothing. The contrast with England is striking. Although English machine builders and craftsmen used technical drawings extensively, they did not learn mechanical drawing formally in schools. They were all apprenticeship-trained.[23] Why did French elites in this period fixate on geometry and mechanical drawing as the principal method for organizing technical education? Why did they teach these techniques to both artisans and engineers? Were the two in fact being taught the *same* techniques? And finally, what difference did this training make to the organization of production?

There are, of course, many different types of technical drawing: from freehand sketches to projective geometry. As "tools" to assist in the organization of the workshop, these different types of drawings imply (but do not require) different social relations. That is because a pencil sketch and a registered blueprint imply very different degrees of discretion for conceivers and builders, and the distribution of discretion in the workplace can serve as a rough-and-ready map of where authority resides. The historian Yves Deforge has read into the new eighteenth-century forms of technical drawing an increasingly sharp separation between the functions of the conceiver of the artifact and its executor. But technical drawing is more than a barometer of such changes. It was also a tool for creating a new, stratified productive order. This stratification, however, was by no means foreordained. There were two contrasting programs of drawing education in the eighteenth century. Mechanical drawing, especially the descriptive geometry, was taught to engineers. Academic drawing was taught to artisans. Both programs contained ambiguities which were only gradually resolved.[24]

Disciplining the Artifact

The descriptive geometry is a mathematicized method of mechanical drawing developed in the engineering schools of Enlightenment France and still taught in technical schools throughout the world. In the words of its creator, the eighteenth-century mathematician and political revolutionary, Gaspard Monge, it was a "[universal] language necessary to all those who work in the mechanical arts" because it allowed one "to represent with exactitude, on drawings which have two dimensions, those objects which have three, and which can be rigorously defined." It accomplished this by portraying a given object from multiple, orthogonal, projective views; that is, by showing a given object face-on from two (or more) sides at right angles. For millennia, architects had drawn buildings in plans, elevations, and facades. And artisans (especially masons and carpenters) had long possessed secret stereographic methods for calculating the various block faces needed to build, say, a Gothic vault. These techniques were generalized by Desargues in the seventeenth century. In Monge's descriptive geometry, however, these views were rigorously interrelated, consistently laid out, and coupled to the world of mathematical analysis. As a result, the descriptive geometry was also a "constructive" technique, allowing its possessor to solve for new shapes and configurations. By permitting engineers to analyze and manipulate these physical objects on paper it helped them solve problems in fortress construction, stone-cutting, and even machine design. This then is the story of what happens to the world of things when people inscribe them on paper. It marks a first step toward understanding how the way things are made has been transformed by the way they are represented, and how the social order reflects the ordering of pictures.[25]

In a sense, projective representations function within engineering culture much the way photography functions within today's popular culture: as a picture of "the way the world really is." But this analogy—like the analogy with the Renaissance techniques of perspective—can be misleading. Engineers and architects use projective views because they avoid the distortions of shape that perspective drawings intentionally introduce to give the illusion of depth. As Descartes had pointed out, perspective is a deception set aright by the judgment of the mind's "inner eye."[26] Perspective drawings are "views from somewhere," and hence, still within the realm of the personal (albeit a readily translatable "personal"). Projective drawings, by contrast, are "views from nowhere." As Abraham Bosse had noted in the seventeenth century, they are the equivalent of perspective views seen from infinitely far away—except that they are close up. Such drawings are "objective" in Lorraine Daston's sense of being aperspectival; they have no "point of view." Or to use the formulation suggested by Daston and Peter Galison, they are the negation of subjectivity. They signal an effort to eliminate all individual and group idiosyncrasies, and obviate any need for judgment, even (supposedly) by the mind's inner eye. In the first half of the eighteenth century, Frézier taught engineers to reject perspective drawings as inadequate if they wished to speak to subordinates with a minimum of ambiguity. Only projective views would do. Paradoxically, projective views do not, to untrained eyes, look even remotely like the world "out there."[27]

Mechanical drawings achieve this effect, in part, by reducing the representation of objects (and their decoding) to a set of formal rules. The goal is to reduce the discretion of both the person drawing the plan and the person reading it. In this other sense, as well, we may say that a mechanical drawing is an "objective" picture of an artifact (though it "looks" nothing like the artifact). As a quasi-public and mathematicized language—what Theodore Porter calls a "technology of distance"—mechanical drawing binds those who use it to a common vision of the object. Here "distance" has at least three layers of meaning. First, as we have seen, mechanical drawing bridges the epistemological mistrust that exists between the inner eye and the external world. Second, mechanical drawing allows for a common intra-group conception of an artifact across space and time, particularly useful for those bureaucratic organizations which must coordinate far-flung activities. And third, mechanical drawing helps bridge the chasm of mistrust that lies *between* groups by providing a common referent. Consider the famous steam engine drawings of

James Watt. Watt first used mechanical drawings (rather than freehand sketches) when he began to contract with clients for the sale of his steam engine. In the 1780s, when he exported his steam engine to France, he used technical drawings to ensure that parts made by subcontractors in France would fit those made in England, and also to monitor and coordinate the work of bricklayers and other workers at the assembly site.[28] French military engineers used mechanical drawings in much the same way: as a sign that their design was approved by the king's ministers, as a master pattern to which far-flung entrepreneurs and suppliers had to conform, and as a method of parceling out tasks to various groups.

In sum, this type of representation is what Bruno Latour calls an "immutable mobile." Images in this form can be moved across physical and cultural distances without undue distortion, and collected at a remote site of power. There, they can be synoptically compared with other images so that discrepancies may be noted and corrective actions taken. To the extent that a cathedral plan coordinates stonecutters and a military map deploys soldiers, an engineering drawing commands workers. Of course, pictures do not in themselves coordinate, deploy, or command. These drawings make possible the exercise of power by enabling their possessors to master phenomena on a scale inaccessible to others.[29] What Latour's analysis slights is the degree to which the authority of these pictures derives from the *self*-discipline necessary to make one. Before engineers can use pictures of this sort to command workers, the drawings themselves must be highly ordered entities. Students spent years learning the self-restraint that enabled them to picture only certain carefully defined characteristics of objects. As Daston and Galison point out, making objective images is a painstaking and laborious undertaking.

This especially applied to the *mathematicized* drawing techniques taught to engineers. Monge always acknowledged that the descriptive geometry often could not be applied to the thick objects common in commercial life. This limitation, however, only increased the moral value of the descriptive geometry as a tool for training students. "[I]f," he said, "from a young age, designers (*artistes*) had been trained in the study (*exercés à rechercher*) of the lines of curvature of different surfaces which are susceptible to exact definition, they would be more aware of the form of those lines and their position, even for objects less [readily] defined; they would [then] grasp them [mentally] with greater precision and their work would be more expressive."[30] In other words, these drawings would enable their creators to perform what Michael Lynch has called the "disciplining of the object"; the process by which the graphical properties of the object are made to embody the "natural object," making the object scientifically knowable and manipulable, much like the docile bodies of Foucault's prison institutions.[31] Such a process involves years of training.

In other words, mechanical drawing may preclude the illustrator's judgment about how to represent the object, but paradoxically one motive for training engineers in this technique is to form their judgment about what are proper objects and how to manipulate them. Indeed, the very rigor of this training suggests that the descriptive geometry is not a "natural" representation, but a cultural convention which arose historically and reflects its creators' view of their place in the broader social order.[32] Technical drafting defined the social role of engineers in late ancien régime France as the conceivers and orderers of artifacts, thereby preparing them to act as intermediaries between state patrons and venal producers.

From the 1760s onward, Monge taught his methods to two generations of engineers-savants (and future republican leaders), including Borda, Bureaux de Pusy, Coulomb, Carnot, and Prieur de la Côte-d'Or. Analogous techniques of projective drawing were also taught at the artillery schools. In the 1740s, the commander of the artillery school at Metz could claim that the importance of drafting for engineering students was so widely recognized as to need no defending. According to Jean-Pierre Du Teil, who directed the Auxonne school when Lieutenant Bonaparte was in residence, technical drawing was indispensable to all artillery officers. Under the guidance of a drawing master, they began with drawings of the natural terrain or strongholds from various "geometric" perspectives. They then moved on to elaborate exercises in drawing fortifications, the basics of civil architecture, and the plans of powder houses, artillery batteries, and artillery parks. They finished with elevations and profiles of the "machines of the artillery," working from current models kept in a special *salle des modèles*. This drawing curriculum provided students with a common body of knowledge, and enabled these sons of petty noblemen and bourgeois notables to direct craftsmen and manage the complex works involved in armaments production.[33]

To reiterate, particular methods of representing artifacts—from technical drawings to today's computers—do not in themselves *require* any particular socio-technical order. As Shoshana Zuboff has shown, the use of computers does not imply a particular form of work organization, though it places a considerable burden on how people are educated to cope with the instrument's new possibilities. As we will see in chapter 8, the legacy of Monge's methods was contested during the Revolution, and much of that contestation centered on pedagogy. Mechanical drawing only made *possible* the separation of the tasks of the conceiver and maker; it did not require their separation. However, by enabling engineers to translate objects into geometric figures, which they could then manipulate and break down analytically, projective drawing made possible a disciplining of artifacts. As Desargues noted in his seventeenth-century treatise on stonecutting, his method "left no shape to chance, or to discovery in the act of making." Engineers could now define tasks and communicate them with sufficient rigor so that the final assembly need not depend on the practical aid of a "fitter." Mechanical drawing did not necessitate interchangeable parts manufacturing, but it is hard to imagine organizing production in this way without some such objective representation.[34]

Gribeauval's famous *Tables de construction* are a splendid example of this disciplinary program (fig. 4.2). Already the 1732 law announcing the Vallière system had been accompanied by official plans for the artillery pieces. From 1764 onward, the Gribeauvalists assembled mechanical drawings of every component of the artillery matériel. These were carefully scaled projective views, with parts specified in dimensions down to 1/200th of an inch. Gauges, jigs, and rulers were also enshrined in the *Tables*. The gigantic five-volume set of drawings was finally published in 1792. With this master plan, artillery officers now had a common reference for all the objects of their technological life. They also possessed an analytical tool for dividing the job into individual tasks, and a disciplinary tool for holding up each piece to an immutable standard.[35]

Palpable Design

The contrasting form of representation is the sketch, or "freehand" drawing. The emphasis here is on the open-endedness of the image—and of the artifact. The "rules" of drawing here are ill-defined, even idiosyncratic. Generally, this is a quasi-private language, used as an extension of the creative process, or as a kind of private notation to oneself or one's immediate colleagues.[36] Such a drawing implies a high degree of trust between the designer and executor of the object. At the limit, they may be one and the same person. As Leora Auslander has pointed out in her study of the French furniture trades, artisans used freehand sketches as a bridge between knowledge and manual skill; their drawings did not exhaust or replace their skills. That is because even when they copied patterns from others or used geometric forms, they still exercised discretion about how to implement their designs.[37]

This was the form of drawing Rousseau recommended to his pupil, Emile. Its purpose was to teach him to see for himself, and to that end Emile was to sketch, not from conventional drawings, but directly from nature. Rousseau carefully distinguished this sort of drawing from geometry. He went so far as to lock away all rulers and compasses, so that his pupil might not become accustomed to these mechanical crutches. The emphasis was on the formation of an independent judgment. For similar reasons, he counseled Emile to learn a trade, as the ability to earn his living would guarantee his independence and the manual labor would teach him virtue.[38] This was a far cry from Diderot's vision of how engineering drawing would organize work.

The difficulty was that before engineering drawings could be of any use, they had to be comprehensible to those further down the productive hierarchy. That was why the artillery taught drawing to its worker-soldiers. Teaching artisans to draw attracted considerable interest from the enlightened elite in this period. The largest part-time drawing school for artisans, the Ecole Royale Gratuite de Dessin, exemplifies the contradictory attitudes of pedagogues to the unenlightened populace—and to the organization of artisanal work. This Parisian scholarship school was founded in 1766 by the energetic

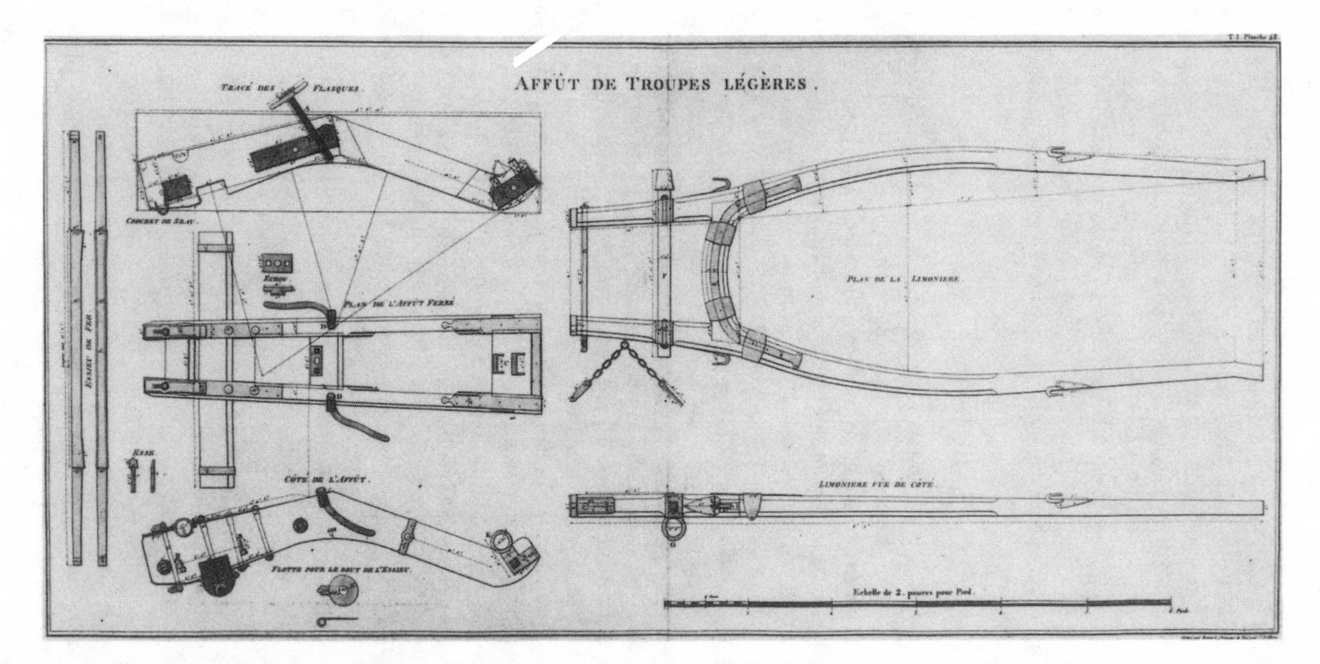

Fig. 4.2. Artillery Carriage, Gribeauval System. This plate from Gribeauval's *Tables de construction* depicts the cannon carriage and limber. Note the use of geometric constructions to indicate the curvature of the cheeks (*flasques*), and the way the orthogonal projections have been interrelated. The elevating screw (*vis de pointage*) is depicted in the upper left. From the Musée de l'Armée (K12793).

Jean-Jacques Bachelier. Some four thousand student-apprentices, most between the ages of twelve and fifteen, passed through his school in the twenty-four years before the Revolution. All began with a common instruction in elementary geometry. Thereafter, depending on their trade, they took one of three classes: architecture, figures and animals, or flowers and ornaments. The student's primary task during his six years of part-time study was to copy twenty-three hundred sequential academic drawings. None pictured mechanical devices.[39]

Why would anyone expect an educational program of drawing of this sort to advance French industry? Louis-Sébastien Mercier, the perpetual critic, ridiculed the idea that academic lessons would accomplish anything other than turn good artisans into second-rate artists. Did lessons in Euclidean geometry and years tracing ornamental flowers really improve young gunsmiths' craft skills? Bachelier believed they did because such lessons transformed the aesthetic and technical *judgment* of young artisans. Technical skills depended on proper mental representations, and conversely, patient hand-tracing formed the mind. But rather than look to nature and the imagination, as Rousseau had advised, Bachelier believed that artisans needed geometry to serve as a "mold for the operations of the mind." He claimed that geometry would make artisanal work more "precise" by teaching students the "exact knowledge of the dimensions of objects considered under various aspects." And he averred that this had practical results: "From certainty in work comes promptitude in execution; [and] rapid execution will unleash the industry of the nation by lowering prices."[40]

Bachelier's real enemy here was the artisan's "ignorant and prejudiced" imagination; only geometry could "prevent the imagination from flying off, and contain it within the bounds of reason." The course aimed to wean artisans from ugly irregular shapes. Catering to the neoclassical tastes of the elite was certainly a smart strategy by which Parisian artisans might increase their share of the luxury market. In this period, fashionable design figured increasingly in the appeal of goods to discriminating consumers—including the buyers of fine firearms—so much so that English producers lured French drawing masters to Birmingham to help with new patterns. But Bachelier's avowed purpose went deeper. The academic style of drawing disciplined and invigorated the ranks of the artisanal community. The neoclassical style expressed virtuous themes, replacing the "tormented" and "monstrous" forms of the baroque with a manly self-restraint and civic service. Self-discipline in taste correlated with self-discipline in the workshop. Bachelier believed his school gave the habit of work to young men who otherwise tended to be lazy and disorderly.[41]

At the same time, however, Bachelier's course played to the artisan's aspirations for autonomy and pride in his craft. The course was to secure for "each artisan the ability to execute by himself and without outside help those different works which his particular genius for his art enables him to imagine." The productive world Bachelier envisaged remained that of the independent handcraft worker governed by the norms of corporatist culture. That artisanal foyer offered a hedge against the "unemployment which troubles the social order and the peace of the family." This model was explicitly paternal—Bachelier pointedly enjoined his students to become virtuous fathers—and as such it reflected the king's paternal rule and the larger corporate structure of the state ("which has done everything for you, while you have [as yet] done nothing for it").[42]

This ambivalent relationship of the school to the particularist and patriarchal world of the corporations became evident with the coming of the Revolution. The abolition of the guilds also presented the school with some very practical problems, since a tax on the corporations had supplied half the budget. In his petitions to the National Assembly, Bachelier now put his emphasis on the universalizing properties of drawing and how it would encourage progress in the mechanical arts by breaking down the divisions between trades. He no longer described his course as an aid to the autonomy of the artisan. With many master artisans suddenly concerned about the loss of control over their journeymen, he now offered "geometry" as a substitute for the authority of corporations. It supplied, he argued, "the subordination which costs nothing to laboring citizens." In 1792, he assured the National Assembly that his course would not give apprentices ideas above their station; but supplied useful knowledge which "recalled them to the estate (*état*) of their father." In this sense, the program of technical drawing taught to late eighteenth-century artisans was ultimately used to reinforce a stratified social order in which one's position in the cognitive hierarchy implied one's position in the social hierarchy. The question to answer is the degree to which this new cognitive and social order transformed the way real artifacts were made.[43]

Of course, an artisan can make *some* kind of artifact from almost any technical illustration; the question is the degree of judgment exercised in carrying out the "instructions" embedded in the picture, and the extent to which it matters whether the constructed object conforms to the plan of the conceiver-illustrator. However, *no* device can be reproduced from a drawing alone. The French learned this when they tried to assemble a working steam engine from Watt's drawings; despite the famous drawings, the pieces did not always fit. Direct personal contact was necessary to transmit the tacit skills needed to replicate a novel technology. And no drawing can capture the set of tactile skills needed to make a working device.[44]

Manufacturing Tolerance

Pictures do not in and of themselves discipline artifacts or coerce labor. Mental representations are essential to work and technological process. But as Diderot understood, one must ultimately grapple with thick things. So to translate their drawings into artifacts, engineers embodied their idealized instructions in physical "tools." Among these mediating devices were gauges, jigs, cutters, fixtures, dies, and (most famously) automatic machinery. Gauges come in two general classes. Go and no-go gauges enable one to judge the "fit" of some dimension of an artifact against some physical standard, and calibrated gauges give a numerical reading of some dimension of an artifact. Jigs, cutters, and fixtures guide the shaping of a piece of work. They transform the general action of a human hand or machine into a specific action. A jig attaches to the workpiece and guides a hand file or cutting tool. This cutter can itself be shaped. Workpieces are held in place by fixtures attached to the machine. All of these devices enable a producer to get a variety of actions out of the large capital investment sunk into a general-purpose machine. Along with steel dies to forge pieces to precise dimensions, these humble devices are the foundation of interchangeable parts manufacturing. They are also essential to "flexible specialization."

Despite their ubiquity, almost no historians of industrialization have examined gauges and jigs. No doubt this is because they are seen as unproblematic agents of industrialization: disciplinary devices to deskill workers. To be sure, gauges define the shape of the artifact by setting the standards for whether a piece of work is acceptable or not. And jigs define the shape of the artifact by limiting the discretion of the worker whose tool is now guided as it shapes the piece. But gauges and jigs are not some external "resource," brought in as the unambiguous agents of a "rational production." Gauges and jigs are the *outcome*, not the precondition of conflict in the workplace. Understanding this process will enable us to historicize the development of machinery more generally. After all, automatic machinery does not simply magnify labor power; it also mechanizes standards of production by means of jigs and fixtures.[45]

To begin with, the use of gauges and jigs long predates the era of industrialization. Before the eighteenth century, watchmakers certainly used these mechanical aids, as did gunsmiths. Almost all the trades used calipers, compasses, and rulers of various kinds.[46] Though not as widely disseminated, calibrated gauges—such as the Vernier micrometers—had been employed by scientific instrument-makers since the seventeenth century. Then there is the question of deskilling. As Robert Gordon has pointed out, the use of gauges, jigs, fixtures, dies, and cutters put heavy demands on the skills of metalworkers well into the nineteenth century. Although gauges reduced the latitude of workers to set their own standards of production, workers (and inspectors) still had to learn the "touch" of a gauge. This required experience. Once a machine had been set up with a jig or fixture, it certainly reduced the machine-tender's discretion over how to shape a piece. But there remained considerable room for expert judgment. Nor was there any *a priori* reason why a metalworker might not set up his or her own machine. Then, too, enormous skill went into the making of these gauges, jigs, cutters, and dies; and these might also be made by the worker who tended the machine.[47]

This should remind us that the social meaning of gauges and jigs depends on the institutional context in which they are deployed. The ability to repeat certain tasks (and check for deviations) is of value to craftworkers in their own shop. Deployed in a modern machine shop, gauges and jigs can be used to separate those who conceive of the artifact, those who design the machinery, and those who operate the machinery. Yet these roles were not distinct in the eighteenth century. At that time, there was ongoing conflict over the use of these devices. This is because gauges and jigs define the *limits* of the agreement between the parties involved in production. In this sense, they are the physical bearers of manufacturing "tolerance," a term and concept first introduced by the engineers of ancien régime France.[48]

Manufacturing tolerance, despite what its name evokes, increases the stringency of the supervisor's control over the work process—albeit in a paradoxical manner. Everyone in a workshop knows that specifying the exact dimensions of an artifact is delusionary. An engineer who demands an ideally cut piece in fact leaves the worker an unspecified degree of discretion about how to shape it. A specific band of tolerance, however, spells out the limits of acceptability. And the use of go and no-go gauges (or jigs) defines this band of tolerance in concrete terms. In an era before automatic machinery, this enabled directors of workshops to set standards of production without leaving all judgment in the hands of the workers.

Workers and inspectors come into conflict as an almost inevitable part of the work process. Where the worker is on piece-rate wages, or an independent artisan who bears the cost of rejected parts, inspections can mean the difference between a living wage and starvation. Even where workers do not directly pay for rejections, they resent inspectors. Gauges devalue old skills in place of new ones, and this upsets the scale of social prestige in the workshop. And for workers under direct management, standards have a human face. This is where gauges come in handy. They appear to deflect responsibility away from the inspector because he can plausibly say that he is only doing his job. Gauges, in other words, are "objective" to the extent that they appear to reduce the discretion of the inspector as much as the discretion of the worker. This form of objectivity is again akin to that described by Theodore Porter in his study of quantification in public life. In Porter's view, there are certain forms of quantification (accounting practices or cost-benefit analysis) which consist of rules which are arbitrary to some degree. Such rules are generated at just those points where mistrust reigns and parties have conflicting interests. Indeed, the elaboration of these rules can be read as a record of the continual effort to prevent their further subversion. Quantification of this sort is objective in the sense of being impersonal, a set of mechanical operations that seem to preclude independent judgment, and hence the discretion of both parties. Porter argues that the reduction of some facet of public life to numbers signals the vulnerability of the relevant "experts" to pressure from powerful outsiders ("power minus discretion rather than power plus truth"). In this view, the prevalence of quantification in public life is a sign of conflict, not of consensus.[49]

Gauges and jigs play a comparable role in manufacturing. They succeed to the extent that they appear to bind workers and inspectors to a common set of impersonal rules at just those points where the possibilities for conflict are greatest. They are objective technologies insofar as they appear to substitute mechanical authority for personal judgment, organizational rules for individual discretion, and verifiable standards for trust. This does not mean that conflict comes to an end or that the parties have equal power in this process. Consider the analogous discussion by E. P. Thompson of the transition from task-time to clock-time in the early modern factories. This was a protracted struggle in which workers generally lost much of their ability to control the work process. Once workers were obliged to labor by clock-time, however, then if six o'clock is quitting time, then when the clock says six it is time to quit. And if a worker does not trust the foreman's clock, he or she can check the time on a pocket watch (as some did in the nineteenth century, and with good cause). Moreover, the worker can now frame an argument about the number of hours he or she will work: the twelve-hour day, the ten-hour day, the eight-hour day. The same pattern holds true in Linebaugh's discussion of the painful transition in English shipyards from customary payment (in wood chips) to wage labor (in specie). The existence of such impersonal standards, then, is both an outcome of past conflict and a site for further conflict. To be sure, the dispute is now fought out on a terrain defined by the supervisor. But asymmetry in power is a matter of degree and circumstance: inspectors may be weak and workers strong. This happened, for instance, during wartime when the demands for skilled arms workers was great.[50]

Gauges, then, define a quasi-public standard. Here, as in Thompson's clock-time example, the worker's understanding of his or her own practices is transformed into a disembodied quantity whose meaning is only apparent at the largest level of organization. Now workers are far less able to complain that their workpiece has been rejected without cause. On the other hand, supervisors may not reject pieces (in principle, anyway) without cause. No wonder gauges are both countenanced and resented. And no wonder parties argue over who will have these standards in their keeping.[51]

The underlying problem, of course, is that only the simplest artifact can be defined with the requisite degree of completion and exactness. Most artifacts of commerce and war are irregular. Tolerances can be defined only for a few dimensions and gauges can capture only a small number of those axes deemed critical. Only in the late twentieth century have published tolerances included a formal definition of shape (e.g., circularity or sphericity). The fact that gauging, moreover, requires "touch" leaves considerable room for disagreement. These disagreements could be minimized—though never entirely eliminated—only by further circumscribing the *practice* of gauging. In this way, the elaborate and minutely choreographed routines which guided the gauging of cannonballs and the proving of gun barrels in the eighteenth century form a record of the conflict over control of the work process. They show the lengths to which Gribeauvalist engineers had to go to forestall all the possible subversions of the rules. The "objectivity" of their standards was the outcome of this process.

The Gribeauvalists' approach to manufacturing tolerance is evident in their claim to have halved the windage of the cannon. No dimension of a gun is more crucial than the fit between the shell and the interior of the barrel. And every historian of this episode has repeated this claim, citing Maritz's revolutionary boring machine as the technological "hardware" that enabled the Gribeauvalists to preserve the cannon's range and accuracy, despite its shortened barrel. Even Rosen or Chalmin who set the Gribeauvalist achievement in a social context, here invoke Maritz's machine as a technological "fact." But what does this claim—"halving the windage"—mean? Precision is relative, not just to what came before, but to other elements of a system. Boring a tube with great precision is meaningless unless a corresponding precision is achieved for cannonballs. What is actually meant then by the phrase "halving the windage" is that the Gribeauvalists tightened from 2 *lignes* to 1 *lignes* (1/12th of an inch) the maximum acceptable difference between the maximum cannonball diameter and the minimum inner barrel diameter. Phrased in this way, the claim about "halving the windage" becomes not a hard technological "fact," but a socially negotiated process.

To begin with, the process of gauging the inner bore of the cannon required judgment and depended on the skill of the individual examiner. Where the Vallièrists had relied on the "eye" of the examining officer, Gribeauval had substituted precise gauges, such as the *étoile mobile*. This portable caliper gauge was fashioned by a scientific instrument-maker to measure the inner diameter of the bore to within 0.025 millimeters. After careful calibration in Strasbourg, one such device was distributed to each of the kingdom's foundries.[52] But as the Entrepreneur of the Strasbourg foundry noted, in fitting this device down the bore of a cannon, an examiner still might easily—"even involuntarily"—tilt it ever so slightly. As someone whose livelihood depended on satisfying the specifications set by the artillery service, the Entrepreneur was understandably concerned about this: "The fortunes of the manufacturer are in the hands of he who makes use of the *étoile mobile*; [and] two examiners almost always obtain different results."[53] This variability could occasion disputes.

To minimize these disputes, the Gribeauvalists developed instruments and practices to substantiate the idealized pictures of their *Tables de construction*. Take for instance their methods of gauging cannonballs. European artillerists had long passed their cannonballs through a circular "go" gauge (a *lunette*) to make sure the shell would fit into the barrel. This left the lower threshold for

Fig. 4.3. Go, No-Go Gauges for Cannonballs. This pair of go, no-go gauges (*lunettes*) for an 8-pound Gribeauvalist cannonball are seen here with inserts to preserve them from damage and distortion. They have diameters of 5 *pouces* 0 *lignes* 0 *points* and 4 *pouces* 11 *lignes* 2 *points*, respectively. This indicates a tolerance of 10 *points* (or 1.8 mm). They were made in the Atelier de Precision and date from the Revolutionary period. From the Musée de l'Armée (K22533).

the size of the ball undefined, and hence dependent on the judgment—the "eye"—of the cannoneer. At the prompting of Choiseul, the Gribeauvalists now introduced a "no-go" gauge with a diameter 9 *points* less than that of the *lunette*. Acceptable balls should *not* be able to pass through this gauge. Applied in tandem with the "go" gauge, this defined a zone within which manufacturers had to operate (fig. 4.3).[54] Defining this zone so clearly, however, immediately made it a matter for negotiation. Du Coudray admitted that the artillery service had initially wanted to set the tolerance for cannonballs at 6 *points*, but the manufacturers had protested, and in the end the service had been forced to settle for 9 *points*.[55] On the other hand, this concept of tolerance enabled the Gribeauvalists to assign different degrees of precision to different parts of the gun. Where precision mattered (or could be obtained without undue cost) they asked that production be carried out to within 2 or 3 *points*; elsewhere parts were acceptable if they lay within a 6 or a 9 *point* spread.[56] Disciplining objects is a matter of diminishing returns; how close is close enough depends on how much effort you are willing to expend (or how much you are willing to pay to have someone expend that effort for you).

In the second place, the Gribeauvalists realized that even these tolerances might fail to capture the variation in *shape* they wanted to control. That was because gauges generally measure only one dimension or plane of a thick object. For instance, an oblong cannonball might pass through the "go" *lunette* and still not fit into the barrel of the gun in a different orientation. (Imagine a football shape.) To verify that all of a ball's diameters were sufficiently small, these flat *lunettes* were replaced by a tube five-calibers long, through which the ball was passed. But what if the examiner simply let an oblong ball drop down the tube? To minimize this problem, the Gribeauvalists developed special workbenches on which these cylinders were affixed at an oblique angle. Now a ball would have to be rolled down the tube. In this way the artifact was disciplined by coupling instruments with procedures in an impersonal choreography.[57]

In the third place, the Gribeauvalists recognized that the gauges themselves might vary from one to another, and that the precision of the same gauge might vary over time with repeated use. The solution here was to centralize the production, dissemination, and verification of the gauges themselves. All gauges were made in Strasbourg by "a single hand" to reduce the variation among them, and distributed from there to the various arsenals. They were then periodically verified against a set of master standards held in each arsenal. This verification was itself defined by a tolerance band, so that gauges were discarded if they varied by more than 2 *points* from the norm. To extend their standards across space and time, the Gribeauvalists created a hierarchy of gauges to match the hierarchy of personnel.[58]

Ultimately, this hierarchy of standards extended to the royal system of measures. In the late seventeenth century, all the French arsenals had begun to rely on the *pied du roi*, rather than local measures. Over time, these rulers had come to diverge, and the Gribeauvalists replaced them with new copper measures. This was part of a broader impetus toward standardization in the eighteenth century by the central administration of the state. In the late ancien régime, the Academy of Sciences took charge of establishing, verifying, and maintaining the metal bars (*étalons*) which constituted the definitive royal standards. After the middle of the eighteenth century, these were increasingly calibrated with reference to measurements taken from "nature." In this way, a hierarchy of measures mirrored a hierarchy of material objects and its carefully ranked administrators.[59]

This elaborate structure, created by the engineers, was meant to constrain them as well. Yet nothing in it *compelled* them to carry out their duties with exactitude. Misjudgment, even corruption, were always possible. As M. Norton Wise's recent volume, *The Values of Precision*, makes abundantly clear, precision is a *value*. For this hierarchy to succeed in disciplining artifacts, the engineers themselves had to accept precision as a moral imperative, as well as an operational necessity. Gribeauval was forever reminding his subordinates that his insistence on precision was not "hairsplitting." He could be severe with subordinates who slipped from the "exactitude" expected. "It is perhaps by lack of attention . . . ," he scolded, "that the [cannon] balls of M. Maritz are too small." His correspondence is filled with such exhortations.[60] Despite a panoply of institutional safeguards (mechanical rules and tools), human resolve remained essential to this program of socio-technical discipline. Within their world of rewards and punishments, pecuniary interest was anathema. The engineers who developed rational production did so in conscious repudiation of the values of profiteering merchants and venal artisans. In the face of the temptation to look the other way (and take a financial cut for their laxity), the engineers legitimated their efforts in the name of a higher form of state service. Artillery professor Lombard noted that for officers working in the arsenals of construction, "precision, solidity, uniformity" were the signs by which they would be judged.

> And in all regards their conduct must be of a scrupulous character. For if there is an error, with whom should the responsibility lie?—With those in whom the government placed its trust and who by either criminal abuse, ignorance, or negligence have delivered to the service materials whose defects may have disastrous consequences.[61]

The price of standards is eternal vigilance.

Artillery Carriages and the Spirit of System

These then were the tools of practical reason which eighteenth-century elites developed to help them reorganize the workplace. Some version of these tools—measures of labor, technical drawing, and gauges and jigs—had been used in artisanal shops for centuries. All are employed today in most modern factories. As employed in the world of the artisanal corporation, however, these tools implied a particular set of social and legal relations, a particular technological life. Hence, the artifacts made with these tools may be said to have possessed a corresponding set of political qualities. In transforming these tools so that they would help reorganize the workshop, eighteenth-century elites sought to impose a different set of social relations on the productive order, and hence a different relationship between the laboring body and the products of its effort, between public technological knowledge and forms of social organization, and between the conceivers of artifacts and those who carried out the tasks. Hence, these Enlightenment elites may be said to have hoped for artifacts with different political qualities.

At the forefront of these efforts were the Enlightenment engineers. The various corps of the French engineers were involved in transforming several large-scale industries in this period: naval construction, fortress building, roads and bridges.[62] Any of these examples might have served to illustrate how these rational tools were actually implemented, but I will focus on the making of artillery carriages. In this setting, the introduction of these tools unexpectedly resulted in the first successful implementation of interchangeable parts manufacturing.

The Gribeauvalists' goal of mobile firepower understandably led them to focus their attention on the carriages that carried the field cannon into battle. From their point of view, the old Vallièrist artillery carriages constituted what Thomas Hughes has called a "critical problem." This phrase does not imply that the failings of the old carriages existed in some real sense. Rather, these "critical problems" are identified and solved within an institutional, cultural, and political context. To be sure, some of the impetus was narrowly technical. Having halved the weight of the cannon, artillerists now also had to reckon with a doubled recoil velocity. To the extent that the carriages were now more mobile (with bigger wheel diameters, etc.), this increased the distance they recoiled and the amount of time it took to run them back into firing position, and hence decreased the rate of fire. The Gribeauvalists reduced this recoil velocity by designing the sides of the carriage to slope more steeply, sending the recoil force into the earth. This, however, strained the structure of the carriage. One of Gribeauval's main concerns at the Strasbourg tests was to determine whether the carriages would be able to withstand the shock of that redirected recoil.[63] In March 1765, he reminded his subordinate Jacques-Charles Manson "not to lose from view our principle, which is to lighten [the carriages] without doing harm to the necessary solidity."[64]

Issues of design and production are entwined here. Throughout the first half of the century, artillerists circulated proposals for redesigned carriages—among them the innovative siege carriage that brought the young lieutenant Gribeauval to the notice of the War Ministry. Innovators, like Gribeauval, justified their plans using geometric constructions. Naval engineers had long used geometric manipulations to trace out the complex curves of a boat's hull with only a compass and straight edge. The artillerists now "constructed" all the dimensions of their gun carriages from a single parameter, typically the caliber of the piece. These Cartesian manipulations never referred to rational mechanics, let alone to the constraints imposed by production. They seem less a sign of rational design, than a mnemonic device to facilitate the ready reproduction of that design in remote locations (fig. 4.2).

Once in power, the Gribeauvalists set about redesigning the carriages with more practical views in mind. They proceeded through design-by-committee: the bureaucratic mode of invention. Proposals for, say, a new axle box circulated by way of Gribeauval from Maritz (the forge-master) to Charles-Robin Chateaufer (the superior officer in charge of carriage design) to Manson (the captain in charge of the worker-soldiers who actually built the carriages).

These engineers did not justify the new design as ideal, so much as operationally superior. Their goal was mobility, and the new carriages were more maneuverable, despite the weight added by the iron axle. They were hitched up for horses in tandem, the wheel diameter was increased, friction in the axle box was reduced, and a second set of supports centered the cannon's weight during transport. Prototypes were extensively tested at Strasbourg in 1765.[65]

Having designed a more mobile prototype, they now had to ensure that the carriages would be produced in conformity with the new pattern. This uniformity was also said to contribute to the central goal of mobility. No longer would each of the six arsenals make carriages to its own specifications. Officers complained that variations in wheel size and axle length made it impossible to hitch carriages from Strasbourg together with those from Auxonne. Even within a single arsenal no two carriages resembled one another, making repairs complicated and time consuming. The Vallière administration defended these variations. Carriages, they argued, were customized for the weight and size of individual cannon. As for the axle length, one officer noted: "It's up to the worker to give [it] the length it needs for the roads of the country." As for the various iron pieces that studded the carriages, "those responsible can give a sketch of what they want [made] to the [local] forge master."[66] In other words, to cope with the thickness of artifacts, the Vallièrists granted considerable discretion to local officers and craftworkers to adapt their matériel to local conditions. The Vallièrist ordonnance of 1732 had deliberately neglected to specify the dimensions of the carriages (limbers and caissons).

What prompted change was pressure from above. During the Seven Years' War, the destruction of artillery matériel made the variations a grave concern. Cannon were being mounted on inappropriate carriages, and carriages of widely different dimensions were being hitched together. With the artillery now under the authority of the War Office, Minister Belle-Isle wrote to the commander of the Strasbourg arsenal: "As it is essential to establish a general uniformity with regard to the [carriages], you will send me a table of the dimensions in use at Strasbourg both for the wood pieces and for each piece of iron for all the carriages, and to make them more intelligible, attach a drawing of each part." The Vallièrist administration never delivered on this request, and in 1762 the War Office under Dubois reiterated that uniformity was "indispensable in our arsenals." Gribeauval made this request the centerpiece of his program; yet another sign that reformers were less innovators on their own account than engaged in bringing their activities in line with the demands of the central authorities. As experts, however, they retained a considerable latitude over how they went about this task.[67]

Bureaucracies adopt uniform rules and precise instruments for eminently practical reasons, typically because central administrators wish to facilitate exchanges of personnel and matériel throughout their domain. For the bureaucrats themselves, however, these goals must come to seem a self-evident proposition, an aesthetic admired for "its own sake." The engineers of the Enlightenment worshipped at the shrine of uniformity and precision. These are actually two faces of the same god. What counts as uniformity depends on the degree of precision sought, and precision can only be measured against a pattern of uniformity. As we have seen, this ethic was intrinsic to professional engineering education with its rigorous program of self-discipline. In 1750, Colonel Claude-Marie-Valenninet Le Duc—later a close collaborator of Gribeauval—advocated the standardization of the carriages, because to the rational well-ordered mind, the absence of uniformity was a "dismal cacophony." He believed the uniformity of matériel to be as important as "the uniformity of religion, laws, weights and measures, and police." A good artillery officer, he said, had "a spirit of order and meditation."[68] In this context the redesign and manufacture of the Gribeauvalist carriages meant the "disciplining" of the artifact.

Here the usefulness of technical drawing was apparent. Gribeauval enjoined Manson to "determine the new dimensions and to prepare the tables and drawings that would render the constructions perfectly uniform in all the different arsenals."[69] The dimensions of the eighty-odd pieces of the carriages were enshrined in the famous *Tables de construction* (fig. 4.2).[70] The Gribeauvalists did not merely "represent" uniformity in pictures, however. They also defined precision operationally, as the ability to replace pieces without refitting them. In a sense, this was a generalization of the idea of tolerance, whereby the range of acceptable dimensions of any part was now set by the fit of that part into the finished whole. At the limit, carriage parts became interchangeable. The engineers understood that achieving this would involve a radical transformation of the means of production.

Arsenals of Construction and Interchangeable Production

The Gribeauvalists were as relentlessly Modern in the workplace as they were on the battlefield. That is, they devalued the relevance of historically established ways of doing things in the face of present-day contingencies. Du Coudray sneered that production in the old arsenals had been based on the "particular proportions which the officers employed there transmitted as if by heredity."[71] He contrasted this servility to tradition with the rational approach of the Gribeauvalists, who had achieved "an extreme precision in the proportions of all the composite parts [of the new carriages], and the exact assembly and rigorous uniformity which results." Henceforth, all carriages would be identical, whether produced in Metz, La Fère, or Strasbourg. Jean Du Teil held up this uniformity as a sign that "nothing was left to chance, nor to blind habit, and that even for the most seemingly minute subject, one was always guided by principles."[72]

The Gribeauvalists succeeded in carrying out this radical program of interchangeable parts manufacturing because of the unique conditions under which the carriages were made, the *régie* system. Unlike the cannon foundries or musket manufactures, the carriages were made in "arsenals of construction" located in the vicinity of the artillery schools in the towns of La Fère, Douai, Auxonne, Strasbourg, Metz, and Nantes. These were workshops owned by the government, managed by artillery officers, and staffed with a workforce drawn from specially recruited soldiers of artisanal background. This form of state monopoly capitalism, in which managers and workers were both subject to military discipline, was the cradle of the uniformity system of production.

Manson, Bellegarde, and Du Coudray were typical of the technically minded artillery captains who commanded battalions of these special troops. Each was assisted by a staff of 15 NCOs, themselves artisan-soldiers promoted from the ranks to be "foremen" in the arsenals. The number of workers under their command fluctuated from 420 to 639, but they were supplemented by local *ouvriers d'état* hired on a contract basis. Both types of workers could moonlight for private employers in town during the slack season. Indeed, the organization of these worker-soldiers, for all its military regimentation, owed much to the corporate structure of artisanal life. The workers were divided into squads by trade: ironmongers, wheelwrights, carpenters, and cabinet makers. Within each trade they were divided into first- and second-class workers, plus apprentices. Promotions followed the same procedures as for artillery NCOs; when a vacancy opened, the members of that rank drew up a list of three nominees from which their superiors selected the successor. As I noted earlier, this corresponds in certain respects to the methods by which master craftworkers elected their confederates. Recruiting was also run along corporate lines, with sons succeeding fathers. Under the Vallièrists, the NCOs were compensated on a piece-rate basis, and they then paid their subordinates as they saw fit.[73]

In other respects, the differences with the artisanal trades seem striking. Artisans working for the military in this period were gradually being drawn into a total institution. Once they passed an initial trial period, workers served an eight-year contract with a substantial bonus for renewal. Gribeauval put everyone to work on a fixed salary, though a first-class worker earned 60 percent more than a first-class cannoneer. The government supplied food and lodging, as well as the necessary tools and raw materials (though workers were responsible for breakages), and guaranteed employment for the duration of their contract. During the Vallièrist era, workers were billeted in town. In 1774, Le Duc built a barracks for the Strasbourg workers, its interior furnished with only that "indispensable for [their] habitation." Production itself was also centralized under a single roof. A morning bell summoned the worker-soldiers to the arsenal and signaled their two half-hour breaks in each nine-hour day (eight-hour days in winter).[74]

To bring the process of production under their control, the engineers deployed the tools of rational production discussed above. This was a matter of conscious preparation, and involved a deliberate attempt to give managers, foremen, and workers distinctive cognitive training. Gribeauval noted that only when young lieutenants were properly trained in school—rather than by workers on the job—would they be able to "correct the work of the sergeants." The *Tables de construction* not only represented the ideal toward which production should strive, they served as a model of social relations in the workplace. The forms of technical drawing which the Gribeauvalists extended to NCOs always stayed within the limits prescribed by their rank and abilities. The Ecole des Sergents begun in the 1760s made sure that NCOs could read, write, do sums, and draw "at least enough to be able to trace the work and understand what one asks of them when they are given a plan or drawing." The worker-soldiers were taught drafting techniques and the use of gauges. In the 1770s, Le Duc added a special room to his regimental artillery school where artisans could trace out the patterns for their work. Regulations promulgated in 1792 (but typical of the ancien régime as well), specified that the director of the arsenal set aside an entire day each week "to exercise the workers in their tracings of [artillery] carriages. . . ."[75]

The engineers embedded these specifications in an array of precision gauges and jigs. Wrights received patterns (*patrons*) which accepted deviations of no greater than a quarter of a *ligne* (1/50th of an inch); these included iron rulers to measure straight distances, mandrels for concave shapes, and *lunettes* for convex shapes. Workers in iron were given dies and jigs to forge the correct curvature and pierce holes "with the greatest exactitude." The Gribeauvalists standardized the tools themselves. All model carriages, gauges, tools, and measures emanated from the Strasbourg arsenal.[76]

Behind these mechanical authorities stood military discipline. Every Saturday inspecting officers verified the dimensions of each workpiece, and marked accepted pieces with their sign. Each worker carried his own "workbook," which recorded his tools, raw materials, and tasks. This sort of daily *livret*, which became universal in the Napoleonic period, documented the history of a worker's behavior, and not coincidentally, determined his pay. Du Coudray admitted that the new standards had been difficult for the workers to achieve at first, but after the officers had demonstrated their "inflexibility" and rejected every piece that did not conform to "spec," the workers had complied. After all, these artisans received a daily wage, not a piece-rate payment. Lost time and wasted raw materials were not their concern. And the range of military punishments available to the engineers made protest difficult.[77]

For all the artillerists' boasting, however, this top-down rationalization could not supplant the skills of the artisans. Gribeauval noted to Manson that "if you are satisfied with the methods the workers have found to pierce the nave of the wheel, then so am I." And he added, "In this sort of thing, the simplest [procedure] is always the best." In this and other ways, the Gribeauvalists acknowledged that they could not analytically capture the tacit skills of artisans, and hence had to defer to their methods. A further proof of this is that the transmission of the new techniques required face-to-face contact. Strasbourg was not just the central repository for gauges, but for skills as well. Gribeauval had to rotate the Strasbourg worker-soldiers through the other arsenals so that they might teach their fellows the new methods.[78]

The Gribeauvalists justified this effort to make the carriage parts uniform by the rapidity and ease with which parts could now be replaced; a critical advantage during a retreat, for example, when a breakdown meant the cannon might fall into enemy hands.[79] Completely unexpected was a further by-product of this interchangeability: a decrease in the costs of production. Du Coudray noted that to his surprise, the exacting precision and rigorous inspections had made the new carriages *less* time-consuming to construct.

> This precision gives rise to another property unknown in works where well-paid artisans labor for private clients. This property, which can be considered superfluous, for it was not sought, is the consequence and proof of that precision so unsuccessfully sought till now, and now so rigorously achieved. It would be natural to believe that these new constructions, executed with so desirable an exactitude, would require much more time [to produce] and would, as a result, be much more expensive than the old. . . . That, however, is not the case.[80]

The engineers' explanations for this result are instructive. For them, the cost savings associated with uniform production turned not on economies of scale or the division of labor, but from the substitution of mechanical for personal supervision. Du Coudray cited the "help given the workers to judge their work for themselves." In the old arsenals, the carriages had been assembled "gropingly," a slow and costly affair. Now gauges and templates guided the workers' hands quickly and surely, and allowed them to evaluate their own labor. After all, Du Coudray noted, even the old carriages had been "precise" after a fashion, in the sense that they had to be hand-fitted together. Obviously, that fitting took time. Now this final costly step had been eliminated. As the economic historian Russell Fries has pointed out, this "system" aspect of interchangeable parts production was responsible for most of its cost savings.[81]

The Vallièrists, however, condemned the method of uniformity production, pointing to the dangerous inflexibility of the system. Even granting that the method was feasible and cheap, the Vallièrists found this "extreme accuracy" undesirable. To begin with, the new carriages would be more, not less, difficult to repair. The Douai tests had shown the new carriages were fragile; some broke down after only ten shots. If spare parts ran out, local artisans in the theater of war would not be able to fashion replacement pieces on the spot. The new iron axles, for instance, could not be refashioned on the spot, as the old wooden ones had been. By moving technical knowledge up to the institutional level, local knowledge could no longer cope with contingencies.[82] Conversely, universal design could not cope with local variations. The Vallièrists had always adapted their axle lengths to local road widths.

In time, the Gribeauvalists realized that the Vallièrists had a point. In 1775 Gribeauval himself confessed that at the time of his initial reforms he had been "too occupied with that uniformity then so greatly desired," and had overlooked the need to distinguish the carriages destined for Italy from those destined for Provence. Manson fretted that to change the *Tables de construction* now would be "embarrassing for the service." But Gribeauval insisted that if the inequalities existed in reality, the *Tables* would have to be modified. This partially explains why even though the drawings were complete in 1767, they were not published until 1792. The formal prescriptions of the *Tables* defined the aspirations of the artillerists to uniform production, yet Gribeauval realized that he would need to bend them to meet unexpected eventualities.[83]

By 1770 some 3,300 new carriages and caissons had been assembled, and in 1783, the government was still spending 600,000 *livres* a year on construction. Shortly thereafter the minister of war ordered across-the-board cuts in the military budget. The arsenals of construction had always been unusual in that they reported directly to the minister of war on budgetary matters. Wholly at government expense, they made an obvious target. From 1784 through 1787, the government spent only 40,000 *livres* annually, 6.7 percent of the former allowance.[84] By then, however, the arsenals of construction had served for twenty years as "model factories" where a generation of artillerist-engineers learned the theory and practice of interchangeable parts manufacturing. Artillery students visited the adjacent arsenals, sketched the matériel, were quizzed on its every particular, and absorbed the ethic of uniformity. The service spread the doctrine of interchangeable parts production to the wider public as well. Du Coudray's 1772 book, *L'artillerie nouvelle*, first publicized the uniformity system. In 1777 Heinrich Othon von Scheel's widely circulated *Mémoire d'artillerie* reprinted Du Coudray's paean to interchangeable production. In 1800, this work was translated into English by Jonathan Williams, the recently appointed head of West Point Military Academy, just as Eli Whitney was petitioning Congress for funds for his New Haven gun factory.[85]

In the view of production I have been developing, interchangeable parts manufacturing represents the next logical step in the attempt to confine the discretion of workers to a carefully circumscribed band of "tolerance." Under such a uniformity system of production, a worker who fails to make objects within the boundaries of this tolerance would encroach on the output of fellow workers and disrupt the entire organization. And conversely the "fit" of the final assembly is a measure of the rigor with which the productive order has been policed. That is why the Gribeauvalists offered *public* demonstrations of this ability to exchange parts, and trumpeted their "discovery" with such fanfare. This showmanship was an attempt to interest patrons and secure funding. As an operational proof of precision it was also a sign of their control over production.

From Carriages to Muskets

Despite all the public claims, we have today no artifactual evidence that the French artillerists ever achieved interchangeable production such that the carriage pieces could be assembled without a final fitting. Given Eli Whitney's exaggerated claims for his interchangeable gunlocks and the controversy over Blanc's claims (discussed below), one might well doubt the engineers' word. Success is probable, however. First, because each carriage was a large object constructed of both wood and metal parts, and therefore its pieces were easier to shape precisely than those of a small all-metal artifact. Second, because the carriage pieces did not need to function closely together in a moving mechanism, and therefore would not need to be as precisely tooled as in a gunlock. Finally, because the special circumstances of the *régie* system and the artillerists' unquestioned military authority over the workers gave the officers the power to impose their vision of production. None of these conditions prevailed in the musket-making town of Saint-Etienne.

Yet the same logic that drove the artillerists to tackle the design and production of the artillery carriages drove them to focus on the design and production of muskets. This was a time, as we saw in chapter 3, when the French army was adopting a tactics that placed renewed emphasis on musket fire. Yet the accuracy of these weapons was derisory, and their role sharply questioned. Even the most partisan advocates of gunpowder weaponry were hard pressed to vouch for their effectiveness. Du Coudray's grim calculus of death had put the kill rate of muskets at no better than 1 shot in 150; whereas cannon shot killed 11 times as many. Within the "killing system" of the new Modern tactics, then, the musket represented a "critical problem" that attracted concern. Du Coudray urged his countrymen to "perfect your muskets as you have begun to perfect your artillery."[86] This is not to suggest that Du Coudray had necessarily found a "real" inefficiency. But as the advocates of Modern thin-line tactics and the supervisors of the kingdom's armories, the artillerists were understandably eager to improve the performance of French small arms. Since 1717, the artillery service had been given responsibility for overseeing the production and purchase of muskets for the regular army, much like the artillery-staffed Ordnance Department in the United States or the Board of Ordnance in Great Britain. However, the involvement of the artillerists in the production of small arms differed markedly in each country, and consequently, so did the organization of production.

This chapter has described how Enlightenment engineers developed various tools of rational production as a substitute for corporatist control over the workplace. And it has described how these tools—measures of work, technical drawing, and gauges and jigs—were proposed by theorists, taught in schools, and implemented in the unique circumstances of a state-monopoly manufacture. I have also emphasized the ambiguous history of these tools, pointing out that their effect on the organization of production depended on the context within which they were implemented. By deliberately leaving this history open-ended, I have eschewed both technological and social determinism. To that extent, I have also left room for a political history of the artifact. In the case of state-monopoly manufacture, the engineers were able to implement interchangeable parts manufacturing. That is certainly one form of the technological life.

But what, one might plausibly ask, does all this have to do with the history of industrialization as we have usually heard about it? Where is the familiar, boisterous eighteenth-century world of putting-out merchants, risk-taking entrepreneurs, collusive artisans, and domestic workers living on the margins? And what about the local social and political context in which these various groups interacted? In the chapter which follows, I examine the attempt of the engineers to impose their production methods in such a setting: the armory town of Saint-Etienne. There, their program encountered fierce resistance from the arms merchants and artisans who produced muskets for the army and for private sale. Examining the complex interplay of engineering management, merchant finance, and artisanal skills will demonstrate how the techniques of rational production were shaped by social struggle and negotiation.

Chapter Five

THE SAINT-ETIENNE ARMORY: MUSKET-MAKING AND THE END OF THE ANCIEN REGIME

PASSING THROUGH Saint-Etienne in 1778 Jean-Marie Roland de la Platière, royal inspector of manufactures, noted down these impressions for his fiancée, the future Madame Roland.

> [Saint-Etienne] is perpetually shrouded in coal smoke which penetrates everywhere and vents at great distances. [The town and its faubourgs] encompass approximately 30,000 inhabitants, almost all of them occupied at the forge: men, women, children, boys, and girls; armorers, ironmongers, metalworkers of all sorts. You cannot have any idea of the number of forges, and their activity: these are the true dens of Vulcan. Each one works with tools proportionate to his age and strength; all are blackened, with white eyes that stare at passers-by even as they busily continue working. You see the heaving breasts of women, everything in them palpitating with the force of their work, the heat of the moment, and the ardor of the fire. . . . They make children to order here: Saint-Etienne is an ants' nest.[1]

Vulcan's forges and the insect swarm, the pagan fire and breeding pit—clearly, the coal-blackened artisans of Saint-Etienne presented the French elite with a disturbing challenge. The administrative elite knew that the kingdom's wealth and military strength depended on the unruly, independent, and prolific artisans of its provincial towns. How could this force be contained and made to serve the state? Saint-Etienne was the kingdom's premier source of muskets, a crucial instrument in the state's monopolization of violence. Roland took particular note of the armorers in their shops forging musket barrels for the army, and alongside them, armorers making gun barrels for private sale. Wistfully, he remarked, "We try to make them all submit their weapons for proving, but aside from those working for the king, they often evade this requirement." He hinted, however, that changes were underway.[2]

Indeed, the year of Roland's visit, the engineer-officers of Gribeauval's artillery corps introduced a new model musket, the M1777, and with it, a new rigor into their administration of the Royal Manufacture of Saint-Etienne. This was the period of controversy over French battle tactics, and the artillerists wanted to heed Du Coudray's call to "perfect your muskets." Until this time, the manufacture had remained a legal fiction: a collection of dispersed ateliers owned by artisans and coordinated by a merchant deceptively called an "Entrepreneur." All this was supposed to change with the introduction of the M1777 and the new engineering management that went with it. However, resistance from both artisans and merchants soon threatened to tear this legal fiction apart from within. When the local engineer-inspector began to enforce the new standards, the number of gunsmiths registered to work for the king dropped in half, and so did the output of the muskets. By the mid-1780s, the premier armory of the kingdom was in ruins. In 1785, at the moment of crisis, the inspector of Saint-Etienne desperately proposed that the state take over the manufacture and run it as a *régie*. As we will see, Gribeauval rejected this program of nationalization. Not that he underestimated the crisis. As he explained to the minister of war, "*Without a prompt remedy, the manufacture of Saint-Etienne will soon be without value to the state.*" It was in this context that Gribeauval proposed a long-term "techno-fix" solution, the method of interchangeable parts manufacturing.[3] In the short run, however, the state's financial troubles obliged the engineers to stand by and watch as the skills and work habits they had built up over several decades deserted the military armory to make guns for private sale. Indeed, the state's failure signaled the triumph of a particular social alliance in Saint-Etienne, one with its own vision of how the production of arms should be organized.

Not long after, Roland faced the consequences of this collapse. In 1792, he was chief minister of the Republic. For half a year, his Girondin party had been aggressively seeking war against the Continental powers, and on 20 April 1792, France declared war on Austria. Despite assurances that France had 200,000 muskets on hand, examination of the stockpiles proved otherwise.[4] With the Prussian army on French soil and rapidly advancing toward Paris, Roland called for 300,000 volunteers and wrote to Saint-Etienne for the immediate delivery of 3,000 muskets. "We could not have a more pressing need of them," he pleaded.[5] Local functionaries duly visited the Entrepreneurs of the defunct manufacture and the town's civilian arms merchants, and urged a patriotic effort on them. Unmoved, the merchants refused to promise future guns at today's prices. The functionaries could scrape together only 68 muskets, with another 198 by mid-September, "not precisely the caliber of the M1777, but they all accept [military] cartridges." The central government had lost all control over the premier armory of the nation at a time when its troops were engaging the Prussians at Valmy and Jemappes.[6]

This chapter, then, examines the struggle over which institutions would define the production of armaments at the end of the eighteenth century. This was a struggle which predated the French Revolution, but which set the local terms for that upheaval. On one level, then, this is the story of how the engineering program of production shaped (and was shaped by) nascent Revolutionary politics. And on another, this is the story of how the politics of production can be read through the physical characteristics of a particular artifact. At the intersection of these two narratives lies a nonteleological history of "mass" production.

Alternatives Revisited

Recent scholarship has questioned the triumphant story of mass production. Not long ago, economic historians dismissed as irrelevant those smaller, secondary firms which either supplied primary, mass-production firms with specialized tools or were forced by them into the margins of the cyclical market. Recent studies by Piore, Sabel, Zeitlin, Scranton, and others have partially reversed this judgment. They have underlined the persistence and vitality of these smaller firms, whose flexible technologies allowed them to retool with shifts in fashion and run off small batches of complex goods. When grouped in "industrial districts," these firms possessed hitherto unexamined local institutions—municipal councils and mutual assistance societies—which coordinated their activities and enabled them to survive market downturns and surmount "free rider" problems. And finally, these firms operated within a code of values that enforced communal behavior, yet went hand in hand with a social mobility that rewarded innovation. In sum, these historians seek to recapture the economic, social, and political world taken for granted by Pierre-Joseph Proudhon, the nineteenth-century artisan-politician who saw no contradiction between economic competition and productive association, market success and small firm size, innovative machinery and the enhancement of artisanal skills.[7]

This is not merely an academic exercise. In large measure Sabel and Zeitlin attribute the demise of flexible specialization to the allure of the *idea* of Fordism. In particular, they denounce the hold of the ideal of mass production on the centralized state, especially the Enlightenment vision of the French engineers. According to Sabel and Zeitlin, these "enthusiasts of mass production" helped shape economic development "precisely by declaring one future inevitable." Against this self-fulfilling prophesy, Sabel and Zeitlin ask us to imagine distinct alternatives, industrial worlds that might have been: either a (better) world in which petty commodity producers sustain their technological dynamism through a quasi-communal property regime; or a (worse) world in which the large magnates and centralized states triumph utterly in imposing mass production upon an unfree peasantry. These alternatives, they suggest, reposition the European "breakthrough" to a dual-economy industrialization as a "historically contingent middle way."[8] At stake is whether there is more than one path to industrialization. Their typology understandably raises important historical questions. In particular, it invites us to reconceptualize the origins of mass production.

While they outline the nineteenth-century heyday of flexible specialization and hint at the reasons for its demise, Sabel and Zeitlin acknowledge that the origin of these districts "remains obscure." In their article on "Historical Alternatives to Mass Production," Sabel and Zeitlin cite Saint-Etienne as one such nineteenth-century industrial district, with its specialty steel firms and volatile silk ribbon trade. They note its coordinating institutions: municipal unemployment insurance, trade schools, a syndicate of small masters, and for armaments, the proof house (*banc d'épreuve*).[9] Yet, as we will see, Saint-Etienne was also the birthplace of interchangeable parts manufacturing. For this reason alone, the eighteenth-century origins of an industrial district such as Saint-Etienne are worth examining. This is all the more worthwhile because other historians have painted a very different picture of the town known in France as the "cradle of the Industrial Revolution." Michael Hanagan, in particular, has described the proletarianization of Saint-Etienne's workforce under the hand of nineteenth-century heavy industry. Up to a point, of course, these two pictures are compatible. Piore's dual-economy model posits the coexistence of both heavy and light industry sectors; and Hanagan acknowledges that artisanal producers led its nineteenth-century labor movement.[10] But the class struggle which energizes Hanagan's Saint-Etienne has little place in Sabel and Zeitlin's socially fluid Saint-Etienne where they find, "properly speaking, no bourgeois ruling class." More generally, the two analyses differ over the role of production in shaping politics. Where Hanagan posits an innate hostility between labor and capital, and hence a confrontational polity, Sabel and Zeitlin emphasize a consensual polity based on accommodation among the strata involved in production. These dissonant pictures make Saint-Etienne a touchstone for how we should conceive of early French industrialization.[11]

Economic historians still debate whether France chose a particularly French path to industrialization, or whether it belonged, in Landes's phrase, to a Continental "emulation." Champions of the emulation thesis argue that France's plentiful, small, labor-intensive firms avoided new techniques, either because of risk-averse family-proprietors or in deference to a recalcitrant workforce. Only later in the nineteenth century, this story has it, were these moribund firms forced aside. This familiar tale of Gallic "catch-up" has always valued the state as the principal promoter of technological innovation and large-scale industrial concentration. As the revisionists have pointed out, however, France did not experience lower rates of overall economic growth than Britain between 1750 and 1850, nor did it innovate less. As for the state, some have come to see its role as obstructionist, imposing the straightjacket of centralized production on innovative small producers.[12]

In this respect, the history of gun production has lessons for us. While there is a vast scholarly literature on the textile industry in this period, few scholars have focused on the metalworking trades. And none have addressed the making of guns. This is not for lack of evidence. Rather, this neglect can be explained on other grounds. First, because the metalworking industry did not employ nearly as many workers as the textile trade—and the arms industry employed fewer still. Second, because the technological changes in metalworking did not exhibit the dramatic saltations we associate with the revolution in cotton production. And third, because social historians have been understandably preoccupied with the relationship between the development of private capitalism and class formation, a process most clearly exhibited in the textile factories. For each of these explanations, however, there is an answer. First, to the extent that gun-making represented the high-precision end of the hardware trade—including the nascent machine-building industry—the techniques developed in this industry were to prove central to the broader pattern of industrialization, in the textile trade and beyond. Second, as Maxine Berg has recently suggested, the pattern of incremental, gradual change characteristic of the hardware trades—what she calls the "Birmingham model"—may in fact be more typical of eighteenth-century technological development than the exceptional revolution in textile production. And third, the armaments industry, no less than the textile trade, was in the hands of private producers. With Sabel and Zeitlin's admonition in mind, social historians should embrace this chance to test their assumptions about the evolving relations between production and politics.

To be sure, gun-making was exceptional, both with respect to the role of engineering design and the intensive scrutiny of the state. Yet even this exceptionalism was a matter of degree; all the French trades operated by the leave of the absolutist monarch. And in a wide variety of industries, textiles included, the state claimed the right to set standards of production. Given the widespread hypothesis that the French state played a central role in the pattern of industrial development, it is precisely this *mix* of small-scale private capital and intensive state scrutiny that makes the armaments industry a valuable site to study the social character of French industrial life. Saint-Etienne is a place where these developments can be observed quite readily.[13]

Second Nature in an Industrial District

The town of Saint-Etienne owed its early development to a bounty of natural endowments: a fast-flowing stream for power, plentiful coal to fuel forges, grindstones to reduce metal, as well as some water transportation to markets. But social institutions profoundly shaped access to these endowments, and hence the pace and character of economic development. For instance, the region quickly outstripped its local supplies of iron ore, and by the seventeenth century, the thriving metalworking trades of Saint-Etienne were dependent on their "daily bread" of iron from Burgundy. In the 1780s, some 9,300 metric tons of iron and 175 metric tons of steel, most of it *acier nature* from Dauphiné, arrived annually by way of merchants in Lyon. The iron was then made into rod by the owners of slitting mills, before being distributed to ironmongers who lived in small towns across the valley. This distribution system had profound implications for the structure of the trade, giving merchants leverage over petty producers.[14] Even access to *local* resources was mediated by social institutions. Water reservoirs to store motive power, privileged access to coal, canals for transport—all these transformed Saint-Etienne's "natural" endowments into "resources" to be exploited. Nature itself must be designed before it can be put to use. What William Cronon has called "second nature" is the socially conditioned environment which makes capitalist exploitation possible. The history of these institutions in Saint-Etienne was paradoxical, however. All were sanctioned by the state, yet all relied on private capital. All were intended to make production more certain and markets more accessible, yet they also made possible the cooperative arrangements characteristic of industrial districts. The institutions which underlie "second nature" are themselves sites of conflict and negotiation.[15]

Consider, for instance, the problem of transportation to markets. The restrictive geography of the region had long shaped trade in and around the Saint-Etienne valley. Only during the eighteenth century did this isolation slowly end—albeit in a highly mediated way. The old province of Forez sits athwart the upper reaches of the Loire River, the main geographic feature of the province. Through it lies access to the rich regions of Orléans, Paris, and the Atlantic. Meanwhile, across forty kilometers of steep ridges to the east, lies Lyon, and by way of the Rhône, the Mediterranean. Transport in either direction, however, was severely limited until the late eighteenth century. Only in 1780 did a private concession build a canal from Rive-de-Gier to the Rhône, vastly increasing coal exports from the eastern part of the province—though loads from Saint-Etienne still had to be carted over the pass to reach the canal-head. And while another private concession had in 1705 dredged the Loire as far as Saint-Just-sur-Loire, ten kilometers from Saint-Etienne, boats could still only be fully laden at Roanne in most seasons. Within the province transportation proceeded at a medieval pace.[16]

Or consider the issue of the fuel supply. The region's main competitive advantage was its proximity to rich deposits of coal, the largest then known in France. Access to this fuel was a zealously guarded local privilege, however, and can serve as an example of how disputes over resources regulated production in the eighteenth century. In theory, extraction required a royal concession, but quasi-legal quarries proliferated up to the Revolution. The outskirts of towns such as Saint-Jean-de-Bonnefonts were pocked with hundreds of digs that tracked the shallow, broken seams. Local workers had customary rights to this fuel. The few larger mines run by merchant families (Neyron, Jovin) were primitive by English standards. Since 1701, the town had enjoyed the formal protection of the "Reserve," which forbade the export of all coal extracted within two *lieus* (eight kilometers) of the central square. This ruling coincided with the opening of the newly dredged Loire to the north, and reassured locals of plentiful supplies. An edict of 1763 halved the protected radius, but in 1782, the Reserve still produced twice the amount extracted outside its boundaries. Then, in the 1780s, the central state intervened decisively on behalf of large-scale mine owners who promised exports to the capital. In 1783 a royal inspector of mines was appointed for Saint-Etienne. In 1780 the canal at Rive-de-Gier on the eastern edge of the region opened. Within five years the Forez became the largest French producer of coal, and exports rose tenfold.[17] The final blow came in 1786, when a royal concession was granted the marquis d'Osmond on the western fields, a region formerly covered by the Reserve. The new representative councils spawned in 1787 warned that the local price of coal had risen 50 percent and had disrupted the activities of ironmongers in towns such as Saint-Priest, Saint-Genest-Lerpt, and Villars. In more passionate terms, the parish *Cahiers de doléances* denounced Osmond's monopoly. Then in the early days of the Revolution—on 24 July 1789—a crowd, apparently led by an alderman of Saint-Etienne, ransacked his buildings and evicted his German overseers. This was only the first of several occasions when the Revolution seemed suddenly to license collective action on behalf of long-standing customary privileges. But behind such dramatic confrontations lay a continual struggle to retain access to factors of production. Such struggles were complicated in the case of the armaments industry by the activities of the state engineers and their legal powers of enforcement.[18]

Two Manufactures: One Common Life

The challenge of the artillery inspectors who supervised the artisanal production of muskets at Saint-Etienne was what might be termed the "problem of command": how to ensure that the guns they bought for the king's troops matched their specifications, and not once but ten thousand times. After all, guns do not simply reflect the technical know-how of ballistic engineers faced with the constraints and possibilities of battle; they take shape in the hands of armorers. Unlike the arsenals of construction, where the state owned the means of production, in Saint-Etienne the artillery service purchased the king's muskets from designated merchants (Entrepreneurs) who subcontracted the work out to artisanal producers. And unlike the arsenals, where the artillery commanded worker-soldiers, here these armorers and merchants conceived of their interests in economic terms. They also possessed legal privileges and were locked in a complex relationship of their own. The goal of the artillery inspectors was to shape this relationship to get the sort of guns the state wanted in the desired quantity and at the mandated price (figs. 5.1 and 5.2). To this end, they established new institutions to define standards of production—notably the proof house to test gun barrels and the reception room to inspect gunlocks. And they staffed these inspection sites with expert

Fig. 5.1. Gunlock for Musket, Saint-Etienne M1777. The gunlock pictured here is mounted on a musket destined for artillery troops. The model type is an M1777, and the lock dates from 1786. From the Musée de l'Armée (#16640).

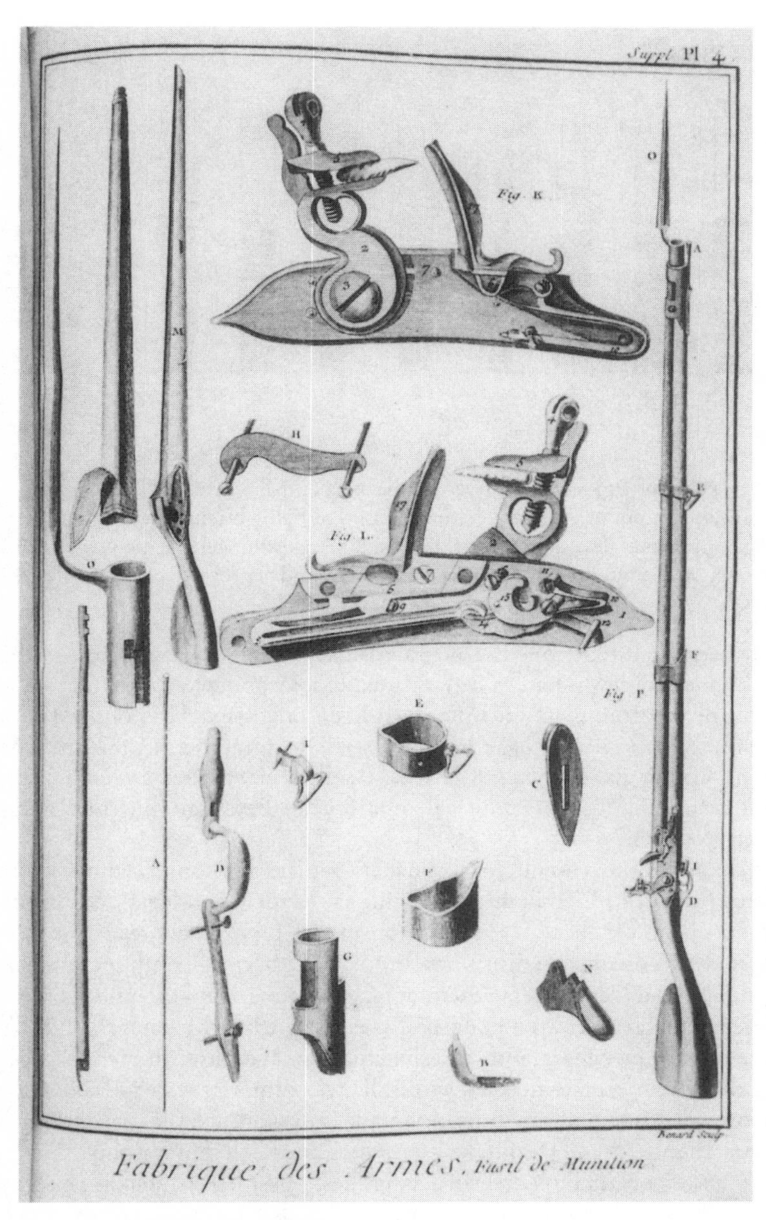

Fig. 5.2. Military Musket and Gunlock. This plate shows an interior view of the gunlock mechanism L, with the tumbler *14* partially obscured by the bridle *15*. On the left are the bayonet and its ring O and G, the stock M, the barrel (unlettered), and the rod A. The various parts of the trigger and mountings are shown at the bottom, while the fully assembled musket is on the right. The illustration is not to scale, nor does it conform to the norms of mechanical drawing. Compare it with the official technical drawings of 1804 in fig. 5.5. [Le Blond], "Art militaire," *Supplément à l'Encyclopédie*, plate 4.

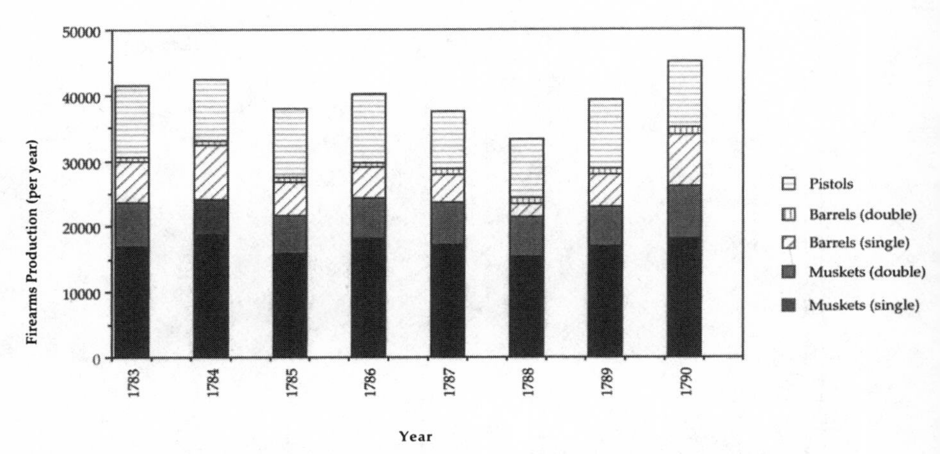

Fig. 5.3. Firearms Production, Private Industry, Saint-Etienne, 1783–90. This graph indicates the output of the private arms industry in Saint-Etienne in the final years of the ancien régime. The data were collected by Augustin Merley, the civilian proofmaster appointed after much controversy in 1782, and are reproduced in Galley, *Election*, 403.

controllers recruited from the artisanal class. But these standards—which had the force of law—had to be continually reformulated in the face of subversion by merchants and artisans. And the engineers were obliged to use coercion to enforce their agenda. In the end, defining the sort of guns which artisans would make, merchants would sell, and engineers would accept, meant reformulating the political relationship between the state and its citizen-producers.

At the root of this conflict was the fact that the division of Saint-Etienne's firearms trade into a civilian and a military sector was a legal fiction rather than an economic fact. The separation of the two sectors rested on a oft-reiterated seventeenth-century law stipulating that: (1) only merchants so designated could contract with armorers to make guns of military caliber (defined as a bore that would take balls weighing 18 to the pound), and (2) in return for tax privileges and other exemptions, certain armorers agreed to work on military guns as needed, while all other armorers were forbidden to do so. However, since military armorers could still work for the private market, and merchants holding military contracts could sell nonmilitary guns privately, inspectors had to cope with an endless quasi-legal traffic in personnel, raw materials, and finished and half-finished goods. One such gray area was the sizable trade in guns of military caliber for the king's colonies and allies. Another was the sale of rejected military barrels for the overseas trade. It was to prevent such trafficking that the artillerists insisted that they be allowed to supervise the proof of civilian gun barrels as well as military ones. The same proof which enabled them to enforce standards of production also enabled

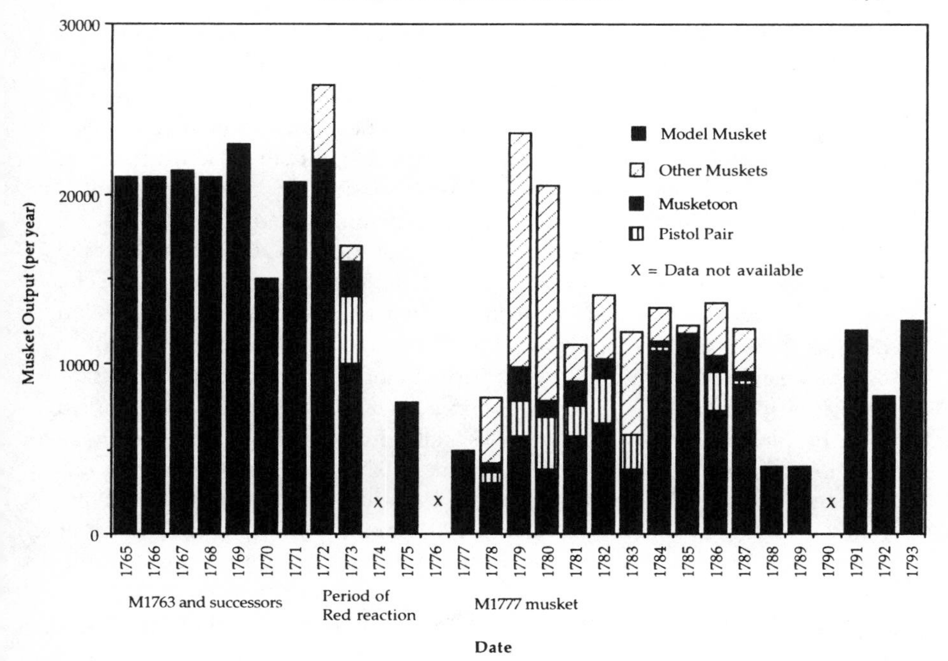

Fig. 5.4. Firearms Production, Manufacture Royale de Saint-Etienne, 1765–93. This graph shows the production of military guns in the last three decades of the ancien régime. It shows the relatively stable production levels of the 1760s, and then the erratic rates beginning in the late 1770s, culminating in a drop in the late 1780s. The data are from Dubessy, *Manufacture de Saint-Etienne*, 102–9, and are based on the number of firearms ordered by the central office in Paris.

them to police the boundary between two types of guns—and, consequently, between two political-economic worlds.[19]

The artillery officers' efforts to police that boundary, however, were frustrated by the fact that the civilian arms-making industry at Saint-Etienne was caught up in an international trade in guns. On the eve of the Revolution the civilian sector produced between 40–60,000 muskets and pairs of pistols annually (fig. 5.3). These varied greatly in quality and were destined for widely different sorts of markets. At the bottom end of the scale, roughly half the total guns were destined for the Atlantic trade and cost as little as 9 *livres*. At the other end of the scale, wealthy customers in France, Switzerland, Germany, Spain, and the Levant bought fowling pieces for as much as 100 *livres*. Saint-Etienne also supplied most of the gun barrels which urban arquebusiers throughout France assembled into muskets and pistols. At the end of the ancien régime the trade brought 1.2–1.6 million *livres* into the town. By contrast, the military armorers produced less than half that many muskets, up to 25,000 guns per year, and these brought in up to 500,000 *livres* a year (fig. 5.4).[20]

The central authorities justified state involvement in civilian gun production by pointing to the need to control the possession of an artifact so disruptive to the public safety and public order. The eighteenth-century musket was a dangerous weapon, especially to its owner. The explosive blast took place frightfully near the vital regions. So since the 1660s, the state had insisted that both military and civilian weapons be subject to a barrel proof (a trial firing to assure the buyer that the barrel would not rupture). At mid-century, some 20–30 percent of barrels failed this test. Without the proof, these barrels would very likely have removed the user's head.[21] Even without invoking this special danger, the French state had more general reasons for regulating the possession of a gun. Like any other aspect of public life in the ancien régime, the right to carry a weapon was a title of an exceptional kind, a privilege granted to individuals by virtue of their membership in a particular collectivity. Guns marked their bearer's public role and signaled social status. The law made careful distinctions about who could carry a firearm, of what sort, and under what circumstances. Exceptions were granted to "gentilshommes" and those "living nobly." These laws were often flouted, of course, but this only shows that French subjects took seriously the social claims implicit in these material objects. Guns were particularly bound up in an aristocratic male code of honor. Sometime in this period, dueling with swords gave way to dueling with pistols.[22]

In such a scheme, the right to make a gun was a privilege as well. Indeed, production in the ancien régime was generally a privilege conferred by the state. As we have seen, the king was the ultimate fount of legitimacy for the trade corporations. And even in towns outside the purview of the guilds—and Saint-Etienne was free of almost all guild restrictions—patents, taxation, and exemptions from tariffs were individually granted. True, state officials in the late eighteenth century became increasingly attuned to laissez-faire ideals. But even minister-philosophes like Turgot or Trudaine insisted that the state retain its role as regulator of production. Paradoxically, they argued that laissez-faire obliged the state to assume even greater powers so as to protect property rights and enforce uniform standards.[23] This argument was not accepted by the arms makers and merchants of Saint-Etienne, however. For them, the lack of local guild restrictions implied a complete freedom to make private contracts without state interference. In practice, however, the entanglement of the military and private sectors meant that the state refused to countenance such arrangements.

Both sectors relied on an overlapping set of some twenty-five hundred armorers, three hundred to six hundred of whom were officially registered for the king's work at any one time. In both sectors, roughly one-third were master artisans owning their own atelier; the rest were journeymen and apprentices employed one or two to a shop. Both sectors also depended on the services of arms merchants, though here a difference emerges. The civilian market was coordinated by a shifting population of some one hundred to two hundred arms merchants, who ranged greatly in fortune. Most also worked as master artisans and shipped fewer than ten muskets a year. These were *fabricants*, a term meaning merchant, but which still implies a close association with production. Among the merchants, some eight to ten families were noted for their disproportionate fortune. These large-scale merchants, known as *négociants*, were themselves only a generation or two from their artisanal origins. This latter group vied for contracts to supply armaments to the king, and were the sometime allies of the artillery engineers.[24] In 1769, with the backing of Gribeauvalists, the leading figure in one of these families, Carrier de Monthieu, won the monopoly privilege of constituting himself the sole Entrepreneur of the "Royal Manufacture of Arms." But even then there was no armaments factory, only a few administrative buildings, and 90 percent of armorers still worked in their own shops, and sold their products for a negotiated price—just as they did in the private market.[25]

In this and other respects, Saint-Etienne differed from the two other French armories: Charleville and Maubeuge. Both of these manufactures had been created *ex nihilo* on France's northern frontier as part of the mercantilist "industrial" policy of the later Bourbons. Each produced roughly the same number of military muskets as Saint-Etienne. However, each had a single Entrepreneur who had long since collected his gunsmiths—mostly immigrant armorers from Liège—under a common roof. And neither made guns for private sale. Saint-Etienne, by contrast, had a three-century tradition of indigenous gun-making. In this, it resembled its rival arms-making towns such as Birmingham or Liège, with their dispersed craftsmen-proprietors. This was something the engineers deplored, and were determined to change.[26]

Artisanal Labor and the Technological Life

In the eyes of the artillery engineers the artisanal mode of production at Saint-Etienne was hopelessly tradition-bound. This ignored considerable evidence that the armorers there lived a technological life as dynamic and innovative as their own. Understanding the artisan's technological life is essential to understanding their resistance to the demands of the engineers. The armorers of Saint-Etienne participated in a complex interplay of productive, familial, and communitarian associations, all set within the disparate and overlapping rhythms of work and life. There was the task-time of forging and reforging iron. There was the daily cycle of firing the forge, eating, and sleeping. There was the social pattern of Sunday masses, festivals, and market days. There was the seasonal variation in agricultural work, with stoppages due to shortages of water power and water transport. There was the lifecycle itinerary from apprentice to journeyman to master, all interwoven with the passage from youth to marriage to child-rearing to old age. Finally, there was the communal succession of generations, whose handed-down knowledge, skills, and capital had to be redeployed in light of changing circumstances. Woven through all these rhythms was the irregular cycle of market prices, which set the constraints within which individuals, families, and communities made choices. And there was another rhythm as well: the rise and fall of military orders that ran to the erratic fortunes of war and peace.

Reconciling these competing and changing demands was not always easy; and the armorers' solution seems to have baffled the engineers. The engineers derided the artisans as excessively venal (for aggressively pursuing their economic interests) and, in a somewhat contradictory vein, complained that they acted collectively to resist new techniques (imposed at the engineers' behest). In fact, these artisans were innovators in their own right, as well as zealously protective of their social and economic autonomy. As William Reddy has pointed out, if anyone was an "entrepreneur" in the ancien régime it was these petty commodity producers. They owned their own shop, bought raw materials, hired a journeyman or two, and sold their goods in full risk of market downturns. They were caught up in the circuits of a complex trade which obliged them to make fine economic calculations. Saint-Etienne's armorers were unique in France in not belonging to a legally recognized guild. Indeed, the social fluidity in this industry (and in the town as a whole) was remarkable in this period. Journeymen were able to set themselves up as independent artisans. Certain artisans were able to set themselves up as small merchants. And merchants branched out from one trade into others. But that is not to say that these artisans operated in some sort of free market oasis set within the regulations of the ancien régime. These artisans had to coordinate their activities with many other commercial agents—merchants, other artisanal specialists, journeymen—and in doing so they relied on family networks, patronage-client relationships, and municipal associations. Well-defined hierarchies of skill and status regulated roles in production and commercial exchange. Life in the town was marked by a constantly replenished set of social and economic relations.[27]

Most of Saint-Etienne's armorers did not reside in the town proper, but were scattered across the valley in ways that reflected the division of labor in the trade. Already in the eighteenth century, a firearm was the joint product of some two dozen subtrades: barrel-forgers, barrel-borers, barrel-grinders, breech-forgers, breech-filers, screw-makers, locksmiths, stock-makers, makers of mountings, bayonet-forgers, assemblers, et cetera. These tasks can be grouped into three clusters: barrel-makers, locksmiths, and gun assemblers. Elite armorers associated with barrel-making lived in along the Furan River, which supplied their power. Those master barrel forgers who were also *fabricants* lived in town in the respectable Chavanel district, near the main offices of the Royal Manufacture. The makers of gunlocks (locksmiths), like the region's ironmongers, lived in some thirty-six small towns like Villars, Saint-

Jean-de-Bonnefonts, and Saint-Priest, which closely tracked the local coal deposits. Many still had agricultural interests. Map 5.1 shows the domiciles of the 124 master locksmiths and 59 journeymen employed by the Royal Manufacture in 1782. (And map 5.2 shows the domiciles of the 587 locksmiths who worked for the National Manufacture in 1799, when almost all of the region's private locksmiths had gone to work for the state.) Finally, gun assembly typically took place in the town proper in the hands of another series of artisans who worked on a piece-rate basis for merchants or other armorers.[28]

A finely graded hierarchy ranked metalworkers above ribbon-weavers, armorers above metalworkers, and barrel-forgers above most other armorers. These artisans derived their status from their family connections, their skill, and the prices they could command. Within the armory, artisan-*fabricants*, such as the members of the Merley, Javelle, Bouillet, and Boutet clans, constituted an elite. Many subcontracted work out to their fellow artisans; others became salaried controllers for the state. As for the bulk of master armorers who owned their own ateliers and earned substantial wages, they too depended on family ties for their status. For example, when town leaders in Tour-en-Fouillouse ranked the 136 resident locksmiths in three categories, they indicated that most of those capable of "fine" work were members of the Reverchon clan. Across the generations, members of the same family plied the arms-making trade, even remaining within the same subspecialty. A genealogy of the Maguin family from 1750 to 1850 shows that half their offspring married either armorers or the daughters of armorers.[29]

Training in the trade was by apprenticeship, which was not formally regulated until the 1770s. Masters generally hired only one or two journeymen. Locksmiths had at most one journeyman to help with the filing. Of the 645 armorers who worked for the Royal Manufacture in 1782, half were masters and half journeymen. Skills were acquired young, usually from older male relatives. One sample of 146 workers shows that 67 percent had begun to work by the age of sixteen, and 93 percent by the age of twenty. These young artisans were trained individually. A sample of 48 elderly locksmiths had trained 127 locksmiths in their thirty-plus year careers, or less than one apprentice per decade.[30] After completing their training, armorers such as Merley "the Parisian" and Jean-Louis Jalambert took the *Tour de France* to see the world and perfect their skills. Many, like Honoré Blanc, Joseph Bonnand, August Merley, and Jean-Baptiste Javelle innovated with their craft. They experimented with "ribbon-wound" gun barrels, multiple-shot muskets, and damasque and engraved decorations. All master artisans made their own tools. Some, like Javelle and Blanc, invented special-purpose machine tools. No formal avenues of instruction were available. In 1766 the municipality permitted the cleric P. Chabert to offer free lessons on "practical geometry"; but the course never met. Artisans could, however, take lessons from fellow *artistes*, such as Jacques Olnier who had studied academic drawing in Lyon.[31]

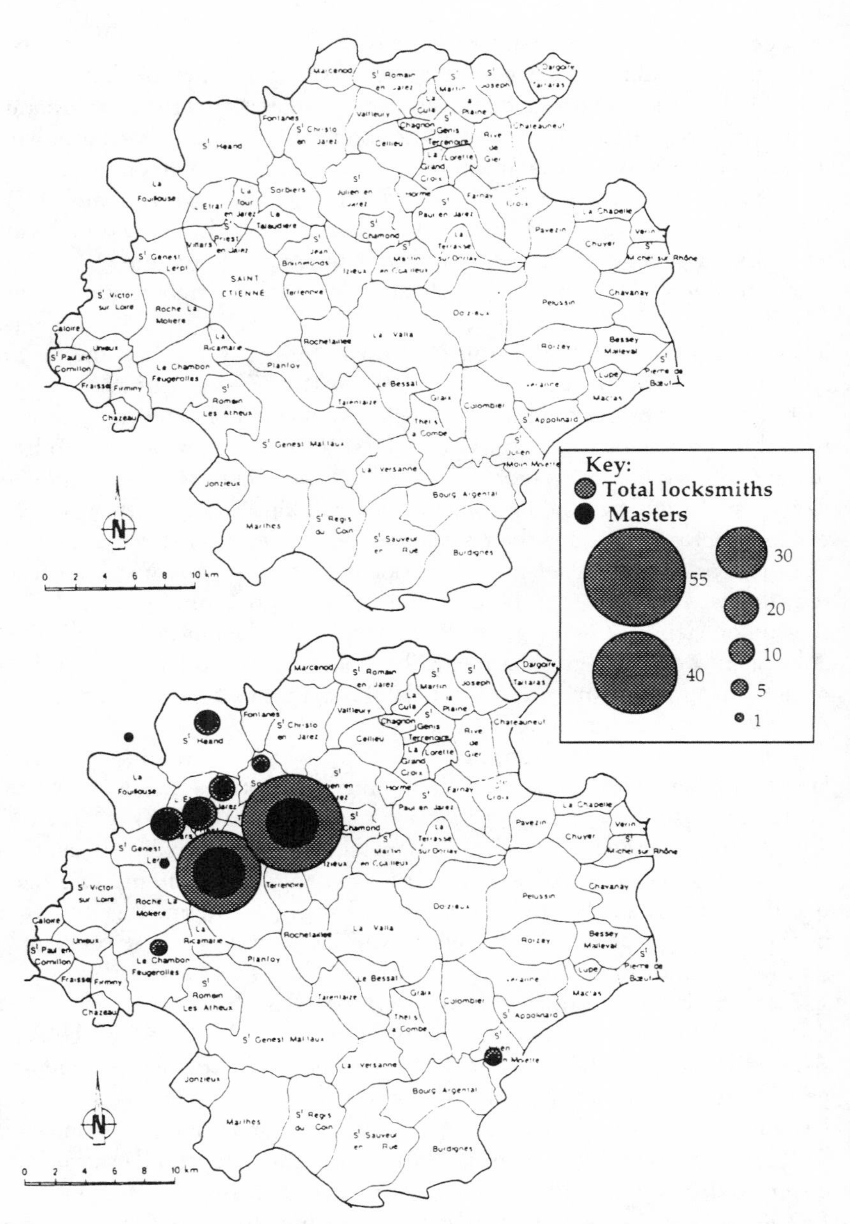

Map 5.1. Distribution of Military Locksmiths in the Saint-Etienne Region, 1782. This map indicates the geographic dispersal of the military armorers involved in the gunlock trade during the ancien régime. The map is divided into districts (a division used after the Revolution), and underestimates the scatter of the locksmiths. The forty-one locksmiths of Saint-Etienne, for example, actually worked in ten separate locales, and the twenty locksmiths ascribed to the town of Villars worked in five different hamlets. Locksmiths were almost equally divided between masters and journeymen. The data are from S.H.A.T. 4f5 Danzel, "Etat et dénombrement," 1 January 1782. Location of the place names was facilitated by Dufour, *Dictionnaire topographique*.

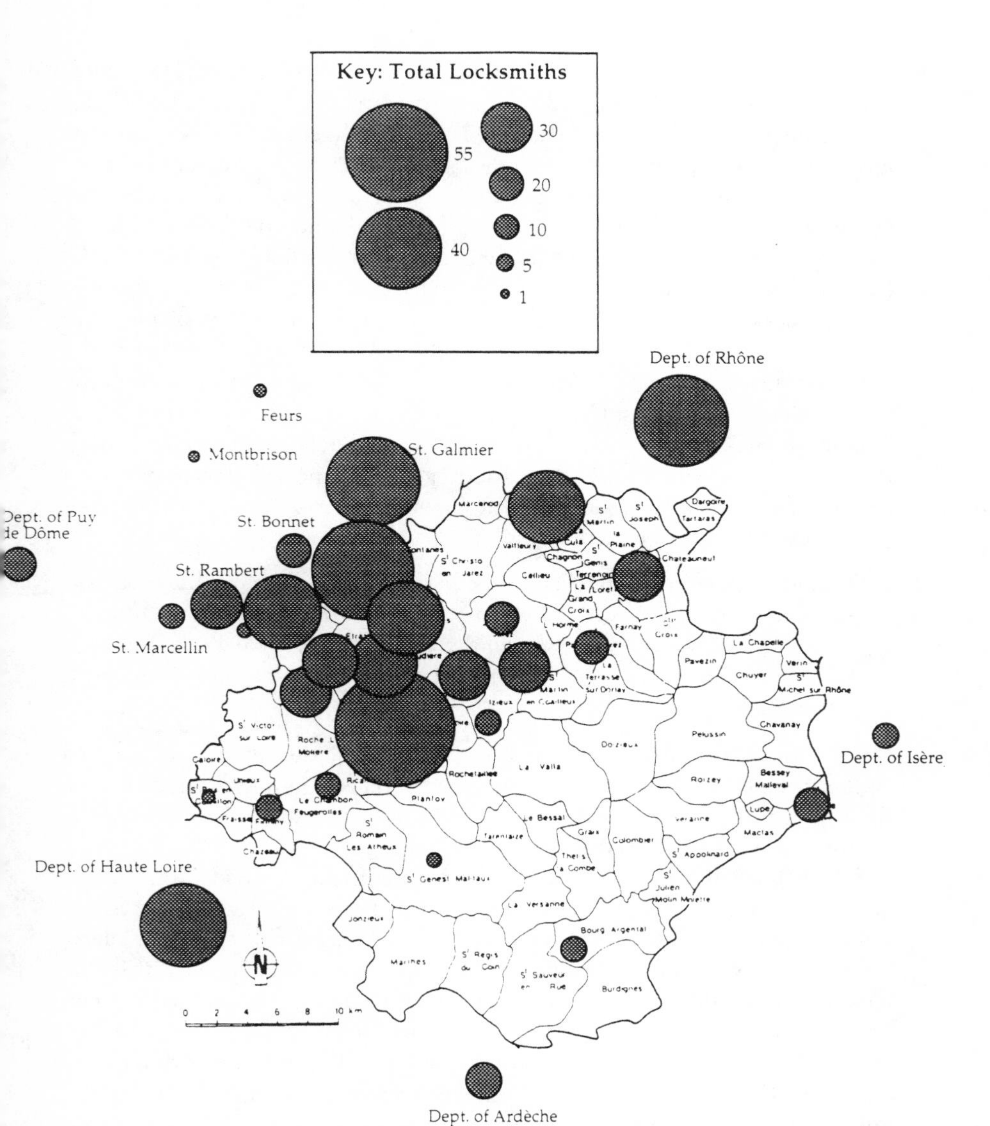

Map 5.2. Distribution of Military Locksmiths in the Saint-Etienne Region, 1799. This map shows the persistence of the geographical dispersion of locksmiths in the post-Revolutionary period, and also the intensification of the military effort. By 1799 some 587 locksmiths were supplying the state, of whom 30 percent worked outside the département altogether. Data are from A.D.L. L942 Berardier-Merley et al., "Etat nominatif," 15 floréal, year VIII [5 May 1800].

For armorers of all ranks, their skill was their main possession. These skills could not be easily acquired, and belonged to a form of bodily, tacit knowledge that could not be exhaustively described. These skills shaped the bodies of the armorers—to the point where an outsider like Roland considered them deformed: "The habit of working early and upright, principally at filing, the body acting with effort back and forth, not exactly in its natural direction, forces them to spread their legs, cambers their thighs, and visibly deforms them."[32] Belatedly, in 1804, one artillerist confessed that "the supervision of engineers, the instructions of controllers, the punishments and rejection of pieces at inspection time are all means; success in the manufacture of arms always depends on the skills (*habilité*) of the workers."[33] But this was not the view of the ancien régime artillerists. One engineer-inspector described the armorers as "the most thick-skulled workers imaginable, and likely to deviate from even the most simple principles."[34] And another scorned the armorers as primitive creatures, incapable of reason, blindly following traditions propagated in ignorance from generation to generation.

> One must not let [the worker] know that he is necessary, that one has need of him. The worker is a kind of stubborn animal who recoils exactly when one wishes him to advance. His needs and those of his family are the sole motives which resolve him to work, and only his fear of being without a job inspires him to work hard.[35]

This was how the engineers referred to some of the most highly skilled craftworkers in France. In fact, it was the very industriousness of the armorers, their "greed" and their interest in their trade which continually frustrated the engineers, who valued service to a higher calling.

The artisans also lived on the other side of a linguistic divide. On the eve of the Revolution, the 120,000 inhabitants of the rapidly growing Saint-Etienne basin (28,000 in the town and its faubourgs) still spoke a dialect-language known as Franco-Provençal. This patois—which varied within the region well into the nineteenth century—raised a tremendous barrier between artisans and the "foreign" military officers who came to administer the armory.[36] Their mentality was also far removed from that of the self-disciplined engineers. Dialect poems of the seventeenth and eighteenth centuries offer a unique insight into the artisans' attitudes toward work, toward bureaucratic elites, and the conditions of their own lives. Verses attributed to Antoine Chapelon, himself born to a nail-making family, feature a hard-drinking, indebted old ironmonger named Denis Bobrum, who recounts his Rabelaisian adventures on his deathbed.

> Si lou savans entre ellou sont en guerra,
> Si lou richards appróvrésont la terra,
> A que sert-ou de tant s'écourpela
> Par avez d'émou et d'argent de tous la?

J'ai mió ama vióre sens tant de scienci
Et sens argent, hazard de prendre à crenci,
Par me tenir l'esprit toujours en jouai,
Lou coeur content, librou, tranquillou et gai.[37]

If the wise ones all make war,
And the rich impoverish the earth,
What's the good of breaking your back
To acquire knowledge (*science*) and money?
I'd rather live without so much knowledge
And money, and in risk of debts,
And keep my spirit joyful,
With a heart contented, free, tranquil, and gay.

At death's door, Bobrum mocks the hypocrites around him: his doctor, priest, wife, and daughter-in-law. Deep in debt and drink, all he has left is his coal-blackened linen, a handful of cooking utensils, and his tools (a mandrel, files, a hammer, a bow-drill, some half-finished swords). That, plus some curmudgeonly advice: "To the smiths, that they should early rise/And always choose the best in wine." His epitaph reads:

Cy gey lou rey dó palengüns,
Que tous sous jours ériant de lüns,
Lou patriarch de le pelles!
Au leissèt rulir son avit.[38]

Here lies the king of good-for-nothings,
To whom every day was [Saint] Monday,
The patriarch of sluggards!
He let his anvil go to rust.

This affectionate exaggeration—though literary in its form—offers a glimpse of the Saint-Etienne artisan's self-image. The rags and wine belly are real enough. In the middle of the eighteenth century, the local civilian administrator estimated the town consumed 60,000 *années* of wine per year; over a liter of wine *per* day *per* man, woman, and child.[39] But the Cockaigne fantasy that every day be "Saint Monday," must be balanced against the evident industriousness of the townsfolk noted by all visitors. Yet even that industriousness was at the workers' good pleasure. One inspector observed with exasperation that when a sudden (and profitable) order arrived, an armorer could produce more in one or two weeks than he had in the previous months.[40]

Other dialect songs convey the subversive side of the eighteenth-century artisan. Georges Boiron, a file-maker who never took to his trade, composed ribald drinking songs that celebrated the petty heists and sexual conquests of his marauding band of roughnecks. Boiron also wrote ditties for the popular *Jeux d'Arc* archery competitions held at various stands in town. Gunsmiths in town to drop off workpieces at the Royal Manufacture could "bend an elbow" at the game-stand and at the cabarets across the square.[41] One popular song captures the resentful pride of the artisan in the late eighteenth century. Composed by an anonymous metalworker at a time of great resentment against the artillery's attempts to reform the Royal Manufacture, this dialect poem takes the form of an allegorical address to the artisan's *basana* (a leather apron worn by ironmongers).[42]

Póra basana,
Toun so ey malhérou.
Ti sey la grana
Que fat lou bionhérou:
Au richou proudus tout,
Au richou fourniey tout.
Vou-ey t-ïn ma de migrana
Que te fara langui, póra basana!

Póra basana,
Ti lou salues tous,
Moussue et dana;
Ti lou respectes tous.
Ti lou veu parmena
Dïn l'ai levant lou na,
Te menaçant de cana;
Vou-ey t'à tet à céda, póra basana!
. . .
Que sant-i faire?
Que gratta lou papie
[D']ïn ai de plaire,
Bion bère et bion migie.
Sant pa cougnie ïn cló
(Vous ririas voutroun só),
Pas nió breÿie de mana.
Hélas! que fariant-i sen la basana?

Quand la basana
Ti ne po plus porta,
Ti pronds la cana,
Ti charches l'hôpita;
D'iqui la Charité,
L'i sei par quóquous meys,
Peusson, vai Notra-Dana!
Hélas, veyquia toun so, póra basana![43]

Poor basana,
Your lot is unhappy.
You are the seed
Which makes the happy ones:
For the rich, you produce everything,
For the rich, you furnish everything.
It's an old war wound
Which makes you languish, poor basana!

Poor basana!
You greet everyone,
Monsieur and Madame;
You pay your respects to all,
You watch them promenade
With their noses in the air,
Threatening you with their cane.
It's for you to give way, poor basana!
. . .
What do they know how to do?
Besides scratch at paper
With a pleasing air,
Drink and eat well.
They don't know how to knock in a nail
(They laugh at their ease),
They can't even bust rock.
Alas! What would they do without the basana?

So when, basana,
You can't carry yourself about,
You'll pick up a cane,
And seek out the hospital,
And from there, the charity hospice,
Then, after few months,
Off to Notre Dame cemetery!
Alas! That's your lot, poor basana!

This song suggests the tenor of social antagonism in Saint-Etienne on the eve of the French Revolution. Directed against elite pretension and bureaucratic despotism, rather than against economic exploitation, it nevertheless cast a cold eye on the rich, for whom the *basana* "furnishes everything." The somewhat plaintive tone of the song ("Poor *basana*!") must be balanced against its contempt for paper-pushers ignorant of even the most basic crafts. Saint-Etienne does not seem to have experienced any of the sort of "league" activity which had to be ruthlessly suppressed in contemporary Lyon. Yet the armorers were also capable of collective action to enforce their rights. In 1761, a group of three hundred gunsmiths threatened to burn down the warehouses of the Entrepreneurs of the Royal Manufacture if they were not paid the money they had coming to them. The merchants dispersed the crowd with promissory notes for bread, and they were paid in short order.[44]

Merchant Capital and Industrial Investment

The merchants who coordinated production in Saint-Etienne were an equally heterogeneous lot. But they were universally averse to investing directly in production. Eighteenth-century merchants ran risks too. They carried inventories and often served as creditors to artisans, who might default. Some *négociants* owned water mills where barrel reamers and grinders rented motive power. But in general, they were content to supply outworkers with raw materials and buy back their semi-finished products for further processing or resale. This was a matter of endless frustration to the artillery engineers who expected the armory to develop along the lines of the northern manufactures or the English mechanized factories of their technocratic imagination. Why were Saint-Etienne merchants so notoriously unwilling to sink capital into production?

The problem was that the rate of return on any investment in manufacturing depended on the stability of demand and the labor supply. Tellingly, the lauded northern manufactures needed constant refinancing in the later part of the eighteenth century. Those Entrepreneurs were at the mercy of an erratic military demand and a mobile population of Liègeois armorers who threatened to run home at the first sign of a slowdown.[45] At Saint-Etienne, military armorers could not flee, but they could (and did) shift production to the private sector. For this reason, the manner in which military demand was superimposed upon the demand for civilian guns was of great consequence to both armorers and merchants.

This raises the converse question: why didn't artisanal producers simply eliminate merchants altogether? Or if their function as coordinators of the trade proved of some value, why not squeeze *their* profits? In other words, why assume that economic power lies in the hands of the coordinating merchant, rather than with the artisanal producer? After all, the knowledge needed to coordinate the different branches of the trade was available to many artisans. Indeed, the boundary between master artisan and *fabricant* was highly fluid. Both groups, however, were often in conflict with the wealthy *négociants* over access to raw materials and the market. Winning control over this access was the crucial source of merchants' power over petty producers. We have already seen how the Lyon merchants and the owners of the local slitting mills controlled the flow of iron into Saint-Etienne. And we have seen how access to coal was jealously guarded. Access to buyers was also contested. When the new proof-house law of 1782 required merchants to list the recipients of all gun shipments, the merchant Jovin protested that this would be disastrous for *négociants* who wanted to keep this information proprietary. "By giving the worker freedom [to know about buyers]," he noted, "one disgusts (*dégout*) the *négociant*, who will abandon this [form of] commerce."[46] In particular, access to military contracts was a great advantage to the *négociants*. Having no capital tied up in production, merchants with a legal monopoly on military contracts could actually take advantage of the irregularity of wartime contracts. This happened because large military orders tended to coincide with a drop in civilian demand (as war cut off foreign markets). In such periods, the 1750s for instance, their control of military orders enabled certain *négociants* to dominate their rivals and bind the armorers to them with debts.

Yet the *négociants* never succeeded in wholly dominating the economic terrain. The third and final possibility, the "sweated alternative," never materialized in Saint-Etienne, at least not in the arms trade.[47] Indeed, when the market was properly aligned (as it was in the early 1780s), the armorers and *fabricants* were able to turn the tables on the *négociants* and undermine the Royal Manufacture. This should remind us that artisans and merchants did not experience the market as a mysteriously impersonal force, but through face-to-face social relations that expressed the relative power of big merchant and little merchant, merchant and artisan, master and journeyman. And although these "forces" were partially stirred up by multiple and faraway hands, they also reflected (in the aggregate at least) the actions of local actors: their efforts to create new markets, their appeals to Parisian ministers for tariff protection and new regulations, and their threats against those who violated the norms of the local economic culture. The challenge, then, for armorers and merchants (both large and small) was the "problem of coordination": how to regulate their activities in the face of these multiple uncertainties, whilst those around them did likewise. The difficulty here was that Saint-Etienne lacked the formal corporate guild structures or *compagnonnages* that usually performed this function in the artisanal trades of the ancien régime. Nor did the law courts or registry lists of journeymen mediate the relations between journeymen and masters, or masters and merchants. Instead, the municipal government was the principal battleground for public disputes over the proper regulation of economic life.

It was against this background that the monarchy introduced the first formal institutions to guide economic life in the arms industry: the proof house in the seventeenth century, and the monopoly on military contracts and the formal work laws in the eighteenth. It was through these institutions that the engineers tried to get the sort of guns they wanted. Consequently, these institutions lay at the center of the antagonism between the civilian and military sectors of the arms trade. The armorers and small-time arms merchants bitterly protested the requirements of the reception room for gunlocks and the proof house for gun barrels. Only gradually were they able to make these institutions serve their own ends. The politics of the proof house are the by-play of a process of conflict and negotiation over who will set the standards of production. For it is here, where the problem of command meets the problem of coordination, that the artifact takes shape.

From "Monopsony" to Management

The history of the Royal Manufacture of Saint-Etienne between 1666 and 1789 can be divided into four phases. In each of these phases the state moved increasingly into a direct managerial relationship with artisanal producers, using legal and economic leverage to reconfigure the productive process. Yet this trend masks a basic continuity: throughout this period, the armorers of Saint-Etienne were largely able to retain control over production.

During the reign of Louis XIV, phase one, the armory was run as a monopsony, with a single Parisian commissioner, Maximilien Titon, granted sole title to buy from local merchants. The centralization of musket purchases and the standardization of design (two sides of the same coin) served the purposes of the king's army by simplifying the supply of ammunition, standardizing drill, and (ostensibly) raising the quality of muskets. The tenfold increase in the number of soldiers, *plus* the War Office's determination to assume responsibility for gun purchases, meant that Titon had to procure over a million arms for the French army during Louis XIV's wars. Although the purchase price of the finished musket was set by the minister of war, Titon had complete freedom to set the price he paid to merchants and they to armorers. This economic power was buttressed by legal privilege. In 1666 the right to produce muskets of military caliber was forbidden to all nonapproved workers. A year later, the regular proving of military gun barrels began. This enforced Titon's monopsony and justified confining his business to a narrow circle of merchants. His rule over the manufactures was near absolute and he died worth over 3 million *livres*.[48]

In Saint-Etienne, this monopsony aroused passionate opposition. Merchants excluded from contracts denounced the "two or three families which grew rich during the last war, and who profited from the work of our artisans and the sweat of our workers." Master artisans and *fabricants* excoriated Titon's monopoly as a monstrous abuse of their right to contract freely with any buyer or seller. Specifically, they denounced the proof as an onerous extension of the government's power. "Liberty and secrecy," they argued, "are the soul of commerce."

> What frightens us even more is that things will not stop there, but that the specious pretext of the interest of the state and the public good can only lead to

a monopoly that will suck up all the profits of our manufacture, which ought to be shared equally between the merchants and workers of our town.[49]

From the beginning, control over standards meant control over the market.

Phase two began in 1717, when the state assigned the artillery service the task of buying its guns and assuring their quality, a task they retained into the twentieth century. As emissaries of the state, the engineers now dealt directly with designated local merchants (Entrepreneurs), who operated under the on-site supervision of an artillery officer (the inspector). Each inspector was assisted by three controllers selected from among the armorers and paid an annual salary: one served as proof master for both civilian and military weapons; one inspected the gunlocks; and one monitored the final gun assembly. New regulations spelled out the dimensions of the firearm, and outlined some procedures for its manufacture. After some controversy, the state agreed to let the price of gun parts be set in private negotiations between armorers and Entrepreneurs. Orders were placed with Pierre Girard, Robert Carrier, Jean-Louis Carrier, and Duchon and Jourjon, and these same families continued to win contracts in the decades to come. Delivery schedules, however, were erratic. Between 1750–55, no orders for military guns were placed. These uncertainties in the military market were exacerbated by tardy repayments; during the 1750s, the state was seven years in arrears.[50]

Although the artillery inspectors were expected to act as disinterested supervisors, they quickly became enmeshed in the financial and familial relationships of the nepotistic Saint-Etienne valley. Inspectors relied on these local Entrepreneurs to guide them through the complex family politics of Saint-Etienne and keep them informed about the technical operation of the armory. From a nineteenth-century perspective, inspectors and Entrepreneurs might appear to have formed a unified social class: both came from *roturier* families who had recently acquired social prestige and ennobling office. The Entrepreneurs had done so through the profits of trade, the artillerists by serving the state. And in certain respects, they forged a mutually beneficial alliance. The Entrepreneurs allied themselves with the inspectors to maintain their monopoly over government contracts. And the inspectors allied themselves by marriage with these exceedingly wealthy merchant families. Moreover, in a town where the dialect was so marked, at least they both spoke French. Yet the alliance was always a troubled one, and they cannot be said to have constituted a cohesive "bourgeoisie." The engineers plainly despised the avaricious dealings of the Entrepreneurs. As a corps which aspired to noble status in a military culture, the artillerists had predicated their merit on the higher calling of loyal service. By contrast, the Entrepreneurs were "interested," their actions driven by immediate personal gain. Vallière, *père* called them "hustlers and greedy to excess (*remuants et avid à l'excès*).[51]

The Carrier clan of arms merchants was particularly successful at marrying their offspring to prominent business associates, collateral cousins, and well-placed government officials. They and a handful of leading families controlled patronage, credit, and contracts in the closed and nepotistic Saint-Etienne valley. They bought up rural lands, lent money to struggling armorers, allotted work to their "own" artisans, and intrigued for control of the municipal council. Josette Garnier's analysis of property sales in the Forez vividly shows their financial power. Purchases of lands among the merchant class of Saint-Etienne rose at an almost exponential rate in the final decades of the ancien régime. They also functioned as the principal source of credit in the town, making large gains during crises (when the price of grain was at its highest). The artisans' collective net indebtedness grew by 71,000 *livres* between 1746–54—even though some 40 percent of artisanal loans came from other artisans. The merchant class, meanwhile, generated a net credit of 114,000 *livres*. While the artisans did not fare as poorly as the peasantry or nobility, all groups saw their economic position slip relative to the *négociants*.[52] In a town virtually without nobility, these merchant families used their wealth to purchase offices. The Carriers collected a long string of royal charges that exempted them from taxes, and enabled them to influence municipal affairs. These families dominated what little high society there was in town. The Carriers were prominent members of the popular *Jeux d'Arc* clubs. But even the Carriers never quite succeeded in fully dominating their partners, rival merchants, or the armorers.[53]

Phase three began after the Gribeauvalists took over the artillery service in 1763. The new inspectors they sent to the town found the manufacture in a state of chaos. This was its normal state. But in the eyes of the young scientifically schooled engineers, the organization of the armory was offensive. For evidence, they compared the Saint-Etienne muskets with those of the northern manufactures. Saint-Etienne's gunlocks were considered particularly horrendous. Sent down to report on conditions, Jean Maritz, the foundry master, cited deficient iron supplies, intermittent water power, archaic machine technology, and lackluster skills. To solve this problem, the state offered both a carrot and a stick. The War Office under Choiseul increased the price of a musket by 60 percent. Equally important, it promised steady levels of procurement (20,000 guns a year) and prompt payment. In return, it demanded far greater control over the organization of production, and far greater authority over work processes. The model here was the northern armories. For the engineers, the higgledy-piggledy competition at Saint-Etienne was the enemy of quality production. Only the orderliness of a centralized manufacture could guarantee quality firearms. So the inspector of Charleville, Fiacre François Potot de Montbeillard, was transferred down to reorganize Saint-Etienne.[54]

The Gribeauvalists began by centralizing financial authority. In 1763–64 Montbeillard formed a consortium of nine merchants. In 1765, their number was reduced to five, and their company received a nine-year monopoly in return for the promise to build a centralized factory.[55] That year, Montbeillard was replaced by his second-in-command, Louis-Alexander Cassier de Bellegarde, who continued his policies. Bellegarde was active, intelligent, and ambitious; he had a thorough knowledge of metallurgy, and had just designed the new M1766 musket. Like Montbeillard, he saw Saint-Etienne as a valley of competition, collusion, and greed. In his eyes, the "jealousies, hates, and private interests" of the arms merchants produced chaos, waste, and imperfect weapons. He had an even lower opinion of the intelligence and honesty of the armorers. The problem was a lack of rational direction: "Hands are not lacking at Saint-Etienne, but they labor without [guiding] principles." Rigor alone would bring to heel the despicable venal practices and underhanded dealings of the inbred, uneducated, half-French provincials of Forez.[56]

Yet for all their contemptuous remarks, both Bellegarde and Montbeillard agreed that Charleville and Maubeuge should be shut down and Saint-Etienne made the kingdom's sole armory. Indeed, the Choiseul Ministry intended to carry through on the logic of the northern manufactures to the point of closing them altogether. Only Saint-Etienne was far enough from the borders to be safe in times of invasion. Only Saint-Etienne employed French artisans and possessed a private trade whose skilled laborers could be tapped when military demand rose during war. And only Saint-Etienne boasted the natural resources to be a truly national armory. The problem remained of how to organize Saint-Etienne. Invoking models of military subordination and engineering rationality, Bellegarde described the armory as a machine whose gears needed to be kept in equilibrium. Only a single Entrepreneur having the necessary funds for the construction of the "buildings and factories necessary for a manufacture" could raise himself above "all objects of division and dispute," and so ensure the "security of the service and place no further obstacles in the way of perfecting our armaments."[57] In sum, the engineers imagined a rationalized military production insulated from the depredations of market competition, yet tapping the skills it honed.

In 1769 Gribeauval was authorized to put this policy into effect. He closed the Manufacture of Charleville, and granted Jean-Joseph Carrier de Monthieu a perpetual monopsony over military purchases in Saint-Etienne, including guns for the colonies and foreign powers. Carrier de Monthieu already dominated the consortium, with 400,000 *livres* in property, whereas his four associates together had only 63,000 *livres*. At the same time, he acquired a central office building on the Place Chavanel, plus two large mills along the Furan. The engineers expected a central factory would soon be built. For the time being, however, work was still subcontracted out to skilled armorers in their privately owned ateliers. As we will see, the rigor of the barrel proof and gunlock inspection was also tightened and new methods of production were introduced.[58]

This consolidation (phase three) was disrupted by the return of the Vallièrist party in 1772. In the battle of Ancients and Moderns, the goings-on at Saint-Etienne played a prominent role. In addition to his duties as inspector, Bellegarde had been asked to comb through the kingdom's stockpile of arms, setting aside good muskets, selling those repairable for shipment abroad at 25 *sols* each, and disposing of the rest for scrap at 10 *sols* each. It was a Herculean task: spring cleaning in the government's Augean stables. Between 1765 and 1769 Bellegarde rejected as defective some 472,000 guns, nearly two-thirds of the entire national stores. The sole buyer of this huge consignment was Carrier de Monthieu. But when the Vallièrists returned to power, they discovered that thousands of these "defective" guns had been sold *back* to the king as new weapons—at a 3200 percent markup! They then discovered that Bellegarde and Carrier de Monthieu were brothers-in-law by a secret marriage. In an intercepted letter Carrier de Monthieu confessed that he was on to a "sweet deal." A widely publicized trial in 1773, which saw both men convicted by a special Council of War, helped bring the Gribeauvalists down.[59]

The motives behind this scandal are illustrative of the agenda of the Gribeauvalists. The Gribeauvalists blamed intrigue: the Entrepreneur of Charleville was angling to reopen his manufacture, and rival arms merchants in Saint-Etienne wanted to recoup their slice of the military market. Bellegarde, they argued, had been ridding the arsenals of faulty firearms that put the king's own soldiers at risk. Bellegarde's wife (Carrier de Monthieu's sister) accused the prosecution of being "the murderers of our troops."[60] In fact, Minister Choiseul had originally justified the project as a cost-saving measure; rather than maintain hundreds of thousands of rusting muskets, it was better to sell defective weapons for scrap, and ship substandard guns to the slave traders or American rebels.[61] That way room would be made for the fifty thousand new muskets produced each year. These new muskets would all conform to the most exacting standards. Simon-Nicolas Linguet, the defense lawyer, recognized "the universal vow of the military man . . . to achieve uniformity in this essential area . . . [where] a disproportion [was] excessively worrisome and even dangerous. . . ." Where the Vallièrist officers saw functional muskets, the reformers saw firearms that did not conform to "spec." Bellegarde was equipped with gauges, and he insisted on real precision.[62] Older weapons were fit for service only after they had been properly reconditioned by the Entrepreneur and had successfully passed inspection, a process which wiped clean the spotted history of the muskets. The engineers had open contempt for the heterogeneity of the past.

The M1777

In 1777, back in control of the artillery service, the Gribeauvalists introduced a new musket model and a new method of managing the armory. This time (phase four), the Gribeauvalists were hopeful that they were at last on the verge of achieving the "perfection" of the musket. They set prices for gun parts themselves, imposed new techniques of production, and sharply increased the stringency of their quality controls. The unintended result was to greatly exacerbate long-standing frictions within the armory—frictions which would ultimately tear it apart.

This new stringency in production cannot be separated from the new program to redesign the musket. The artillery's efforts to design small arms followed the same logic (and raised the same problems) as their design of cannon. The M1763 had been considered too weighty, and Bellegarde's M1766 had been slightly modified in 1768 and 1770. Then in 1774 the return to power of Vallière, *fils*, had resulted in yet another musket. Under these circumstances, colonels continued to custom-order muskets for their troops, undermining the artillery's monopoly over the supply of firearms. And troops in the field were notorious for personalizing their firearms. In 1775 many soldiers were still shortening the gun barrel which they found too heavy to aim easily. Just as he had stepped in to redesign the French cannon after Marshal Broglie began to have pieces rebored on his own, Gribeauval was obliged to redesign the musket to reassert the service's authority over the supply of armaments. In this context, "uniform production" meant artillery control.[63]

Gribeauval laid the ground carefully. To ensure the musket would be of superior quality—and accepted as such—he held an "open" competition. The judges were Montbarey (the minister of war), Du Châtelet (commander of the Régiment du Roi and long-time patron of Gribeauval), and Gribeauval himself. Five prototypes were considered. After tests held in January and February 1776 at the Château de Vincennes, the jury awarded the decision to Honoré Blanc, chief controller of gunlocks at Saint-Etienne. From the beginning, he had been the inside candidate. He had designed his prototype while working on a special commission for the duc Du Châtelet's Régiment du Roi. Under state patronage, Blanc assembled an entire experimental workshop, where he spent almost 3,400 *livres*, tapping the assistance of talented armorer-inventors such as Javelle, Bouonnet, and Jacquet.[64]

In its deliberations the committee emphasized the role of the new gun in the new thin-line tactics. In the official three-man line, a soldier had to load his musket while holding it upright. So to prevent powder from spilling out of the pan, Blanc set it an at angle. The pan of one competing design, the examiners noted, would have "inconvenienced the soldier." On the other hand, Blanc's barrel, while one *pouce* shorter than the old, retained its basic ballistic characteristics. The guns were not tested for accuracy of fire—Gribeauval had sufficient authority to dispense with public demonstrations—but in 1788 and again in 1790 the artillery was obliged to defend the M1777 against rival designs.[65] At that time, Professor Lombard examined the performance of the M1777 in expectation of Guibert's mixed tactics. Using data on the initial velocity of the ball, he calculated its range for different angles of fire. From this, Lombard suggested how soldiers should use the musket's *but en blanc* to aim their weapons more accurately. This was something every hunter learned by experience, but soldiers needed rules of thumb, which Lombard supplied.[66]

The engineers meant to leave no doubt that their new design was an optimal match of design and military need. Blanc bragged that only with the 1777 musket "had anyone seriously concerned themselves with perfecting the firearm since every piece, without exception, has been thought about and discussed, and whenever there remained the least uncertainty, we had recourse to experiment."[67] The implication was that nothing had been left to chance or tradition. We should not, however, take such claims too seriously. Between 1700 and 1840, the design of the flintlock musket remained essentially constant. Despite their efforts, the Gribeauvalists only delivered marginal improvements in the safety, reliability, and accuracy of firearms. The pace of change was evolutionary, not revolutionary. Radical modifications in gun design were hostage to the repository of gunsmiths' tools, skills, and interests. Firearms production at Saint-Etienne remained a handcraft industry until the end of the nineteenth century; and reliable and accurate guns had to wait for percussion locks and rifling.[68] The best the Gribeauvalists could hope to accomplish was to make a slightly more effective musket by producing a more precise one. As in the case of cannon, cannonballs, and artillery carriages, this meant defining a band of manufacturing tolerance. As we will see, the new regulations laid out a minimum and maximum for the gun barrel and lock pieces. Indeed, Gribeauval delayed the introduction of the M1777 by a year so that all the procedures, measures, and gauges for each piece would be ready before sending out the new pattern guns.[69]

At the same time, the artillery service moved into a direct managerial relationship with the armorers by setting a price for each gun component. This had been tried and abandoned in 1759–60, when the War Office had first seized control of the artillery back from the Grand Master and Vallière. But local arms merchants had protested; a manufacture, they objected, was not an army battalion, and "not susceptible to this same degree of perfection." "Military formations (*évolutions*)," they argued, "depend on the will of a single person, [but] everyone has influence on the permutations (*révolutions*) of commerce." As a third party, "uninterested" in the outcome of the contract, the state lacked the necessary information to set prices. Indeed, these prices were not calculable. Did the engineers know that all armorers were not equally skilled? And that not all orders—even from the government—were identical? Setting prices by command, they implied, could never replace the dealings of interested parties.[70]

Yet the Gribeauvalists asserted that they *could* substitute rational calculation for these market negotiations. The newly appointed Inspector Pierre-André-Nicolas d'Agoult—formerly Bellegarde's second-in-command—calculated the gun's component prices by breaking the manufacturing process down into dozens of analytical tasks. For each he assessed the wage and material costs on the assumption of an annual output of twenty thousand muskets. Inaccurate as his estimation proved to be, it marked another stage in the engineering analysis of production used by Perronet for pin-making. It was a milestone in the evolution of modern management—an innovation that owed more to French bureaucratic rationalism than to the private sector's drive for profits.[71] The role of the merchants had not been entirely eliminated, however. A company of Entrepreneurs was formed, including Carrier du Réal (brother of Carrier de Monthieu), Carrier de Thuillerie (cousin of Carrier de Monthieu), and Benoît Dubouchet. Gribeauval granted them a 9 percent fee per gun, plus a 5 percent interest on funds advanced workers, plus a 5 percent fee for the approximately 550,000 *livres* in fixed capital the Entrepreneurs had sunk into the manufacture. This brought the price to 22 *livres* 10 *sols*—a 14 percent increase. But the authority of the Entrepreneurs over production was much diminished and henceforth in the inspector's eyes, they served simply as local financiers ("*bailleurs des fonds*").[72]

The Gribeauvalists also sought to define the qualifications of arms workers and establish fixed work rules—what eighteenth-century elites called the "police" of the manufacture. As far as the engineers were concerned, the cycle of shoddy production could only be broken by direction from above. Their model for instilling these "simple principles" bears a superficial similarity to the corporate structure of the urban trades. Up until this time, armorers who wished to work for the king had simply registered with the inspector. That was now to change. In 1773—only a few years before Turgot (temporarily) abolished the guilds—Vallière, *fils*, decreed that candidates for the mastership had to submit to the inspector's "theoretical" examination on the firearm, and perform a piece of work (a "master piece") under the supervision of the controllers.[73] The Gribeauvalists amplified this corporate structure when they returned to power. Henceforth, master armorers needed the inspector's approval to take on apprentices. Journeymen were forbidden to change shop without their master's *and* the inspector's consent, and in any case a substitute had to be found. Private arms merchants were emphatically forbidden to "seduce" armorers employed by the Entrepreneur. (Master armorers, however, continued to move freely from private contract work to military production.)[74]

Steven Kaplan has shown how the Parisian authorities of the late ancien régime reinforced the authority of guild masters in order to serve their mutual interests: the guild masters acquired legal controls over their sometimes unruly journeymen, and the state assured itself of social stability in the seething urban scene. But the engineers' work rules went even farther, giving them nominal control over the title to mastership, the quality of production, piece-rate prices, and patterns of employment. As with their "democratization" of artillery promotions, this blend of corporatism and bureaucracy was intended to reinforce the hierarchy, not level it.[75] The Gribeauvalists' purpose in all this was to transform the processes by which guns were made. The northern and southern manufactures still employed distinctive work practices going back generations. The service now insisted on a "perfect uniformity in the three manufactures." The task of preparing the technical means for this synchronization fell to Honoré Blanc. Promoted in 1777 to the new job of chief controller of the three manufactures, Blanc was charged with seeing that all the armories were "provided with the various tools and instruments necessary to assure the uniformity of the work, acceleration of production, and economy in price." After years in Saint-Etienne, he toured the northern manufactures in 1781–82 to standardize their tools and methods too.[76]

The artillerists' commitment to quality over quantity had immediate results. In 1782, Agoult crowed that the musket was "at last achieving the desired perfection." But "perfection" came at a heavy price. Both the Entrepreneurs and military armorers resented the new standards. After all, the cost of rejected weapons came out of their pockets. The output of the manufacture fell from twenty thousand to twelve thousand arms per year between 1776 and 1781 (fig. 5.3). And the workers quit the king's service in droves. In Charleville and Maubeuge, they left for Liège. In Saint-Etienne, they turned to the private market, where prices were rising, thanks in part to the American Revolution. The artillery tried to keep armorers registered for the king's service by increasing the price of guns and by tightening legal sanctions. Gribeauval raised the price of the M1777 in 1781, and again in 1784, for a total increase of 28 percent since 1776. However, the Entrepreneurs complained that their returns had fallen to 6.5 percent, far below what they said they could earn in private commerce. The evidence suggests that the armorers had lost even more. The resulting battle over standards of production—and the artillery's attempt to enforce them with legal sanctions—produced a political revolt that foreshadowed the outbreak of the Revolution in Saint-Etienne. By 1785 the prize armory of the kingdom was in ruins, and the artillery faced a stark choice: lower standards or nationalize the armory.[77]

In each of the following three sections of this chapter, I take up the three principal stages of musket-making: (1) the forging of the gunlock, (2) the forging and grinding of the gun barrel, and (3) the barrel proof. Each of these was a site of considerable friction between armorers and engineers, and each illustrates a different aspect of how the engineers sought to manage production *without* assuming direct ownership of the means of production. My purpose in each example is twofold. On the one hand, I aim to give content to that ever-vague term "the rationalization of production," so as to show how it operated (or failed to operate) in practice, and on the other, to demonstrate how the battle over standards of production set the terms for Revolutionary politics in Saint-Etienne.

The Problem of Command: Gauges and Gunlocks

As we saw in the previous chapter, the Enlightenment engineers—when confronted with conflict in the workplace—sought to replace personal authority with objective measures. To this end, they prepared mechanical drawings and substantiated these designs in physical tools. I also noted that the implementation of these tools was fraught with ambiguity. The individuals who actually wielded these tools in the armories were the controllers, the state-employed "foremen" who ascertained the quality of the finished gun and disciplined those artisans whose output did not come up to grade. These controllers were the crucial links in the chain of command that ran from the state to petty producers. Consequently, the men who filled these posts (including the proof master) were at the center of any conflict between the engineers and artisans. The engineers conceived of the controllers as the NCOs of the manufacture: men elevated from the ranks of the "soldiery" to relay the orders of their superior officers. Just as Enlightenment military reformers curtailed the entrepreneurial role of the army NCOs, so too the artillerists sought to make the controllers subservient to an established hierarchy. Montbeillard called them "the eyes of an inspector"—but added, "that one could say, with the proverb: *You shouldn't always believe your eyes*." Despite a state salary of 1,000 *livres* a year they were entrepreneurial agents in their own right, involved in under-the-table arrangements, long customary at Saint-Etienne. Before 1763, for instance, the proof master collected an illicit 4 *sols* "premium" for proving a military gun and 10 *sols* for a civilian gun. Another 3 *sols* went to the controller of gunlocks. The position had become something of a family legacy, as the list of proof masters between 1717 and 1765 makes perfectly clear.[78]

One of these men ("more interested in his pocketbook than his duties") was a member of the Bonnand family, who had long had quasi-hereditary rights to the position of controller of gunlocks. Vallière, *père*, had winked at such arrangements, and simply asked that the controllers limit their "take" to 2 *liards* per lock.[79] Bonnand informally ranked armorers into three classes and paid them according to his estimation of the quality of their work. This qualitative judgment depended less on the individual lock than on reputation, informal personal ties, and a pattern of delivering good locks in the past. A sign of this personal judgment is that Bonnand obliged locksmiths to deliver the work-pieces to his own house. This discretionary power made Bonnand an important patron in his own right. Control over standards is control over access to the market.[80]

The Gribeauvalists found these activities intolerable. "Premiums" sapped the controllers' willingness to reject faulty arms. It also highlighted their discretionary power. Montbeillard deplored their authority over the workers.

> Also one commonly sees them be mercenary, brutal, cruel to the workers, and demand from their former equals a deference and blind submission to their wishes. . . . The former inspectors allowed the controllers an authority that they often abuse, setting themselves up as the tyrants of the workers and the blood-suckers of the Entrepreneurs.[81]

The challenge for the reform-minded engineers was to recruit controllers loyal to *their* interests. On taking charge of the armory, Montbeillard made a

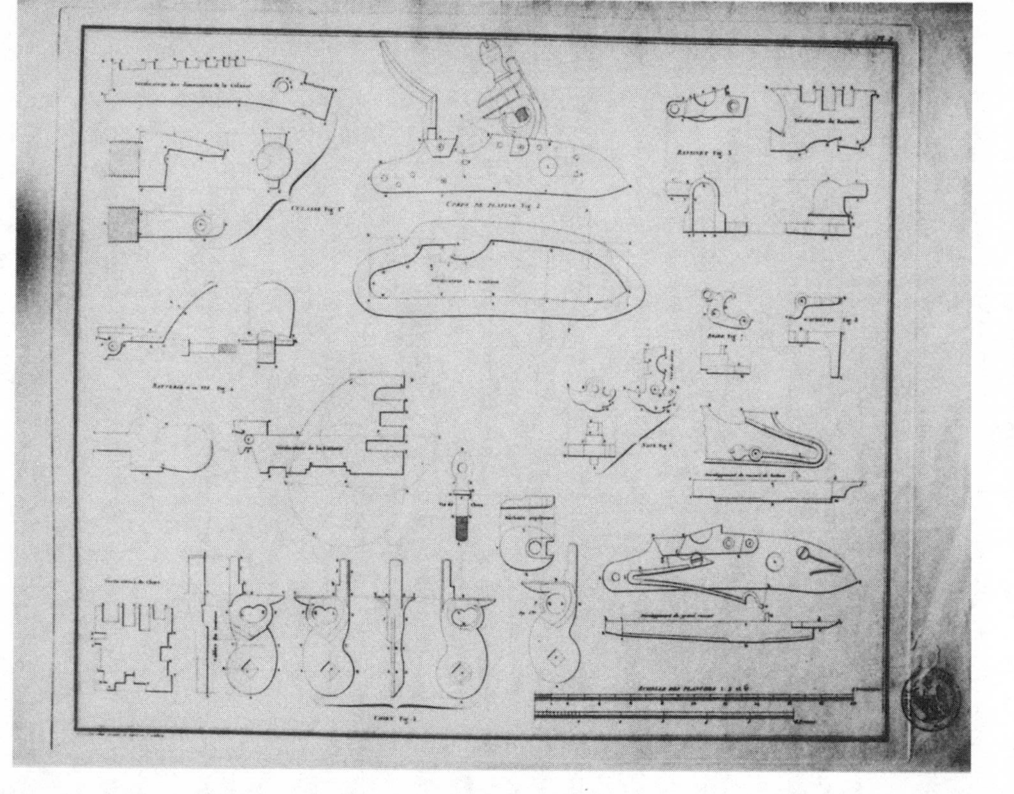

Fig. 5.5. Technical Drawings for the Modified M1777 Musket, Year IX [1802–3]. This series of official drawings accompanied the *Règlements* for the musket model of the year IX [1802]. They show the way orthogonal views were rigorously interrelated. See, for instance, the three views of the breech (*culasse*) in the upper-left corner. They also show how the artillery officers used geometrical constructions to try to define the shape of lock parts. In this regard, see the drawings of the lock plate (*corps de platine*). The two views of the tumbler (*noix*) in the center right are accompanied by a top view of the gauge for the tumbler. Note the notches of the tumbler, its small pivot, the large pivot (arbor), and the square (flats). The final plans were not published until 1805. The drawings were once held in S.H.A.T. 4f22 Min. War, *Règlement fixant les principales*

clean sweep of personnel and increased their salary to 1,200–1,500 *livres*. In the place of Bonnand, he promoted Honoré Blanc, the young master arquebusier he had brought with him from the Charleville armory. When the Vallièrists returned to power in 1772, Bonnand resumed his post and signed up his old armorer-clients. He was then removed and Blanc resumed his post when the Gribeauvalists returned.[82] The deeper problem, according to Inspector Agoult, was that controllers shared the passions, rivalries, and personal interests of the workers "to whom they are bound by blood, by marriage and by friendship." These familial, personal, trade, and class loyalties frustrated the artillerists. They tried transferring the controllers far from their families and friends, but the unique methods at Saint-Etienne meant that only local controllers had sufficient knowledge of the idiosyncrasies of local craft practices to follow the process closely. More generally, the extensive division of labor in the arms trade meant that few armorers were qualified to supervise other subspecialties. Finally, the Entrepreneur usually managed to promote those controllers loyal to him, and these men could hardly be trusted by the inspector.[83] The problem of command, then, was caught up in the social relations of the armory. So in the place of social relations, the engineers sought to substitute objective measures.

This was not an easy undertaking. No master drawings of the M1777 were ever made, only a manuscript list of the dimensions of the gun's parts. Not until 1804 were official technical drawings ever compiled for small arms (fig. 5.5). The challenge in compiling these specifications reminds us of the difficulty of mastering thick objects. Consider the tumbler, the piece which the artillerists referred to as the "brains" of the lock. It was the key piece of the gunlock mechanism because it transferred the force of the spring to the flint when the trigger was pulled. Even after assigning numerical values to nine of the principal dimensions of the tumbler, its contour remained ambiguous. Officers debated whether the length of the "claw" should be 4 *lignes* 9 *points* or 5 *lignes*. This distance mattered; it was critical to whether the tumbler engaged the spring. The officers also encountered great difficulties defining the contour of the claw, which also affected the action of the lock. In 1804, the engineers tried to express this contour using geometric constructions, inscribing it within a circle traced from the pivot. But the claw was not perfectly circular; indeed, it must not be if the lock were to function properly. Blanc had always admitted that lock-making required "much intelligence" from the worker; yet he also claimed that the goal of production was "perfection." The engineers resolved this tension between the goal of uniform manufacture and the unspecifiability of the artifact by defining a tolerance for the tumbler's dimensions with gauges and jigs.[84]

A surviving set of master gauges for the M1777 is housed in the Musée de l'Armée in Paris (fig. 5.6). It is almost certainly the work of Honoré Blanc. These instruments are themselves a remarkable achievement. There was no

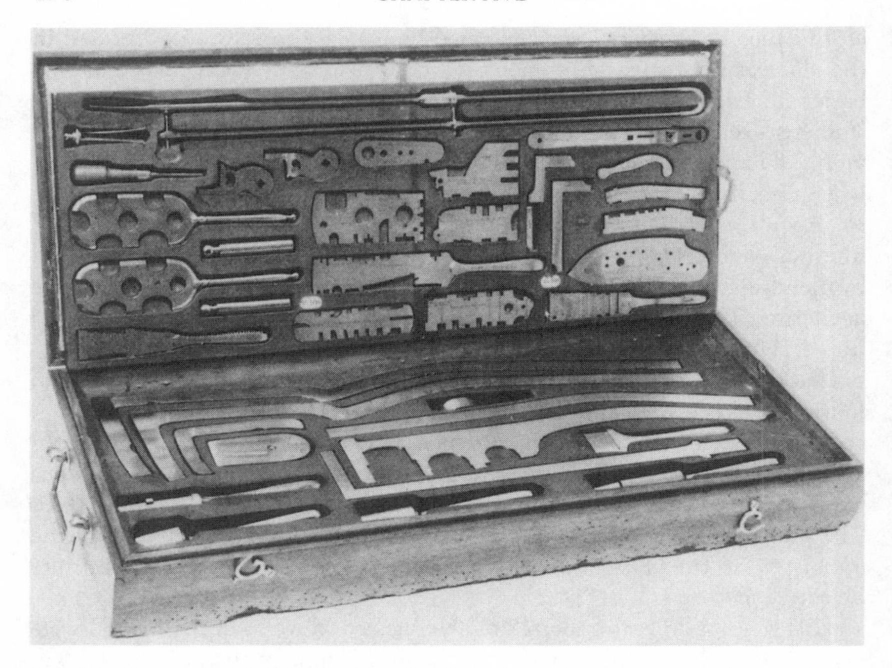

Fig. 5.6. Gauges for M1777 Musket. This boxed set of gauges was used to verify the dimensions of the M1777 musket. This is one of the finest examples of precision manufacturing in the ancien régime, and is probably the work of Honoré Blanc. The gauges are primarily of the go, no-go variety and therefore left some judgment in the hands of the examiner; the calipers allowed for a quantitative reading. From the Musée de l'Armée (P108).

equivalent in the United States (the presumed birthplace of interchangeable parts manufacturing) for another forty years; and in private industry such gauges were still rare in France in the late nineteenth century. Moreover, these master gauges were only the beginning. A set of pattern guns (*types*) was distributed to each manufacture, accompanied by strict orders from Gribeauval to "watch with the greatest exactitude that everything concerned with production . . . conforms exactly [to these regulations]." These models were then used to calibrate a set of gauges and jigs supplied to each artisan by the Entrepreneur. From these, the artisan was to make a set for his own daily use.[85] These sets of gauges defined the physical shape of the gun, and limited the artisan's freedom to control the production process. They also reduced the discretion of the controller. In this sense, Blanc built his supervisory duties directly into these instruments. Since 1763, he had been traveling to the scattered workshops to instruct the locksmiths in the use of gauges. When he became chief controller in 1777, he expanded these trips to include all three manufactures. His goal was to proceed until it seemed as if each worker had "the same gauge." Bemoaning a shipment of inadequate muskets that had somehow got past the controllers, he announced that the gauges are "our guides and ought to be our laws." In September 1782, he returned to Saint-Etienne to begin experiments with the "dies and tools proper to rendering [gunlock] pieces perfectly exact and uniform." In the place of the controllers' discretion, the lock's pieces were now judged against an impersonal and public standard.[86]

To further discipline the shape of the lock, the engineers changed the organization of production. Originally, all the lock pieces had been forged and filed by a master locksmith and his apprentice like any other piece of ironmongery (figs. 7.6–7.9). After 1763, however, the tumblers for military gunlocks were made by two specialists who worked "as near as possible to the manufacture," and were paid by the piece. There, the tumbler's large pivot ("arbor") was rounded with a tool which "had the effect of a lathe." This was almost certainly Blanc's hollow milling machine, the first of its kind. This device, which used the workpiece itself to guide the cutting tool, was later adapted by Eli Whitney in the U.S. The workers also used a filing jig, called a *calibre double*, which Blanc introduced in 1765 to place the square "flats" exactly in the center of the arbor. This filing jig also guided the worker in shaping a small shoulder around the arbor which reduced the friction of the tumbler against the lock plate. Four different gauges then checked the thickness and circumference of the tumbler and the diameter and placement of its axes.[87] The engineers' conceit that the tumbler was the "brains" of the gunlock is revealing. Their decision to remove the production of this piece from the hands of the artisans and place it under their direct authority is an apt metaphor for the way that the engineers hoped to transfer knowledge of the productive process from the atelier to their own offices.

The skills of the tumbler-makers were not thereby made superfluous. The fact that they could be fired for work that failed to pass inspection suggests their job still involved considerable skill. Nor were the other locksmiths suddenly deskilled, though their discretion was now hemmed in by gauges, jigs, and the need to make their pieces function in concert with the supplied tumbler. At the reception room in Saint-Etienne a controller examined the "play" of the lock, and then visited each piece individually with a set of gauges. He then reassembled the lock to see that it still functioned. If so, the worker was paid, and the pieces were marked and taken to be case-hardened.[88]

This new rigor, of course, came at a price. The number of rejected locks rose appreciably after 1777, as did the effort that locksmiths had to devote to production. Both costs came out of the armorer's pocket. Many tried to quit military work and turn to making locks for civilian muskets. The Gribeauvalists increased the price of the lock by 16 percent, but the desertions continued.[89] In 1785, with production plummeting, Agoult ordered his subordinates and the controllers to ease up on their inspections. In fact, the regulations of 1777 had always distinguished between the different degrees of exactness necessary for different parts of the lock "because the liberty given in this regard to the worker was judged useful to ease the production of the piece." But the gauges had proved too tempting to pass up. Dealing day to day with armorers whose livelihood depended on a sale, junior officers and controllers had taken refuge in their independent arbitration. Now Agoult ordered them to see beyond the gauges. "No more tedious minutia," he declared, "no more maximum, no more minimum." He asked his officers not to enforce standards beyond the limit of what might do harm to the manufacture. The gauges did *not* do away with the need for judgment, he reminded them. On the contrary, the "just and enlightened" officer, assisted by the observations of an "intelligent" controller, needed to allow deviations which did not affect the essential proportions of the lock. This put a premium on the scrupulous judgment of the supervisory staff, on "the justness of the *coup d'oeil* of the officer and the know-how of the controller." And yet (for judgments always lie between the two branches of that phrase: "and yet"), the officer must not let too many locks pass inspection because "it is essential that the worker be bounded (*contenu*) [by the gauges]."[90]

Professional status is synonymous with the authority to make judgments. We saw this in the actions of gunners on the battlefield. Whether they aimed by eye or with a mathematical table, the variety of circumstances and the thickness of things meant that only art could help them hit their targets. Similarly, artisanal skills cannot be reduced to routines, nor can the art of inspection. Gauges and tolerances, it would seem, belong to an asymmetrical form of objectivity, an objectivity which appears more stringent the lower down the hierarchy you are. For the artisanal producer they define the limits within which he must work. For the controller, they define the zone within which he is not likely to be called to account. For the manager, they are guides about which he can exercise judgment. Policing work with objective measures only *seems* to substitute impersonal for personal relations. "Objectivity" looks different from different places in the social hierarchy—and so do objects. To the armorer, the gun is the embodiment of a thousand skillful strokes of the file, a batch of iron that had a different odor, a week of quarreling with his journeyman, and means to get a livelihood. To the controller, the gun is a set of specifications that must be met. And to the artillery officer, the gun is an imperfect replica of a thousand other guns that must be supplied to the king's armies, for honor's sake.

The Problem of Productivity: Gun Barrels and Mechanization

The Gribeauvalists also tried to transform work practices through machines, such as the hollow milling machines Blanc developed to cut the tumbler. Can we assimilate the development of machinery to the sort of analysis we just applied to gauges? After all, machine tools are not simply instruments for amplifying labor power, they also mechanize standards of production. Indeed, a main difference between a tool and a machine is its ability to operate automatically; that is, to multiply and reiterate the same standardized action, and thereby make production even less subject to individual discretion. The efforts of the Gribeauvalists to mechanize the making of the gun barrel largely failed, however. Not that the armorers of Saint-Etienne were averse to machinery *per se*—they themselves innovated with tools and methods—but they objected to machinery that shifted control over production.

Making a musket barrel was a complex task involving thirteen armorers and four supervisors. In Saint-Etienne, barrel-forging alone required four artisans: a master forger, two journeymen hammerers, and a worker to insert the mandrel around which the "skelp" (*maquette*) of iron was hammered. In the northern manufactures of Charleville and Maubeuge, the barrel-forger superposed the two edges of the "skelp," then used a triphammer to weld the seam shut. The artillerists thought the advantages of this method decisive. Yet according to the artillery's own data, the armorers at Saint-Etienne forged barrels 18 percent more quickly: in two hours and fifty minutes, as opposed to three hours and twenty minutes. The machines did, however, dispense with the labor of one journeyman. One of Blanc's tasks in the years after 1777 was to introduce this superposition technique to Saint-Etienne. He did not have much success.[91] So in the 1780s, the artillery service brought Charles Pierrotin and his son, barrel-forgers of Liège, down to Saint-Etienne to introduce the use of mechanical triphammers. Larger triphammers had long been used by foundry masters to prepare bar iron for distribution, but this was a machine that directly transformed the process of artisanal production. Armorers at Saint-Etienne acknowledged that the northern technique possessed a finer finish, but they asserted that the method diminished the malleability of the iron, and therefore the durability of the barrel. The Liègeois encountered vehement opposition. In July 1789—a month of industrial action at Saint-Etienne—their workshop was pillaged. Five months later, they complained that the city aldermen had given them six weeks to leave town. National authorities ordered the municipal authorities to protect the Pierrotins, who remained in Saint-Etienne. But in the 1840s, the welding was still accomplished manually, with three artisans wielding handheld hammers.[92]

More telling is the conflict over grinding the exterior of the barrel. This was perhaps the most labor-intensive step of the barrel-making process. It took place in a water mill owned by an Entrepreneur or another merchant, and where the grinder rented space and motive power. Barrel-grinding was dangerous and required great strength. Grinders lay on a high plank and forced the barrel down against the grindstone. They checked their progress against a wooden bar called an *échantille*; its metal edge offered five cut profiles to verify the shape of the barrel at different points along its length. In the eyes of the artillery officers, the barrel-grinders took a perverse pride in their ability and courage. Their hands swelled painfully. Grindstones sometimes exploded. The grinders contracted white lung disease from millstone dust, and few survived into their forties. These hazards and skills needed also kept them in short supply. Consequently, they held a formidable economic position in the manufacture. By the inspector's admission, this one group controlled the pace of production. In part, they did so by controlling the standards of production. By mechanizing new objective standards, the engineers hoped to break the grinders' stranglehold on the production process. As Blanc noted in 1781, the uniformity of the gun barrel's thickness was crucial to the durability of the gun. At the time, the Saint-Etienne barrels were unacceptably irregular. Unfortunately, the proper exterior dimensions of the barrel had not been specified by the regulations of 1777. This needed to be immediately rectified. But rather than use "vague terms which left so much arbitrary," he proposed a new caliper gauge (a *compas pour l'extérieur*). Placing these in the hands of a new category of workers, called *dresseurs*, would expose the "secrecy" which had allowed "a small number of armorers" to flourish.[93]

This caliper gauge was not developed. Instead, the engineers devoted considerable effort to mechanizing this task. Already in the 1770s, the northern manufactures used simple lathes to turn the barrel. In the late 1780s, a Saint-Etienne armorer-mechanic named Jean-Baptiste Javelle invented three machines for turning and milling the exterior of gun barrels (figs. 5.7–5.9). Javelle had been a controller since 1777, collaborating closely with Honoré Blanc. Their state salary freed both men to experiment with building new machine tools. Javelle's machines reproduced the barrel grinders' skills as a three-step process. The first lathe mechanically scored the gun barrel at various points along its length to indicate the depth for milling. This replaced the need for an *échantille* or caliper gauge. The second lathe reduced the iron of the barrel with a cutting tool that moved laterally in a slide rest via a rack-and-pinion feed. In a number of respects, it resembled the 1818 lathe of the American machinist Sylvester Nashe, although the rotation of the barrel and cutting tools were not geared together. Like it, however, the machine was not self-acting and the workman checked his progress against the marks left by the first lathe. Nor did it cut to a variable distance to give the barrel its tapered shape. The third machine tool milled the octagonal facets at the breech of the barrel. This was a genuine milling machine, not like Blanc's hollow milling machine. The barrel, fixed in a metal chassis, slid back and forth along a short measured distance while a toothed cutting tool stripped it flat. To bring the barrel in closer contact with the cutting tool, the entire chassis was raised with a large screw mechanism. One novel feature of the machine was its notch-and-pin guide that helped the worker reposition the barrel to cut eight symmetrical facets.[94]

In his report on these machines, the Ponts et Chaussées engineer, Griffet de la Beaume, noted that a dangerous job monopolized by skilled artisans could now be carried out year-round inside a factory. Although the person tending the machine still needed considerable skill to turn the barrels, the grinders' monopoly would now be broken. For his part, Inspector Agoult expected the lathes to lower the cost of each gun barrel by 15 *sols*, a reduction of about 5 percent. Javelle's lathes received the first patent granted any resident of Saint-Etienne, and during the early years of the Revolution, he briefly opened his own workshop. He could not make a success of it, however. By 1793, he had gone to work for Jean-Baptiste Jovin, a wealthy *négociant* in the process of dominating the national armory. Even then, Javelle's machines were only gradually adopted. The machines themselves were far from perfect; Javelle and his son worked to improve the design for decades. According to artillery officers, resistance to the machines ("jealousies") among the barrel-grinders still prevented them from employing the new technology at Blanc's Roanne manufacture in 1800, even though it operated away from Saint-Etienne. In the 1820s, barrel-grinders still used millstones to finish the barrel.[95]

In sum, the evidence is mixed on the role of mechanization in armaments production in this period. Individual armorer-mechanics experimented with machine-tool design, but they aroused more enmity than imitators. And the engineers' interventions proved insufficient to break the hold of the grinders. But what is intriguing here is what their effort reveals about the engineers' version of rational production. Reproducing the grinders' skills with machinery may have had the net effect of undercutting their wages. But the engineers appear to have been less interested in lowering production costs than they were outraged that the grinders had dared to take advantage of their position to make economic demands at all. This was part of the artillerists' larger assumption that armorers served under their martial authority.

Soldiers or Citizens?: The Politics of the Proof House

During this period, the legal status of an arms worker was in dispute as much as the proper methods of production. According to the artillerists, the few privileges granted armorers who signed up for the king's service—some tax relief and an exemption from the *corvée* and from quartering soldiers—meant that those armorers were also subject to their martial authority. This claim, of course, was intended to give the engineers economic leverage over the arms workers. And the artillerists' claims went farther. In their additional role as guardians of the proof house for civilian guns, they asserted that their martial authority extended to *all* armorers in Saint-Etienne. No aspect of the gun-making trade aroused more contention than the barrel proof. The state justified the proof as necessary to regulate the public sale of this dangerous weapon. Control of the proof house enabled artillery officers to force armorers to use better quality raw materials and take more care in production. The armorers complained that this made their guns uncompetitive with those of

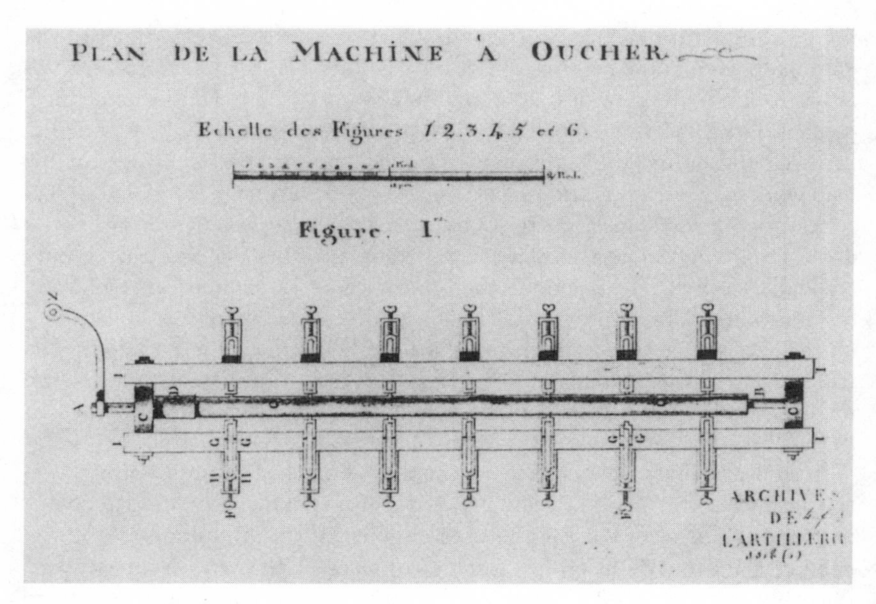

Fig. 5.7.

Figs. 5.7–5.9. Javelle's Three Machines for Turning Musket Barrels. These three machines of Jean-Baptiste Javelle turned the exterior of musket barrels, and were developed in the mid-1780s. The first lathe (fig. 5.7) scored the gun barrel *Q* to the proper depth for grinding using seven pairs of cutting tools *GH*. The second lathe (fig. 5.8) cut the barrel to the indicated depth by means of the cutting tool in slide-rest G, which was advanced laterally using the rack-and-pinion feed *N*. Both lathes were driven by a large human-powered rotating drum. The milling machine (fig. 5.9) cut the octagonal flats at the barrel breech by advancing and returning the entire chassis *H* with a rack-and-pinion feed (crank *OK* and screw *Q*). The pressure of the cutting tool increased as the machine frame *V* was elevated. The notch-and-pin index *E* was then reset to present a new facet to the cutting tool. From S.H.A.T. 4f8 C. P. Griffet de la Beaume, "Trois machines inventées par le Sr. Javelle," 18 January 1787.

Liège, particularly for second-rate guns destined for the Atlantic trade. From the point of view of the arms workers and merchants, then, the proof house was an economic burden. But they also considered it something more: an infringement on their rights.

Local merchants had long asserted an unlimited right to make contracts with whomever they pleased. Since the days of Titon, they had argued that commerce was a private transaction, involving only a seller and buyer. That is why "secrecy" was the "soul of commerce." It shielded the contract from third parties who were not "interested" in the outcome. In such a view, state inspection was necessarily specious. This is a familiar notion of laissez-faire: the free market as the sum of private exchanges. As Steven Kaplan has pointed out, however, this idea of a market *principle* must be distinguished from the ancien

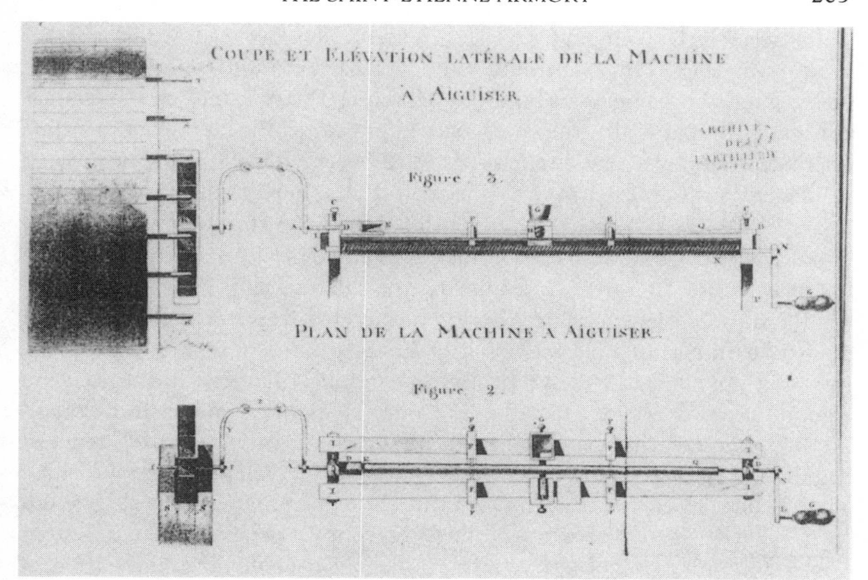

Fig. 5.8.

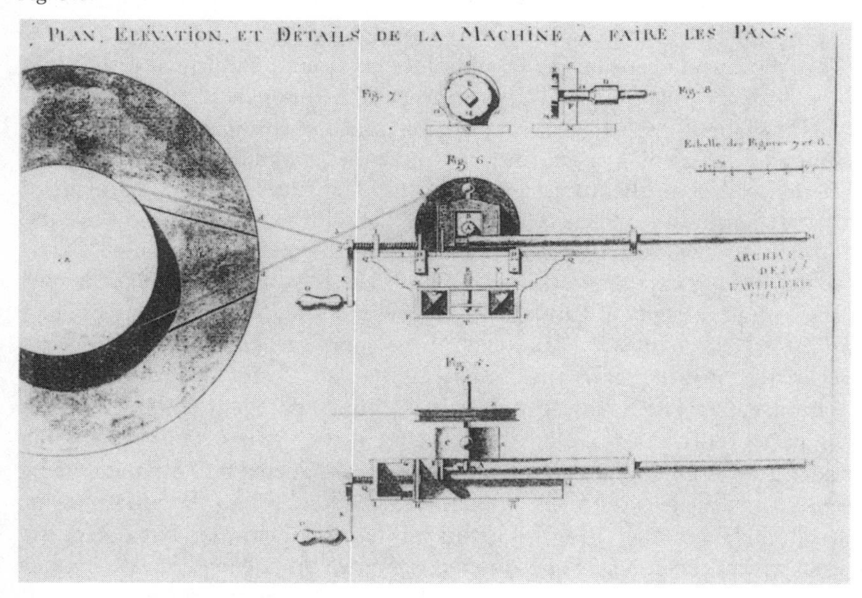

Fig. 5.9.

régime notion of the market as a *place*. In such a marketplace, the state—acting through what in France is called the "police"—could regulate exchanges in the name of the public good. One goal might be to ensure that everyone had equal access to trade. In the case of gun barrels, the proof house was a marketplace which secured the public safety.[96] The challenge was to enforce these transactions. In the early 1780s (as military armorers began to shift production to the private sector) the engineers began to tighten up the strictures on the civilian arms proof. In particular, they criminalized practices long customary in Saint-Etienne. In response, the armorers appealed over the heads of the artillery to civilian authorities in Paris, invoking their legal rights as "citizens."

On 19 December 1780, the armorer-*fabricant* Augustin Merley was thrown in jail. He had spent five weeks in hiding, accused of violating the statutes of the proof house. Returning to Saint-Etienne to protest his innocence, he had been confronted by the *maréchaussée* and ordered to pay a 200 *livre* fine (3 months' earnings for a master artisan). When he refused, he was imprisoned. Inspector Agoult contended that Merley's unusual gun breech disguised the amount of powder loaded into the weapon, halving the amount in the proof. Merley protested that he had been experimenting with a gun of "reduced recoil" that imparted twice the motive force to the ball, and that one of his laborers had "by accident" submitted this barrel for proving. He was, he said, "perfecting his art." (New forms of "false" breeches *were* being introduced from England at this time.) Merley further objected to being accused without a hearing. The proof master who made the accusation of malfeasance and took the suspect barrel to Agoult was Merley's own brother, the proof master Jean-Baptiste Merley.[97]

A new law concerning the proof house had been promulgated earlier that year, on 10 February 1780. Mostly, it repeated the litany of requirements from older laws, but for the first time it gave the inspector strong powers of enforcement over the civilian proof: the right to impose heavy fines, imprison offenders, and call upon the *maréchaussée*. For the first time, the law also required civilian gunsmiths to break rejected barrels, as military armorers had since the 1760s. The law further stipulated that all arms shipments be accompanied by special passports. Merley had already petitioned the district judiciary in Montbrison and the regional Parlement. Now he protested his innocence to chief minister Necker himself. On behalf of his comrades, he averred: "We have always regarded the proof as voluntary on our part."[98]

In fact, the state had been imposing ever more stringent requirements for the past twenty years. Earlier in the century, the civilian proof, nominally under the authority of the inspector, had been conducted on a separate site by someone delegated that task. The armorer paid a "fee" to the proof master, despite his state salary. Ostensibly this was for his gunpowder, but it kept the proof within the compass of economic exchange that defined social relations in the civilian sector. Complaints against defective guns had long ago led Vauban to plead that "at the very least the proof must be carried out by someone not [financially] interested in production." Standards had to be impersonal. After 1763, civilian guns were to be proofed by the same personnel and on the same site as military firearms. They were also subject to the same "double proof" which uncovered many more faulty barrels. These moves occasioned vehement protests from "some 90 families, all having the same rights to commerce." Petitions addressed to the controller-general of finances, and cosigned by the mayor and aldermen, conceded that without some sort of proof certain *négociants* would undermine the reputation of the town and sell shoddy guns. What they wanted was an independent civilian proof under the authority of their syndics. But Inspector Montbeillard dismissed the "syndics" as a fiction. The town of Saint-Etienne, he noted, had no formal guilds. "Humanity," he said, "demands the proof." In response, the civilian authorities conceded that only the military inspector could prevent abuses. And Choiseul, the minister of war, had the town's petitions burned in front of the town hall.[99]

The armorers and merchants, however, had continued to evade the proof in other ways. And in response, the artillerists had devised methods to prevent cheating. So, over time, the proving of a gun barrel became a highly choreographed activity, a set of precisely mapped procedures. These rules can be read as a codification of mistrust, the accretion of years of conflict between economic agents and state bureaucrats over the terms of a transaction. Imposed by the state as the conditions for acceptability, these procedures reflect the state's lack of direct control over the work processes in the armory. (If the engineers had controlled production, they would not have needed codified standards.) Reluctantly accepted by armorers as the conditions for sale, the standards forced changes in the way they made guns. (The state was sometimes the only customer in town, and one with particular powers to enforce its demands.) This elaborate ritual, then, defined the artifact as an object—an impersonal commodity which could be exchanged for specie. A brief summary conveys the elaborate nature of this test.

Prior to the test, the proof master examined the barrel for defects. He dried the interior with a rag "in case the worker has maliciously slicked it with grease or garlic." He then tested the barrel with gauges. After 1777, for a military gun to pass inspection, a short steel cylinder of 7 *lignes* 9 *points* had to slide easily up and down the interior, while a cylinder of 8 *lignes* could not enter at all. Nonconforming barrels were milled flat along one side for 15 *pouces*, which did not prevent them from being submitted for the Atlantic trade. For the first test the proof master introduced a quantity of powder equal to the weight of the ball (18 to the *livre* for military guns). Controllers verified that the test shot was round and correctly wadded. Under the close supervision of an officer, the armorers then loaded the barrels, tamping the shot by letting the rod fall "of its own weight" three times. The barrels were set horizontally along the proof bench, their breeches touching a metal band, and fired simultaneously. Afterward, each was rod-rammed in the presence of an officer to ascertain whether it had actually fired. If not, it was fired again. The barrel was then examined for fractures. After a second proof—with one-fifth less powder—the officers and controllers again examined all barrels, rejecting those with any evidence of cracks. If the cause was due to the poor quality of the iron, a better batch was procured. Surviving barrels were then washed in the river and branded with the official proof mark. Military barrels, however, were not to be let out of the sight of the proof master—"it being of the greatest importance never to leave them at the disposition of the workers." An assistant of the proof master, called an examiner, then dried the barrels against a large stove and checked them for pits and crevices. He again used the two gauges to check the caliber. He also checked that the breech plug and chamber engaged well, and did not rattle. After another series of operations by workers (filing the breech, trimming the muzzle, attaching verges, and polishing the exterior), the barrel was stored for a month in a locked humidity chamber. The final exam always took place in the presence of an officer. If the humidity had highlighted any flaws, the barrel was marked so that it could not be resubmitted. Accepted barrels received a proof mark bearing the current year and were distributed to the finishers who assembled the musket with the stock, mountings, and gunlock.[100]

By Gribeauval's own admission the stringency of the new proof increased the number of rejected barrels from 15 percent to between 20–25 percent. One barrel-maker, Jean Lullier, spelled out how this made the difference between a living wage and starvation. If he forged twenty-four barrels in a six-day week at 4 *livres* 5 *sols* each, after deducting the cost of iron, coal, tools, and 12 *livres* for two journeymen, he was left with 3 *livres* 10 *sols* a day—a substantial wage. But because the Entrepreneur retained 10 percent per barrel against rejects, he actually earned only half this; an adequate 1 *livre* 15 *sols*. This, however, assumed a rejection rate of 20 percent. (Half the cost was born by the armorer and half by the Entrepreneur.) Now that rejection rates were well above 25 percent, he would take home less than a *livre*. Worse still, because he did not yet know the rejection rate, he might well end up owing the Entrepreneur money. *Fabricants* who supplied complete barrels were even less happy. Their twenty-four barrels a week netted them 156 *livres*, but out of this they had to pay 44 *livres* 8 *sols* to have them ground, bored, and finished, plus buy twenty-four breech plugs for 9 *livres* 6 *sols*. A rejection rate of one-in-six (17%) meant a take-home pay of 26 *sols* a day. But at the 25 percent rejection rate, a *fabricant* would earn 5 *sols* a day, less than the least skilled ironmonger.[101] Establishing the same standards for the military and civilian guns prevented armorers from shifting their labor to the civilian sector.

Not until the law of 10 February 1780 was the full rigor of this examination felt by armorers, both military and civilian. That April a "sting" operation by Agoult's second-in-command caught royal armorers secretly working on barrels for private sale. Some hid the goods; others had them forcibly confiscated. The barrel-maker Jean Lullier complained: "We made those barrels because we are starving while working for the Entrepreneur of the king. . . . Where is the ordinance coming from his Majesty which treats us like slaves? We are citizens, fathers of families, subject to all the charges of the state."[102] Lullier's and Merley's petitions brought to light a whole series of imprisonments imposed by Agoult since 1777. They included fourteen prominent master armorers from the Lyonnet, Javelle, and Gonon clans, "and there were others, but the inspector is very lax about legal formalities." Armorers had been jailed for shoddy workmanship, for refusing work, for selling military muskets on the private market, for seeking work elsewhere, for absence from work, for drunkenness, for laying off a journeyman, and for simply talking back. Pierre Gizbach and Antoine Gourgoillant were jailed for having tried to seek work at Tulle: "But if their labor does not furnish them with the means of subsistence, humanity speaks in their favor; they ought not to have been treated like criminals." A few months later, on the last day of the carnival, Lullier was thrown in prison for fourteen days when an artillery officer overheard a "witticism" he made to one of his comrades.[103]

At the heart of the problem was the artillery officers' conception of the workers as soldiers and therefore subject to military discipline. This discipline enforced economic relations: proper work procedures, proper hiring practices, proper commercial transactions. By criminalizing customary practices—as Peter Linebaugh has remarked—the law stripped laborers of control over their products.

> It is common that when a worker is in debt to the Entrepreneur, Inspector Agoult puts him in prison. The prisoner then has to pay the jailer. Workers should not be treated like soldiers.[104]

These men, however, were neither soldiers, nor helpless day laborers. They were elite artisans, among the most prominent persons in town. Enraged, the local municipal council took their part. It sent the *négociant* Jovin to Paris to appeal over the heads of the military engineers to the civilian authorities and ask them to "tear the thick veil that covers the manufacture." Armorers—even military armorers—paid taxes, had standing with the civilian courts, were domiciled in town, "and consequently belong to the body of citizens (*font corps avec les citoyens*)." The municipality thereby invoked a notion of citizenship tied to the traditional privileges of city dwellers, but which also pointed to their long-standing demand that economic freedoms not be shackled by legal impediments. This time they got a hearing.[105]

As was often the case in the ancien régime, the different branches of the central state worked at cross purposes. The civil authorities went to battle with the War Office. They agreed that the state must regulate a trade which "exposes the life of a citizen." But in demanding the "greatest perfection" for civilian as well as military guns, Agoult had gone too far. The army had no right to extend a "military discipline" over Saint-Etienne and imprison workers without judicial hearings. In his own defense, Agoult wrote that the law had been passed by the civilian intendant at Lyon. Each article had been read at the town hall before the mayor, the aldermen, and the majority of armorers—and approved. In reply, the armorers objected that neither the judge, the fiscal procurer, nor the municipal secretary had been present at the town hall meeting, and most of the prominent arms merchants had been away at the Bordeaux fair. Only Agoult had been allowed to speak. The new law had neither been posted or published. Since when did the military have the right to control commerce?[106]

In August 1781, the minister of war ordered Agoult to cease his surveillance of the civilian barrel proof and to moderate his discipline of the armorers. Then on 17 January 1782, the new controller-general, Joly de Fleury, passed an ordinance definitively separating the military and civilian proof. Henceforth, civilian barrels would have to submit to only one proof; the amount of powder would depend on the caliber of the barrel tested; and those barrels which failed the test would be returned to the worker. Most important, an independent civilian proof master would be selected from a list proposed by a new municipal body called the "syndics of the proof house."[107] This body would be composed of arms merchants and leading master armorers. The first set of syndics were prominent arms merchants such as François Jovin, Benoît Penel, Fleury Royet, and Jean Carrier du Molard. And the proof master they selected was none other than the aggrieved barrel-maker Augustin Merley, "whose honesty is known to all the assembly." He retained his functions until 1811, and the proof of civilian guns remained in municipal hands throughout the nineteenth century.[108]

The civilian gun trade thereby acquired the sort of coordinating institution which Sabel and Zeitlin have pointed to as a central feature of industrial districts. The proof house was not the vestige of some corporate heritage stretching back into the hazy mists of "tradition." Nor did it emerge from a spontaneous association of property owners seeking to better their lot through formal ties. The proof house was born in conflict. It emerged at the intersection of the state's interest in regulating trade *and* the desire of local producers to seize those regulations for their own purposes. The triumph of the civilian proof house was a victory for a particular class of producers, the *fabricants* and master armorers, as against the military engineers and their sometime allies, the large merchant *négociants* who ran the manufacture. It represented the first time in Saint-Etienne that local notables had seized control of economic institutions from the central state. In doing so, they had articulated the legal rights of economic agents, as against the administrative interests of the state. They thereby acted upon a logic which was soon to be enshrined in Revolutionary rhetoric: that citizenship followed from productivity. That victory was to prove the harbinger of the municipal revolution which was to sweep Saint-Etienne—and much of urban France—by the end of the decade.[109]

Nationalization and Machine Production

But before turning to that municipal revolution, we must first consider the consequences of this economic rebellion for the Royal Manufacture. Gribeauval accounted the creation of an autonomous civilian proof house a disaster for the state's ability to procure military arms. No longer could the inspector prevent armorers from being "seduced" by the tiny price advantage of civilian production. What would happen, he asked, when the state needed to increase production during war? "All our plans are based on this principle [of] subordinating commerce to the military manufacture," he complained.[110] The question was what to do about it.

Gribeauval promoted Inspector Agoult to become director-general of all manufactures. His replacement, André-Charles-Emanuel Danzel de Rouvroy was a man of "the greatest exactitude" yet also "gentle and humane." Yet he too resorted to martial law to enforce standards. Those who failed to furnish promised work—"and done with the requisite precision"—or who ignored reprimands, were jailed. In a display of its authority, the artillery had Louis XVI himself sign a three-day arrest warrant for the notorious Joseph Bonnand (from the family of former controllers known for taking "bribes"), because he refused to release a manual laborer and had "raged against the regulation and authority of the officers." By Danzel's own admission, the town prison soon overflowed with "drunk and debauched" armorers, and he proposed building a special detention room for them.[111] But a new edict of 7 July 1783, which seemed to expand the inspector's police powers, further undermined them. Armorers had to obey inspectors and controllers in "all that is relative to their work." And artillery officers could detain for twenty-four hours all "workers who have the gall to insult or menace an officer or controller. . . , [or] who incite revolt or a uprising." But lengthier punishments had to be approved by the minister of war, and workers had the right to appear before local judges. In the place of arbitrary martial authority, this was a cumbersome procedure.[112]

It was at this junction that Danzel proposed running the armory on the *régie* system. The state would buy out the Entrepreneurs, borrowing money at 5 percent, far less than the Entrepreneurs' profits. A nationalized industry would operate from a centralized factory; armorers could be hired on a contract basis; the account books would be properly supervised; and artillery inspectors would manage the whole business professionally. This was the logical outcome of a 150-year history which increasingly put the state in direct contact with armorers. In the Entrepreneurs' counterproposal, they agreed to sell their buildings to the king and function as suppliers of raw materials, money lenders, and administrators. For this they asked a fixed commission. This was the logical outcome of an equally long effort to achieve risk-free profits; known today in military-industrialist circles as "cost-plus."[113]

In a lengthy memorandum to the minister of war, Gribeauval rejected both approaches. Not that he underestimated the chaos at the manufacture. The rise in standards (and drop in production) had caused misery among the workers. "The perfection demanded and necessary in all the parts of the firearm only permits the production of a small number [of guns], which can only produce a small profit." Not that he condoned any relaxation in standards; that would be "dangerous" for the soldiers and force costly repairs down the road. But nor would he recommend a *régie*. With only a few superior muskets to be made in the future, the king would be unwise to buy out the Entrepreneurs now. Besides, Gribeauval did not want artillery officers to negotiate the sort of commercial transactions a *régie* would entail, such as selling rejected gun barrels to the slave trade. The government's experts should retain their authority over the design of the firearm, the registry of workers, the techniques of production, and the standards of reception—but the capital and commercial responsibility must remain in private hands.[114]

The problem was to keep the Entrepreneurs in the business. Gribeauval's stopgap solution was to decouple the quality of production from their rate of return. He proposed a "capitalization" fee (15% of their 550,000 *livres* of capital), plus a 10 percent profit on each gun. This shows the steep price the French state was willing to pay to avoid nationalization. But what about the armorers? Gribeauval confessed the army would soon reach its full complement of muskets and revert to replacement levels, perhaps halving purchases yet again. Under those circumstances, the armorers would starve. All the state could do was reaffirm their tax exemptions, reduce their raw material costs with tariffs against the export of iron, enforce their monopoly on selling military weapons abroad, and hope that once they were habituated to the new techniques the number of rejects would drop—as they had in the arsenals of construction. But in the long run, Gribeauval could only offer a techno-fix solution that would replace the need for workers' skills altogether: the manufacture of firearms with interchangeable parts. He therefore ordered a trial development of "the means recently invented by Sieur Blanc"—the subject of chapter 6.[115]

For all its sophistication, Gribeauval's solution adheres closely to the corporatist reflex of the ancien régime (and of military procurement officers everywhere). Gribeauval's main concern was to insulate his supplier from the free market, that ruinous force which cut corners and cheapened goods. He recognized that the two sectors were bound together, that civilian producers served as a repository of skills and capital for times of crisis. But he did not want his officers corrupted by the commercial world. And he was only tangentially concerned with price, and then, mainly in deference to ministerial pressure.

The new regulations proved disastrous. The price of iron was rising: between 1778 and 1784 it had gone up 15 percent; between 1784 and 1788, it rose another 50 percent. Not that the Entrepreneurs cared; they received an annual payment of 75,000 *livres* even when no guns were produced. And they had never lost their right to trade in civilian arms. In the last five years of the ancien régime, Saint-Etienne produced essentially no military muskets—though the number of armorers officially registered for the king's work had only dropped in half. In 1786 Danzel de Rouvroy resigned as inspector. His replacement, Augustin de Lespinasse, presided over an empty shell.[116]

The Municipal Revolution

In many recent histories of the French Revolution, modern politics burst unexpectedly upon the scene in 1789. Revisionist historians have argued that, until that date, dissatisfied French men and women filed grievances with the bureaucracy or engaged in philosophy, but did not "formulate their interests, organize to demand changes, and mobilize to achieve their goals." When it finally did occur, this flurry of revolutionary political activity first began in urban centers. Townspeople first tasted the maneuverings of representative government when they elected delegates to the Etats Généraux. Then all through the summer of 1789, rising food prices and rural unrest inspired urbanites to form emergency committees. These supervised a local militia and monitored bread distribution. This "municipal revolution," as Lynn Hunt has described it, had occurred in twenty-six of France's thirty largest towns by August 1789. But even though this local seizure of power fed upon the crisis of legitimacy in Paris, the phenomenon did not simply radiate outward in imitation of the capital. The actions in some towns (Marseilles) predated the revolution in Paris; others occurred in towns far from Paris before those near the capital. The municipal revolution merits attention precisely because it signals the first break from the administrative practices that governed political life in the ancien régime.[117]

Hunt argues that wherever it occurred, the course and intensity of the municipal revolution was shaped by local socioeconomic structures. In particular, she cites the relative openness of the traditional elite to the new elites (professionals, merchants, and master artisans), and the relations between the new elites and those further down the economic order (other artisans, outworkers, and peasants). In almost all the municipalities she studied, the new leaders claimed legitimacy in the name of the commune, but "stopped short of committing themselves to active support of the crowd." Yet Hunt mainly emphasizes the political nature of the rupture: how the new municipal institutions created a "space" for new elites to participate in public life and gave at least muted voice to certain members of the lower orders (other artisans, wealthy peasants). And in her later work she has stressed cultural processes as much as structural explanation, symbolic representations as much as the play of material interests. She has eschewed both the disembodied politics of Furet's rhetorical rupture and the interest-driven politics of Soboul's nascent class struggle. Instead, she stresses the unstable relationship between the various factions of the bourgeoisie and the contingent process by which new municipal elites periodically renewed their membership with groups of ever lower social standing. She concludes that "insofar as the bourgeoisie made the Revolution, they made it as the agents of the town rather than as the agents of a new capitalist order."[118]

Yet there were also important continuities between the economic conflicts of the late ancien régime and early Revolutionary municipal politics. Public discussion may have acquired a constitutional cast only after 1788; hardly a handful of speculative philosophers questioned royal authority before then. Nor was participation in public life—as either an elected representative, a voter, or a member of a crowd—conceived in participatory terms before 1789. In emphasizing the political rupture of 1789, Hunt has been shying away, as have the revisionists, from exploded materialist explanations for the origins and unfolding of the French Revolution. In a series of papers in the 1960s, George Taylor effectively undercut the Marxists' assumption that there existed an industrial capital-owning class in the late ancien régime. His successors have pointed out that the republican leaders cannot be easily reduced to an identifiable economic interest. Increasingly in the past decade, the revisionists have made politics largely independent of economic position.[119]

But what if the error of the materialists lay not in drawing a connection between politics and economic structures—a connection for which Hunt holds out hope—but in our image of what constituted economic development and material culture? From this perspective, the "missing" class of industrial capitalists becomes irrelevant. The Revolution may not have been initiated in the name of the factory, but that does not mean that it was not fueled by those seeking to create new institutional forms to regulate production. Freed from teleological assumptions about the necessary direction of industrialization, we can see how conflict over the terms of material culture shaped both Revolutionary social agendas and their political expression. That was certainly the case in Saint-Etienne.

Though Saint-Etienne does not quite make it onto Hunt's list of France's thirty largest towns, it was typical of the larger pattern. In Saint-Etienne, too, constitutional questions were first debated by local notables in the regional assembly of 1787. The creation of an "intermediate committee" allowed men allied to the educated scions of the merchant class to retain considerable authority over municipal public security through July 1790. Meanwhile, the aldermen seized control of the town council from the appointed noble mayor early in 1789. Yet these groups were soon under pressure from those below them on the social scale. How they handled that pressure shows continuities with the allegiances formed in the battle over the barrel proof.

Almost immediately, popular movements in Saint-Etienne took the guise of industrial conflict, rather than a food riot.[120] The protests that rocked the town in the second half of 1789 were a continuation of the conflicts of the 1780s, with the standoff in Paris suddenly licensing the reversal of that decade's grants of state-led monopoly privileges. The royal concession granted Osmond's large new mining operation in 1786 was rendered null and void by the crowd action of 24 July 1789; an action in which the municipality was complicit. That same month, a crowd of armorers, again with the municipality's consent, threatened to run the Pierrotin family out of town for having introduced mechanized barrel-forging with triphammers. And in September, Sauvade's fork factory was destroyed.

In demolishing Sauvade's mechanized fork factory, the petty commodity producers of Saint-Etienne defended the existing structure of the trade. Like the contemporary Luddites of western England, their objection was not to machinery *per se*, but to machines which caused "redundancy."[121] Jacques Sauvade's factory on the Furan River used stamping dies to punch-cut a variety of small metal goods on sheets produced by three rolling mills. By means of slight modifications, he hoped to manufacture forks, the *rosettes* of knives, the frames of buckles, shields of locks, and flat bolts. Then, just as the factory was ready, a crowd of local fork-makers gathered outside on the night of 1 September 1789. Municipal authorities arrived to calm the populace. Sauvade himself offered to "delay perfecting his establishment until the people believed it offered some hope of employing workers, and if not, then desisting [from his innovations]." In the meantime, however, he let the cylinders be taken to the mayor's office for safekeeping. Satisfied, the crowd disbanded. However, the precious cylinders were somehow "lost" along the way, and the next day another crowd systematically dismantled the factory's power train and mechanical devices, and threatened Descreux—an artisan who had helped Sauvade. This incident should not be viewed as an isolated act. Though the crowd's actions were licensed by the Revolution, they belonged to a local tradition of artisanal resistance to mechanization. And according to Sauvade, elements within the merchant class had silently encouraged the destruction. As for the municipality, it equivocated, praising Sauvade's mechanical ingenuity, but refusing to punish the perpetrators.[122]

The next target was the Royal Manufacture. During the Great Fear, the beleaguered aldermen had withdrawn 2,633 firearms from the stores of the Royal Manufacture. The brigands never materialized, nor did some 800 muskets.[123] More serious was the outburst of November 1789. Then, a group of arms workers, led by the armorer Claude Odde, accused the Entrepreneurs of shipping muskets abroad to the enemies of the people. The day after the municipal government arrested Odde, a large crowd demanded his release and confronted the bourgeois militia, whose commander, the Baron J.-B. Bernou de Rochetaillée, was fatally trampled. The next day, they pillaged the manufacture, making off with the entire stock of 5,612 muskets. Municipal officials, state officials, and artillery officers were all "proscribed," and fled town. Odde returned to Saint-Etienne in triumph. Inspector Lespinasse noted that "the workers are the masters now and they know it." Within a week, however, the authorities had returned and the rioters turned in the stolen firearms. According to Lespinasse, "The journeymen of various fabriques had risen against their masters. As a result everyone had seen the need to disarm the people."[124]

This analysis of class politics in Revolutionary Saint-Etienne is borne out by the composition of the new municipal government. It consisted of elite master artisans and small merchants, a coalition which headed off industrial conflict at Saint-Etienne for the next few years. Faced with this sort of "industrial action," the aldermen convoked syndics of merchants and master artisans. These syndics were filled by the same personnel that had seized control over arms production in the 1780s. With Paris in disarray, they seized the political upper hand, and moved quickly to assert their economic interests.[125] The arms industry controlled 30 percent of the members; and the armorer Noël Pointe, later Saint-Etienne's representative to the National Convention, made his first political appearance here. Their authority was such that Lespinasse needed their permission to ship the rest of the manufacture's guns to Lyon.[126] Needless to say they took a dim view of the manufacture. The rhetoric of laissez-faire was at full voice in Paris, and could now be turned directly against state institutions. In September 1790, the municipal council stopped payment for lodging the inspectors and controllers, and demanded that the manufacture be closed definitively.

> Considering that the Royal Manufacture of arms in this town is one of those corrupt establishments which concentrates in only a few exclusive hands a branch of industry and commerce, [giving rise to a situation] onerous to both the People and to the State, such a regime cannot be continued under the new order of things.[127]

The municipality also took aim against the standards so carefully erected by Gribeauval. They repeated their long-standing accusation that such standards infringed their latitude as free citizens to make economic decisions in their own best interests. Under pressure from local producers, the National Assembly ordered its Military Committee to review the design specifications of the M1777, and ease the requirements in the hopes of boosting production. The Committee agreed to reconsider all those ". . . minute details whose execution has nothing to do with the making of the firearms, [and which] have so greatly hindered manufacturing that [the workers] can only make a third the quantity possible if those shackles were removed."[128]

A new musket, designed the No. 1, was now authorized. It adapted elements from various earlier models, and cost about 15–20 percent less than the M1777. Shortly after the declaration of war in April 1792, Saint-Etienne grandly offered to furnish eighty thousand to one hundred thousand of these guns a year, but only if the War Office relinquished its exclusive contracts with a few monopolistic Entrepreneurs, recalled its "perfectly useless" inspectors, and renounced its obsession with precision manufacturing.

> What does it matter if the stock is more or less fashioned, more or less thick or light, so long as the barrel, the gunlock, and the breech offer the former solidity. . . ? Why hesitate to give commissions to zealous citizens whose reputation for the production of arms is not in the least doubtful?[129]

That summer, with war underway and the price of guns rising exponentially, the state conceded defeat. The municipality of Saint-Etienne was authorized to appoint a Commission of Verification to purchase armaments "from any individual," but only at a fixed price and in conformity with published standards. In theory, the guns would still have to submit to two proofs, and have their locks disassembled and tested to see that they functioned properly. (The army was composed of citizen-volunteers now, after all!) But even this relaxed examination (no gauges, no humidity chamber) was circumvented. Composed of two local functionaries, an artillery officer, and two armorers, the Commission of Verification was largely ineffectual. So began a five-year hiatus in the artillery's control over the armory.[130]

It hardly mattered. The manufacture was deserted, its monopoly meaningless. The government price was half the black market price. Every day emissaries from Languedoc, Provence, Burgundy, and elsewhere contracted for muskets for their national guards. The king of Sardinia even attempted to purchase ten thousand muskets to arm the émigré army massing against France. The central government had lost control over its local administrators, and hence over their demands for armaments. The army's arsenals were scandalously empty, and political factions began to look to the War Ministry for scapegoats.[131]

In discussing this first federalist phase of the French Revolution, it is tempting to conflate the growing demands of local elites for economic autonomy (the quest for decentralization) with the revolutionary attacks on the ancien régime's system of royal privilege (the call for the rule of public law). In practice, these two goals went together in towns such as Saint-Etienne. There, arms merchants and armorers resentful of the monarchy's legal restrictions on trade were also those groups which had been forcibly excluded from lucrative military contracts. From their point of view, legal privilege and economic monopoly (and the objective standards which enforced them both) emanated from Paris. In this context, local political control meant local control over standards, and hence, relative commercial freedom. That is not to say that these merchants and armorers were the shock troops of industrial capitalism. Rather, they sought to reduce the central state's role in the regulation of exchange in order to substitute local supervision. In doing so, they translated their economic grievances into a form of political action that the central state had to take into account. The net result was something akin to capitalist decentralization.

Toward Patriotic Production

In 1792–93, the artillery service's worst nightmares were realized. The armories on the northern frontier were overrun by foreign troops, and Saint-Etienne's armorers and merchants refused to supply guns to the state. In September 1792, Minister Roland could only procure three hundred muskets with the fortress of Verdun on the verge of surrender. Despite thirty years of effort by the engineers, gun production had remained in the hands of merchants and petty commodity producers driven by commercial considerations. Up until 1782, the state had been gaining leverage over these groups. It had created institutions to control its purchases, the supply of skilled labor, and the standards of production. And it had backed all this up with formidable coercive powers. After 1782, the revolt of the armorers and the penchant for free trade dominant in elite government circles had allowed those institutions to fall under local control. By 1785, the royal War Office had lost control over the production of war matériel, and the republican state could not get it back in 1792. At the same time that Gribeauval's artillery officers drove the Prussians from the field at Valmy, they had been routed in the industrial battle at Saint-Etienne.

Yet over the next few years, the French state did regain some measure of control over the production process. It did so, firstly, by reasserting its authority over local organs of government and monopolizing all purchases of guns, both private and military. And secondly, it did so by aligning itself with the social class of artisans and *fabricants* now pivotal in municipal affairs. To consummate this alliance, the new emissaries of the Revolutionary state translated democratic decision-making into the realm of production. Ultimately the state was to demand that production be a patriotic act. Experimenting with new quasi-consensual institutions to organize production enabled a new national armory to expand the scope of its operations, eventually enveloping both civilian and military gunsmiths in a national regime. But the price of doing so was to foreclose the artillery service's ambitious plans to mechanize and centralize production. It signalled the end of Gribeauval's attempt to impose uniform production from the top down.

The first step toward this mode of patriotic production was initiated by Gilbert Romme, the Jacobin mathematician sent to Saint-Etienne as a representative-on-mission in late September 1792. Upon arrival, he decreed that all private trade in firearms cease, that the central state have sole power to purchase muskets, and that all muskets conform to the M1777 or No. 1 model (with barrels properly proofed and locks verified). "The citizens of Saint-Etienne," he announced, "are enjoined to stop the brigandage that has entered into the manufacture and the trade in arms; they must silence the egotism and greed of certain bad citizens." He also devised a novel democratic mechanism to set the price of muskets, control the quality of production, and select the responsible personnel. Each local commune with resident armorers elected deputies to a Council on Armaments. After some haggling over who should be president of the Council—Romme proposed artillery officer Claude Gabriel de Fyard—Thomas Berthéas, an arms merchant was elected. Etienne Bonnand was elected first controller, and Jean Bonnand, first examiner. Jean-Baptiste Merley-Berlier and Augustin Merley (the former civilian proof master) were given responsibility for gun barrels. A few weeks later, the guard of the storehouse was replaced by Jean-Baptiste Bonnand. Thus, the central government had conceded control of the manufacture to the coalition hostile to the Gribeauvalist program. Not surprisingly, the commissioners used their control to prevent their rivals' guns from coming to market. To control standards is to control economic power.[132]

The crux of the problem, however, was that the government's price for guns was uncompetitive. Before Romme left town he set the musket's price at 42 *livres*. But when the Council met that month to distribute this price among the various arms-making specialties, rebellion broke out. Democratic deliberations and patriotic exhortations would not induce the armorers to accept prices below market rates. On October 30, the locksmiths' delegates from Saint-Héand, Villars, and the other surrounding towns demanded piece rates the Council found excessive. The next day, the same thing happened with the breech-makers and reamers. On November 2, no workers were invited, even though the sessions were supposed to be open to the public. The Council pushed through fixed prices. Consequently, the government received only 6,835 muskets from Saint-Etienne in 1793.[133]

The central government had now made almost every concession; it had ended the local monopsony; it had pretended to let the price be set by the armorers themselves; it had relaxed its production standards. And still no weapons were forthcoming. Merchants refused to sell arms except at the market rate on the day of delivery. The state's latest representative, the onetime local armorer Nicolas Bouillet, suggested offering the merchant "whatever their avidity desires." The only other option was the use of superior force, which he warned was both repugnant and impractical. The artisans were geographically dispersed—and they were obviously well armed. With their own shops and irreplaceable skills, the armorers and merchants of Saint-Etienne could bend the government to their terms—so long as they had alternative buyers. For the time being Romme's patriotic solution failed.[134]

Then in the summer of 1793, the war turned disastrous, Dumouriez defected, the allies again crossed into French territory, the Vendée rose in revolt, and everywhere political agitation increased. Within the army, the need for guns was desperate. Troops on the front line complained about the diversity of calibers. Army generals began sending representatives to Saint-Etienne to procure guns on their own authority. On the political front the central administration faced imminent collapse. The federalist revolt—a rebellion of municipal "moderates" against the increasingly populist Jacobin policies—had important consequences for Saint-Etienne. On May 30, forces led by the property-owning class seized the Lyon government, the provincial capital from which Saint-Etienne was administered. Within days, the National Convention in Paris fell definitively into the hands of the Jacobins. Defiant, the Lyonnais sent a battalion of twelve hundred guardsmen to Saint-Etienne to secure a supply of guns and to prop up the municipal government there, now under growing pressure from its own local Jacobin clubs. Three days later, the mayor of Saint-Etienne welcomed in the Lyonnais, who occupied the town without significant resistance. The manufacture had definitively ceased to be of use to the national forces.[135]

Chapter Six

INVENTING INTERCHANGEABILITY: MECHANICAL IDEALS, POLITICAL REALITIES

WHY did the Gribeauvalists turn to interchangeable parts manufacturing for muskets in the mid-1780s? After all, there was nothing new about the ideal of interchangeability in the late eighteenth century. The artillerists already had two decades' experience with the method in the arsenals of construction. And there had been attempts even earlier. In the 1720s, Christopher Polhem, a Swedish inventor, built a wooden clock composed of interchangeable parts. For this feat, Polhem has been often described as a lone technological visionary, a man far ahead of his time.[1] But at the same period in France, an armorer-inventor named Guillaume Deschamps manufactured gunlocks with interchangeable parts. His experience is instructive of the obstacles Honoré Blanc would face sixty years later as he tried to realize this mechanical ideal.

In 1723 at the Hôtel des Invalides, in front of august witnesses, Deschamps disassembled fifty flintlocks and recombined their parts to produce fifty functioning flintlocks. The minister of war ordered Vallière, *père*, to supervise the inventor's attempts to expand his system. By 1727 Descamps had manufactured 660 locks judged interchangeable by Vallière's own inspectors, "all properly reassembled without a single stroke of the file." Each lock, however, had cost 20 *livres*, five times the current price. Undaunted, Deschamps proposed a larger manufacture to produce five thousand identical gunlocks, which he expected would cost only 2 *sols* 7 *derniers* each (or one-twentieth the current price). This proposal anticipates many of the technical and institutional features of interchangeable parts production. Tasks were to be divided among specialist workers, who would use dies, gauges, and filing jigs to shape parts precisely. Master pattern locks (*types*) would be distributed to the War Office, Grand Master, inspector, controller, and examiner. Deschamps noted that with six hundred locksmiths in the Saint-Etienne region, he could locally staff a manufacture to produce forty thousand gunlocks a year (more than the current national output).[2]

This proposed manufacture never opened. Inspector Guérin reported on the hostility toward Deschamps among the armorers and merchants of Saint-Etienne. The artillery bureaucracy was no more favorable. Not even pressure from the War Office could persuade a service which still answered to the Grand Master. That same year, however, Vallière, *père*, did decree that henceforth a single lock design would suffice for all military muskets. He expected this "simplicity and uniformity" to be achieved "once the master locksmiths of the different manufactures are all supplied with pattern guns." He hired Deschamps's confederate, Bicot, to instruct the locksmiths of Saint-Etienne in their use. But Vallière distinguished between this desirable degree of standardization and Deschamps's "superfluous, impractical, and expensive" interchangeability.[3]

What Robert Gordon has called the "mechanical ideal" has long appealed to artisans devoted to perfecting their craft. This might or might not produce valuable operational advantages that would appeal to administrators (translatability across space and time, ease of repairs in the field), or generate economic gains that would appeal to cost calculators (cheaper unit prices, reduced costs for repairs). As a set of procedures, however, the uniformity system required the coordinated use of skills and tools, and something else as well: a new form of the technological life. Making interchangeability a technical reality proved a daunting institutional challenge.[4]

This chapter is devoted to the efforts of Honoré Blanc to make this ideal practical in the years 1785–93. This uniformity system capped the development of the rational tools discussed in chapters 4 and 5—the process by which artifacts were to be disciplined and the workshop organized to produce "objective" commodities which could be exchanged for king's coin. I argue that interchangeable parts manufacturing can be understood as a political solution to a set of current dilemmas. To make this clear, I will set this French engineering program against the familiar story of entrepreneurial innovation, particularly in the hardware trades of the British Industrial Revolution. Certain of Blanc's techniques owed their inspiration to Birmingham manufactures, yet the method of interchangeability did not catch on in England. Why did the country that gave us the Industrial Revolution not adopt the methods of uniform production? And what does this say about French "industrialization" in the late eighteenth century?

The Uniformity System: Costs and Controls

The French artillery service underwrote wide-scale experiments in interchangeable parts manufacturing from 1763 to 1807 without hard evidence of a clear-cut cost advantage over traditional methods. Its advocates claimed that it would save on skilled labor and thereby lower production costs in the long run, or they asserted that the system would reduce the lifetime cost of the musket by eliminating expensive repair work. At other times, advocates claimed that interchangeable parts manufacturing would facilitate the repair of guns in the field. And sometimes, in their private correspondence, they touted the advantages of the method for controlling recalcitrant armorers so as to assure a reliable source of guns for the French state.

The immediate logic of the artillery's push toward the ideal of interchangeability is clear enough; interchangeability was a public sign that the engineers had achieved uniform standards in the armories under their authority. Preparing the technical *means* for this program was the task of the armorer-inventor Honoré Blanc. Blanc was born in Avignon in 1736 and apprenticed to the gun-making trade at the age of twelve. He was a master armorer at the Charleville Manufacture when Inspector Montbeillard brought him to Saint-Etienne in 1763 to serve as controller of gunlocks. There he applied himself "with such knowledge, zeal, and gentleness that the Entrepreneurs and workers alike are surprised at the prodigious changes he has wrought." Blanc's technical abilities were not limited to a single aspect of gun-making. He experimented with new methods of tempering steel and forging gun barrels, invented a new pistol gunlock, and fashioned presentation muskets for officers. He was literate, but without a formal education. However, his private memoranda and *Mémoire important* of 1790 display a keen understanding of political, economic, and social conditions in contemporary France.[5]

Blanc resembles nothing so much as the mechanic (*mécanicien*), a social type crucial to European industrialization in the nineteenth century. Each country has had its illustrious members. Among many others we might cite Henry Maudslay, John Wilkinson, and James Watt in England; John Hall and Thomas Blanchard in the U.S.; and Jacques Vaucanson, Honoré Blanc, and George Bodmer on the Continent. These men designed and built the machine tools at the core of the textile and metalworking industry. Many, like Blanc, spent some part of their career under state patronage with varying degrees of contentment. Many, like Blanc, tried their hand at running their own manufactures as entrepreneurs with variable degrees of success. Unlike the military engineers, these men tended not to have extensive training in higher mathematics or rational mechanics—though many appreciated quantitative analysis of machinery. Apprenticeship trained, they placed a premium on craft skills. Already in the eighteenth century, state officials recognized that these men possessed a crucial form of technical know-how, and they began to believe that a small number of such men might make a great difference to the economic fortunes of the nation. England forbade their emigration. France attempted to recruit them.[6]

From the beginning, Blanc attached himself to the reformist party of Gribeauval. Already in the 1760s he was producing specialized gauges, jigs, and machinery for the gunlock. With the adoption of his design for the M1777, Gribeauval appointed him chief controller of the three manufactures, with a salary of 1,800 *livres*. He was given authority over the controllers and armorers in all three manufactures, and charged with equipping them with the "tools and instruments necessary to assure uniformity [of output], acceleration of work, and economy of price." To this end, Blanc was given a workshop in

Saint-Etienne staffed with a dozen highly skilled armorer-inventors. There, at government expense, he produced the "dies and tools proper to render the pieces [of the gunlock] perfectly exact and uniform." By 1780 he had turned his attention to bringing the tools of Charleville and Maubeuge in line with their sister armory.[7]

In this context, the principle of interchangeability represented operational proof that the engineers had pushed production to the highest degree of precision. In the same letter of 1785 which Gribeauval wrote to the minister of war to lay out his financial plan to save the Saint-Etienne armory from ruin, he also noted a surprising result just achieved by Blanc.

> We have just performed an experiment here with unexpected success; after having disassembled and mixed the pieces of 25 gunlocks of muskets for the Régiment du Roi, they were taken up at random and reassembled with an ease that was highly satisfying.[8]

Already, Blanc had two hundred interchangeable gunlocks on hand. The astonishing demonstration had been repeated before Du Châtelet, and Gribeauval asked the War Office to fund the further development of this technology with a view toward instituting it in all the armories.[9] The artillery offered several rationales for this program. As we saw before, this period was one of deep conflict in Saint-Etienne. By taking control of the price of gun parts, the engineers had taken on a direct managerial relationship with armorers. And Gribeauval's stopgap financial solution, while it eschewed nationalization, had further sidelined the Entrepreneurs. The engineers were now preparing themselves to take direct charge of the organization of production.

At the same time, however, the Gribeauvalists were under pressure from above to cut costs. The price of the M1777 had already risen 40 percent over the previous model, capping a doubling in the price of a musket since the Gribeauvalists had assumed power in 1763. The outlay for armaments still represented a small fraction of total military expenditures, but they had risen to half the artillery's budget in 1783. Then, with the war in America winding down, the War Office cut the budget for the purchase of firearms between 1783 and 1785 from 1,200,000 *livres* to 100,000 *livres* where it stayed until the Revolution. The artillery leadership hoped to mitigate these cuts with "the techniques of Sr. Blanc." On the other hand, developing these techniques was not cheap. Funding for Blanc's efforts took up nearly 15 percent of the annual budget for firearms in the years 1785–90, for a total of nearly 80,000 *livres*. These appreciable sums were given begrudgingly. With each change of minister, Gribeauval was obliged to reassure the War Office that he expected to recoup this investment.[10]

Hence, it is not surprising to see the artillerists claim that interchangeability would cut the costs of production. These savings would result from replacing skilled artisans with semiskilled laborers, generating a "real and considerable reduction in the price of arms . . . due to the infinite abridgment of labor costs."[11] According to Danzel de Rouvroy, the beleaguered inspector of Saint-Etienne, "If a full success crowns [Blanc's] aims, one will be able, with ordinary laborers, to construct all the parts of the soldier's musket much better than one can at present with the most skilled hands."[12]

As a practical matter, this assumed, of course, that the price of engaging highly skilled mechanics to construct and maintain new machinery would not outweigh the savings expected from substituting cheap laborers for armorers. The problem again was that mechanizing any single task provided only marginal savings at best. Only systemwide interchangeability would save substantial amounts by eliminating the need for a final, expensive hand fitting. This required a unified production process and a significant outlay of capital. The armorers and *fabricants* understandably opposed such a change. And now that the Entrepreneur-merchants no longer set prices for gun parts, they would not realize those savings either. Hence, the logic of cost savings through uniformity production appealed only to the state. But in the late 1780s, the crown was in no position to make that kind of investment.

The artillerists also argued that by facilitating repairs, interchangeable parts reduced costs over the *lifetime* of the musket. Salvaging and refitting old locks—either in the storerooms or in the regiments—constituted a significant drain on the arms budget. Bellegarde's disastrous triage had been primarily intended to reduce this expenditure. After the Revolution, Aboville asserted that musket repair had cost 20 million *livres* between 1792 and 1795, and another 5 million *livres* between 1795 and 1800. In 1786, Gribeauval authorized another round of triage in the stockpiles; this time without arousing any objections. Here again, of course, economizing on these repairs did not interest producers, only the state.[13]

Just as often, however, interchangeability was justified with reference to operational motives. In the face of its economic shortcomings, historians have usually cited the military's desire to ease repairs on the battlefield. The French artillerists certainly made this argument. On the battlefield, in the colonies, aboard ships, in case of siege: interchangeable parts would speed repairs and hence augment the fighting ability of troops. This ignored several problems, also raised by the Vallièrists with respect to artillery carriages. Either the spare parts had to be readily available on the site, or cannibalized from other guns. But the number of interchangeable weapons was never more than a tiny percentage of the total, nor were they marked as such. Even in the early 1800s, when Blanc's Roanne factory produced roughly ten thousand interchangeable gunlocks a year (perhaps 5% of the imperial total), he was not allowed to print the name of his manufacture on the lock plate.[14]

Each of these rationales explains why the state, and not private investors, would be interested in interchangeable parts production. But none fully explain the timing and character of the artillerists' program. The historical circumstances outlined in the previous chapter suggest how the ideal of uniform production emerged from a larger program aimed at consolidating the artillerists' authority over the armory. The engineers' goal was to minimize the human discretion interposed between their conception of the artifact and its realization. From the artillerists' point of view, the fount of this discretion was the greed of both armorers and merchants. One challenge they faced at Saint-Etienne stemmed from the disruption to military gun production caused by competition for skilled labor in the civilian gun sector. From this point of view, the uniformity system, by transforming the sort of labor used to make guns, was intended to prevent artisans from selling their wares to the highest bidder and so hold the state ransom with their skills and capital. That is, interchangeability was intended to undercut the ability of workers to operate as entrepreneur-*fabricants*. The other challenge was the Entrepreneur, whose interest in profits led him continually to cut corners on quality. In their private memoranda several artillerists suggested that the state get rid of the profit-hungry Entrepreneurs altogether. They saw the interchangeability system as implying a new government-owned manufacture along the lines of the arsenals of construction. Danzel argued as much. So did others who suggested that the worker-soldiers in the arsenals of construction were "happy," and that the *régie* system would solve the problems at Saint-Etienne.[15] The commissioners who investigated Blanc's methods in 1792 argued that Gribeauval had always intended ultimately to shift musket production to state manufactures in the arsenals of construction.

> And so the state would cease to be dependent on the Entrepreneurs of the manufactures and would no longer have to pay that multitude of supervisors known as controllers, examiners, et cetera.[16]

The "worker-less" factory has been part of the engineering mentality since the days of the French military engineers. What has been less often noted is that it was also conceived as an "entrepreneur-less" factory. The close tolerances necessary to achieve interchangeable parts would oblige the armorers to work to the engineers' specifications. Building command into the production process would solve the problem of finding controllers to honestly supervise their fellows. And the rational analysis of production would obviate any need for merchant coordination. This technocratic ideal presupposed that relationships between professional engineers and semiskilled workers would be mediated solely by technical drawings, gauges, jigs, and machinery. Against our familiar vision of an Industrial Revolution fueled by the entrepreneurial drive for profits, we need to consider this *engineering* vision of industrial modernity. It may not have been a model destined to succeed—but even its failure was to have important consequences.

Certainly, the armorers and merchants of Saint-Etienne read the artillerists' program of interchangeability as antagonistic to their economic interests.

Blanc denounced them as greedy and self-interested, and he found their hostility palpable. At every turn he encountered "prejudice, laziness, ignorance, and irrationality (*passion*)," and hinted ominously at the "powerful enemies" trying to stop him. In his *Mémoire important*, he noted, "[t]he invincible repugnance which the proprietors of these manufactures have manifested against everything which aims at the perfection [of gun-making]."[17] This opposition was grounded in a simple economic logic. Interchangeable parts manufacturing required substantial outlays for capital which the *fabricants* had little hope of securing. Consequently, the *fabricants* and master armorers of the town were allied against any moves to raise the barriers to securing lucrative government contracts. Blanc's techniques, to the extent that they tightened standards and reduced the need for skilled labor, would have divided the gunsmiths into a group of haves and have-nots, employers and employees.

As Merritt Roe Smith has pointed out in his study of Harpers Ferry, mockery and derision are potent weapons in a community's effort to combat technological innovation, or indeed novelty of any kind. And the armorers of Saint-Etienne had proved themselves capable of acting on their derision. For this reason, the artillery service decided to transfer Blanc's experiments to "an isolated locale, so as to no longer be the butt of naysayers, who, according to him, had not ceased to heap ridicule on his operations." To be on the safe side, Gribeauval asked Du Châtelet to sequester Blanc's workshop in the dungeon of the Château de Vincennes, where recent prisoners had included Diderot, Mirabeau, and the marquis de Sade. This isolated locale behind the three-meter-thick dungeon walls may seem to some historians a fitting metaphor for the view that technological innovation emerges from the creative minds of innovators acting from outside society. But in fact the artillerists proactively transferred Blanc and his dozen assistants to this protected site precisely in order to insulate them from social strife. Or so they thought.[18]

Mutiny and Tolerance

Blanc's transfer to Vincennes was prompted by an episode earlier that year at the Royal Manufacture of Bayonets at Klingenthal. What happened there offered the artillerists an insight into the social and political effort that would be needed to make interchangeable parts manufacturing succeed. The disturbance began when artillery officers introduced high-precision molds to stamp the metal ring, called the *douille*, which held the bayonet fixed to the muzzle of the musket (figs. 6.1 and 5.2). These stamping methods were akin to those being developed by Blanc to forge interchangeable gunlock parts. The rationale behind their introduction was also similar. The artillery wanted to define precisely the bayonet ring to ensure that it would attach to the muzzle without an individual fitting. In his *Mémoire important* Blanc noted that precision-forging the bayonet ring with dies and calibrating them with gauges would

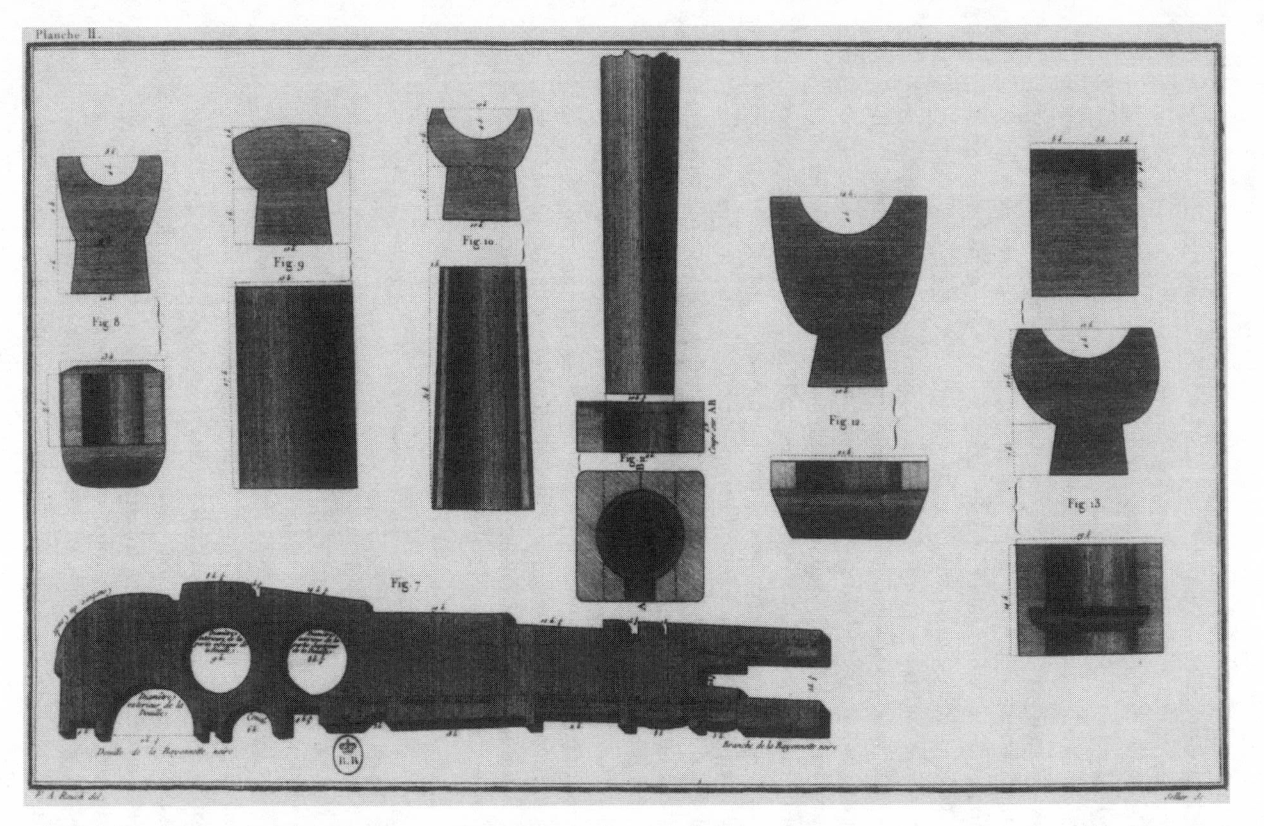

Fig. 6.1. The Bayonet Ring and Its Gauge, Manufacture Royale de Klingenthal, 1780s–1790s. The bayonet ring (*douille*), the piece which affixed the bayonet to the muzzle, is here seen along with the gauge used to verify its dimensions. The two holes in the gauge are for "the interior diameter of the lower part of the ring" (9 *lignes*) and for "the interior diameter of the upper part of the ring" (8.75 *lignes*). Again, compare this technical drawing with the view of the ring G in fig. 5.2. The drawing comes from Vandermonde, *Fabrication des armes blanches* (1794), but the gauge dates from the conflicts of 1782–84.

save the state the expense of transporting bayonets across France from Klingenthal to Saint-Etienne to be hand fitted.[19] But when the engineers introduced these techniques they met with violent resistance. This mutiny of the Klingenthal armorers, though far from Saint-Etienne, alerted the artillery to the social repercussions of the uniformity system. Of the five military commissioners who evaluated Blanc's techniques in 1792, two had intimate involvement with the mutiny. Their experience explains the cautionary tone of that otherwise favorable report, and accounts for its conclusion that it would be "impolitic, even dangerous" to introduce interchangeable parts manufacturing in "such times as these."[20]

To the historian, the episode also reveals how achieving the standards of production which made interchangeability possible meant *resolving* conflicts between economic rationalists (artisans, merchants) and administrative rationalists (engineers). The embodiment of those standards in material tools (such as gauges and jigs), and the use of those tools to define the shape of artifacts (here, the bayonet of a musket) was a social process. Artifacts in this sense are political. As the negotiated outcome of conflict over the terms of an exchange, they reflect the nature of the social settlement between distinct forms of technological life. In this case, we are dealing with a three-cornered conflict involving state engineers, a private merchant, and artisanal producers. In their complex interaction, no one party had a clear upper hand: the artisans could shift their efforts to other forms of work; the merchant could quit this form of commerce; and the engineers could cut (or raise) prices. Moreover, all parties could (and did) call on powerful patrons, appeal to certain legal rights, and muster some kind of physical force. At stake was the status of French subjects: did the artillerists have the right to imprison artisans who refused to work on terms which they considered unacceptable? Also at issue was the boundary between state interests and private capital: should the government take direct ownership of the means of production? The answer to these questions cannot be understood in narrow economic terms, but only within the political framework of late ancien régime France. Resolving these questions played a key role in defining the shape of the bayonet, not simply as an idealized drawing, but as a "thick" object produced in bulk at a set price. Hence, in defining a tolerance for this artifact, the state defined its relationship to its citizen-producers.

In the 1730s the hamlet of Klingenthal, thirty kilometers outside Strasbourg, was created *ex nihilo* as the French army's premier source of swords, sabers, and bayonets. Staffed by skilled German artisans brought in from the steel-working town of Solingen, the armory encompassed language and caste barriers that mirrored the fault lines at Saint-Etienne. The artisans were relatively prosperous, compensating for fluctuations in military demand with private orders for fancy swords for officers and civilians. A merchant (Entrepreneur) provided local finance and coordinated production, and in return received tax exemptions and a quasi-monopoly. Only in 1765 was an inspector actually housed in the town. Then, in the early 1780s, Klingenthal was the site of disturbances instigated by the Entrepreneur Louis Gau. Gau, a well-connected financier from Strasbourg, was agitating for an increase in the price of bayonets. The town had just lost its monopoly on the supply of sword mountings. And the new inspector, the able Amiable Marie Givry, had set out to enforce rigorous standards of production.[21]

The trouble began in the winter of 1783, when Givry announced a new standard model for the bayonet ring. The placement of the *bourrelet* (the bulging circle at the end of the ring) was still done by eye. Positioned incorrectly, the *bourrelet* impeded the correct attachment of the bayonet to the musket. Complaints on this score had been heard in Saint-Etienne where the fitting took place.[22] Moreover, the ring-filers, Givry asserted, were complaining of the excessive filing needed to shape sloppily forged rings. The ring-forgers, however, refused to conform to this new model. The artillery, they responded, was altering the terms of its own contract; of the eight thousand rings they had sold to the artillery since the regulations of 1777, none would be acceptable under the new model. Where was the law authorizing the new model? Now the ring-filers complained as well. They said the new standard, by increasing the number of rings rejected at inspection time, had halved the number they could file in a day. In rebuttal, Givry denied that he had altered the regulations. The minister of war, however, had notified Gribeauval that the Entrepreneur Gau was bringing pressure to bear at Court. So Givry went to Paris in March to calm the waters.[23] In his absence the confrontation took a violent turn.

Upon leaving, Givry turned over the operation to his second-in-command, Captain Villeneuve. Villeneuve ordered one of Klingenthal's controllers, François-Antoine Bisch—"the only controller in this place capable of carrying out his duties"—to construct a new stamping die and companion mandrel. He further proposed that every ring-forger be provided with his own copy of the new model, along with "the gauges, mandrels, and dies conforming to the models produced for each piece." According to Villeneuve, the filers were happy, but "when [the forgers] learned that forging [the rings] to this degree of perfection required an additional stamping and heating, they murmured [against the change] and ended by mutinying to excess. . . ."[24]

In the face of the forgers' vocal refusal, Villeneuve insisted that Bisch fashion the new gauges, and he issued an order—translated into German—that all workers gather on the morning of April 2 to receive them. Disobedient workers, he added, would be imprisoned. At that morning meeting, a group of forgers, led by Jean Schmidt, threw their workpieces down in disgust and stormed out. Brought back, the forgers scoffed at the young officer, telling him (in Villeneuve's words):

> . . . that they had no need to take instruction from me; that they all knew their trade; that they would continue to make the rings just as they had up till now; and

> that if we didn't want them, we should [get them elsewhere]; and that it was no business of mine to interfere with whether the rings were well or poorly made for the filers.[25]

Brought back a third time by fear of imprisonment, they demanded to see the new regulation which proposed this standard, and if it were true, demanded a price raise to compensate for the additional work. If not, they would quit the manufacture. The forgers—the most prestigious and highly paid group of armorers—felt that their autonomy and wages were under threat.[26]

Villeneuve denied the new procedure incurred costs for the forgers. It might slightly reduce the number of rings forged in one day (from 23–25 to 20–22), but this would be compensated for by the decrease in the number of rejected rings at inspection time. Translated into German, Villeneuve's stonewall caused the artisans to storm out a final time, threatening Bisch and his wife with violence if they continued to assist with the new models. Bisch, whom Givry had recommended for promotion to first controller, broke one of the mandrels out of fear. The workers then filed a petition of complaint with the marquis de Lasalle, the provincial commander, and met in a cabaret to swear their defiance. They insisted that all discussion of the new tools and procedures be postponed—at least until this present work order was over. Two days later, Villeneuve, with the assistance of the *maréchaussée*, sent Schmidt and his associate Hiet to prison. When their comrades demanded to be arrested as well, Villeneuve ordered the police to bind in irons the arms and legs of anyone who presented themselves at the jail.[27]

This episode came quickly to the attention of the Comité Militaire, the high commission then entrusted with reforming the military, which included Gribeauval's patron, Du Châtelet. The Comité took note of the complaints of the workers, as relayed by Lasalle, and they learned from the Entrepreneur Gau that Inspector Givry kept the master patterns locked up in his office. Gau further accused Givry of seeking to nationalize the manufacture. For his part, Givry denied that he had altered the standards. In a somewhat contradictory vein, however, he conceded that certain artisans ought to be paid more since "their prices had been calculated on the basis of a tolerance [that had proved to be] inimical to the king's interests." In fact, the 1777 regulations had said nothing explicit about tolerances for the ring. Givry now announced that he would henceforth accept ring diameters forged within 1 *point* of the final filed size. But workers must not be allowed to judge his "greater or lesser exactitude." Armorers who questioned the authority of military officers should be imprisoned. Indeed, all armorers under contract to the army—in return for their tax exemptions and privileges—were subject to military discipline.[28]

What was to be the state's role in production? What powers did it have over economic agents? In its ruling, the Comité Militaire attempted to adjudicate the boundary between administrative and economic rationality, and between the interests of the state and the interests of local merchants and producers. At the root of the problem, the Comité acknowledged, was the "natural animosity" between Entrepreneur and inspector. The Comité's solution, therefore, was to demarcate their respective spheres of authority. The inspector's powers did *not* extend to the "interior" of the manufacture. Power over the hiring of workers, their pay, or the use of their time "belonged to the Entrepreneur." The inspector, however, set the standards for raw materials and finished goods, and for "general work." The Comité then proceeded to define the border between these realms of economic and administrative rationality in physical terms. The Comité pointed out that the manufacture could not be expected to turn out perfect bayonets, only ones of "good quality up to a certain degree." If the state wanted more perfect weapons, it would have to pay for them or let the manufacture fall into ruin. If willing to accept less than perfection, the state would have to "fix the degree of rigor with which the inspection was to be carried out, according to the amount the king is willing to pay." To that end, it insisted that the tolerances for acceptable artifacts be specified as exactly as possible. It asked the artillery to draw up a *procès-verbal*—a contract—indicating as exhaustively as possible the tolerances to which bayonets would have to conform and the procedures by which the degree of conformity would be judged. These necessarily had to be public standards, and the Comité ordered the artillery service to give the Entrepreneur a copy of the new master patterns. In a final effort to resolve the dispute "impartially," the Comité sat through a demonstration which showed that most of the bayonets were acceptable, with only minor repairs.[29]

Givry accounted this a victory—though he warned Villeneuve that in the future they would have to document their every decision. A year later Givry did in fact spell out the tolerances exactly; the thickness of the ring, for instance, was to be 10 ± 2 *points*. New patterns were distributed to the War Office, the inspector, and the Entrepreneur, plus new gauges, jigs, and mandrels for the Entrepreneur and controllers (fig. 6.1). Yet when these standards were conveyed to the armorers, they balked again. Givry offered them what he estimated was a 10 *livres*/week profit ("not bad for men housed by the state")—but they refused. Only when Givry received authorization from the minister of war to strike recalcitrant workers from the rolls did the artisans give in.[30]

It was not, however, a personal victory. While Bisch received his promised promotion to first controller, Givry was unable to parlay his managerial success into the kind of reward that was the best an (honest) engineer might hope for. Givry—the very model of an Enlightenment engineer—was denied premature consideration for the Croix Saint-Louis. Born to the minor nobility, he had chosen the technician's route to high office in the ancien régime. Within the artillery he had specialized in technical tasks, commanding a company of worker-soldiers before becoming an armaments inspector. His library consisted of more than professional volumes (Bézout, Du Coudray, Vallière, Du Puget, Guibert); his wider reading included plays by Racine, Diderot, and Beaumarchais, as well as tales by Voltaire, Sterne's *Voyages*, and Diderot's

Bijoux indiscrets. Among his philosophical and historical books were Helveltius's *Esprit*, Rousseau's *Confessions*, and Mercier's *Bonheur de campagne*. Educated, capable, rigorous—a card-playing ladies' man without fortune or high connections—he was a lieutenant-colonel when he emigrated in 1792. Such was the lot of a talented officer in the ranks of the ancien régime.[31]

In the meantime, his antagonist, Entrepreneur Gau, had carried out his threat to quit the business. In 1786, he sold the business with the Court's approval and continued (successfully) to pester the crown for additional reimbursement. The new Entrepreneur, Jean-François Perrier, did not stand in the way of the new forging methods. In 1794, when the patriotic savant Alexandre-Théophile Vandermonde reported on the Klingenthal Manufacturing process, he noted that four stampings were needed to forge the bayonet ring and that extensive gauging was performed throughout the process. At the time, Vandermonde was the director of the Atelier de Perfectionnement, an experimental workshop where interchangeable parts techniques were being developed by the Revolutionary state.[32]

On a corporate level, then, engineers had triumphed. But they had succeeded only by renegotiating the terms of the exchange. Confronted with recalcitrant producers and under pressure from state authorities, the engineers were forced to make the terms of that exchange explicit and public. Givry had vowed that he would not allow artisans to question his judgment. In fact, his discretion had been reduced. Objective criteria, enshrined in material tools and public law, henceforth defined the bayonet ring. For their part, the mutiny of the artisans and the recalcitrance of the Entrepreneur had enabled them to translate their greater work effort into greater payments from the state. In doing so, they had also made the engineers acknowledge their rights as citizen-producers, free to place their labor and capital where they thought it would pay best. Why were the engineers willing to pay this price? Behind their engineering program to discipline artifacts lay an attempt to coordinate production not just within a single manufacture, but across dispersed manufactures as well. Unlike the artisans or Entrepreneurs, the engineers thought in national systemwide terms. At that level of analysis, they saw how interchangeable parts production would enable them to save the cost of transporting bayonets from Klingenthal to Saint-Etienne. My point, therefore, is not that engineering standards lack their own economic logic, but that deciding *who* will pay for them and reap their advantages is a social process. The same logic explains why interchangeability never materialized in industrializing Britain.

English Means and French Ends

The machines which drove the Industrial Revolution are presumed to be British and the motives are said to be entrepreneurial. But despite the phenomenal growth of the hardware trades of eighteenth-century Birmingham, British manufactures never organized their workshops to produce interchangeable parts. Why not? And why did eighteenth-century Frenchmen think they had? According to Inspector Danzel, Blanc's method of manufacturing "solely with the use of dies" had "already raised in England several types of manufactures to a very high degree of splendor."[33] This was not an idle comparison. The artillery service kept close tabs on developments across the channel. Yet the engineers' view of the Birmingham phenomenon was selective, and they hoped to apply its successes to realms the British never contemplated.

In the latter half of the eighteenth century the hardware "toy" trades of Birmingham's industrial district acquired an international reputation for high quality and relatively low cost. As Maxine Berg has pointed out, this transformation was not driven by mechanization—Arthur Young in 1791 complained that metalworkers lacked the equivalents of the power loom and spinning jenny—but by a diversification of products, subtle redefinitions of skills, and more stratified relations between masters and "men." The late eighteenth century in Birmingham did not see the replacement of small firms with large ones, so much as changes in the relationship of both to the growing consumer market. Matthew Boulton, Birmingham's most illustrious industrialist—famous for his association with Watt's steam engine—was also vitally engaged in producing and marketing a veritable cornucopia of small metal goods: buttons, jewelry, buckles, and all the appurtenances of the age's fashion. And this multiplying diversity, rather than the sheer bulk of Birmingham's output, was as dramatic a shift in eighteenth-century British industry as the famous innovations in factory organization in the textile trades.[34]

The French state fingered this complex and dynamic productive world for transfer across the Channel. Trudaine, the activist commerce minister, especially wanted Saint-Etienne to develop the vitality he associated with large-scale industry and mechanization. He saw no incompatibility between his laissez-faire ideology and state-led efforts to promote a particular version of industrial development. In this sense, Trudaine and his bureaucrats sought to define *prospectively* the direction of "modern" industry. Trudaine was particularly concerned about the cost and quality of Saint-Etienne's ironmongery as compared with Birmingham and Sheffield. To this end, he recruited an English button manufacturer named Michael Alcock to come and diagnose Saint-Etienne's deficiencies. Alcock was critical of Saint-Etienne's relative backwardness, which he blamed on technical factors (the supply of water power was inadequate); on the social organization of work (each French metalworker insisted on having his or her own forge); and on state policy (internal tolls and taxes which made the economic climate uncertain). The result was that ironmongers had little hope of escaping the poverty near universal among the locals. According to Trudaine's English analysts, Birmingham's success depended on its greater liberty; which did not just mean liberty from state tariffs and privileges, but also freedom from the violence which threatened those who dared to mechanize production. Certain changes were underway in

Saint-Etienne, however. In the 1760s, some masters began to operate workshops with thirty to forty journeyman, all engaged in different tasks. The Eustache knife and a plethora of low-cost cutlery produced in this manner brought in some 500,000 *livres* in the 1760s. But Trudaine was not satisfied, and he tried to get Alcock to build a factory there in collaboration with Carrier de Monthieu.[35]

Instead, Alcock first established a factory for "ironmongery, cutlery, and jewelry in the English style" at La Charité-sur-Loire. The factory succeeded in a narrow technical sense, employing two hundred workers who made three hundred gross of buttons daily (or a theoretical rate of 13 million buttons a year). Yet the firm lurched from crisis to crisis, and in 1764, Alcock formed a manufacture in Roanne, at the other end of the Forez province from Saint-Etienne. Roanne possessed excellent access by the Loire River to northern and western France. Coal and other raw materials could be procured from Saint-Etienne. What is more, the municipality eagerly welcomed the new manufacture. By 1771 Alcock and his sons owned a factory in town and a water mill just outside. The firm diversified into buckles and jewelry and did an annual trade of between 84,000 and 146,000 *livres*. Part of their success depended on military orders and the establishment was well known to the artillery service. In 1777, the firm received the contract to make buttons for Du Châtelet's Régiment du Roi, and the fastidious Anglophile duke found them comparable to those of London.[36] The factory was a local marvel, using stamping techniques to produce buttons by the millions. And during the Revolution, Blanc bought out Alcock's buildings, water mills, and tools.[37]

Whatever his inadequacies as a businessman, Alcock had a firm grasp of how to organize production. He divided his factory into separate workrooms for separate tasks, most of which he mechanized. The mill at Beaulieu employed waterpower to hoist the weight which would then forge blanks in jumper dies ("monkey" dies). This process resembled the drop forges used by Blanc for stamping out lock plates. All told, each button passed through forty hands. Factory work was draconian—nothing like conditions in the arms industry. Workers at La Charité labored fifteen hours per day in summer and thirteen in winter. Most of the workforce was composed of women. All were forbidden to be absent for more than a quarter-hour. "Leagues" or complaints about work conditions could be punished with prison.[38]

At the same time, the operation depended on highly skilled machinists and artisans, in line with the dual labor market that characterized Blanc's operation. Alcock's machines needed constant repair. And the firm took great pains to keep up with changing fashions; an inventory mentions 1,057 different engraved dies for buttons, suggesting that Alcock was using production flexibility to cultivate the same sort of diverse market as Boulton had tapped. Alcock brought talented English assistants with him and dispatched his wife and sons on expeditions to entice others.[39]

Alcock thereby shaped the French state's image of the industrialization then taking place in Britain. In 1784, the artillery service dispatched two expert artillery officers to spy further on the English metals industry: Klingenthal's Inspector Givry and Ignace Wendel, founder of Le Creusot. The two men reported on the sixty or more "fly-presses" used in Birmingham to stamp all sorts of metalwork out of copper, tin, silver, and pinchbeck. In France, they noted, commercial use of these presses was banned for fear that they would aid in the counterfeiting of coins. Were this restriction lifted, France might enjoy this method "to which the town of Birmingham owes its rapid and astonishing prosperity." Yet as the French spies admitted, the Birmingham *gun* trade was nothing like this. And they noted too that its military guns did not equal the French in their quality or fit.[40]

Industrially advanced Britain never attempted the interchangeability system for gun-making. A few British gunsmiths did adapt stamping techniques to gunlock production (though not as early as Blanc). Isaac Mason was pressing out gunlock parts with a fly-press in 1796. And in 1818 John George Bodmer, who had manufactured interchangeable gunlocks for the French government during the Empire, visited "Mr. Little's shop where I saw the locks made according to the [interchangeability] model. But I do not think they are a great success." Certainly this "failure" cannot be ascribed to technical incompetence. Compared with the artisans of Saint-Etienne, the armorers of Birmingham had access to superior raw materials (including Swedish ore), plus a thriving machine-tool industry (then forming under Maudsley et al.). By contrast, French metallurgy was backward, its machine-tool industry in its infancy, despite a few exquisite pieces by Vaucanson, Senot, and other Parisian *artistes*.[41]

Rather, the explanation lies in the relationship of the state to the structure of the Birmingham gun trade, which differed from that of Saint-Etienne in one respect. In both towns, production was in the hands of artisans who lived in town and the surrounding region, their access to markets and resources mediated by merchants who ranged widely in their wealth. Similarly, both towns sold shoddy muskets for the Africa trade, fancy sporting guns, and muskets for the military. In the military trade, however, a crucial difference emerges. In England the Board of Ordnance bought unfinished gun parts on contract, which were then assembled by state-employed armorers in the Tower of London.[42]

This meant that Birmingham gun merchants did not invest directly in mechanization until the late nineteenth century. Russell Fries has made their rationale clear. The savings a private producer might realize by mechanizing one individual step in the process was small (given the low price of skilled labor), and the loss of flexibility was risky if orders fell. Substantial gains *might* have come by mechanizing the entire process if that obviated the need for expensive hand fitting. But the contract system of buying gun parts meant that only the state would benefit from this savings.[43] This left the state to take the initiative. After all, the French artillery service had established the same incentive structure in Saint-Etienne after 1777 by setting prices for gun parts. But only in France was a powerful government agency willing to commit resources to a "mechanical ideal," for its own eminently practical reasons.

The exception proves the rule. In 1805–8 in Portsmouth, a manufacture was founded to produce wooden pulley blocks using a series of 45 single-purpose machine tools. This manufacture initially turned out 1,420 blocks a day; by 1808, the rate was 130,000 blocks a year. The sponsor (and sole customer) for this undertaking was the royal navy and its methods were masterminded by three engineers: Samuel Bentham (inventor of the panopticon and inspector-general of naval works), Marc Isambard Brunel (a French-trained military engineer), and Henry Maudslay (master machine-builder). On superficial examination, this system went far beyond the French methods for achieving interchangeability. Portsmouth finely divided the manufacturing process into a series of analytical tasks, each of which was handled by a single-purpose machine tool. However, interchangeability was not a goal here, so much as an unintended result. It could only have occurred with the relaxed tolerances of woodworking, as it had for the French artillery carriages. But whereas the French artillerists then took up the principle of interchangeability as the rallying cry of rigorous production standards, no one in England took up cudgels for the *ideal* of interchangeability. The Portsmouth case remained an isolated instance.[44]

The comparison with England, then, is paradoxical, both for contemporaries and for us. Out of a diverse and mutating British economy, the French state selected certain methods of manufacturing, and tried not only to import them into France (with mixed success), but also to extend their domain to areas the English never attempted. François Crouzet has illustrated the somewhat whimsical French views of English industrial superiority in the eighteenth century.[45] Technology transfer is invariably a process of cultural selection and local adaptation. For the historian, the note is cautionary. It suggests that "mass production" is not a monolithic technique with a single, simple lineage. There are as many versions of mass production as there are versions of flexible specialization. Diverse factors have shaped any particular version. We can see these influences at work as the French engineers' version unfolded in Revolutionary Paris.

Out of the Dungeon

In 1789 Gribeauval's health failed. He was seventy-four and in too much pain to sleep. His memory became unhinged. As he sunk into senility, his lifelong obsession with uniformity possessed him more and more. On 3 April 1789, he wrote to his former patron, the duc de Choiseul, to assure him that "he would continue to occupy himself daily with all those portions of the artillery confided in him . . . and in particular with everything necessary to establish uniformity in all the different parts of the musket . . . , so desirable for the good of the service and to economize on the king's finances."[46]

Choiseul never answered that letter; the former minister had been dead for four years! There is no more fitting symbol of the deep fixation on order, duty, and control than this senile letter. According to a report, the first inspector-general had spoken of Blanc's work "in the last moments of his life." In his final days, "nothing [remained] in that so well-organized head of exquisite discernment but a singular stubbornness (*entêtement*) and despotic egotism." In May 1789 Gribeauval died.[47]

The Revolution quickly severed Blanc's remaining lines of patronage. With Gribeauval dead and Du Châtelet disgraced, pressure began to mount against Blanc's operation. In 1790 a new minister of war ordered Blanc to send his assistants back to Saint-Etienne and to transfer his experiments to an arsenal of construction. Funds were drying up; the government was behind in the payroll. But Blanc proved adept at mastering the new forms of publicity the Revolution made possible. In early November he addressed a twenty-page *Mémoire important* to the National Assembly, urging them to establish a manufacture along the principles of interchangeability. Critics within the War Ministry complained that Blanc's pamphlet was printed "in profusion," and written "with much artistry and in the most seductive manner." In fact, it offered a cogent appraisal of the ills confronting the armaments industry. Then, on November 20 at the Hôtel des Invalides, Blanc staged a public demonstration of interchangeability before the latest minister of war (the engineer Duportail), plus legislators on the Military Committee, officers of the artillery service, and members of the Academy of Sciences. Selecting pieces at random from bins containing the parts of five hundred gunlocks, the gunsmith assembled a number of working gunlocks. At that time, Blanc had in stock over one thousand gunlocks, all of them tempered, polished and fully interchangeable. What he had accomplished on this small scale, he now proposed to do for the entire nation.[48]

Blanc promised legislators that a single new manufacture based on interchangeable production would lower the cost of armaments. In particular, he emphasized the savings in lifetime costs of repair . He acknowledged, however, that the implementation of these new techniques faced serious obstacles, especially at Saint-Etienne. The artisans there had been "imbued with bad principles which they abandon with difficulty." More to the point, rival private arms merchants constantly "debauched" workers. And the armorers' "natural intractability (*indocilité*)" disposed them constantly "to deliver themselves to the highest bidder." Blanc deplored these artisans who placed their venal interest above their duty to perfect their craft or serve the state. The Entrepreneurs, too, were more interested in their purse than in the good of the service. They too had thwarted Blanc at every turn. What was needed was a set of tools to transform production. In place of what he called the "ancien régime of the manufactures," Blanc called for the creation of a national *Encyclopédie-pratique*. This *Encyclopédie*, however, would not be a picture book for display in elite salons, but an actual workshop where superior craftsmen would perfect their skills building machine tools (*gros outils*) to transform the mechanical arts. Blanc did not embrace abstract rationalization. Theory counted for less than practical skills; only the hand could guide the hand. He was an armorer, not an artillery officer. Technically proficient but ignorant of rational mechanics, Blanc's utopian workshop of artisan-inventors would supply the techniques by which the mechanical ideal would be made real. These machine tools would then be employed in his new manufacture, providing jobs and social discipline for the indigent.

> By the grace of these same tools, one can employ the poor day laborers that come from all classes, even children of 10 years or more, who can be usefully occupied, be it in acquiring the habits of good work hours, or in enabling them to earn, at least at the beginning, a portion of their nourishment, which will comfort many poor families who themselves have almost no means of subsistence.[49]

In the years after the Revolution, Blanc went so far as to suggest that deaf-mutes and blind students from the hospices at Picard and Abiny be employed in his factories. By attaching the poor house to the factory, society's marginal population could be gainfully employed and the social utility of large-scale manufactures demonstrated. Diligent children could toil alongside disciplined paupers; the factory offered both social control and social regeneration. The year 1790 was a time of widespread hunger and want. Urbanites felt threatened by the roving bands of beggars, impoverished by poor harvests. Blanc offered the National Assembly a means of ordering this destabilizing element.[50]

Blanc's publicity prompted two investigations. The minister of war asked the Academy of Sciences to study Blanc's claims, the sort of task which that body had long performed for the state. Three of the four members of the Academy's jury were connected with the *armes savantes*, including: Coulomb, the officer in the Génie who had investigated the laboring body; Borda, the naval engineer who had worked on ballistics; and Laplace, the mathematical physicist hired by Gribeauval as examiner for the artillery schools.[51] The fourth member and the chief author of its report, Jean-Baptiste Le Roy, came from the celebrated family of clock-makers, and had often investigated new technologies. Through his intimate scientific friendship with Benjamin Franklin, he had become one of Thomas Jefferson's guides to the French capital, and had presumably alerted the ambassador to Blanc's achievements. Le Roy also advised Jefferson to seek out the skilled machinist François-Philippe Charpentier to construct his new portable copying machine, and Charpentier later became the chief mechanic of the Atelier de Perfectionnement, the experimental workshop that developed interchangeable gunlocks during the Terror. While we cannot track the exact flow of information among these men, their mutual contacts show how widely word of Blanc's achievements had traveled among French savants, administrators, and machinists.[52]

Blanc received a ringing endorsement in the Academy's *Rapport* when it finally appeared in March 1791. The full Academy also witnessed a repeat demonstration of interchangeability, watching as the tumbler of a recently manufactured gunlock was fitted into a gunlock held in safekeeping since his earlier demonstration six months before.[53] On receipt of this favorable report, Blanc rushed off another petition to the National Assembly, which forwarded his claim to its Military Committee. The War Office then agreed to appoint a five-man artillery commission to assess interchangeability and its potential cost-savings. It included Givry and Villeneuve, experts on the use of dies at Klingenthal; and Guérin and Guériot, who commanded the making of interchangeable artillery carriages. From 16 September to 30 October 1791, these men dogged Blanc's workers, recording every action as they stamped, milled, filed, and case-hardened thirty-seven gunlocks composed of twenty different pieces each. Their conclusion was that Blanc had achieved true interchangeability.[54]

Evaluating Interchangeability

How did Blanc achieve interchangeability? The answer requires us to pay close attention not only to the machines and tools Blanc invented, but also to his procedures and organization. Blanc came to divide the manufacturing process into some 156 analytically distinct steps. Of course, these were not apportioned among the workers so finely. He also mechanized several operations, though considerable skill was still needed to use these flexible machine tools. As the tools and procedures for each piece were similar, I will again restrict my discussion to the tumbler, the "brains" of the lock. My description will also take note of changes introduced by the various successor methods in France, both at the Revolutionary Atelier de Perfectionnement (1795–96) and at Blanc's Roanne Manufacture (1796–1807). Understanding these procedures is important for understanding the *kind* of "mass production" which Blanc inaugurated.[55]

The manufacture of the tumbler took place in five successive stages, with a total of twenty-five distinct operations, totaling an average of two hours per tumbler. The first set of operations rough-shaped the tumbler. After forging, the blank was placed in a steel mold (the "jumper" die) of a drop forge. The weight was hoisted by four apprentices; when released, its violent impact stamped the tumbler with great precision. The tumbler was then removed, reheated, trimmed, and returned for a second stamping. At the Atelier de Perfectionnement, one of these drop hammers was fifteen feet high, the weight guided by rails, and the mold securely held in place by four adjustable screws. It resembled the drop hammers which excited so much comment in the rifle works of the American inventor, John Hall, in the 1810s. A smaller drop press was mounted on a heavy cabinet of wood, where the mold could be finely positioned with a crank (fig. 6.2). The Atelier de Perfectionnement also used a giant screw-operated coining press, procured by the Revolutionary government from the mint at Rouen (fig. 6.3). With it, various punches pierced holes in the lock plate or cock. For every operation there was a corresponding precision mold. Without proper precautions, stamping could produce fragile and brittle pieces, but Blanc avoided these by employing a high grade of iron.[56]

The second set of operations employed a "large milling machine" invented by Blanc to fashion the tumbler's two pivots. This was undoubtedly a *hollow* milling machine where the cutting tool itself supported and guided the piece. Although it lacked the precision of the universal milling machines of the nineteenth century, Blanc's machine could be adapted by changing the milling bit. One bit removed the seams and milled smooth the large pivot side. A different cutting tool then milled the flange at the base of the large pivot and the other face of the tumbler. Finally, a third cutting tool milled the little pivot and the surface of the tumbler. After each operation, the cut was checked with a specialized gauge. The flexibility of this tool must be weighed against the time lost in setting up fixtures. At Roanne, the milling machines were dedicated to a specific task.[57]

The third set of operations rough-shaped the bursts on the tumbler's lever and claw, thereby setting the rest and cocked positions of the lock. This was a hand operation, guided by a filing jig which Blanc had introduced into Saint-Etienne in the 1760s. The fourth set of operations milled the tumbler to its final dimensions using a "double milling machine" that advanced two opposing hollow cutting tools onto the working piece. The fifth set of operations gave the pivots and notches their final form. A worker sawed and filed the pivots to their proper length. Another worker cut the square of the pivot, guided by a specially constructed filing jig. And an apprentice drilled a hole in the center of the square; this operation was carried out with a vertical drill press that was guided by a drilling jig. At the Atelier de Perfectionnement (and probably at Vincennes as well) the drill was mounted on the wall, and included a sight to check the accuracy of the work, and (most important) a fixture to position the piece precisely (fig. 6.4). Following this, a master finished and tapped the hole. Then a worker finished the square of the pivot, the flange, and the "rest" notch with a hand file.[58] Last of all, the tumbler was case hardened. Blanc's tempering process differed slightly from that employed in the other manufactures. According to Le Roy, this process "did not alter the form of those pieces." Blanc also took precautions against the wear of his dies; precisely tooled steel patterns (*types*) enabled him to reconstruct new dies exactly.[59]

Fig. 6.2. Drop Press, Atelier de Perfectionnement, 1794-96. This drop press was used to die-forge pieces of the gunlock. The guided weight fell from the height of one meter. A ratchet mechanism positioned the working piece precisely. Another drop press of twice this height was also used. C.N.A.M. Drawing 13571-425(23) "Découpoir et mouton, servant à la fabrication des platines," [1794–96].

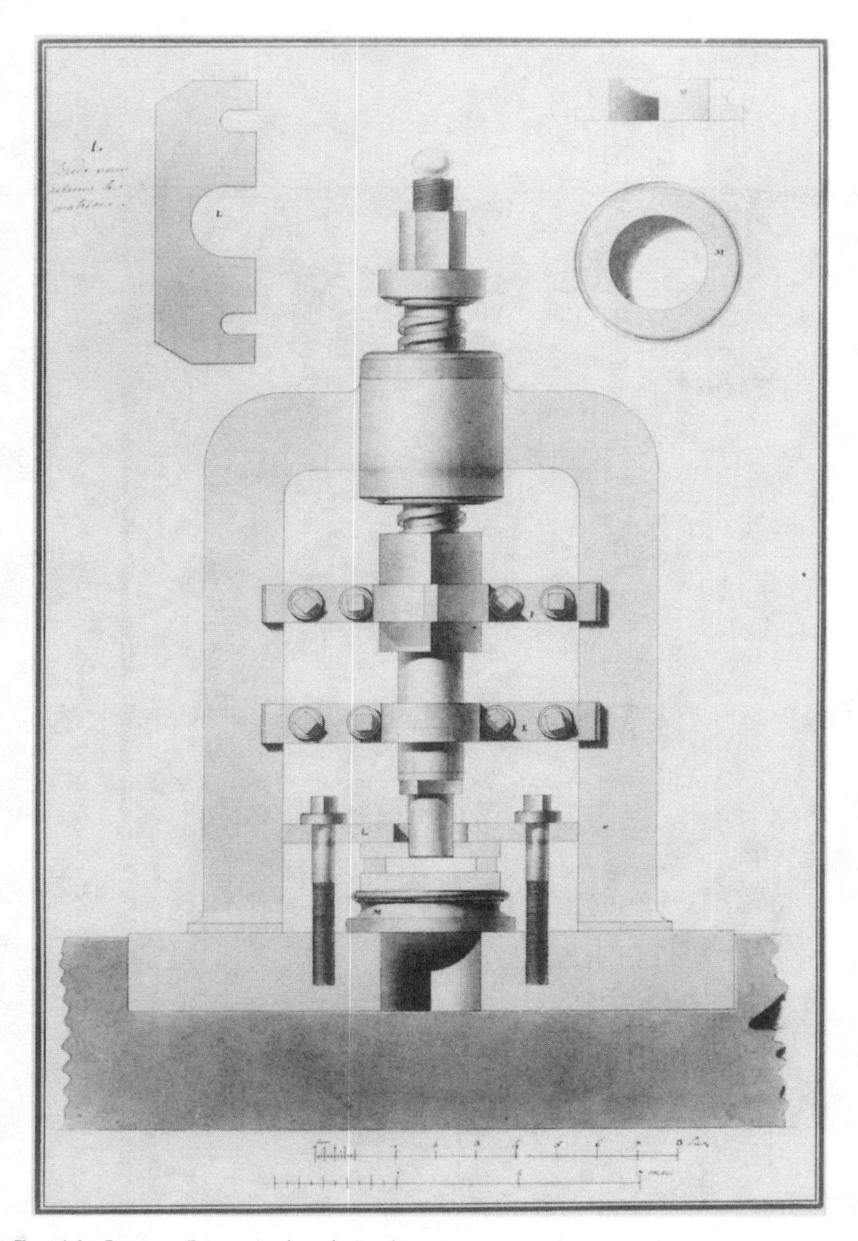

Fig. 6.3. Coining Press, Atelier de Perfectionnement, 1794–96. This massive coining press, over six meters high, was used to forge lock pieces and pierce holes in the lock plate. The press was tightened by a giant windlass (not pictured). Various molds could be inserted. Compare with the coining press in fig. 4.1 C.N.A.M. Drawing 13571-426(4) "Cinq dessins d'un découpoir ou emporte-pièce servant à la fabrication des platines de fusil," [1794–96].

Fig. 6.4. Drill and Fixture for Lock Plate, Atelier de Perfectionnement, 1794–96. This drill was braced against the wall and tightened by means of a screw. It was fashioned with greater precision and solidity than the drill depicted in fig. 7.10, but its main advantage was the fixture, pictured at the bottom left, which precisely positioned the lock plate and guided the placement of the drilled hole. C.N.A.M. Drawing 13571-418(1) "Machine à percer les trous de la platine," [1794–96].

As this summary shows, Blanc still relied on hand methods at Vincennes; nearly 75 percent of the time there was taken up with hand filing. However, mechanized tasks were not carried out with single-purpose machinery. Instead, Blanc's methods relied extensively on fixtures. Paradoxically, Blanc's method can be described as *uniformity production with flexibility*. Given the limited capital available, his short production runs, and the large number of distinct operations needed to make a gunlock, this flexibility is perfectly understandable. In such a situation, fixtures—like gauges—represent a compromise. They reduce the worker's discretion about each particular operation, but only by preserving a large degree of discretion about how the machine itself would be configured. In theory this made *possible* a separation between a skilled "setup man" and a semiskilled operator. But Blanc did not divide his labor pool this way, at least not at Vincennes. Later, in Roanne, he did try to substitute day laborers for skilled artisans. Yet even then he needed skilled hand filers. In this sense, interchangeable parts manufacturing in the late eighteenth century was a far cry from modern machine-driven mass production.[60]

Recently, Sabel and Zeitlin have returned to their alternative histories of mass production. They now suggest that their idealized distinction between mass production and flexible specialization does not account for the "hedging" strategies by which entrepreneur-practitioners mix and meld techniques associated with the two patterns. Here we have a clear example of how this hedging strategy was substantiated in the physical design of machinery. In this sense, the method of production Blanc developed for the engineers was a suitable match for their middle epistemology. The engineers realized that uniform methods of production (like uniform rules of calculation) needed to be tempered by sufficient flexibility to account for real constraints (limits on resources, the diversity of local conditions). System-builders, if they want their systems to thrive, must provide them with the flexibility to respond to changing circumstances.[61]

Although none of the locks made by Blanc have survived, we do have good evidence that his method produced interchangeable locks, at least in limited runs. Both before and after case-hardening, the thirty-seven gunlocks had been repeatedly assembled, disassembled, mixed randomly, and reassembled. In all instances, the lock mechanisms functioned "without requiring the least adjustment." "Blanc," the commissioners said, "is infallible in execution with regard to uniformity, at least when closely observed and using a small number of workers." And given the accusations of later years, it is worth noting that the commissioners also found the quality of Blanc's locks satisfactory.[62]

With regard to the time needed for production, Blanc did not fare so well. His average of twenty-eight hours and forty-eight minutes per lock was 30 percent more than that required at Saint-Etienne and twice the time required at Charleville and Maubeuge. He had, however, achieved considerable savings in the production time for the cock, frizzen, and lock plate, and hoped to do so for the bridle, the sear, and the pan. Moreover, the commissioners noted that the workers at Vincennes were neither accustomed to their jobs, nor assigned to a single task. In a larger manufacture Blanc would be able to take full advantage of a division of labor, so that each worker might "labor constantly on the same object and necessarily become more expeditious." Of course, total production time was even then a naive way to calculate costs. It did not take into account the substitution of cheap labor for skilled craftsmen. Blanc used "apprentices" for the 27 percent of total time devoted to forging and filing with dies and machines, and skilled armorers for the 73 percent devoted to hand filing. At Saint-Etienne the situation was somewhat reversed. There, master artisans forged pieces, which occupied 6 percent of the total time, and masters *and* journeymen spent the rest of the time filing, milling, drilling, and adjusting the piece (75%), tapping screws (6%), and case-hardening and polishing (13%). So partial savings might be expected from this substitution. On the other hand, the commissioners did not consider the additional capital outlay for machines and tools. These costs were estimated at 20 percent of the price of the gunlocks at Roanne. The bottom line was that Blanc there delivered gunlocks at a cost about 20–30 percent higher than those of Saint-Etienne.[63]

In fairness to Blanc, the commissioners noted that the Vincennes dungeon was not the most favorable site for an arms manufacture. Raw materials cost more in the capital, and water power was unavailable. In principle, the city offered the "highly knowledgeable workers" necessary for machine building, but Blanc had been obliged to bring his own assistants with him, including Javelle. There was another unexpected problem with the capital. Blanc's output of gunlocks had been reduced since 1789 by the "frequent interruptions occasioned by the insurrections."[64]

In the event, not even the moat and medieval stonework of the royal fortress protected Blanc from the upheavals of the Revolution. On 28 February 1791 a vast crowd of perhaps ten thousand men and women from the nearby artisanal quarter of Saint-Antoine assembled outside the Château. In the preceding month, the Parisian government had decided to restore the keep at Vincennes as a prison. To some citizens, this signaled the founding of a new Bastille. Had the people torn down one bastion of despotism only to see another raised in its place? This fortress, moreover, contained its own gun factory. As one pamphlet was to note, "The repairs to the fortress, along with an arms factory, caused the most animated disquiet." With the assistance of a small renegade party of National Guards, the crowd seized the tools of the workers repairing the drawbridge and began to demolish the castle keep, smashing glass, prying apart the bars of the windows, and destroying cots suspected of being destined for prisoners. They also smashed machinery. In his *Ami du peuple*, Marat noted the discovery of the atelier for gun-making inside the dungeon: "Soon the alarm spread, 150 citizens arrived at the Château; they found old cannons, about 200 cots, gunlocks for muskets, and all the tools of a workshop." The National Guards, eventually joined by Lafayette himself, halted the demolition and made some sixty arrests. The Parisian municipality assured the population that it had no intention of "resurrecting one of those prisons that despotism raised against liberty."[65] Though not a principal Revolutionary *journée*—such as 14 July 1789 or 10 August 1792—this crowd action marked a significant stage in the deepening confrontation for control of the capital. We will probably never know whether the crowd thought they were razing another tower of despotism, making one of the period's common forays for firearms, or engaging in a bit of wanton machine breaking. We do know that Blanc's machines and tools were damaged in the attack.[66]

In the end, economic, technical, and military factors had little to do with the rejection of Blanc's proposal. In their report, the commissioners worried principally about the *social* effect of substituting cheap labor for skilled artisans in the highly charged climate of Revolutionary France. A new manufacture, they feared, would reduce the activity of the old armories to the point where "the work would languish and the former gunsmiths fall into poverty (*la misère*)." Once the armorers got wind of the proposal, their reaction was sure to be violent.

> Independent of the indigence to which 500 or 600 individuals would be reduced, almost all of them fathers of families, the commissioners believe that it would be impolitic, even dangerous, to have only one such establishment of this type in a state such as France. What's more, one ought to consider that such a manufacture could not be erected with sufficient silence that the suspicions of the locksmiths would not be aroused; and what a stirring up (*fermentation*) this would then excite in the manufactures when the armorers saw preparations afoot to deprive a great number of them of their livelihood (*existence*), above all in the present circumstances in which France finds itself.[67]

Blanc, realizing that state support was not forthcoming, asked permission to set himself up as a private entrepreneur, and promised to supply fifteen thousand muskets with interchangeable parts gunlocks annually. In return, he asked for release from his military obligations, propriety rights to all his machinery, his salary as chief controller, and an interest-free loan of 30,000 *livres*. The commissioners concurred, with the caveat that the government reserve the right to make copies of Blanc's machines. Yet when the minister approved the project on 3 April 1792, Blanc made no move. This was no time to transport large machinery. That month France went to war against the First Coalition. The next spring the campaign turned disastrous, and the French Revolutionary government took extraordinary measures to increase the production of armaments. In that summer of 1793, while Blanc delayed, the Committee of

Public Safety scrambled frantically to organize the mammoth Manufacture of Paris. Finally, in November the Convention authorized his purchase of a Ursuline convent in Roanne, along with the water mill previously owned by the Alcock family. In return, Blanc was to manufacture thirty thousand interchangeable gunlocks per year, plus English-style steel files. But not until March 1794, with the third war season underway and the Committee supreme in Paris, did Blanc leave for Roanne under direct orders from Prieur de la Côte-d'Or. This was at the height of the Terror and the frustrated Committee was already setting up its own interchangeable parts workshop in Paris.[68]

Uniformity—In Parts

At times, the scholarly debate over the origins of mass production has had a peevish quality: which industry—clock-making or gun-making—gets credit for inspiring interchangeable parts production and the machine tools which made it possible? Advocates for each side tout their respective Renaissance visionaries, eighteenth-century forebears, and nineteenth-century success stories.[69] But behind this "precedentitis" lies a more contentious issue: the proper form of social organization. At stake is a presumed contrast between economic rationality and administrative rationality. Did interchangeable parts production follow from a desire to reduce the costs of production or from a logic of command and control? Two value judgments are intertwined here. First, there is the question of whether mass production should be associated with material progress and consumer prosperity—or with the demise of innovation and labor subjugation. Second, there is the question of which agent, the entrepreneur or the state, has done more to aid and abet the methods of mass production. As is often the case in contemporary intellectual life, there is no combination of views that does not have its defenders. David Landes and Alfred Chandler celebrate the techniques of mass production, which they ascribe to the efforts of private American corporations to satisfy the emerging nineteenth-century mass market. Others, like Lindy Biggs and Stephen Meyer, condemn private Fordist control over workers and blame it on the insatiable demands of private capital for profits. Sabel and Zeitlin denounce the state for imposing mass production on vital industrial districts, which they associate with small-scale private capital using flexible specialization. And finally, there is even a (dwindling) number of scholars who celebrate state-sponsored programs of mass production.[70]

My aim with this history of eighteenth-century arms production is to caution against parsimonious explanations. Plainly, the achievement of twentieth-century mass production depended at a minimum on *both* economic and administrative "rationality." Indeed, an entire school of business history has grown up around the concept of "transaction costs," the rational basis upon which administrative procedures are substituted for market mechanisms. In these Chandlerian histories, the rise of an articulated middle management is made subservient to the technological imperative of coordinating vast networks of railroads (or assembly lines). But this cannot be the whole story. The Fordist achievement required social innovations as much as strictly technical and administrative ones: consumer credit and advertising campaigns to prime the mass-market pump, the five-dollar day and social services to control the labor supply, and professional schools and suburban housing to sustain middle management. It also required a cadre of engineers who could build the machine tools, analyze work-effort, and plan production runs, all while touting interchangeable, assembly line manufacturing as laudable technological progress. Mass production emerges from this matrix as an innovative solution to a set of early twentieth-century problems.[71]

Eighteenth-century uniformity production in the armaments industry involved a similar process of negotiation among groups with divergent interests. Resolving the interests of state administrators, provincial merchants, and artisanal producers depended on the relative strength of the parties, itself a function of the social and political context in which the conflict unfolded. Interchangeable parts manufacturing was a step in an ongoing negotiation about the terms of an exchange between engineer-managers, merchant-capitalists, and artisan-workers. What were the tolerances of work to be? How would those tolerances be monitored? Who exercised judgment about whether the standards were met? Who would enforce them? From this perspective, the increasingly objective standards which the engineers enshrined in new measures of labor, technical drawings, gauges, fixtures, and machine tools were all part of an effort to discipline artifacts and the artisans who made them. But they were also a *response* to the continual subversion of those standards by artisans and merchants. Interchangeable parts production emerged from this matrix as an attempt to maintain fixed standards of production for an artifact essential to the authority of the state. My argument in short is this: that an artifact is not the fixed expression of the interests of one social class, but the negotiated outcome of a contested and constantly subverted social process. Artifacts are "objective," then, in the sense that they are the sum of a set of standards agreed to as the terms of an impersonal exchange. And artifacts are "political" in the sense that the outcome of this negotiation depends on the relative strength of the parties involved. Where the managers are state engineers with powers of legal sanction, the conflict becomes explicitly political because its resolution defines the limits of the state's authority in the realm of production. With the coming of the Revolution this resolution was undermined—and reinvented.

PART THREE

Engineering Society: Technocracy and Revolution, 1794–1815

War, which is a horrific barbarism for kings, is just for a people who take up their rights and liberties. War has become for the French Republic a joyous occasion to develop all the powers of the [mechanical] arts, to exercise the genius of savants and inventors, and to consecrate their utility with ingenious applications.
—Chemist and legislator, A.-F. Fourcroy

Chapter Seven

THE MACHINE IN THE REVOLUTION

In March 1794, at the height of the Terror, government posters in the Parisian armaments workshops were found marked with graffiti. Beneath the signature of Robespierre someone had written "anthropophage" (eater-of-men) and beneath the names of Prieur and Lindet someone had written in red pencil: "stupid brutes" and "deceivers of the people," and in black pencil: "thieves, murderers" (fig. 7.1). The employees of the musket workshops had been in revolt for several days. Two arms workers were arrested, and all the workers were made to copy these words to compare their handwriting. The culprit, however, was not identified. The Committee of Public Safety blamed the revolts on "intriguers" and "aristocrats."[1]

In the preceding months, the economic situation in the capital had been worsening: the *maximum*, which limited prices, had taken a beating; the black market was rampant; meat had been rationed. Political opposition to the Committee was heard from the Hébertist sans-culottes spokesmen on the left and the Dantonists and moderates on the right. The marriage of convenience between the Jacobins and the urban masses was unraveling. To reunify the nation, the Committee decided to put on a grand display of fraternity. On March 20—at the same time that the Hébertist sympathizers and Dantonists were being crushed—the government organized a triumphant procession to celebrate the makers of the weapons of war. The Commissioners of Armaments led eighty gunsmiths, bearing "a stack of firearms topped by a cap of liberty." Following them were some of the nation's leading savants and the professors of the Revolutionary Course on the Methods of Armament Production: Monge, Fourcroy, Berthollet, Guyton-Morveau, Hassenfratz, and Vandermonde. They led three groups of students, twenty-five across, carrying their freshly printed instruction booklet and hauling a basket of saltpeter, a mechanical tun for gunpowder, and a cannon they had forged themselves a few days before. Young women and mothers marched with them, as well as city notables, mounted soldiers, and two musical bands.[2]

The rift was not healed. The communal orgy of republican devotion did not usher in a utopian consensus. A week later a large delegation from the rebellious atelier appeared before the Convention. They railed bitterly against the "tyranny of the administration." Candlelight work had resumed, adding two hours of work. Wages, however, had not risen. For some time they had been eating nothing but bread and cheese. They pledged devotion to the Committee of Public Safety, but hinted ominously that someone might seek to take advantage of their "discontent."[3]

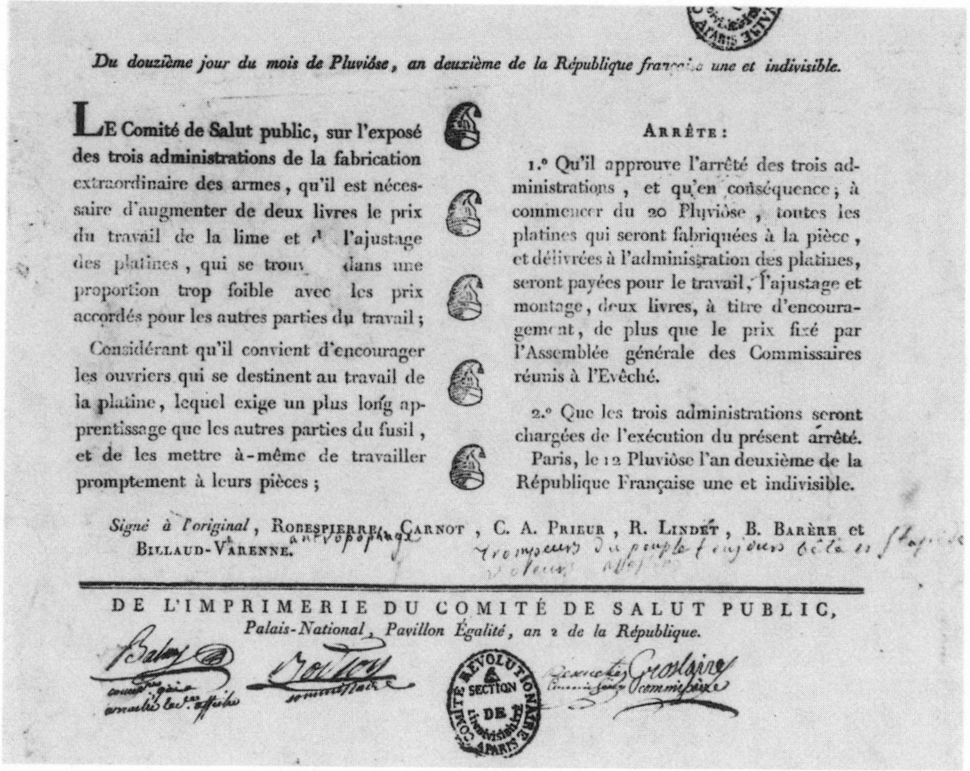

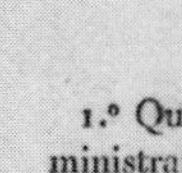

Du douzième jour du mois de Pluviôse, an deuxième de la République fran[illegible] une et indivisible.

LE Comité de Salut public, sur l'exposé des trois administrations de la fabrication extraordinaire des armes, qu'il est nécessaire d'augmenter de deux livres le prix du travail de la lime et [illegible] l'ajustage des platines, qui se trou[illegible] dans une proportion trop foible avec les prix accordés pour les autres parties du travail;

Considérant qu'il convient d'encourager les ouvriers qui se destinent au travail de la platine, lequel exige un plus long apprentissage que les autres parties du fusil, et de les mettre à-même de travailler promptement à leurs pièces;

ARRÊTE:

1.° Qu'il approuve l'arrêté des trois administrations, et qu'en conséquence; à commencer du 20 Pluviôse, toutes les platines qui seront fabriquées à la pièce, et délivrées à l'administration des platines, seront payées pour le travail, l'ajustage et montage, deux livres, à titre d'encouragement, de plus que le prix fixé par l'Assemblée générale des Commissaires réunis à l'Evêché.

2.° Que les trois administrations seront chargées de l'exécution du présent arrêté.

Paris, le 12 Pluviôse l'an deuxième de la République Française une et indivisible.

Signé à l'original, ROBESPIERRE, CARNOT, C. A. PRIEUR, R. LINDET, B. BARÈRE et BILLAUD-VARENNE.

anthropophage

trompeurs du peuple toujours bête et stupide

voleurs assassins

DE L'IMPRIMERIE DU COMITÉ DE SALUT PUBLIC,
Palais-National, Pavillon Égalité, an 2 de la République.

COMITÉ RÉVOLUTIONNAIRE SECTION DE L'INDIVISIBILITÉ A PARIS

Fig. 7.1. Placard for Arms Workshop, with Graffiti, Manufacture of Paris, February 1794. This placard was posted in a lock-making atelier of the Manufacture of Paris in early February 1794. The title was "AVIS AUX OUVRIERS" ("Notice to workers"), and it promised a 2 *livres* increase in the piece rate for gunlocks. Under the name Robespierre someone has written "anthropophage" (eater-of-men) and under the signatures of Prieur, Lindet, and Barère on the right, someone has written: "trompeurs du peuple toujours bête et stupide/voleurs assassins" (deceivers of the people, forever stupid brutes, thieves, murderers). The culprits were not identified. From A.N. W77 plaq. 1, pièce 15, Section de l'Indivisibilité, 22 ventôse, year II [12 March 1794]

April 1794 marked a turning point in the Revolution. For the first time, the Committee used violence to forestall further popular insurrection in the capital. In the north, the French armies were poised on the edge of victories that would make her mistress of the Low Countries. Speaking on behalf of the government, Billaud-Varenne declared that the days of trial and error and spontaneous popular movements were over; henceforth everything would proceed according to plan, according to a "system." In his words, the revolutionary state had become "a machine."[4] Over the next few months the state reasserted its monopoly on violence both on the battlefront and on the urban streets. But beneath the surface of the totalizing language of the nation-in-arms—a rhetoric central to the arrogation of Revolutionary power—divergent social interests threatened to reassert themselves. April also marked a turning point for the Manufacture of Paris, a colossal industrial project undertaken by the Revolutionary government to create *ex nihilo* Europe's largest center of musket production.

The previous summer, the neophyte Jacobin government had been on the verge of collapse. The manufactures on the northern frontiers lay in enemy hands and Saint-Etienne had fallen to the Lyon rebels. Meanwhile, the republican armies were begging for weapons. Desperate, Robespierre and the other lawyers on the Committee brought on board two military engineers from the Corps du Génie: Lazare Carnot and Claude-Antoine Prieur, usually known as Prieur de la Côte-d'Or (fig. 7.2). For their labors, the two men, aged forty and thirty, respectively, have been heralded as the "Organizers of Victory." The epithet has stuck. Under their protection, a whole bureaucracy of savant-technicians went to work for the state. It was a kind of eighteenth-century Manhattan Project, even including a secret site for the testing of exotic new weapons. In the event, these men and their coterie of "techno-Jacobin" savants proved themselves the apogee of the Revolution—expanding the boundaries of state power—and simultaneously the Trojan Horse that ended the popular revolution on the Paris streets. These technocrats carried through on the project initiated by their colleagues in the artillery service and the military reformers of the ancien régime. This cadre of military engineers directed both the armies in the field and the production of war matériel. They were the first to try to implement the logic of total war. Single men between the ages of eighteen and twenty-five were to serve in the army as citizen-soldiers; the rest, in the words of the Committee, "constituted an army occupied with the production of arms."[5] Article 1 of the *levée en masse*, drafted by Carnot, and proclaimed on 23 August 1793, signaled a new relationship between the state and its citizenry.

> Until such time as its enemies are chased from the territory of the Republic, all French people are permanently requisitioned for army service. The young men will go to combat; married men will forge armaments and transport provisions;

Fig. 7.2. Claude-Antoine Prieur, Known as Prieur de la Côte-d'Or (1763–1832). Prieur de la Côte-d'Or was trained as a fortification engineer at the Ecole du Génie at Mézières in the ancien régime. He won election to the National Convention and served on the Committee of Public Safety during the year II, with primary responsibility for the production of war matériel, for engineering education, and for the promulgation of the metric system. The portrait comes from the B.N.

> women will make tents, clothes, and serve in hospitals; children will tear old linen for bandages; old men will be carried to the public places to stimulate the courage of the warriors [and] preach the hatred of kings and the unity of the Republic.[6]

For two years, French leaders had been calling for mass armies to emancipate Europe. The concept of an invincible citizen-army was no longer an Enlightenment daydream. In the words of the *levée en masse*, "In a free country, *every citizen is a soldier*."[7] The French Revolution founded a new juridical state, one based on the absolute right of property and the universal rights of man; it also founded a new national state, one which initiated twenty years of warfare across the Continent and superintended an unprecedented "crash" industrial project to manufacture guns. Overnight, Paris was transformed into a public arena where the struggle for political power took place against the backdrop of state-sponsored industrial workshops. In 1794, the year of the Terror, the capital rang with political speeches and the blows of hammers.

At its peak, the Parisian manufacture employed over five thousand citizen-workers in thirty different government-run workshops. It was the largest "crash" industrial project Europe had ever seen. Its goal was to produce one thousand muskets per day, more than six times the combined arms output of the ancien régime's three armories. But the Committee's reliance on patriotic fervor in the workplace was in tension with its engineering program of efficient production and rational administration. The Committee might silence political dissent, but it found that it could neither co-opt skilled armorers, nor "revolutionize" Parisian metalworkers into skilled armorers overnight. This was the context in which the Committee founded the Atelier de Perfectionnement—the "Workshop of Perfectibility"—an experimental workshop to construct machine tools to manufacture firearms with interchangeable parts. The growing rift between the government and arms workers would no longer be slicked over with pledges of fraternity and festive parades. For the technocrats, the utopian community was not a procession of garlanded youth. They dreamed of a prosperous future enriched by the bounties of uniform production and populated by disciplined, rational, and innovative citizens. Jean-Henri Hassenfratz, a Jacobin savant who participated in the procession of the Arms Makers, spoke of the festival with contempt:

> Let us be careful lest while we are busy organizing our festivals, our neighbors may organize their industry. . . . It was not with festivals that the English have been able to acquire a great preponderance over the political balance of Europe. It is not with festivals that the United States of America became a flourishing people.[8]

For these techno-Jacobins, the utopian community was a workshop, and the Atelier de Perfectionnement was the seedbed for a new kind of precision-minded industry and a new kind of technician. They imagined a community of skilled machinists devoted to a common program of innovation—along the lines of Honoré Blanc's *Encyclopédie-pratique*. In the long run, perhaps all French citizens might share in this program. In the short run, however, the machinery developed in the Atelier would help the technocrats impose discipline on the great mass of arms workers. As the sign of this discipline, they aspired to produce firearms with interchangeable parts. In place of fraternal vows and organic consensus, social harmony would be achieved by mechanical routine. The great machine of the state would run to the rhythm of mass production. Of course, this vision did not come to pass, anymore than Robespierre's Republic of Virtue.

This chapter, then, reintegrates the story of a (failed) technological revolution into the familiar history of a political revolution. It examines the attempt of the engineers to guide the French Republic toward technocracy: first a form of democratic technocracy, and then a hierarchical technocracy. The Manufacture of Paris mobilized social and technical resources on an unprecedented scale. Out of the rubble of the ancien régime emerged a state capable of organizing the manufacture of over 140,000 guns in a year. This was a vast creative act, involving the effort of thousands of French laborers, artisans, savants, and administrators. Whatever their differences, these citizens bent to their respective tasks with a ferocious energy. This was also an act of great violence, not only because it forged the instruments of mass war, but because it engendered radical social forms that threatened to alter the political equation in the state. So despite its brief existence, the manufacture posed a question central to the new régime: what was to be the relationship between republicanism and the productive order?

Patriotic Production

We know shockingly little about work practices during the Revolutionary period, and the dominant school in Revolutionary historiography is hardly promising on this score. François Furet's thousand-page *Critical Dictionary of the Revolution* is silent on the subject of how the common people earned their daily bread. Hence, it says nothing about the relationship between changes in working conditions and the course of Revolutionary politics.[9] The closest thing we have to such a study is still the forty-year-old account of the "sans-culotte movement" by Albert Soboul, *Les sans-culottes parisiens en l'an II*. This was followed by its companion pieces: George Rudé on the Revolutionary crowd and Kåre Tønnesson on the defeat of the movement in the year III. These histories attempted to recapture the popular face of the Revolution through the proclamations of the "sans-culottes," political activists who put popular pressure on their erstwhile Jacobin allies in the state. Soboul associated this group with the artisanal classes, and thence with the aspirations of the undifferentiated urban masses. But in making this association, Soboul and his followers were caught in a circularity. Soboul always acknowledged that the sans-culottes included both master artisans and their journeymen-employees, and therefore could not constitute a Marxian social class. What gave these employers and employees grounds for social cooperation, Soboul believed, was the intimate and cooperative nature of artisanal work as celebrated in sans-culotte lore. And what gave them a coherent political program was their common vulnerability as consumers in the face of escalating food prices—hence, their demand for a *maximum* to limit food prices, and the success of the food riot as a political force in Revolutionary Paris. It was this social program whose political face was called the Terror.[10]

Recently, this circularity has been thoroughly exposed. The problem with Soboul's analysis is that it mistakes a rhetorical construction useful in the pursuit of political alliances for an expression of "authentic" social interests. Rather than connecting social rank and political action, Soboul has painted a brilliant tableau of the political culture of Revolutionary activists. Richard Mowery Andrews has shown that the sans-culotte activists were socially indistinguishable from the Jacobin leadership they initially supported and later opposed. Michael Sonenscher has stressed the diverse experience of working life in the ancien régime—which included both intimate ateliers and large workshops—and hence that the image of the "sans-culotte" functioned as a rhetorical meeting ground for artisans and activists in a turbulent city. Finally, William Sewell has pointed out that bread prices were not particularly high in 1793 when the demand for the *maximum* was loudest, and that what counted here was the way the militants expanded the popular equation of dearth with hoarders into a broader demand to extirpate secret and selfish *political* conspirators.[11]

These critiques remind us that the relationship between social standing and the articulation of interest is a fluid and mediated process. This is certainly true of the relationship between one's position in the workplace and claims about one's place in the social order. As we saw in chapter 4, William Sewell and Michael Sonenscher have suggested some of the ways in which artisanal producers articulated their interests in this period before the language of class became predominant. It has long been clear that work became invested with new moral imperatives during the French Revolution. The Le Chapelier and d'Allarde laws of 1791 abolished the corporations and journeymen's *compagnonnages* as the public referents of work. What I wish to explore here is how, in the new climate of the nation-in-arms, work became (temporarily) identified with patriotism. This equation has recently been hinted at by Haim Burstin, who quotes the observer Parrière shortly after the *levée en masse*, "Love of work was united with love of the *patrie*. . . . [W]e could say perhaps even more precisely that it was love of *patrie* that excited the love of work."[12] In such a context, the collective interests of workers *qua* workers quickly came into contradiction with their membership in a presumed "people" whose sovereignty lay in the state. Wage demands were seen as selfish if they challenged the *maximum* on salaries, and hence politically subversive. Yet who were "the people" if not these workers?

Nowhere were these contradictions more evident than in the workshops for military weapons. There, the duty of the citizen-worker to produce was explicitly tied to the citizen-soldier's duty to fight. The Committee asserted that the armorers were soldiers recruited for industrial work, and like their brothers in the field, "owed the Republic all the time which their strength permits." Workers "formed part of the army" and "constituted an army occupied with the production of arms." Tying their lot to that of their brothers-in-arms served both as patriotic exhortation and as veiled threat: younger workers could be sent to the front, and all were subject to military discipline on the job. For example, as the Christmas season approached, the Committee warned that workers, like soldiers, would not be allowed to take off Sundays or fêtes, but only the *décadi*, the last day of the republican ten-day week (*décade*).[13] The artillery service's long-standing claim that the arms workers of Saint-Etienne were subject to military discipline was now subsumed into a larger obligation to the state.

Yet, as in Saint-Etienne, the arms-making ateliers were a source of continuous unrest. And, as in Saint-Etienne, this unrest was no trivial matter. Workers contested wages and conditions, and organized collective protests against the Jacobin Committee. This was a material threat to the state because the army was desperate for muskets. It was also a direct political challenge to the legitimacy of Jacobin rule because if the work was an expression of patriotism, then protest about the conditions of work was political protest. The logic of state Terror made opposition from such a collectivity impossible to countenance. Hence, as we will see, the meaning of work had to be continuously renegotiated during this period as the political equation changed. This had practical implications for wages and working conditions. It also transformed the ways guns were made. In this brief period, the production of weaponry became *explicitly* politicized; that is, it partook of a public contestation over the meaning of republicanism.

This transitory (but illuminating) development has escaped previous investigators. Previous accounts of this episode have all relied on Camille Richard's meticulous and monumental *Le Comité de Salut Public et les fabrications de guerre sous la Terreur*, published in 1921. Richard, however, adheres closely to Mathiez's concurrent effort to glorify the Jacobin dictatorship for saving France from invasion. The specter of World War I haunts every page. And like Mathiez, Richard never squares his admiration for the Herculean Committee with the opposition they faced from the "sans-culotte" armorers he also admires. He accepts on face value the claim that the disruptions in the ateliers were caused by "adversaries" of the Revolution. In part, this is because almost all his evidence comes from official government sources. Tellingly, he never mentions that the graffitist of March 1793 scrawled "Robespierre anthropophage," noting only that Barère, Prieur, and Lindet were impugned. The sole study to appear since then is the 1976 unpublished master's thesis by Françoise Gaugelin, written under the direction of Albert Soboul. In the spirit of Soboul's program, it distinguishes between the movement of the sans-culottes and the interests of the Jacobin dictatorship. Like his work, however, it takes the sans-culotte representations on face value. Again, this is partially a methodological problem; the police reports that she uses (valuable as they are) tend to magnify threats to the state.[14]

These studies equate words and deeds, rhetoric and social structure. In the pages that follow, I will offer a more plausible and complex picture of the Manufacture of Paris, based on new findings in the archives. Only after this social structure has been mapped out can we address the contest over republicanism. Offered by the Jacobin state as a patriotic framework for a common endeavor, trumpeted by the sans-culottes as a call for democratic participation, republicanism was (tentatively) taken up by a heterogeneous group of workers in defense of their interests. The negotiations between the techno-Jacobin state and the Parisian arms workers over their mutual obligations in the workplace (wages, working conditions, and technology) also involved a negotiation over the terms of republicanism and the meaning of the French Revolution.

The debate over Revolutionary France's "industrial policy" was refracted through a conflict over its political culture. One important aspect of that debate concerned the role of the state in encouraging large-scale manufactures. As in the contemporary United States, there was considerable debate over the effect large-scale enterprises would have on urban life and the fledgling republican experiment. Some French men and women positively welcomed large-scale manufactures as a potential source of employment and discipline for the poor. Others feared that manufactures would create a threatening concentration of workers in an already troubled city. As one former policeman put it in 1791: "Politically, it is often a dangerous strategy to assemble a massive number of men deprived of property, thus conferring upon them the kind of independence that almost always accompanies the sense of force."[15] A central question of French political life in the early 1790s was the degree to which the state should sanction, encourage, or take the lead in this effort. The Jacobin leadership, as is well known, sought to uphold the sanctity of private property. Generally hostile to state control of the economy, they acceded to the *maximum* and state-sponsored workshops only as a political expedient. On this question, the sans-culotte populists had a distinct point of view.

The Manufacture of Paris lurched into existence as part of that social program and its attendant rhetoric which has become known as "the Terror." This program was first enunciated by the sans-culottes before being appropriated by the state. The Convention first heard the call to found "national ateliers for the production of arms" on 31 May 1793, the day of the first insurrection against Roland de la Platière's Girondin party. It was announced as one of the seven demands of the Parisian sections, along with a domestic *armée révolutionnaire*, a fixed price of bread, aid to widows and orphans of the war dead, and the cashiering of all noble-born military officers. Three days later a tremendous crowd, assembled under the banner of this program, enabled the Jacobins to defeat the Girondins and purge the Convention. All that summer, populists continued to demand that all workers be placed in the service of the state. Hébert's *Le Père Duchesne* announced: "We must requisition all workers who work with metal, from the farrier to the goldsmith, establish forges in all the public places, and labor night and day to manufacture cannons, muskets, sabers and bayonets."[16] The manufacture belonged root and branch to the radicalization of the Revolution.

By that fall the Jacobin leadership echoed the sectional militants, recruiting workers as part of the mobilization of the *levée en masse*. On behalf of the Committee of Public Safety, Billaud-Varenne and Collot-d'Herbois declared: "Let locksmiths cease to make locks; the locks of liberty are bayonets and muskets."[17] Soon the Committee had acquired the authority to requisition all metal workers, and exempt from the army all citizens capable of producing arms. For the time being, the Jacobin leaders suppressed their doubts about the wisdom of state-sponsored production. The Jacobin state portrayed itself as the fullest expression of the nation-in-arms. Republicanism, in such a situation, was predicated on an ideal of universal service, including the nationalization of production. As the Committee put it:

> It is necessary that each citizen render unto the Republic all the services which she has the right to expect of him; of two services, that which is the more important and not within the capacity of all citizens must be preferred; the need for firearms is the most pressing of all those which in this moment of crisis faces the Republic.[18]

This duty was backed up by extensive powers of enforcement. Those who evaded the requisition for military service or the manufacture of arms, or who produced a firearm not destined for the army would be placed in irons for two years. When artisans employed in the Manufacture of Paris were arrested for avoiding the draft, the Committee of Public Safety turned around and imprisoned the arresting officer.[19]

Of course, nothing obliged the state to concentrate this arms-making effort in Paris. Paris possessed no indigenous armaments industry, the prices of raw materials were higher there, and water power scarce. Yet the Committee acceded to sans-culotte demands that the manufacture be located in the capital. More than anything else, this political decision shaped the fate of the endeavor. It meant that the manufacture emerged over the course of the year II as a hybrid beast, designed to meet a variety of interests and purposes. It was part state-owned and part private capital; it was composed of skilled armorers, Parisian metal workers, and occasional laborers; and it produced both new weapons and repaired old ones. It was an armory, and simultaneously a vast public works program.

Municipal leaders had always feared the great masses of urbanites. This fear was magnified with the Revolution. Not only did the Revolution dismantle the traditional mechanisms for maintaining order in the workplace, it disrupted the normal circuits of commerce (particularly in the luxury trades).

This economic calamity was also an affront to the fraternal generosity of the Republic. In 1790–92, the state organized charity workshops for the poor. But as Alan Forrest has pointed out, these schemes became financially untenable as the military needs of the nation escalated. An armaments manufacture in Paris solved both problems; it put the capital's underemployed to work combating the greatest threat to the nation. Carnot acknowledged that the motive for establishing the armory in Paris was to provide work for restless urbanites. A police observer noted that the prospect of national arms workshops in the Faubourg Saint-Germain seemed to have "revitalized and encouraged" the young men of the *section*. The citizenry expected even more jobs to be created. The manufacture, then, cemented the new political alliance between Jacobin leaders and the urban populace.[20]

But even as the government sought to mollify the mass of city workers, it also needed to supply its troops with adequate weapons at the lowest price and as quickly as possible. The solution adopted by Carnot and Prieur, the organizational heads of this leviathan, was to divide the workers into three general categories: those supplying parts under private contract at their own shops, those working in the national ateliers on piece-rate wages, and those in the national ateliers receiving a fixed daily wage. These three types of workers were not all equally laudable, however. The technocrats clearly would have preferred to leave as much of the operation as possible in private hands. Like Gribeauval, they spurned the *régie* system because it was inefficient, unreliable, liable to fraud, hostile to innovation, and more concerned with bureaucratic aggrandizement than the public good. They also worried about the "anti-democratic" threat posed by large state-run enterprises; that is, they worried that the concentration of workers could be mobilized to threaten the state. In April 1794, when the Committee finally felt free of populist pressure from the sans-culotte militants, Prieur refused to countenance the creation of yet another nationalized foundry.

> Whereas *régies* are onerous to the nation for reasons of economy and prejudicial to the army because they do not deliver at regular intervals; whereas the administrators of *régies* are more preoccupied with their own convenience and adding to their establishment than in the actual work processes and products; whereas in the *régies* new processes are never introduced and improvements (*perfection*) in the works make no progress; and whereas the *régies* accord officials the disposition of appointments, giving them a dangerous power in a democratic state.[21]

Yet this hostility to state ownership was mostly an assertion about the *limits* of government intervention. The technocrats realized that only the state—and only a state of a particular sort—could command the resources to accomplish this extraordinary task on such short notice. Where Roland's Girondin government had disastrously failed to supply the army with weapons, the technocrats proved willing to do what it took. Throughout the year II, the vast majority of guns made in the Manufacture of Paris came from factories owned and operated by the French government. On the same day he lambasted the system of *régie*, Prieur awarded state moneys to Lakanal's state-run armaments community in Bergerac, and within a fortnight he had founded the Atelier de Perfectionnement.[22]

The Manufacture of Paris: Managers and Workers

From the point of view of the desperate French armies, the Manufacture of Paris lurched into activity with frustrating torpor. When Carnot reported on two months of progress on the manufacture to the Convention on 3 November 1793, he had only six new all-Parisian muskets to show. In mid-December, the manufacture was producing 200 muskets a day, but these were mostly repairs. In the meantime the army had grown from 300,000 to 500,000 men and was still expanding. The daily and seasonal rhythm of battle demanded that guns be placed immediately in the hands of soldiers. All that winter and spring, complaints of acute shortages of muskets and bayonets streamed into the offices of the Committee of Public Safety.[23]

From the point of view of the city, however, the manufacture mushroomed into existence almost overnight. Within a matter of months, some 5,000 armament workers were laboring in large workshops of 200–300 each, all run by a single employer: the state (fig. 7.3). This meant that roughly 25,000 men, women, and children depended on the manufacture, a substantial proportion of the 250,000 persons in the capital who lived by manual labor. Drawn from a variety of trades and possessing a wide range of skills and interests, they nonetheless formed a considerable mass in the heart of the capital. The ateliers themselves were established in a variety of state-appropriated buildings: the mansions of émigrés, former monasteries, and even the university. By the spring of 1794, they included 17 factories devoted to gunlocks and musket assembly, and 14 devoted to gun barrel production (6 for grinding, 5 for reaming, and 3 clusters of 156 forges). Eight factories were mounted on river boats where waterpower from the Seine helped ream and grind gun barrels—despite the lowest water level in decades.[24]

Supreme management of this vast operation fell to Prieur de la Côte-d'Or. Almost every order emanating from the Committee concerning armaments bears his signature, but he worked mostly behind the scenes. Saint-Just would later lash out at his colleague who, he said, was "enslaved in the bureaus." Meant as a bitter insult—Prieur as the antithesis of the Revolutionary ideal of openness and accountability—this characterization contains a great deal of truth. Prieur was a Corps du Génie engineer uncomfortable with oratory and literary forms. After graduating from Mézières in 1784, he worked on chemical research with his older cousin, Guyton-Morveau. Like Guyton, Prieur was elected to the Legislative Assembly in 1791 where he had a relatively undistinguished career until he took Guyton's place on the Committee of Public Safety in the summer of 1793. There, Prieur exercised considerable power by means of the bureaus, arguably more than Saint-Just did at the bar of the Convention.[25]

During the first half of the year II (from September 1793 to April 1794), Prieur, Carnot, and the rest of the Committee gradually usurped the War Office's authority over weapons production and vested power in a civilian National Commission on Arms and Powders. In September 1793, provincial artillery inspectors were replaced with "intelligent republicans, *artistes* who perfectly understand all the details of production." In February 1794, the Committee took over the provincial armories, and gathered the army's arsenal into the Manufacture of Paris. The absolute rule of the Committee over military matters (both army command and supply) was consummated at the high tide of April.[26] By then the Commission on Arms and Powders included a bureau devoted to musket production, itself divided into sections for gun barrels, small pieces, and accounts. This administration supervised government ateliers, supplied raw materials and tools, recruited the work force, assigned production quotas, signed private contracts for armaments, and assured "uniformity in the price of all the parts of the manufacture and repair, as much as the different locales will allow."[27] Yet even as this administration usurped the artillery's authority, it employed personnel trained in the military offices of the ancien régime. The three chiefs of the armaments administration were Dupin (a War Office adjunct), Capon (navy), and Bénézech. And Carnot and Prieur (themselves military engineers) were seconded by a coterie of savants, many of whom had long-standing ties to the army. As we will see, these technocrats remained in office under the Thermidoreans, a striking confirmation of Tocqueville's hypothesis of the continuity of administrative organs across the Revolutionary divide.

This top-down bureaucracy, however, was also subject to popular pressure, which varied with the political situation in the capital. The Manufacture of Paris was governed by a quasi-democratic Council, which included worker-representatives as well as administrators. Members of an oversight board included: Hassenfratz, Régnier (an armorer-inventor), and Mégnié (Lavoisier's instrument-maker). These administrators were to combine technical know-how with patriotism, so that they would not "mislead the people." This combination was one which the savant-engineers seemed to fit most readily. When the Committee needed a roving inspector who was "knowledgeable, honest [and] republican," it turned to Vandermonde.[28] The internal organization of the Parisian workshops also reflected this compromise between administrative rationality and popular pressure. Each government atelier was placed under the authority of a technical supervisor (the *directeur*) nominally selected by the workers. He was, however, subordinate to an administrator (the *agent comptable*) who was appointed by the government, sat on the board of the central administration, and coordinated relations between workshops.

But all problems paled before the shortage of skilled labor. Workers were recruited from three main sources: (1) skilled armorers from the former royal armories (especially Maubeuge), (2) metalworkers from Paris and the provinces, and (3) indigent day laborers from the capital. Within days of the *levée en masse*, the Committee called to Paris some twelve hundred armorers from occupied Maubeuge (including some from Liège). These skilled artisans formed the core staff in many of the government ateliers, acting as chiefs of production, particularly for gunlocks. In his report to the Convention, Carnot painted a picture of these artisans toiling happily for the Revolution, bringing the ringing sound of hammers into monasteries formerly encased in silence, inactivity, and ennui. In fact, these workers presented a number of special problems. They complained bitterly about the shortages of tools, coal, and waterpower. They complained about the high infant mortality rate in the city. Their wives were bored and homesick. Many badmouthed the administration, refused to train other workers, stuck together, and worst of all, stood up for their rights. They had the skilled craftsman's contempt for the bumbling city-apprentices around them. The Committee tried to accommodate them. It found them lodging in the former monastery of the Miramiones—which later became a hotbed of worker revolt. But when they complained, it denounced them as foreigners (Liègeois) and purged them from the ateliers.[29]

The bulk of the labor force was recruited from the Parisian metal trades. Immediately after the *levée*, each section drew up a list of potential workers, noting their "civicism" and whether they possessed workshops or tools. Lists from 12 (out of 48) sections survive, and of these 1,734 workers only 7 percent were gunsmiths and 7 percent swordsmiths; evidence that the bulk of the workers had no skills in the gunsmithing trades. Most were in "allied" metal trades: 30 percent were makers of door locks and associated hardware (*serruriers*), 16 percent jewelers and goldsmiths, 14 percent cutlers and related ironmongers, 7 percent watchmakers, plus a smattering of machinists and instrument makers. These gross statistics, however, mask enormous disparities. The workers lived unevenly across the city; for instance, nearly half the armorers lived in one section (Amis de Patrie). They also differed greatly in their wealth. In Arcis (a section with mostly jewelers and cutlers) roughly half the metal workers had access to their own "boutique" or "room," and their journeymen were evenly distributed among them. In Poissonière and Beaurepaire a few masters employed the bulk of journeymen, while a majority of masters employed none or only a few. These new data confirm Michael Sonenscher's reanalysis of Braesch's famous census of the urban trades during the Revolution. The diversity of work cultures included intimate ateliers *and* large workshops. To that extent, the scale of the government ateliers would not have been entirely novel to some artisans. Unfortunately, these lists do not indicate which workers actually took jobs in the state ateliers—although employment lists from a few ateliers do confirm these patterns. We also know that eligible workers had to have two years' experience in an "analogous trade," pass before an examining board, and undergo a two-day trial period. Nevertheless, many workers without any metalworking skill were hired as well. These indigents and day laborers performed some of the low-skill tasks associated with gunmaking.[30]

This heterogeneous assemblage of workers was distributed among the different aspects of musket production. To illustrate this, I will examine three different types of ateliers. The first is the three barrel-reaming factories located in large boats moored on the banks of the Seine which employed a total of 126 workers. Each of the two shifts on each boat consisted of a core of three skilled barrel-forgers from Maubeuge or the provinces assisted by 18 day laborers without experience in any relevant trade. These coiffeurs, butchers, coachmen, masons, and men *sans état* performed those low-skill tasks associated with barrel-reaming and grinding operations.[31]

Next consider the atelier of Chartreux (also called Observatoire) where some 240 locksmiths from Maubeuge made gunlocks in a former convent. Of the 759 new gunlocks produced in nivôse, year II (21 December 1793–19 January 1794), 80 percent came from the Chartreux atelier. Retaining these skilled artisans was not easy. Private entrepreneurs in Paris competed with the national ateliers for capable workers. Losses occurred whenever a new state atelier was established, either in Paris or the provinces. This demand for skilled workers quite naturally gave these armorers leverage to demand greater wages.[32]

Finally, consider the Quinze-vingt atelier, the only workshop for which we have extensive daily records (records not consulted by Richard and Gaugelin). Located in the famous artisanal quarter of Saint-Antoine, this atelier was typical in that it made use of Parisian metalworkers. Yet it was itself a heterogeneous operation. Before the Revolution, it had been the steel-goods factory of the entrepreneur Jean-Joseph Dauffe. It had then been nationalized in February 1794, with Dauffe remaining as its salaried *directeur*. The *agent comptable* was a M. Ciraud, "mathematics" professor at a pre-Revolutionary technical school in Paris. One portion of the atelier was devoted to the sharpening of bayonets, another to making gunlocks. It also supplied other ateliers with tumblers for gunlocks, and with steel files. The 21 men sharpening bayonets were paid a collective piece-rate wage. As their productivity rose to 4,000 per *décade* in the early fall, their wages rose from 4.5 *livres* per day to an impressive 15 *livres*. Even this did not satisfy everyone; after the summer, 18 superior workers shared one pot, and the 3 inferior workers divided another. They were assisted by 12 unskilled wheel-turners, paid 3 *livres* per day. Eight horses also supplied power. All told, they sharpened over 75,000 bayonets in 11 months.

While the number of bayonet-sharpeners was held constant, the number of locksmiths in the Quinze-vingt atelier rose steadily: from 37 in mid-February, to 76 in mid-April (when the crisis in gunlocks hit home), to a plateau of 175 in August. Another 60 workers forged lock pieces and steel files. The locksmiths (filers) worked in groups of 9–15, each headed by an armorer-instructor, who was paid a set wage (10 *livres*/day) to train them in the trade. A survey in early April showed them to be typical of the Parisian metalworkers surveyed above: 20 percent were armorers, 51 percent were makers of door locks and ironmongery (*serruriers*), 20 percent were in sundry metallurgical trades (clockmakers, jewelers, machinists, founders, and cutlers), leaving 3 percent in unrelated fields, and 6 percent apprentices. Although these workers were paid a flat daily wage, this varied from 4 to 7 *livres* per day depending on their previous output. Their productivity varied widely from worker to worker. In mid-May, when the atelier's 83 locksmiths turned out 123 locks, 16 workers made 3; 28 made 2; 19 made 1; and 20 didn't even finish one. The same pattern repeated itself *décade* after *décade*. Workers who consistently made three locks were promoted to instructor or sent to work at Chartreux for piece-rate wages. At the other end of the scale, incompetent workers who had not completed a single gunlock in the *décade* were held to 4 *livres* per day, and their names forwarded to the central administration. Yet in July, three workers had not produced a single lock in the past six decades, and four had done so only once. Of the 33 requisition-aged workers in mid-May (roughly one-third the total), 6 made none. Average productivity, after rising steadily from 0.8 locks per worker-*décade* in February to 1.7 in June, began to jump erratically, rising and falling between 1.0 and 2.2 locks per worker-*décade*. At no time did it even approach the levels of 5.0 that would have been typical of skilled armorers during the ancien régime. Overall, the atelier produced 5,356 gunlocks, plus 3,850 tumblers.[33]

The muskets produced in these ateliers reflected this heterogeneity. We know from the sketches for Hassenfratz's unwritten memoir on gun production that the methods superficially resembled those of the ancien régime. Precision manufacture was considered a luxury, however, and arms workers employed only crude gauges to check the diameter of barrels and the basic dimensions of lock parts. The main test was functional. In the span of eight months, only 7 percent of gunlocks in the Quinze-vingt atelier were rejected, an extremely low rate even for experienced armorers, which the Parisian workers were not. The gun proof was conducted alongside the now-demolished Bastille, and it required only one shot, and none of the elaborate procedures of the ancien régime. Barrels which the artillery engineers would certainly have broken were sent back for repairs. On those rare occasions when arms were rejected, workers complained about corruption, blamed the quality of the iron, and demanded to be paid anyway. Speed of production, not quality, was now the paramount concern. And the semiskilled workers of Paris could not match the standards of the ancien régime in the best of cases. One of the central roles of the administration was to distribute master models of the gunlock and other parts (pattern guns) to all the ateliers, both privately run and government owned. But as we will see, the state was not willing or able to enforce higher standards until after April 1794.[34]

This heterogeneity has been obscured by previous accounts of the manufacture. The arms workers have been lumped together by terms such as "artisan" (Soboul, Richard) or "proletariat" (Guérin), when such categories are precisely what was fluid in this period. These workers differed in their geographical origins, their levels of skill, their familiarity with large-scale manufactures, their labor mobility, their role in the production process, and no doubt in their political views as well. As Charles Sabel has recently noted, this heterogeneity may well be a persistent feature of the lives of working people. What is all the more remarkable is that—to the horror of the technocrats—they proved themselves capable of collective action under the banner of republicanism.[35]

Discipline and Negotiation

Insurrections in the ateliers occasionally coincided with rising prices and city-wide mass movements. But rather than responding to bread prices (as urbanites had for centuries), conflicts between the arms workers and administrators revolved around work conditions and pay. In addition to engaging in crowd actions and petitioning the Convention, the workers also staged work stoppages, slowdowns, and other forms of industrial rebellion. To defuse these protests, administrators deployed the rhetoric of revolutionary fraternity, praising the patriotism of their comrade-workers, invoking the higher national purpose, and blaming the troubles on subversive elements who "adroitly exploited the private [pecuniary] interests" of workers. Workers who followed these "intriguers" were unwittingly in league with the enemy. The language of the Terror could not allow that citizen-workers would willfully oppose their own representatives, any more than citizen-soldiers could defy their military commanders. Yet the language of patriotic production also implied genuine consultation with workers—much as citizen-soldiers were also granted some say in who would command them. This had practical implications for the material incentives and punishments that the workers faced. On the negative side, these included the threat of transfer to the army, imprisonment, and (rarely) the guillotine. On the positive side, these included negotiations over wage rates, work rules, and a social insurance scheme.[36]

Take the delicate question of compensation. Time and again, Prieur and Carnot expressed their preference for piece-rate wages because they "increase production, diminish [the need for] surveillance, and give the worker a better sense of liberty." The Republic, they argued, ought to foster the development of the self-reliant artisan. In the tradition of Rousseau, the independence that came from earning one's own livelihood through the exercise of one's skills also made citizens more capable of independent political judgment. Here, the image of the independent sans-culotte dovetailed nicely with the administrators' expectation that piece-rate wages improved productivity. Unfortunately, even experienced metalworkers—clock-makers, say—were unable to earn a living wage making guns on the same piece rates as skilled armorers. Hence, they demanded to be employed in the state ateliers at a fixed daily wage. But as this was the least desirable form of payment from the point of view of the state, administration insisted that daily wages were a temporary expedient and that ultimately all workers would be shifted to piece-rate work.[37]

Consequently, setting the relationship between these two forms of compensation meant balancing different values associated with the working life. So rather than determine wages unilaterally, the administrators negotiated with the workers about their salary, while carefully controlling the process by which agreement was reached. This form of "democratic production" had been pioneered by Romme the previous year at Saint-Etienne. In Paris, the Council of the Manufacture was composed of thirty artisanal delegates elected by the sections and thirty government delegates from the ranks of the administration. The proceedings were monitored by three representatives of the National Convention, including Noël Pointe, gunsmith of Saint-Etienne, and presided over by Hassenfratz. This Council met in October for an extended arbitration hearing. Initially, the administration had adopted the expedient of setting wage rates to equal those of the armory of Charleville. As the Council noted, however, the metalworkers were not willing to work for what now amounted to apprenticeship wages; many had wives and children to support. At a series of meetings, the Council raised the daily wage from 3 to 4 *livres*. These daily wages were only to last three months, however, and the administration wanted to encourage workers to go onto piece-rate compensation. The process of setting these incentives was adroitly manipulated by the administration. On the third day of hearings, a Maubeuge worker proposed that the *embouchoir* (a metal band on the strap of the gun) be compensated at 45 *sols* each. This prompted a tirade from the administrators who claimed this rate would earn selfish workers the "exorbitant" salary of 30 *livres* per day. As a punishment, the name of this selfish worker, Jean-Joseph Vernelle, was marked down in the *procès-verbal*, and the price set at 40 *sols*. This happened several times. Overall, piece-rate wages rose one-third (the same as the daily wage).[38]

At the same time, the Committee expanded its rudimentary program of social assistance. In 1792, the Girondin government had announced pensions "proportional to service" for all armorers with thirty years' experience and over fifty years of age. The law had also promised compensation for work-related injuries. Now the elected Council added leaves of absence and material assistance to injured workers "if they have served the public zealously." The citizen Fernet received the sum of 60 *livres* for an injury received during work; Dauré received 100 *livres*, as did Hubert Galand. On rare occasions, sick pay was granted.[39]

These quasi-democratic negotiations and proto-insurance schemes were meant to allay the mistrust between administrators and workers. Workers were being urged to see their economic exchange with the state as part of a patriotic effort. When workers in one atelier demanded higher wages, Hassenfratz reminded them that: (1) wages had to be the same for everyone and were already higher in Paris than Charleville, (2) loyal sans-culottes would want to do everything they could to defeat the enemy, and (3) the wage package had been agreed to "by the workers themselves."[40] This was part of the government's larger policy which matched the *maximum* on commodity prices with rigorous enforcement of the wage *maximum*. In fact, commodity prices were rising in spite of the law, and the Paris Commune in the hands of the Hébertistes had allowed wages to rise for workers in private industry. This meant that the arms workers (who answered to the national government) saw their wages eroded over the course of the winter.

Arms workers grew increasingly dissatisfied that winter. In early December, disturbances shut down two workshops. After an insurrection over wages in the Capucins atelier, the Committee of Public Safety had the six worker-appointed chiefs of production arrested (all of them armorers from Maubeuge and Liège). At the end of the month, the armorers there still complained about a salary cut. At the same time, Le Faure, the director of the Marché-aux-Poissons gunlock atelier, was arrested because his workshop had mutinied. Hassenfratz reported that a majority of workers had been absent when he made a spot inspection there.[41] Nervous, the Committee instituted "police measures" to reduce the threat of collective action. These decrees amplified the Loi Chapelier that forbid workers' associations, and they were backed up by that month's law of Revolutionary government. The decree forbade all assemblies of workers and warned that all "mobs" (*attroupements*) of workers would be dispersed and the instigators would face two years in irons. Individual complaints would still be heard, but communication between ateliers had to be effected by intermediaries. If workers wished to voice a collective grievance they had to ask permission to assemble and designate two representatives to bear their petition. Chiefs of the workshops were accountable for preventing counterrevolutionary *mouvements*.[42]

Despite these admonitions, the rebellion spread. Government ateliers lost two hundred workers (over 10%) in early December 1793; a rare dip in an otherwise steady increase (fig. 7.3). Arms workers left their forges and promenaded down the avenue. On December 24, the Capucins atelier was still in rebellion; workers refused to stay until eight o'clock in the evening. Two days later, a mutiny erupted in the atelier of the Maison d'Aine. In early January, other ateliers joined the rebellion. Five ateliers held general assemblies in spite of the prohibition. Rumors circulated in early February that arms workers and their chiefs might rise up against the administration. The agitation intensified in the spring of 1794 as the Jacobin campaign against the Hébertistes reached full force. While selected workers made oratorical devotions to the Convention and marched in the festival for the Makers of the Weapons of War, other workers in the Place Fabrication-des-Armes voiced support for Hébert. A police spy overheard an arms worker say, "It's all screwed up! Our bosses are windbags . . . they deceive the Convention." One arms worker was overheard complaining about the *maximum*, "It's all finished now, [and] we are worse off than ever, because we can't buy anything with this money; we're dying of hunger, and they mock us with pretty speeches."[43] Plainly, Revolutionary rhetoric was no longer able to paper over the political and economic interests that divided administrators from workers. With the erosion of the *maximum*, the wage rate in the ateliers was not keeping pace with rising wages in the private sector. Again, that spring the Committee resorted to force and compromise. When an armorer in the Arcis section muttered "a worker must live just like anyone else," he was thrown in jail. Others were arrested in early April, when they refused to work a fourteen-hour day. At the same time, the Committee reconvened the Council for a new round of arbitration, this time with each atelier directly electing six worker-delegates, who were compensated for attendance. However, they acted as observers rather than deliberators, and the composition of the governing board heavily favored administrators (17 officials versus 5 worker-representatives and 5 chiefs of the ateliers).[44]

This combination of repression and concessions did not, however, put the manufacture into equilibrium. The total number of arms workers had reached four thousand in April, and the output of musket barrels regularly surpassed five hundred a day, nearly half the goal (figs. 7.3 and 7.4). However, the output of gunlocks was nowhere near this target. Even with one thousand locksmiths, output was somewhere around eighty per day, one-fifth of the rate typical of the ancien régime (fig. 7.5). Even assuming the minimum wage of 4 *livres* per day, this meant that each lock cost 60 *livres*, or ten times the price in the 1780s. Moreover, many were of poor quality. Finding qualified personnel for this part of the gun was particularly difficult. And the limits of repairing old locks were being reached. Prieur noted that unless the rate of production were improved, "the manufacture would soon be brought to a halt for lack of gunlocks."[45]

The Atelier de Perfectionnement

The Atelier de Perfectionnement was created to resolve this disequilibrium, reimpose standards, and bring an engineering rationality to the unruly manufacture. Historians have wondered why state officials during the Revolutionary decade suddenly came to see machine production as the central factor behind

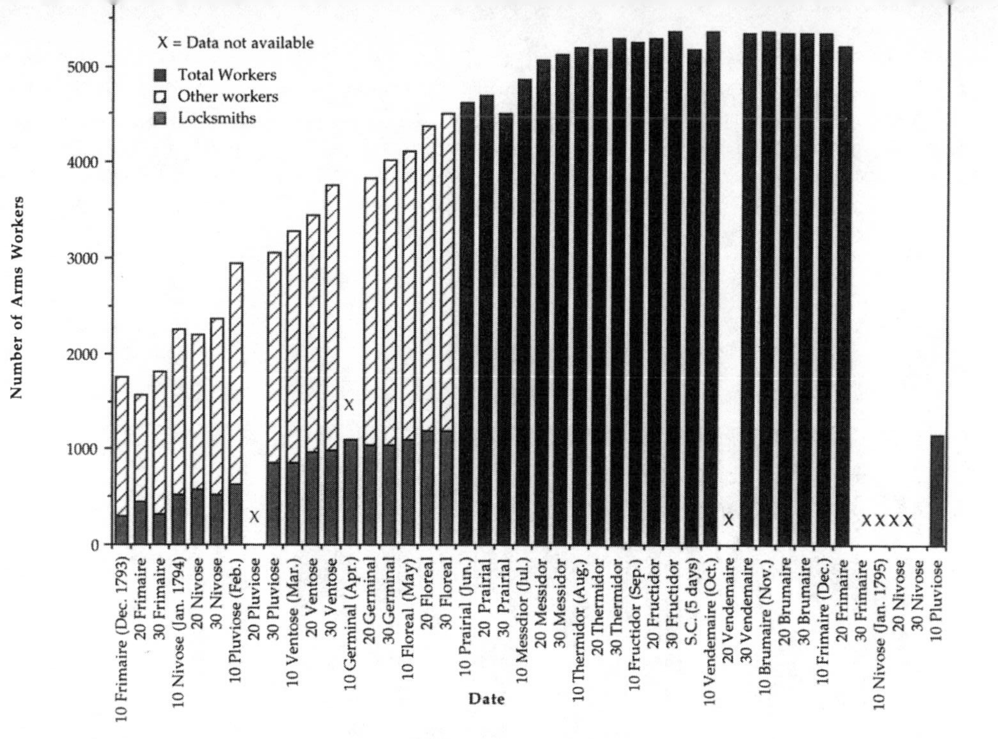

Fig. 7.3. Arms Workers, Manufacture of Paris, December 1793–January 1795. This graph illustrates the growth and decline of employment in the Manufacture of Paris over its one-year history. Employment steadily climbed from 1,600 in December 1793 to 5,000 shortly before Thermidor, seven months later. Then, after a period of stability, there followed a precipitous decline that began in December 1794. Where data were available, I have noted the number of locksmiths employed: their number rose from less than 400 in December to over 1,000 by April. Thereafter, they were not counted separately. The data are from A.N. AD VI 40 *Situation de la Fabrications Nationale des Armes*, 10 frimaire, year II–20 nivôse, year III [20 December 1794–9 January 1795].

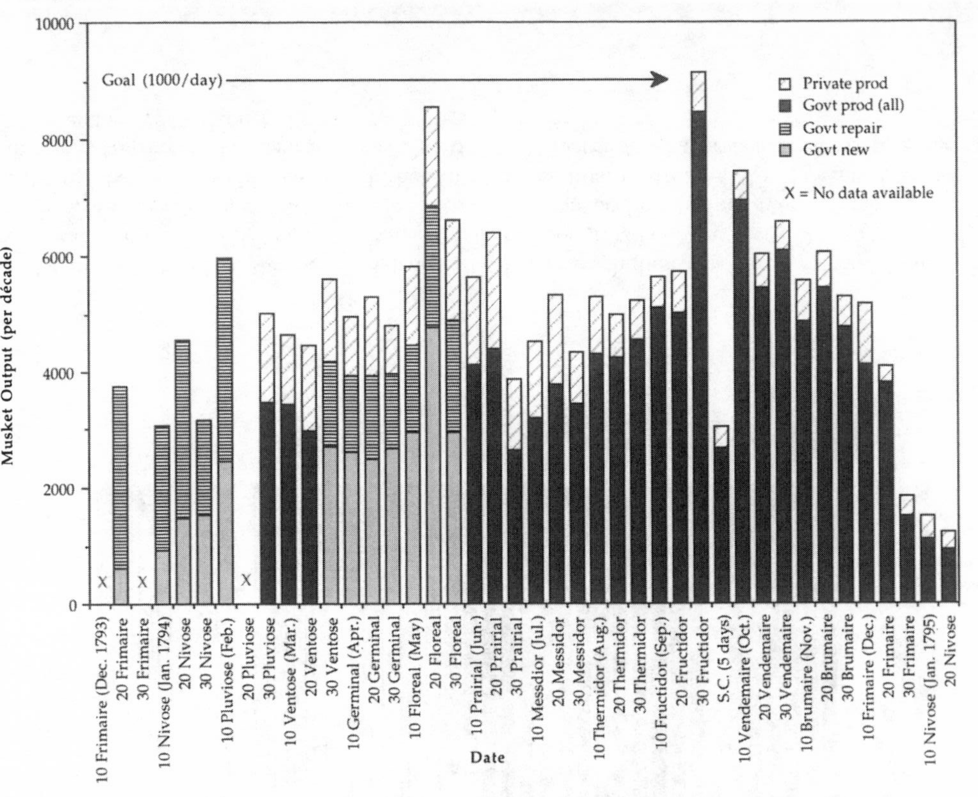

Fig. 7.4. Musket Production, Manufacture of Paris, December 1793–January 1795. This graph shows the progress made by the Manufacture toward the target of 1,000 muskets per day. Output typically fluctuated between 500 and 600 per day. The target was not achieved except during two brief periods: the second décade of floréal [May 1794], shortly after the Committee took a firm hand over discipline and increased pay, and the last *décade* of the year when the accumulated inventory was released. Where data were available, I have distinguished between newly produced muskets and the repair of old firearms. The contribution from private contractors seldom exceeded 15 percent. The data are from A.N. AD VI 40 *Situation de la Fabrications Nationale des Armes*, 10 frimaire, year II–20 nivôse, year III [20 December 1794–9 January 1795].

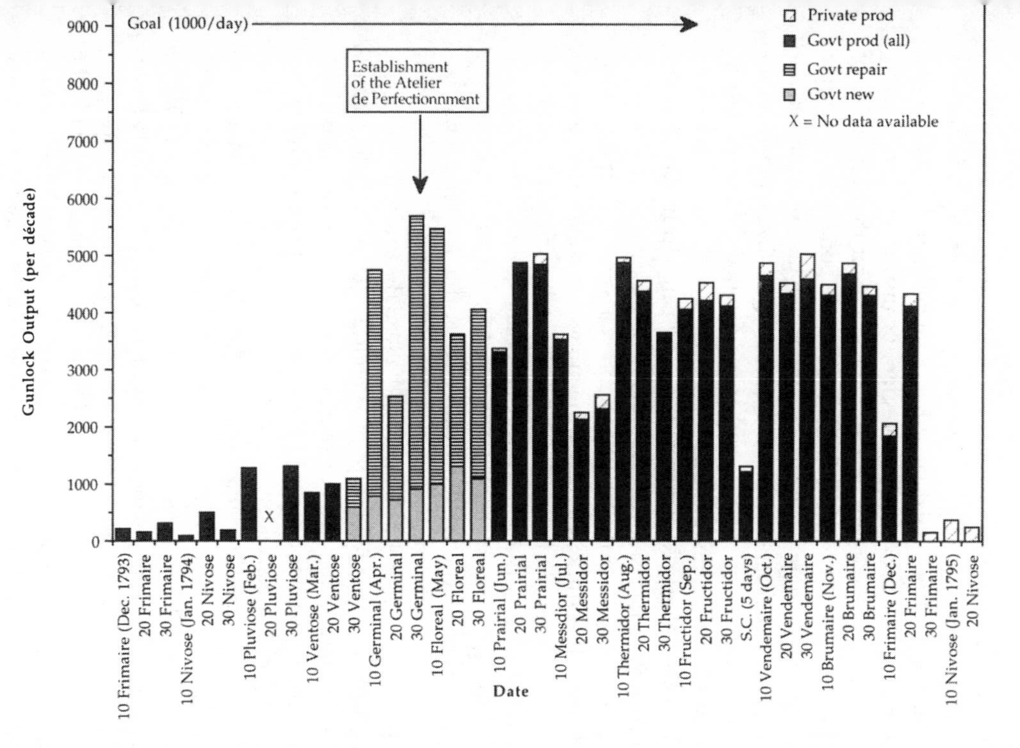

Fig. 7.5. Gunlock Production, Manufacture of Paris, December 1793–January 1795. This graph shows that the production of gunlocks lagged far behind the target of 1,000 per day, and had yet to surpass 100 per day in March 1794, when the output of muskets was five times as great. Prieur's initiatives of April prompted a sharp rise in output, but the production rate for *new* gunlocks hardly improved at all. Most of the increase came from repairs, and even then never surpassed half the target rate. Production of gunlocks by private contractors was negligible. The data are from A.N. AD VI 40 *Situation de la Fabrications Nationale des Armes*, 10 frimaire, year II–20 nivôse, year III [20 December 1794–9 January 1795].

labor productivity. After all, the doctrines of the pre-Revolutionary "Economists" denied that manufacturing produced wealth. And there are many ways to increase output beside machinery. William Reddy, for one, has speculated that an impetus may come from the needs of the state during the Revolutionary wars. This conjecture is borne out by the Committee's solution to the crisis in the manufacture.[46]

The Committee calculated that to meet its goal of a thousand muskets a day, it needed to double the number of locksmiths *and* achieve a fivefold increase in productivity. But the pool of skilled labor had long ago run dry, and so had Carnot's hope that clock-makers, makers of scientific instruments, and "young men less habituated to a single type of work" would master the techniques of gun-making "in a few days."[47] Nor did the locksmiths respond to an incentive pay scheme designed to "excite [their] zeal." And no wonder. Workers supplying one gunlock per *décade* received 4 *livres* a day, with a 1 *livre* supplement for each additional lock. However, once a worker had reached the rate of 3 locks per *décade*, he was given a onetime bonus of 36 *livres* and put on piece-rate wages. This deceptive scheme lowered his pay by 39 percent! In any case, few semiskilled locksmiths could produce more than one lock per *décade*.[48]

This left mechanization and the division of labor. To get production underway quickly, Carnot had initially ordered the ateliers to adopt the handcraft techniques of the old armories (figs. 7.6–7.9). Each locksmith was to make all the pieces of his lock. But at the same time, Carnot had explicitly reserved the right to exploit novel, money-saving techniques which might spring from the genius of an enlightened citizenry. On 2 April 1793, the War Office had announced that it would examine "all inventions, all processes which might simplify, accelerate, or perfect the firearm," and spread those worthy "throughout the Republic." And that winter, the Committee of Public Safety established a Jury des Armes, composed of expert armorers and technicians (Hassenfratz among them) to judge and reward all such inventions. The results were not encouraging.[49] Like Diderot and Adam Smith before him, Carnot had expected that repeating the same task day after day would stimulate the inventiveness of the artisans themselves. The situation, however proved to be exactly the reverse, and the state set out to build machinery that would *impose* the division of labor. Dividing the tasks involved in making the twenty pieces of the gunlock mechanism meant coordinating these activities. And the technocrats had come to realize that this could only be accomplished with the creation of mediating instruments—gauges, fixtures, and machines—which embodied standards of production. The techno-Jacobins were technological determinists.

In early April, Prieur took the first step in this direction by establishing an atelier to manufacture hand tools for the production of gunlocks. Tools were in short supply and workers typically spent the first months on the job making a full set.[50] The Atelier de Perfectionnement was created a few weeks later to change the *nature* of these tools, and hence, the organization of production. Some private contractors had already offered to produce gunlocks mechanically. The firm of Jouvet and Sueur promised "100 locks a day" using coining presses and other machines in the Orangerie of Chantilly near Paris. Nothing came of their efforts. And a citizen Huet from the Museum section of Paris fantasized about a three-cylinder rolling mill that would crank out one million "perfect" gunlocks per year. As for Honoré Blanc—the person with the most practical experience in this line—he proved unwilling to share his discoveries and stalled his departure for Roanne.[51]

Finally, the frustrated Committee decided to mechanize production on its own. The Atelier de Perfectionnement was to construct machine tools and "research methods of manufacturing gunlocks in a more exact and rapid manner."[52] Hassenfratz and Vandermonde, appointed to supervise the Atelier, immediately held up interchangeability as the emblem of rational production.

> Until the present, each worker forged his piece following some arbitrary model, and never were two pieces perfectly similar. It is imperative that we employ procedures so certain that each piece of the same type taken at random can be used as a replacement.[53]

By referring production to mechanical standards, the technocrats believed they could simultaneously reduce the need for skilled labor and enforce a division of labor. The Atelier's chiefs hoped that machinery would "diminish the work . . . , and simplify it in such a manner that all sorts of citizens would be perfectly able to produce the separate pieces of the gun." Immediately, they put this method to the test. They seeded the Atelier with twenty locksmiths chosen from the Manufacture of Paris for "being good workers and the most tranquil," and outfitted them with precision gauges and calibrating devices, all made "of the exact same size" by a group of expert artisans within the Atelier. Then, for six weeks early in the summer of 1794, these workers were assigned separate tasks making pieces and allowed to assemble the gunlock only at the end. The aim was to determine what arrangement of tasks would optimize productivity. Though a number of locks were produced in this manner, the experiment was not judged a success.[54] According to its technical director, Pierre-Bernard Mégnié, "The French genius, especially in a town such as Paris, little lends itself to such a large concentration [of workers], so that our trials did not produce the results we had expected." Defeated, the administration returned to ordinary methods of production until such time as the machine tools were completed. Only then would they attempt to reintroduce the division of labor rejected by the artisans.

> Consequently, our locksmiths have since followed the ordinary procedures until such time as we complete the machines for abbreviating [the work], and achieving the perfection realizable with lathes, dies, stamping machines, and coining presses.[55]

Fig. 7.6.

Fig. 7.7.

Fig. 7.8.

Fig. 7.9.

Figs. 7.6–7.9. Craft Methods of Musket-Making in Revolutionary Paris. This series of sketches from the year II [1794] was supposed to accompany a memoir on musket-making to be written by Hassenfratz. The sketches show the methods used at the Manufacture of Paris. However, they apply equally to the craftwork performed in armories such as Saint-Etienne during the ancien régime. Note the leather apron worn by the smiths (known in Saint-Etienne as the *basana*), their wooden shoes, and the sans-culotte dignity of the artificers. From C.N.A.M. Drawing #135771-424 Anon., "36 croquis représentant des ouvriers de divers états occupés à leurs travaux," [1794]. In fig. 7.6 a master armorer holds the working piece (a jaw screw) in its mold over an anvil and presses a two-headed hammer to it, while his journeyman strikes with his hammer. In fig. 7.7 the lock plate is held in a vise. The artisan is filing the piece by hand, and the caption indicates that he is performing the final filing to gauge. He presumably had a jig clamped behind his working piece. In fig. 7.8 we see a drill typical of those used by skilled craftsmen in a variety of metalworking trades. The downward pressure is applied by the weight hung to the left. The drill was turned by a crank and was stabilized by a metal guide just above the bit. In fig. 7.9 the worker is using a "conscience drill," so-called because the plate is placed near the heart. The drill is powered by a bow, and the working piece (a screw) is ground against a filelike surface to flatten its head. The same method was used to tap holes in the various pieces of the gunlock, particularly the lock plate.

Once again the French state had aspired to interchangeable parts manufacturing at a time of worker unrest and rising wages. Prieur spoke of producing gunlocks in "a simple and expeditious manner . . . with the use of machines to prepare the pieces of the gunlock, so that there remained little work left for the artisan to prepare them to be mounted, adjusted, and finished." The technical means of achieving this outcome was heavily influenced by Blanc's precedent. The Atelier became the center for gauges for the Manufacture of Paris—just as gauges had been the sole responsibility of Blanc's experimental workshop. The government had reserved the right to make use of his procedures. Vandermonde knew his methods. And Le Roy, author of the Academy's report on Blanc, later became governor of the Atelier in his capacity as president of the Conservatoire National des Arts et Métiers.[56]

During its two-year existence, the Atelier was located in the Hôtel Montmorency, the three-story mansion of an émigré now filled with twenty workshops, twelve forges, and seven furnaces. In the central courtyard, Hassenfratz installed fly presses shipped in from the Rouen mint. Daily administration fell to the *agent comptable*, Antelmy, who was later accused of running the workshop as his private fiefdom, bilking the government and tyrannizing workers and staff. The technical director for day-to-day operations was Mégnié, the son of a Dijon locksmith who had headed perhaps the finest scientific instrument shop in France before the Revolution. The two men quarreled bitterly. After a violent dispute in May 1795, Mégnié was replaced with François-Philippe Charpentier, the machinist-inventor who had built Thomas Jefferson's portable presses and invented a horsepowered lathe to mill and ream eight gun barrels simultaneously.[57]

The activities of the Atelier were divided among different groups of machinists, each with its own chief. The Lyonnais watchmaker Jacques Glaësner, with "distinguished talents in the manufacture of gunlocks," headed a group of machinists working on machine tools for the tumbler. Savart, Monge's *aide de physique* at Mézières, led a group of machinists concentrating on the lock plate. Penel, an artisan from a prominent family of Saint-Etienne armorers, led a group of engravers who made precision molds, dies, and stamping tools. And a group of a dozen expert locksmiths constructed model gunlocks, gauges, measures, and other precision hand tools. The artisans themselves were selected for their skills in machine-building. "Many of the *artistes* who work in the atelier," said one report, "can be classed as inventors or improvers." One was Simon Beguinot, a thirty-seven-year-old artisan, who had labored for twenty years as a worker-soldier in the artillery's arsenals of construction, forging the "dies and molds, tools for tapping screw holes, lathes, [and] triphammers" to produce artillery carriages with interchangeable parts.[58] Altogether, these craftsmen-machinists represent the vital links by which the technical accomplishments of the ancien régime leavened the institutions of Revolutionary France. In the eyes of the French technocratic elite, these men represented a valuable resource, machinists who would vault France into industrial competition with Britain and supply her armies with the most up-to-date weaponry.[59] Some economic historians concur in their belief that a small number of skilled machinists can have a wildly disproportionate effect on economic development.

In general, much as in Blanc's Vincennes workshop, the basic plan was to die forge the lock pieces with drop hammers, shape them with coining presses, mill them to size with milling machines, and then hand-finish them in filing jigs (figs. 6.2–6.4). Promising results were obtained in December 1794 stamping out gun cocks. But after tempering, many proved too fragile. Mégnié blamed these poor results on the low quality of the available metal. At least five different machine tools were prepared to produce the lock plate: a drop hammer for die forging, a horizontal grindstone, a hollow milling machine, a vertical drill press, and a final stamping machine. The tumbler was to be shaped by as many as four different machine tools. One milling machine cut the square. Then a milling machine (never completed) shaped the curves and bursts. A lathe turned the tumbler pivots. The piece was finished with a double milling machine, which could be accurately positioned by a screw. This last tool also came equipped with six different cutting tools, which, as the machinists themselves noted, had the property of "adapting to different gauges." This suggests that the Atelier did not focus exclusively on single-purpose machine tools. It does, however, remind us of the considerable effort placed in developing the sorts of heavy machine tools which were to make interchangeable parts production successful in America in the coming decades.[60]

All this work took place against the background of a city in revolutionary turmoil. The fixed wages of the machinists eroded as inflationary pressures mounted in 1795. That March, a worker announced that he and his comrades would leave if they did not get paid "in proportion to the cost of living." In July, the workshop chiefs reiterated this threat. Turnover was extremely high as the Atelier rapidly expanded and then contracted. The April after its founding, not one of the locksmiths who had been hired a year before remained. From fifty machinists at the time of Thermidor (July 1794), the Atelier reached its peak of ninety-five in January 1795, just when the rest of the manufacture was being dismantled. Thereafter, the number of personnel fell to forty in July 1795, and thence to twenty-five through 1796.[61]

Under these circumstances, discipline was a perennial problem. The Atelier was organized along a hierarchy of skill and experience, typical of machine shops in the nineteenth century. But the unrest of the period upset this ranking. "I cannot establish control," Mégnié admitted. There were many personal confrontations. A worker "publicly insulted" Antelmy's wife because he was being cheated of his pay. Workers "more often than not" were away from their posts. Shortly after Thermidor, two insubordinate locksmiths were replaced with two more compliant workers. Mégnié was instructed to lay off those "often absent without a legitimate excuse" or who "interrupt the work by their insubordination." At the same time, Antelmy had apparently bilked the government of tens of thousands of *livres* which he paid back with 8 inflated *francs*. For all of these reasons, many of the novel machine tools remained on the drawing board or were left half-finished.[62]

Nevertheless, the technocrats on the Committee of Public Safety defended the Atelier, and with it a particular role for the state in the realm of technological innovation. Private interest alone was not sufficient to ensure the adoption of unpopular new technologies. In the post-Thermidorean political climate, however, they no longer advocated that the state assume outright ownership of the means of production. Rather, the state should serve as a "strong power" to overcome the social resistance that technological change inevitably aroused, and to supply the capital that private investors were reluctant to put up. With the rest of the manufacture collapsing, Prieur defended his Atelier in just these terms in 1795.

> There is no one who does not sense that these new workshops and ingenious new machines, both for speeding up the production and making it more perfect, will not take root without the help of the government. We know that even the greatest inventions and most obvious improvements, when they depend on a large number of people, need to be encouraged before they will succeed. Inertia and ignorance give way only little by little; and often a strong power is necessary to overcome and conquer them.[63]

To preserve the Atelier, the techno-Jacobins on the "new" Committee of Public Safety—Guyton and Fourcroy—transferred it to their newly created Conservatoire National des Arts et Métiers, an institution whose mission was to instruct citizens in the possibilities of machine production. Henceforth, in addition to its work on gunlocks, the Atelier was to "perfect tools for the majority of the mechanical arts." This was portrayed as a logical extension of its earlier mission. As Vandermonde pointed out, "almost all the arts and trades have a connection to the armaments industry."[64] The aim was to bring artisans into line with the revolutionary possibilities offered by machinery. Unlike the recent year of turmoil and darkness, this was a revolution devoutly to be wished. The Atelier was to initiate a "happy revolution in the practice of the mechanical arts" by "inspiring in a larger number of workers a taste for precision without which the most well-reasoned products become useless and unsuccessful." In the view of its founders, "the imperfection of the mechanical arts belonged to the practitioners and not to science itself." The Atelier would function as a training center for a new breed of worker-machinists. Workers skilled in these new arts might then spread their knowledge to the provinces. As we will see below, the Conservatoire National des Arts et Métiers was to become the living embodiment of Honoré Blanc's *Encyclopédie-pratique*, itself the embodiment of Diderot's vision of a general and public science of machinery.[65]

Dismantling the Machine

In this way the Atelier de Perfectionnement lingered on (though in an altered form). But the Manufacture of Paris could not survive the political shift in the capital. The roots of the Thermidorean coup go back to April 1794 when the Committee of Public Safety executed the more vocal of the sans-culotte activists, suppressed the majority of the provincial rebellions, and overcame the external military threats to the nation. All that summer, even as divisions appeared within the Committee, the state took an increasingly authoritarian position vis-à-vis the armament workers. Typical of this new tone was the pamphlet of Representative A.-L. Frécine, entitled *Death to Tyrants! In the Name of the French People*. Frécine announced that it was time to forsake "fraternal dialogue" to talk "severe reason" to the workers. He denounced their demands for higher salaries as cupidity, implied that strikers were army deserters, and threatened them with the law of suspects (i.e., imprisonment and the guillotine).[66] The administration began to direct the language of the Terror against the workers.

> A worker must honestly earn his living from his work, but anyone who hides unreasonable pretensions under a hypocritical sans-culottism is a bad citizen. His mouth offers his arms to the Republic; his greed refuses them.[67]

This rhetorical construction was part of a larger effort to create the Republic of Virtue. That utopian and authoritarian vision met active resistance from ordinary people, whose interests were not always identical to that of the government. This is not to say that they were not devoted to the Revolutionary cause, but that they had a different conception of what republicanism entailed. For the armorers, wages and working conditions lay at the center of their grievance. The overall rate of production was now falling. In late May, the delegates to the manufacture's governing Council again offered an incentive for increased production: 9 *livres* per extra lock delivered. For the first time, workers who provided their own tools received an extra 2 *livres* per lock. The scheme, however, did not increase wages for the majority of arms workers without their own tools and on the daily rate. In early July, the Committee warned Parisians that the *maximum* for salaries would be held constant through August.[68]

The decision was not greeted warmly. In June, the Committee had preemptively suppressed "counterrevolutionary" insurrections in the arms factories which would otherwise have disrupted the festival of the Supreme Being. Absenteeism, the most common form of worker resistance, increased dramatically and insurrection seethed in the workshops.[69] Having located the manufacture in Paris to provide workfare for the urban masses who were their main political allies, the Jacobin government now felt strong enough to bring those programs under control. But the crackdown on the sans-culottes and armorers cost the Jacobins their links to populist support. Unable to muster the popular demonstrations that had brought them to power, Robespierre, Saint-Just, and Couthon fell on 9 thermidor (27 July 1794). The historian George Rudé has described the bumbling incompetence of the political leadership in July as it suppressed the very workers who provided its political muscle.

> [O]n the 25th [of July] Hanriot was warned that several arms workers, "doubtless led astray by the enemies of the people," had left their workshops; and the mayor, Fleuriot-Lescot, on the morning of 9 Thermidor (27 July), obviously unaware of the drama that was already unfolding inside the government Committees and in the lobbies of the Convention, ordered a military force to keep the workers in check on the following day, which was a public holiday.[70]

The fact that arms workers failed to rally in support of the Jacobins should not be taken to imply that they actively supported the Thermidoreans. In the confusion of that summer, the character of the new regime was not immediately apparent. As one delegation of workers put it later that autumn: "Occupied seriously in serving the cause of all virtuous patriots, our laborers' attention does not permit us to see [through] the chaos of which we are today the victim." Indeed, on the surface, the government's "industrial" policy seemed unchanged. The Convention left control over the French army and its armaments production in the hands of the "new" Committee of Public Safety. Indeed, Prieur and Carnot sat on the "new" Committee until that autumn, when they were replaced with Guyton and Fourcroy, members of the same technocratic circle. The Committee did relax some of the harsher penalties and release several armorers from prison. They also offered limited wage concessions; workers in other Parisian trades were awarded a new *maximum* on August 9 which increased their salaries about 50 percent. Arms workers, however, were specifically excluded from the deal. Moreover, arms workers caught tearing up posters of work rules signed by the beheaded Robespierre were informed that the "police laws" of the ateliers still applied in full force. In mid-August, Prieur reminded arms workers that they were still subject to the requisition, and workers leaving their atelier without permission would still be treated as suspect. In this sense, the "revolution" of Thermidor actually continued the effort, underway since April, to establish discipline in the workplace.[71]

At the same time, the Thermidoreans mistrusted the large concentrations of workers in the manufacture. The defeat of the Jacobins had been possible because of the disaffection of the urban masses. This disaffection might yet turn to insurrection. For such local political reasons, as much as for broadly ideological ones, the Convention denounced state-sponsored manufacturing. The manufacture had to be dismantled. A "temporary expedient" while the enemies of the Republic were on the move, the manufacture was now unnecessary. And the proof that this deviation from the principles of good government must cease was the enormous inefficiency of the state's foray into production. "[Formerly], the need for armaments was very great . . . , and calculations of cost were less important than the [number of] muskets produced." But now that normalcy had returned, the state would do better to rely on the provincial armories where costs could be controlled (because they were not subject to the same political pressures). "The cost of firearms at Paris is incalculable; in the provincial manufacture it is fixed." The difference in cost certainly seemed compelling. A bayonet that cost 15 *livres* in the Miramiones atelier where workers were on a daily salary, only cost 4 *livres* in the provincial armories.[72] The difficulty was shutting down a vast social program in the midst of a volatile city. And then an ideal opportunity presented itself.

On August 31, at 7:30 A.M. a tremendous explosion rocked the gunpowder factory of Grenelle in the heart of the city. A large crowd watched wagons bear four hundred dead workers to the Ecole Militaire. Immediately, rumors gave the event a political cast. Some blamed malevolent instigators: "This is the result of the release of prisoners." Others smelled conspiracy: Why had certain workers spent the day in the cabarets? Others blamed their bosses: Why had officials visited the facility by candlelight the evening before? That afternoon the Convention reassured the public that "all is calm, all is in order." It offered aid to the wounded, honored the victims as "defenders of the *patrie*," and assured the nation that the supply of gunpowder was adequate. A large shipment of powder had gone out just three days before the blast. But if true, had someone had prior knowledge of the destruction? Representative Lakanal, at the podium of the Convention, was interrupted by murmurs when he hinted that phosphorous wicks might have been found on the site, suggesting arson. And what was the connection between this explosion and the violent fire twelve days earlier at the Atelier de l'Unité, the city's largest saltpeter refinery, where again, "almost all the saltpeter had been saved"? The Thermidorean leader Fréron had blamed that fire on a "natural accident." But the coincidence of these two episodes led Carrier in the Jacobin club to implicate the Thermidoreans. A fierce political struggle was then raging over control of the Committee of Public Safety. At the end of the day Carrier succeeded in expelling Tallien, Fréron, and Lecointre from the Jacobin club—though it hardly mattered anymore.[73]

Whatever its cause, the explosion shifted public opinion against the armaments industry in the capital. Anxieties concerning arms facilities intensified in the upcoming days, particularly with regard to the secret facility at Meudon, which some speculated was in the hands of the Austrians. As the immediate fear of further explosions diminished, however, the people seemed willing to leave the investigation in the hands of the Convention.[74] The dismantling of the manufacture could now proceed. Without public support, the protests of the dismissed arms-workers—though on an unprecedented scale—never had much chance of toppling the government. The day after the explosion, the Committee ordered the cessation of all public works in the Manufacture of

Paris, and moved the proof house to Versailles. Dismissing the manufacture as a "prodigy" of the Revolution, the Commission declared that over the next two months the personnel of the manufacture would be dispersed: eight hundred to the provincial armories, twenty-four hundred to the shops of private Parisian arquebusiers, twelve hundred to private contractors, and nine hundred who would remain in state-owned ateliers.[75] Naturally the troublemakers were fired first.

> We must conserve those whose needs, morals, and talents are well known to us. For a long time, we have desired order and peace in our ateliers. Now, quiet reigns there; the father of the family no longer excites his child to complain, [and] the work is executed without dissent as it ought to be where all true republicans are brothers and friends.[76]

But when the layoffs began in earnest that November, they generated the first autonomous workers' movement that Revolutionary Paris had seen. The rebellion spread from atelier to atelier. Worker-delegates from the atelier of the Ile de la Fraternité addressed the Convention on November 17, Réunion followed the day after, the Quinze-vingt two days later, and Jemappes and Bonnet-Rouge the day after that. They objected to starting at 6:30 in the morning in winter, when an artisan was "drenched or shivering, his tools in his numb fingers." Their pay had not kept pace with the rising price of foodstuffs and was inadequate for fathers of families.[77] They blamed the exorbitant price of muskets on their unqualified supervisory personnel who used work standards to persecute those who offended them, yet "could not distinguish between minor mistakes and serious flaws in the product." And they lashed out at the managers of the manufacture with some republican rhetoric of their own.

> Do they come to visit the ateliers? Obviously, the workshops are no longer nationalized. Fraternity is only a word, equality a chimera, and liberty a phantom. We see them, their hats set menacingly on their heads, trampling on their slaves at their whim, intimidating and dismissing those unfortunates who do not please them. . . . You [legislators] are our fathers, and justice has no asylum but here. We demand an administration knowledgeable about work procedures and metallurgy, and whose principles are embodied in the republican virtues, a well-arranged and stable organization where prices are set in proportion to the needs of life, bringing consolation to the old and fathers of families, and emulation to the young, so that with the passage of time, all will remain proportional. A reasonable discipline (*police*) should suffice to return an *artiste* to his duty; the workshops are composed of honest men rather than brigands. . . . It is not with tyranny that one leads men to put their heart into working to defend liberty and equality and respecting the republican virtues.[78]

The rhetorical gloss that had for so long covered the deep divisions between the state and workers had nearly worn through. In the dawning light of the emerging order, a large-scale industrywide workers' movement briefly supplanted the food riot as the primary expression of urban discontent. In the 1930s, the historian Daniel Guérin argued that the Revolution marked a historic turning point in modern worker-employer relationship. Guérin has been properly criticized by post-Jacobin historians for equating Parisian artisans with an industrial proletariat; he overlooks the heterogeneous nature of the arms workers and the fact that this type of movement would not be repeated for over fifty years. Yet, unlike the post-Jacobin historians, he sees through the tendentious Revolutionary language. The workers' mistrust of republican rhetoric—and their adoption of that language for their own purposes—tells us, more than any failed *journée*, that the Revolution is over.[79]

For its part, the government stonewalled. It allowed workers to begin winter work at 7 A.M. and appointed six knowledgeable *artistes* to look into the complaints. But at the same time, it abolished the Council with its worker representatives, and refused to budge on wages. In the ateliers, workers were reminded to make up for lost production due to the protests—or they would be docked pay.[80] Immediately, the protests took a violent turn. Candles were blown out; excrement was left on anvils. In the Quinze-vingt atelier, three instructors were accused of spreading insurrection among the workers. More ominously, 100 instructors from nine different ateliers signed a petition demanding a monthly salary. Then four days later, on 22 November some 350 workers from the Pantheon atelier marched on the Convention. The government arrested the four purported ringleaders.[81] Workers from the Marat atelier allowed to address the Convention on November 24 were harshly rebuked by the legislators.

> Citizens, those men who agitate in the ateliers of armaments, and who lose time in making reclamations, these men are the creatures of Robespierre; we know them for what they are; they [should] tremble; they will soon be returned to the nothingness from which they should never have emerged.[82]

The representatives twisted the armorers' complaints against "ignorant" administrators into an attack on the existence of the manufacture itself. On the assumption that the "people" could never have rebelled on their own, the Convention blamed the movement on privileged young men who had latched onto office jobs to shirk military service. The administration threatened to return to the army "incompetent" administrative personnel, intensifying the purges it had been carrying out since Thermidor. Unproductive workers would also be dismissed. Armorer Philippe Marchand had already been sent back to his regiment for bad conduct. A police spy darkly hinted that "two-thirds" of the workers who had presented themselves to complain at the Convention were of requisition age.[83]

Within a fortnight, the Committee's announcement that the manufacture would be abolished precipitated the largest movement of arms workers yet. It was also the last. On December 10, a throng of four hundred arms workers gathered in Vaugirard and dispersed at nine o'clock "in gaiety, singing patriotic couplets." After failing to coordinate their actions the next day, they succeeded on December 12 in rousing several ateliers and a crowd estimated at six thousand. (Others said it totaled ten or even twelve thousand.)[84] A delegation of twenty workers admitted to the Convention the next afternoon denied their movement was seditious; they simply worried that without state workshops they would starve that winter. To this, the President of the Convention gave a stern paternal answer: even though the state owed work and subsistence to all its children, it would not tolerate "dangerous assemblies during time devoted to the public good." To the very end, the government insisted that the workers did not even know what they were protesting about. Representative Legendre claimed that workers had been forced to join in the demonstration by a small number of malcontents. The government characterized the insurrection as misguided rather than hostile. They ordered the chiefs of the ateliers to calm their workers (simultaneously promising to find the chiefs new jobs). That same day the Convention voted to approve the decision to place the manufacture on a private enterprise system.[85]

These were the last movements in Paris motivated by wages; thereafter, with inflation doubling and then redoubling prices, the urban masses reverted to food riots. The thousands of workers assembled the previous year were discharged. The commissioners Capon and Bénézech ordered workers "who show no talent for armaments" to "return to their proper professions." A list appeared every five days of those workers ordered back to their battalions. A few workers with talent for arms-making were sent to the provincial manufactures. Maubeuge was reconstituted. On the germinal *journée* of April 1, a small group of arms workers joined in the march. Two days before, the Commission had closed the rebellious atelier of bayonet-makers at Miramiones "in the interests of economy and order." The Manufacture of Paris was defunct.[86]

The Guns of the Republic

In February 1795, Guyton-Morveau summarized the achievements of the manufacture to the Convention. His was a stirring justification of the engineers' activities on the Terrorist Committee, and part of their Thermidorean exculpation. He noted that prior to September 1793 arms production in the capital had totaled 9,000 muskets per year. Over the next 13 months, Paris had produced 145,600 muskets—more than all the other manufactures of France combined, and twice the output of all the armories of the ancien régime. Peak production exceeded 600 new gun barrels and 190 new gunlocks a day, for a total of 157,249 barrels and 44,012 gunlocks (shortfalls in gunlocks had been made up with repairs). Over 1.8 million *livres* had been spent on the Manufacture of Paris alone, another 2.0 million in the provinces. This massive effort was now at an end, but it had arguably saved the Republic.[87]

Of course, there soon surfaced other interpretations of this period. With the ebbing of the Committee's power, authority over armaments production gradually reverted to the military. That May the Committee of Public Safety created an Artillery Committee of twelve artillery generals to advise the government on armaments production. They established their own Atelier de Précision to assure uniformity of projectiles, construct verifying gauges, and experiment with new production techniques. By the year VII (1798–99) they had reestablished the ancien régime's Entrepreneur system of production in all the provincial armories. The reign of the legislators—and the system of *régie*—had lasted for five years.[88] According to these artillery experts, such as the reinstated Inspector Agoult, the muskets produced in this hiatus did not measure up to the standards of the ancien régime. And they were contemptuous of the Revolutionary No. 1 gun made in Paris during the year II, denouncing it as a "bastard arm," and sneering that "it is difficult to determine the nature, dimensions, and price of an arm whose production is arbitrary and which is composed of [elements of] so many different models."[89] These artillery experts read the physical properties of these guns as a kind of material script of the Revolution. From their Napoleonic distance, the social chaos and political insubordination of the year II could only have produced guns of shoddy workmanship and outlandish price. They deplored the fact that many of these muskets were still in circulation, requiring constant repair and adjustment. One suspects that they had the same aversion to the heterogeneous soldiers of the Revolutionary *levées*, or to populist Revolutionary ideas altogether. Napoléon's crony and later first inspector-general, Jean-Jacques-Basilien Gassendi, derided the muskets of the year II.

> It is a composite that the circumstances tolerated during the Revolution. . . . And that is where the marvelous ateliers of Paris have led us: the weakness of the governors, the ignorance of the supervisors, the rapacious greed of the contractors, the destitution of those armories where they do know how to make arms, [and] the leveling of the officers of the artillery.[90]

In one sense, these engineer-officers were right: those 145,000 muskets, heterogeneous as they were, represent in material terms the conflicts and negotiations of a frantic year of Revolutionary effort. They also testify to the common purpose the Revolution engendered. In the hands of citizen-soldiers, the guns contributed to the defense of France and forestalled the restoration of the monarchy. Both on the battlefield and in the workshops, this achievement was undergirded by a new ideology that placed all citizens in the service of the nation, but which also took seriously the assent of those citizens. This process gave rise to new institutional forms—such as quasi-democratic wage negotiations and the election of military officers—which were themselves subject to revision as the political equation changed. The guns of the Manufacture of Paris can truly be called republican.

In this sense, the history of the Manufacture of Paris underscores the central theme of this book: that the technological life and political struggle are mutually constitutive. Neither is fixed. Just as this period saw the development of new methods of organizing production, so too was the nature of republicanism under intense scrutiny at this time. Historians have recently pointed to the rich political culture which emerged during the Revolution. What now deserves our attention are the ways in which this political life permeated material culture. This was not merely a matter of iconography: dress codes, theatrical displays, and the like. Political values were also invested in those material objects which we associate with deep patterns of knowledge-making and social function. As we saw in chapter 3, Revolutionary weaponry was invested with political values. The pike was the sign of popular insurrection; the bayonet-musket the sign of the élan of the Revolutionary soldier and his self-disciplined flexibility in battle. These representations had real consequences for citizen-soldiers, however, shaping social relations in the army and the way battles were fought. Similarly, the claim that these Revolutionary muskets should be produced by citizen-workers was not merely a rhetorical exercise. It had practical consequences for the way guns were made. It also shaped the ways in which those citizen-workers could articulate their interests, and hence, emerge as a political force.

We can see an analogous process at work quite vividly in contemporary America. During the 1790s the new United States also began a half-century debate on the compatibility of republicanism and the factory system. John Kasson has noted how various groups of Americans reconceived both republicanism and industrialization in defining the new nation. This began as a debate among elites. Thomas Jefferson feared that democracy would be despoiled by a class of dependent wage earners; political autonomy could only be exercised by landed proprietors. Alexander Hamilton responded that the preservation of America's national autonomy—military and political—required that she shake off economic dependence on Britain and develop her own manufactures; and that moreover, those who worked in the manufactures would acquire a valuable discipline there through a strict regimen of work. This debate was by no means concluded when the first large-scale factories opened in Lowell, Massachusetts, in the 1820s. Hence, the mill owners (whatever their profit motive) dressed their factories in republican garb. They staffed their machines with women operatives who were neither voters, nor likely to become a permanent proletariat. And they located the factories in the countryside, where these workers would not destabilize the restless urban population. Yet by the 1830s, these workers had themselves adopted the language of republicanism to assert their dignity as citizens and organize for better wages and working conditions.[91]

So too, in France, the struggle over the organization of Revolutionary production was folded into a political debate over the character of the republic. In part, this was an elite debate about the political implications of industrialization. Despite their doubts about state-sponsored production, the Committee of Public Safety created the Manufacture of Paris. Intended to satisfy the nation's desperate need for firearms and cement the Jacobin alliance with the city artisans, the purpose of this hybrid manufacture was expressed in the language of patriotic duty. This republican language implied both social discipline and a degree of popular assent. The limit of this rhetoric was reached in the spring of 1794. Then, with the nation reunified and open political dissent crushed, the Committee increasingly emphasized discipline over assent, and mechanical coercion over patriotic enthusiasm. But neither Robespierre's Republic of Virtue nor the engineers' vision of uniform machine production was acceptable to the mass of Parisians. The Jacobin alliance with the urban populace began to unravel and the Thermidorean coup soon followed. Now these diverse artisans and day laborers found in republicanism a common language to express their dissatisfaction. Their invocation of a different vision of republicanism, one which assured them a living wage and benevolent managers, led to the first workers' movement of the modern era.

Revisionist accounts of the French Revolution have properly emphasized the political nature of the transformation of the years 1789–95. Reacting against the materialist hypotheses of Soboul's generation, they have noted the disjuncture between class structure and political allegiance and have exposed the ways in which older historians have mistaken rhetorical constructions for social reality. This study confirms that social class is too fluid a category in this period to explain the alliances of the various factions contending for political power in the years 1792–95. But this fluidity ought not become an excuse for ignoring changes in material conditions. Beneath the language of republicanism, historians may still uncover divergent interests and changing practices. Central to this recovery of buried conflict is an ability to see through the "technocratic pose" of the engineers who exercised power during the year II. This pose was intrinsic to the efforts of engineers to make themselves invisible by cloaking themselves in the mantle of "service"—as if the crucial question were not: "service in whose name, with what means, and to what end?" The engineers thereby obscured the manifold ways in which the relationship between the machine and society may be configured. As we will see in the next chapter, the development of this apolitical engineering ideology can be traced historically. In doing so, this book returns to its earlier subject: the relationship between engineering knowledge and social structure. It will thereby put us in a position to understand the seeming divorce of politics and technology in the modern era, and the rise of capitalist relations of production in the nineteenth century.

Chapter Eight

TERROR, TECHNOCRACY, THERMIDOR

To an unprecedented degree, the French Revolution integrated scientists and engineers into the governing elite. Parisian academicians had long served as royal advisers on policy relating to technology, but even under Turgot's Ministry they never wielded power in their own name. The Academy of Sciences mirrored the corporate world of Colbertist statecraft.[1] In the new Republic, however, many savants (including military engineers) were elected to the national legislature. At a time when public life became the arena in which personal and national destinies were worked out, men such as Condorcet, Bailly, Prieur, Carnot, Fourcroy, Guyton-Morveau, and Bureaux de Pusy spoke with both political *and* technical authority.

"Scientific" men were never more than a small percentage in the legislature, but they figured prominently in the political life of the nation. Scores enlisted as high-level administrators and sat on government committees; some like Monge, Laplace, and Duportail served as ministers. These were educated men of talent, just like the provincial lawyers who held the majority of political posts. Why shouldn't the savants serve as well? After all, in the span of a few short years, millions of French men and women redefined themselves as citizens, volunteered for the national army, participated in elections, joined crowd actions, and began to conceive of their lives in political terms. Jean-Sylvain Bailly, the astronomer and mayor of Paris after 1789, expected the savants to lead the Revolutionary transformation. And Condorcet hoped the new age would fulfill his prophesies for a scientific politics. Didn't science have a republican ethos of its own? Was it not a meritocratic polity, open to all qualified practitioners, regardless of their social standing? Weren't these the principles upon which the Revolution was being established? Then, too, it was an ideal opportunity to make a new career. Monge expressed these hopes to his protégé, the young mathematics professor Sylvester-François Lacroix, stuck in the provincial artillery school at Besançon: "Certainly the aristocratic officers will be furloughed if they haven't already deserted or refused to swear [allegiance to the Republic]. The lot of a meritorious professor will be glorious. You are young and it is for you that the Revolution was made." The republic had need of savants, and they answered the call.[2]

But what are we to make of the fact that their self-appointed task during the year II was largely military, and involved industrial management and secret research into engines of destruction? Weren't these the antitheses of the ostensible values of the "Republic of Science" and the French Revolution as well? And what of those other savants, the majority perhaps, who kept a low profile during the Terror, or were condemned, imprisoned, or executed? In posing this question, I do *not* presume to pass moral judgment on the republican savants. Instead, I want to lay out the broader context in which politics and technology became seemingly divorced in the post-Thermidorean world. Roger Hahn has richly documented the many ways in which the French Revolution recast the institutional life of science. And in a provocative essay, Dorinda Outram has argued that the trauma of the Terror marked a decisive rupture in the life of the scientific community, confirming many savants in their choice of a vocation engaged in a solitary study of nature (a retreat which paradoxically reconstituted the scientific community).[3] I want to examine the reverse side of this same coin: how, after a brief period of coexistence, was public political activity seemingly decoupled from the technological life? And what was this "technocratic pose" meant to cover up? Answering this question will help us understand the emergence of the social and epistemological structures which governed France in the nineteenth century.

My argument will proceed in several steps. First, I will expand on my previous demonstration that these men made a difference to the organization of technological life during the years II–III. Second, I will explain why they later *denied* that these activities had involved them in political choices, and how this denial has blinded many historians to the engineers' significance. And third, I will map out what practical effect this denial had, continuing the discussion begun in chapter 2. How did the hierarchical structure of engineering sketched out at the end of the ancien régime survive a Revolution ostensibly carried out in the name of democratic egalitarianism? As we have seen, this period saw an extensive debate over how to organize the technological life. Among the contentious issues at stake was the relationship between technical knowledge and the social structure, and the relationship between the republican state and capitalist production. Each of these debates was fundamental to defining the kind of polity France would be. And each involved difficult choices out of which no definitive solution emerged. But together they mark the French Revolution as a struggle over technocratic values, as much as a struggle over republican principles.[4]

Technocrats at War—With Themselves

Recent scholarship has reminded us that one cannot easily map the relationship between a particular techno-scientific orientation and a particular political affiliation. For instance, the pat correlations once made between "the left" and "empiricist science" are rife with exceptions. There is no *necessary* transhistorical connection between any two forms of the political and the scientific life.[5] The more tractable and interesting question is how scientists and engineers—including those with pronounced political views—have positioned themselves to achieve political results from a position of *seeming* neutrality. This strategy—what I call the "technocratic pose"—belies the diverse ways in which the technological life can be organized. During the French Revolution deep divisions over this topic split the community of savants.

The Revolutionary government of the year II enlisted technical experts for a variety of military tasks. As we have seen, these engineers and savants served in the administration of the Manufacture of Paris. They toured the capital and provinces to report on industrial conditions, taught in the Revolutionary courses, and gathered and disseminated technical information. In 1794, Monge compiled his famous *Description de l'art de fabriquer les canons*, and he published with Vandermonde and Berthollet a path-breaking pamphlet on steel production, which urged those in the "rear ranks" to devote all their energies to the "battle of metallurgy." And at the secret laboratory at Meudon, a few kilometers southwest of Paris, the savants directed armaments research into incendiary shells, balloons for military surveillance, and a semaphore telegraph to connect Paris with the northern battlefields. They also played the principal role in founding the post-Revolutionary institutions of technical education.[6]

This book has made clear that there was nothing new about savants lending a hand on secret military projects, teaching in military schools, and serving as administrative go-betweens. Indeed, these activities had been central to their social identity under the ancien régime. Even those savants most committed to "pure" research in the ancien régime—Monge, Laplace, Lavoisier, Coulomb—were all amateur scientists in that their livelihood was earned in some other career, typically involving technical aid to the French crown. But the monarchy had also legitimated the savants' corporate life as natural philosophers through its sponsorship of the Academy of Sciences. By abolishing this institution in August 1793, the Revolution precipitated a reevaluation of the relationship between the scientific and political realms. The pressure for this destruction undoubtedly came "from below." As Gillispie has noted, the populist rhetoric of 1793–94, as encapsulated in the accusations of Marat, expressed an active hostility to scientific elites and their monopoly on technical judgments. Populists demanded that science be useful, and denied that elite savants could properly judge practical inventions and patents.[7] Hahn has demonstrated that the abolition of the Academy of Sciences was more an attack on royal privilege and institutionalized elitism, than on science *per se*. He goes on to acknowledge, however, that from the savants' point of view, the dismemberment of the Academy destroyed their scientific life. While pleading to save the Academy of Sciences, Lavoisier put it like this:

> Science is not like literature. The man of letters finds in society all the elements he needs to develop his talents. . . . He depends upon no one. The same is not the case in the sciences. Most of them cannot be pursued with success by isolated individuals.[8]

With the Academy abolished, those in the physical sciences had only two courses open to them: the armaments industry and the tasks relating to the metric system. Thanks to their connections to Carnot and Prieur, certain savants maintained their positions on these projects. There they could demonstrate their usefulness to the Republic and hence be shielded from political accusations. Carnot praised the devotion of France's "most celebrated savants and *artistes*" to the Manufacture of Paris, "despite the aversion some ascribe to them." Prieur protected Monge so that he might continue to work for the manufacture. For similar reasons, Hassenfratz secured the release of Vandermonde from jail. These rescue efforts, however, followed from long-standing patronage relationships. Monge, Vandermonde, Guyton, Fourcroy, Berthollet, and Hassenfratz had all collaborated with one another and with Lavoisier. Guyton and Prieur were cousins. Carnot and Prieur had been at Mézières under the tutelage of Monge. During the early days of the Revolution, they formed political fellowships as well: Monge, Hassenfratz, Vandermonde, and Lavoisier were all members of Condorcet's liberal "Club of 1789," and Monge, Hassenfratz, Fourcroy, and Vandermonde were members of the Jacobin club.[9]

Those outside this techno-Jacobin circle understandably resented the powerful insiders. Once excluded, they could not easily demonstrate their "usefulness" by laboring in a capacity that served the state. This made them even more vulnerable to political accusations. But in fact persecuted savants were condemned for reasons having nothing to do with their science. Lavoisier was executed as a tax farmer. Several savants imprisoned for their "aristocratic" leanings actually *were* from the nobility. And those, like Bailly and Condorcet, who lost their lives, did so because they had become closely identified with a particular political party. I say this *not* to justify the persecution, but to point out that the attack on *science* was largely a perception.[10] No doubt the upending of patronage networks confirmed this impression. At the height of the Terror, Prieur purged Borda, Delambre, Coulomb, Brissot, and Laplace from the metric system commission. This minor scientific figure, now an all-powerful politician, pronounced his illustrious colleagues "unreliable." His motives for this action have been a matter of some dispute, but the important point is that he was later *accused* of persecuting men whose scientific talents he allegedly resented and whose political views differed from his own.[11]

That there were important political differences among these savants was to be expected. They did not all take to the Revolution with equal passion, nor read its meaning in the same way. Hassenfratz was an ardent Jacobin of humble background, active in the journées of 14 July 1789, 10 August 1792, and 2 June 1793. After Thermidor he was obliged to flee the capital. A similar fate befell Monge, who became increasingly pro-Jacobin during his tenure as minister of the navy under the Girondins. Fourcroy eventually became identified with the Thermidoreans, but like many of their number he began as a radical member of the Jacobin club. Outside the techno-Jacobin circle lay savants, like Laplace who despite a humble background identified with the elite institutions of the ancien régime. Purged from the metric system commission, he ultimately served as Napoléon's minister of the interior. Others, like Cassini or Lacepède, were royalists, and during the year II, the targets of slurs.[12] Of course, all the savants' political views varied over time, as they coped with a succession of régimes. In this they resembled thousands of other prominent French men and women. Monge became a friend of Napoléon, accompanied him to Egypt, and later served as an imperial Senator. Hassenfratz took a job in the Napoleonic Ecole Polytechnique. Yet whatever their later peregrinations, their allegiance during the Terror defined their political views in the eyes of their colleagues. In the gathering recriminations of the Thermidorean "White Terror," this was a dangerous association. The astronomer Lalande, for instance, denounced Fourcroy, Monge, Hassenfratz, Berthollet, and Arbogast for not having done enough to prevent Lavoisier's execution, and some historians have repeated the charge.[13]

It was against this searing *accusation* (regardless of its accuracy) that the techno-Jacobins deployed what I am calling the "technocratic pose." This "pose" served their immediate needs by sheltering them from retribution, and by creating a neutral ground upon which the scientific community could be reformulated. However, it also implied a repudiation of the political choices the Jacobin savants had made in the year II. This reformulation was part of a larger effort to rewrite the history of that period.

The Thermidorean Exculpation

Interpretations of this period of French history have been controversial from the time the events themselves unfolded. It is hardly an exaggeration to say that an allegiance to a particular interpretation has signaled one's political identity in France ever since.[14] Yet amid all the accusations and counteraccusations, the engineer-savants have come in for little critical analysis. Historians of widely divergent political sympathies have wrangled over the Jacobins' role in the year II—"despots," "saviors," or both?—but they have almost unanimously lauded the role of the savants in directing the war effort and refused to accord them any role in the Terror. This blindness to the engineers' political and social import is borne out by the treatment of Lazare Carnot. As the individual most often associated with the military effort of the year II—both on the battlefield and in the ateliers—Carnot can serve as a touchstone for how the historiography of the Revolution has obscured the connections between politics and technology. *Le grand dictionnaire* of Pierre Larousse offers a quintessential nineteenth-century characterization.

> Carnot endures in our modern history as the most perfect type of civic purity, of patriotic devotion, of disinterestedness, of rigid probity; and no one better merits this title of "a man of Plutarch". . . for his attainment of the ideal of modern democracy, modesty in heroism and simplicity in grandeur.[15]

This combination of Roman patriotism and scientific disinterestedness provided a usable Revolutionary past for the Third Republic. But it divests Carnot's achievement of real political significance. Carnot, as we have seen, was directly involved in the development of mass war fought with bayonet–muskets, and he was instrumental in trying to lead the Manufacture of Paris from patriotic production to machine production. Both required him to make hard political choices; and these were overlaid by further choices as he served on the Directory and as minister of the interior under Napoléon. Yet despite the wide gyrations of his career—in part, perhaps, because of them—Carnot remains one of the few major figures of this divisive period to find favor across the political spectrum. In his final judgment on the Revolution, the nationalist historian Michelet proudly takes his seat beside Carnot and Chambon, the organizers of war and state finances. The "Dantonist" historian Aulard exonerates Carnot for signing the death warrant of his hero, arguing that winning the war was his only motive. The "Robespierrist" historian Mathiez quibbles about Carnot's flirtation with *ci-devant* officers, deplores his role in the Thermidorean coup, and bemoans his subsequent association with Napoléon, but credits him with reviving the army and saving the Revolution. And the socialist historian Jaurès exonerates Carnot, while turning him into a mere "administrator." Historians on the right condemn Carnot's association with the Terror, but they likewise praise him for saving the nation. Sorel explains that his republican ideals blinded him to the "horrors of the present." Even Taine judges Carnot's soul only "half lost," commends his charity toward *ci-devant* officers, and likewise credits him with winning the war. Apparently, political divisions among historians of left and right count for less than the drums of 1870 and two world wars.[16]

Recent historians have continued to portray Carnot as a technician caught up in trying circumstances. Reinhard's biography describes a loyal military engineer, who abhorred the barbaric violence of the new warfare even as he insisted on every citizen's patriotic duty to fight. Gillispie's approach is more nuanced. He acknowledges the political implications of the war, Carnot's participation in the administration of the Terror, and more important, the "intimate relation" between the two. Yet, Gillispie delicately disassociates Carnot from the personal venom of the other Committee members, and finds in him a patriotic, stoic moderate caught in the crossfire between left and right.[17]

What invites suspicion about all these readings is not simply their implausible unanimity in so contentious a field, but their uncanny echo of Carnot's own justification in the period just after Thermidor. His contemporaries were not so easy on him. After the fall of the "Triumvirate"—as Robespierre, Saint-Just, and Couthon were quickly dubbed—the position of the techno-Jacobins was a tricky one. To be sure, they had helped initiate the Thermidorean coup. But they had also sat on the dictatorial Committee, arrogated power over the army and economy, and signed the death warrants of Danton, the Hébertistes, and other political enemies. In the new climate, these acts appeared unwarranted and tyrannical. Pressure began to build from the resurgent moderates of the Convention to purge Carnot and Prieur from the "new" Committee.

The techno-Jacobins answered the challenge in two ways. First, they denied that they had participated directly in Robespierre's bloodthirsty regime; the members of the Committee, they asserted, had always divided their responsibilities. Second, they hid behind their patriotism; every time the French armies scored another victory, Carnot took to the legislature to silence his opponents. Already in September 1794, while announcing fresh French victories, Carnot asserted that Robespierre had refused to commit himself to military affairs, "The success of our armies was a torture for him."[18] In pursuing this double track defense, the technocrats attempted to decouple their activities from party politics.

The chemist Fourcroy (having joined Carnot on the "new" Committee) first trotted out this dual strategy in his speech of 3 December 1794: "Those Arts which have Served in the Defense of the Republic." There, he outlined the many ways in which "the leading men in the exact sciences" had assisted the war effort with "discoveries more illustrious by their real utility than the noise they have made in the world." And all these contributions—machines to make guns with identical pieces, incendiary shells, observation balloons, a semaphore telegraph, gunpowder production, and technical-military education—had been realized by the "pure" members of the Committee over the violent opposition of Robespierre and his cohort.

> [P]ursued by the usurpers of the rights of the people, [and] proscribed by the tyranny which feared its useful influence, the genius of science and the [mechanical] arts had been forbidden and in a sense hidden from the eyes of the triumvirate, even in the very seat of the Committee where they cruelly exercised their horrific despotism.[19]

This rhetorical strategy succeeded largely because it fed the ruling elite's wider effort to justify their activities of the previous year. After all, the same legislators still sat in the Convention. Among the few lasting achievements they could safely boast about were their technological accomplishments. They could thereby distinguish the creative acts of the elite from the destructive acts of popular ignorance. As Dorinda Outram has pointed out, Fourcroy was here expanding on the themes of Henri Grégoire's famous denunciations of the "vandalism" of the year II. Grégoire asserted that these barbaric attacks, condoned by Robespierre himself, had been directed not only against artistic treasures, but against treasured "men of talent." Grégoire's list of victims reached its crescendo with Lavoisier, an example, he proclaimed, "which must be translated to history." Whereupon he related how the Terrorist Dumas had dispatched France's greatest scientist, saying, "The Republic no longer has need of savants." Seeking to rescue some good amidst this barbarism, Grégoire pointed to the technical and scientific achievements of the year II: the metric system, the military arts, victory. These were the great *permanent* contributions of the Revolution. From this vantage point, Grégoire, Fourcroy, and the Thermidoreans went on to rehabilitate an official scientific life; they established the Institut, created the Conservatoire National des Arts et Métiers, and completed the metric system. Around the neutrality of science, the mechanical arts, and reason, they attempted to forge a new political consensus. But this consensus reposed upon a lie; Dumas, we now know, never made any such statement.[20]

Grégoire and Fourcroy's exculpation was not enough to silence the critics and those seeking revenge, however. During the year III the accusers eventually turned against the remaining members of the Committee. Typical was a pamphlet excoriating Prieur's involvement in a manufacture which had "nourished. . . thousands of sans-culottes." Ominously, it concluded, "Do we need anything more to judge and execute him?"[21] So in consecutive speeches on 23 March 1795, Carnot and Prieur took to the floor of the Convention in their own defense, giving enduring shape to the myth of a Committee of Public Safety divided into patriotic engineers and despotic politicians. They argued that given the immensity of the task and the Committee's multiple responsibilities, each member had been preoccupied with his distinct duties. Robespierre and Saint-Just had had sole responsibility for the police, Carnot for the war effort, Prieur for the supply of ordnance, Lindet for the food supply, et cetera. It was not right to blame everyone on the Committee for the misdeeds of the "Triumvirate." Carnot had always opposed, he said, the violent measures of Robespierre, Saint-Just, Couthon, and the "conspiratorial municipal government" of Paris. And he insisted that he had opposed the execution of Danton. For his part, Prieur confirmed the split between Carnot and Saint-Just on the Committee, although this division had been covered over at the time for fear of giving comfort to France's enemies. In short, the Committee had divided executive authority and the engineers had achieved all the useful results, or so they claimed. As for the means chosen—the state-sponsored workshops and the mass army—these, they now asserted, had been dictated, not by ideology, but by Revolutionary circumstance, by "reason of state." The engineers ought to be above censure because, in the words of an anonymous shout from the floor of the Convention, "Carnot organized victory!"[22]

There remained the tricky question of why the engineers' signatures could be found on the various Terrorist orders, including the death warrant of Danton. Carnot dismissed these scribblings as a bureaucratic formality.

> They were and always had been simply evidence that the documents signed had been *seen*; evidence of a purely mechanical operation, proving merely that the reporter, that is, the first signatory of the minutes, had acquitted himself of the formality prescribed by submitting the document in question to the examination of the Committee.[23]

The "mechanical" act of signing his name, then, was simply an efficient act in the machinery of the Revolutionary state, part of that division of executive labor which constitutes effective administration. In fact, both Carnot and Prieur had themselves handwritten many police orders, particularly in the disciplining of the ateliers and troops. In saying this, my purpose is *not* to belabor their culpability, but to understand the *type* of exculpation they produced under pressure. The Carnot of the year III denied that he had made judgments during the year II. The division of labor and the mechanical signatures signaled the impersonal acts of an organization man, one whose service to the nation superseded any local responsibility. This republican nationalism was an extension of the meritocratic professionalism of the ancien régime engineers. It was the reverse side of the "civic probity" and "disinterestedness" lauded by *Le grand dictionnaire*. And it was the camouflage of the "mere administrator" Jaurès dismissed. The Revolutionary engineers denied that the war effort had involved them in politics. This was a perfectly understandable strategy. They detached technology from politics in order to keep their heads attached to their bodies. Shortly before his execution, Bailly had pleaded that "as a public figure, I was never entangled with any faction, or party to any intrigue; I always walked in the straight path of duty." Plainly, he was not believed.[24]

Again, what made Carnot's exculpation successful was the larger absolution it made possible. The central challenge of this period was what Bronislaw Baczko has called the problem of "how to end the Terror." Against the gathering White Terror, Carnot's "mechanical" model enabled the elite to shelve the self-accusations that threatened to tear apart the Convention. In his address, Carnot pointed out that by the logic of his accusers *every* legislator and government official had participated in the Terror, if not by active cruelty, then by passive cowardice. After all, the Convention had nominally controlled the Committee of Public Safety. Once this was accepted, then the accusations must continue until every legislator was held accountable. The only way out of these endless recriminations, Carnot argued, was to assert that the legislature embodied the will of the people, who might be in error, but could never be guilty. In this way, its members could say that they had only done what "necessity" required—"mechanically," as it were. And what was this "necessity"? Even before he took his seat on the Committee, Carnot had made his Machiavellian allegiance clear. "Every political measure," he had said, "is legitimate, if it is required for the safety of the state."[25]

The evasion here is that in a republic the "political measures" which preserve the safety of the state are supposed to be publicly debated. In this sense, Carnot's formulation reflects the larger sense in which the French Revolution itself was a *denial* of modern (interest) politics, a vast technocratic pose in which "Reason" was to speak with a single voice about the national destiny. As Lynn Hunt has noted, this was the founding paradox of Revolutionary politics: that legislators had always to deny that they acted politically, in the sense of acting for particular interests, while they continually sought to speak in the name of the general will.[26] This "technocratic pose" has beguiled many historians, and obscured the important role of the technocrats in engineering the Revolution.

Each plank of this exculpation has had its adherents. Liberal historians of the post-1945 era, such as Gillispie and Fayet, have cast the Jacobins as the enemies of "pure" science, citing their anti-intellectual rhetoric and the standstill in scientific progress during the Terror.[27] They have thereby made the savants victims of the Terror rather than its prosecutors, not recognizing that the savants were *both*. Moreover, their overly fine distinction between pure and applied science—an attempt to clear the *intellectual* life of science from military taint—ignores the considerable practical technological achievements of this period.[28]

My contention is that the events of the year II must be seen as a single piece. This is not, of course, a new observation. Aulard long ago exposed the collective nature of the activities of the Committee of Public Safety. But the post-Jacobin historians who followed in his wake—Mathiez, Soboul, and Camille Richard—used this argument to assert that the Robespierrists deserved as much credit for the Revolution's scientific and military successes—for saving the Republic—as the savants on the Committee.[29] But Aulard's insight now needs to be pointed in the other direction as well. What both liberals and post-Jacobins elide is how the activities of the technocrats meshed with the broader patriotic and social program that lay behind the Terror. As we have seen, their participation in the Terror was not passive. Indeed, as the directors of a vast military and industrial program founded on republican fervor and state-sanctioned discipline, they possessed as much power as "politicians" like Robespierre (if by power we mean the ability to realize one's political views). One may indeed say the technocrats were *central* to the state's arrogation of the Terror—and this involved them in political choice.

A technocratic interpretation of the French Revolution is not entirely new, of course. So far, however, it has only appealed to some very odd bedfellows. The anarcho-syndicalist historian of the 1930s, Daniel Guérin, makes the technocrats the prosecutors of a virulent class warfare, wielding state power to crush the proletarian workers of Paris. And François Furet hints at a technocratic critique when he calls the Committee of Public Safety, the "revenge of the technicians, the era of the organizers." More promisingly, the "contradictory" trajectory of Carnot's career has been recently noted by Jean-Paul Charnay.[30] But all these writers still imply that there exists a single "technocratic" approach to the relationship between technical knowledge and social structure. In this sense, Guérin and Furet, each in their own way, still assume that technocracy is apolitical, except that they condemn where others have praised.

I have here been arguing that the Thermidorean exculpation was something of an *ur*-event in the relations of science and politics in the modern era. As many historians have noted, science as a profession and politics as a public activity both came of age in France at the end of the eighteenth century. Yet after a brief period of intense involvement, scientists (with very few exceptions) have generally shied away from formal party politics. This retrenchment was itself a political act, part of the conservative retrenchment signaled by Thermidor and which gathered force in the years which followed. This meant abandoning hopes of transforming society through a Revolutionary exertion of human volition, and turning instead to the patient cultivation of technical means by licensed experts. Around this "depoliticized" science, a scientific community could be reformulated. As Outram has pointed out, this community was built around an explicitly apolitical consensus. She quotes Georges Cuvier, who came to preside over the new Institut:

> Fortunately there exists in the midst of political associations an association of another kind, which tries to serve them all, but takes no part in their continual conflicts. The true friends of science, while as loyal to their country as the next man, are also united amongst themselves by the generous lines which attach them to the great cause of humanity.[31]

So universalist science and conflicted politics were to go their separate ways, and the fact-value distinction given institutional form. By this apparent separation of means (technology) and ends (politics), the technocrats hoped to configure the relationship of the state to its citizens in amoral terms, and return authority over the technological life to the bureaus where they served as administrators. Not that savants and engineers lacked political convictions, then or since. On the contrary they have often attached themselves to political causes and occasionally to political parties. Carnot, Prieur, Monge, and their coterie were committed republicans. Indeed, my argument depends on recognizing that their political convictions informed their technological decisions. What I am arguing here is that in recasting their actions as following from technological necessity and administrative logic, rather than from political choice, the techno-Jacobins minimized their own political import. This exculpation was not merely a rhetorical ploy, it papered over real divisions about how to organize the technological life. Moreover, this "pose" had real consequences for post-Thermidorean France at a time when state institutions were in rapid formation under the direction of those same engineer-savants. The result was to give shape to the relationship between the emerging industrial order and the French state, a relationship pioneered by the artillery engineers of the ancien régime.

Within the broad technocratic spectrum to which almost all the savants subscribed, I want to pick out two divergent tendencies. The first is "democratic technocracy," and the second is "hierarchical technocracy." Both are political, although only the first seems so. Both are technocratic, although only the latter is explicitly so. And both are devoted to technological progress based on the development of machinery. Democratic technocracy, however, presumed that technological knowledge would be more equitably shared, and did not value theoretical mastery above practical skills. Both were devoted to a mix of private capital and public intervention, in which the state encouraged new technologies but did not own the means of production. Hierarchical technocracy, however, accepted a wider divergence in the distribution of wealth and technical knowledge, and hence gave more latitude to private control over production. The Thermidorean exculpation played into the hands of those forces (such as the military) who, in the face of the populist upsurge in 1793–94, were determined to reestablish a hierarchical technocracy. The burden of the rest of this chapter, then, is to explain the reconstruction of this hierarchical structure, with its seemingly "natural" correlations between social rank, educational institutions, employment prospects, and cognitive abilities. Doing so will enable us to return in the next chapter to the question of production and the social settlement of post-Revolutionary France.

From the Infinite Universe to the Planned Meritocracy

The nation's legislators recognized that the design of educational institutions was crucial to the relationship between the distribution of technical knowledge and social structure. Hence, proposals for school reform were hotly debated in the new republic, and resulted in several new types of technical schools. In 1792–93, Condorcet and Lakanal proposed a scheme of universal popular education (with a heavy technological component); their plan, however, failed to pass the Convention. Instead, the Committee of Public Safety established other schools, also populist in their orientation, but more in line with the military needs of 1794. Fourcroy justified this conjunction, saying that arms production involved "nearly every branch of human knowledge," and so could serve as the focus of public education. But as the army reasserted its authority after Thermidor, these schools were transformed into a vehicle to reproduce a stratified social order.[32]

Take for instance, the Ecole Normale system, in which provincial students, trained as teachers in Paris, would return home to spread knowledge exponentially. The origins of this system lay in the pioneering armaments school of the year II. In February 1794, eight of France's leading savants (the same techno-Jacobins who directed the Manufacture of Paris) offered a crash course in weapons production to "sons of workers." Two young men from every department were paid 4 *livres* a day and granted free lodging in Paris. Eventually, the "Revolutionary Course" swelled to eight hundred pupils. The lessons were directed toward practical knowledge (the collection of saltpeter, the production of gunpowder, the reaming of artillery cannon), but included relevant scientific background in natural history, physics, machinery, and chemistry. Students also witnessed demonstrations and visited working ateliers around the city. They then returned to their home districts to open schools and spread the knowledge they had acquired. This course had a formative impact on French education. It set a precedent for enlisting the nation's leading savants as instructors. Second, it made a powerful claim about the contribution a scientific education could make to the military and economic fortunes of the nation if it were intensive and popular. Finally, it provided a template for how centralized training could be spread by the pupils themselves. Speaking for the Committee, Barère boasted of the Revolutionary acceleration of learning; patriotism and science, he said, had accomplished in thirty days what would have taken three years under the pedantry of the ancien régime. The efficacy of these methods depended on a democratization of technical know-how, not through a leveling mediocrity, but by raising the ordinary citizen's ability to produce useful knowledge. Barère expected the course to "transform all citizens into physicists and chemists carrying the elements of the thunderbolt against the brigands, priests, and kings."[33]

The Jacobin Committee expanded on this model, first for the Ecole de Mars (which again included instruction on gun production), and then for an analogous two-month course for general public instruction. This latter proposal formed the basis for the Ecole Normale de l'An III, itself the model for the Ecole Normale Supérieure. Originally intended to train primary school teachers, the normale system was hijacked after Thermidor to serve as a method of elite education in high philosophy and science. Although some professors at the Ecole Normale de l'An III stuck to the intentions of the year II, most of the instruction was pitched far above the comprehension of most students.[34]

Similarly, the Ecole Polytechnique, initiated by Prieur and Monge in the year II to train engineers for a wide range of public services, soon became the apex of nineteenth-century technological education. We have already seen that this hierarchical engineering culture—with its mathematical markers of merit and its ideal of state service—had its origins in the ancien régime. In fact, the continuities of state structures across the Revolutionary divide, which I identified in chapter 2, actually masked a bitter debate over how to teach engineering. In practical terms, the years 1789–94 devastated the engineering schools of the ancien régime. The artillery schools, Mézières, and the Ecole des Ponts et Chaussées all faced a dearth of qualified candidates, the departure of teachers, and political attacks on their privileged role as gatekeepers to their respective corps.[35] The manner in which French engineering emerged from this breakdown has recently been studied by a number of historians—most notably Bruno Belhoste, Janis Langins, and Antoine Picon. Their researches have pointed to competing conceptions of the engineering profession advanced during the Revolution. One of these was the vision of Gaspard Monge.[36]

Part of Monge's ambition during the year II was to see technical knowledge disseminated to all French workers. To this end, he suggested that Condorcet's plan for universal education include schools for descriptive geometry in every district. Incorporating this subject into primary and secondary education would ensure that youngsters destined for the mechanical arts would acquire "the habit and sentiment of precision." Here, as in Bachelier's Ecole Gratuite, mathematical drawing was to form the *judgment* of the student. For Monge, this encompassed political judgment as well as technical judgment. By familiarizing students with the use of evidence, the descriptive geometry would make them less susceptible to the predations of charlatans. In this sense, the technological life made good republicans.[37] And although this education would not overturn all prior endowments in talent and capital, it would enhance the autonomy of all students, and thus signal a partial equalizing. In his introduction to his *Géométrie descriptive*, a text based on his lectures at the Ecole Normale de l'An III, Monge argued that the technique should be taught to

> . . . all young people who have the [requisite] intelligence, [both] to those who are well off (*une fortune acquise*), so that they will thereby be one day able to put their capital to better use for themselves and the state; and to those who have no other fortune than their education, so that they may one day raise the value of their labor.[38]

Monge meant his program to be liberatory, teaching universal techniques that would free artisans from old routines. In place of the rote drawing methods secretly passed on by the masonic guilds (for a price), the descriptive geometry would be openly taught to all students, and without a fee. This program was not only personally liberating, it would liberate the nation as well. More precise and uniform French products would compete more effectively with those of other nations. It would thereby free France from its dependence on foreign, especially British, imports. This would be achieved, in part, because the descriptive geometry would help in the development of machine production. Monge frankly admitted that the descriptive geometry made possible a chain of command from "the man of genius who conceives a project" to "those who direct its execution" to "the artisans who must themselves execute its various parts." But this structure involved close interactions between the different roles and social mobility among them. The descriptive geometry was to inculcate theoretical techniques among artisans and manual skills among engineers. It thereby partakes of democratic technocracy.[39]

Under the impetus of Lacroix, the secondary schools known as the Ecoles Centrales (1795–1806) began as a partial fulfillment of this Mongean vision. Lacroix was the young artillery professor for whom Monge had said "the Revolution had been made." In 1800 he argued that just as great wealth amid general poverty could be accounted a failure of governance, so was the existence of savants isolated amid an ignorant populace. Not only was such inequality "useless," it gave credence to dangerous (i.e., revolutionary) ideas. Like economic fortune, technical knowledge should be held along a continuum, and in both cases the intermediaries guaranteed the strength of the whole. To this end, the curriculum of the Ecoles Centrales would link the mechanical arts to drawing, "which is at the base of so many arts." Indeed, a study by Catherine Mérot indicates that technical drafting was the most popular class, typically attended by more than half the students. It was particularly popular among the sons of artisans. In Isère, where young Henri Beyle—later, the author Stendhal—attended school, about 60 percent of students took drawing, and of these, 75 percent were "artisans, servants, and workers in town and country," even though they made up only 20 percent of the school's students.[40] Yet, these schools, like the Ecole Normale, quickly became the preserve of the elite. In the last years of the Directory, even before the Ecoles Centrales were replaced with the restrictive Napoleonic lycées, the curriculum shifted from technical subjects to purely academic fields. Destutt de Tracy, the Ideologue who formulated this new education policy, commended a two-track course of studies: one for the members of the "savant class," who were to rise on the basis of scholastic merit, and another for the members of the "laboring class," who needed to learn "the habits and morals of the painful work for which they are destined." The former would now receive a rational training based on the analysis of grammar. The latter were no longer welcome at the Ecoles. Such drawing as was still taught there was increasingly freehand sketching taught by academic drawing masters, and was no longer coordinated with mathematics as Monge had wished.[41]

In parallel fashion, the Ecole Polytechnique itself was transformed from a practical school for generalist engineers into an elite preparatory school for the military Ecoles d'Application. Monge was the guiding spirit of the "first" Ecole Polytechnique, then known as the Ecole Centrale des Travaux Publics. The school did not formally sit in session until the year III, but as Janis Langins and Bruno Belhoste have shown, its origins lay in the revolutionary impulse of the year II.[42] During this period of preparation, Monge operated under the authority of Prieur de la Côte-d'Or, who had wrested control over engineering education away from the military. When the school opened, training encompassed both theory and practice under the rubric of the "descriptive geometry." Students spent 48 percent of their time on the descriptive geometry and still had a separate course on drawing that took 16 percent of their time. Chemistry lab took 24 percent. Physics (analysis, mechanics) was notably slighted.[43] This education, however, was far from monolithic. Janis Langins has pointed out that the approach of the first school was "Encyclopedic," not just in the range of its activities, but in its commitment to practical training. In the first year students studied stonecutting, carpentry, and machine design. In their second year they focused on roads and bridges, and in the third year on military and naval works. The descriptive geometry provided a cognitive and social framework for this program: as an approach to mathematics, it combined the methods of synthesis and analysis; as an approach to practical engineering, it tied mathematics to the design of physical objects; as an approach to the technological life, it gave engineers a common language with which they might communicate to others in the productive order; and as a pedagogical program, it emphasized the acquisition of knowledge as a *process*.[44] Monge himself considered the descriptive geometry a tool "which [the students] must build themselves." The drawing exercises which "constituted the ostensible lesson" were actually meant to teach future engineers about the nature of *built* things.

> These drawings, these constructions, demand thought (*méditations*); but there is to be no time devoted to thought as such; that is to take place during the construction, and the pupil who at the same time has exercised his intelligence and the skill (*adresse*) of his hands will win for this double labor an exact rendition (*description*) of the knowledge he has acquired.[45]

In good Lockean fashion, then, the student's judgment was trained by exercises that coordinated the hand and mind, with the resulting pictures serving as a record of his mastery of thick materiality. Monge took great pains to ensure that students did not treat this as an abstract exercise, nor mechanically trace out the solutions. Pupils worked with plaster models and actual machines. As Prieur put it in the school's *Journal*, "Without this [insistence on manual labor, students] will have only superficial ideas and be incapable of pursuing a profession."[46] Attached to the school were draftsmen, stonecutters, and carpenters with skills in layouts so that students would observe objects being built on the basis of drawing-encoded instruction. This hands-on learning was reinforced by education in small study groups known as "brigades."

In sum, this first Ecole Polytechnique had an activist/interventionist orientation that epitomizes democratic technocracy. Although it was necessarily an elite institution, it was open-minded in principle. For instance, students were admitted on their potential, rather than on their formal mastery of mathematical material. They also were given full scholarships. And they elected their brigade leaders. No wonder the first-year class, screened in the year II for their pro-republican beliefs, led by Monge, and lodged in Parisian homes, showed a propensity for political action, generally directed against a government rapidly becoming more conservative. That activity—such as their involvement in the *journée* of 14 vendémiaire (6 October 1795)—gave powerful institutional interests, particularly within the military, the excuse they needed to reshape the Ecole Polytechnique to their needs.[47]

Re-Ranking Technocracy

Even as Prieur renamed the school "Polytechnique" in 1795, the school's generalist orientation was under attack. The shift toward a hierarchical technocracy began to gather force within months after Thermidor. The key development was to make the Ecole Polytechnique into a two-year preparatory school for students who then acquired specialist training at the Ecoles d'Application, separate engineering schools run by the Artillery, Génie, Géographes, Ponts et Chaussées, and Mines. The old specialist schools—whose continued existence Prieur had considered temporary—were put on a permanent basis in September 1795.[48]

Even as a preparatory school, Prieur and Monge sought to keep the Ecole Polytechnique a relatively open institution. The problem was that these techno-Jacobins were simultaneously presenting themselves to the nation as loyal servants of the state. Prieur was assiduously avoiding the Convention and any political controversy. Hassenfratz was denying his early political activism. As mere administrators who repudiated their own political activities, these men were hardly in a strong position to insist on their own democratic technological vision. As a practical matter, the current political climate also obliged Monge to flee the capital, and temporarily landed Hassenfratz in jail. Fourcroy who had advocated an all-encompassing Ecole Polytechnique only a year before, now announced legislation establishing the specialist schools. Only this sort of narrow training, he said, could produce those arts which helped in the defense of the nation. "Even the tyranny [of the Revolutionary government]," he noted, had been obliged to respect this fact; and "[the schools] had been a refuge from the decimating axe." Henceforth, students were to be admitted to the Ecole Polytechnique in proportion to the needs of the services, and were to consider their abilities as belonging in some manner to the state.[49]

As the political mood swung even further to the right, the Ecole came under increased attack. The technical arms of the military sought to reassert control over the training of their cadets. They attacked the school as "privileged" because it exercised a monopoly on admissions to state service. In 1799, they tried to disband the school altogether.[50] Their attempt failed, albeit by a narrow margin, but only because the school's governing board of savants justified the school as providing the various engineering corps with the common education they needed to work together. Guyton, the temporary director of the school, described the advantages this had for the functioning of the state.

> [The school] is to bring together, through the memory of a common education, [absorbed] at an age most proper for forming enduring contacts, those who, assigned to different corps or called to other professions, would not otherwise recognize one another [because of] the differences in their duties (*fonctions*) and the obstacles that is all too liable to produce.[51]

The school survived, then, but only by acquiescing to the military and to the demands of the new First Consul. This process was consummated in 1804–5, when Napoléon militarized the school outright, housing the students in barracks, and subjecting them to martial discipline. This definitively ended overt political action by the students against the government. Whereas 45 percent of the first class had gone into the military, the number had risen to 61 percent within a year, and by 1807 topped 95 percent—the majority going into the artillery. At the same time, the students increasingly came from the wealthier portion of the population. Run initially as a scholarship school, the class of 1799 still included 131 sons of artisans and farmers out of 274 students. After 1805, only thirty scholarships were granted and fewer than one-tenth of the students came from the popular classes. Meanwhile, the entrance exam became more rigorous, and left less room for examiners to judge a student's potential. Objective admission criteria masked a shift to a narrower base of recruitment. Napoléon made no secret of his desire to reserve the school for his new notability. "It is dangerous," he said, "where people who are not wealthy are concerned, to give them too great a knowledge of mathematics." In a bold gesture, Monge, back in Paris, threw his teaching salary into the scholarship pot, but the sacrifice did not make an appreciable difference.[52]

These trends toward elite recruitment and military vocation were consummated in the Ecole Polytechnique's role as a mathematical gatekeeper. As Janis Langins has shown, the military's capture of the school dovetailed with the agenda of Laplace, who used his status in the Institut and his position as artillery examiner to transform the school's curriculum. The artisanal assistants were fired in 1797. By the early 1800s, the descriptive geometry and practical courses had been cut to only 30 percent of class-time—an anomalous situation for an engineering school—whereas time devoted to mathematical physics rose from 25 percent in 1799 to 50 percent in 1812. Laplace expanded on the ancien régime tradition of training analytical engineer-designers, using the Polytechnique to recruit a handful of high-level savants in mathematical engineering and physics. This was an approach which would yield rich fruit in the fertile research programs of Navier, Poncelet, and the other engineer-theoreticians of the nineteenth century. But it had less satisfactory consequences for ordinary engineer-practitioners because it divorced them from practical training.[53]

Mathematics at the Ecole Polytechnique was not political simply by virtue of its difficulty. Laplace's hostility to the geometric curriculum was entwined with its tainted revolutionary origins and his distaste for Mongean methods. Eduard Glas has argued that Laplace was antagonistic to Monge's conception of geometric mathematics as a map of *reality* in its most general guise of spatiality; whereas his own conception of analytical mathematics was of a formal language directed toward the *explanation* of natural phenomena from a few regular principles. Monge, moreover, taught by example, and valued hands-on manipulation. Laplace taught by formal exposition, and valued theory over phenomena. We shouldn't overstate the case, however. Both approaches were taught in the Ecole Polytechnique, and were used to solve problems ranging from the theory of machines to fortifications to bridge-building. There can be no doubt, however, that mathematical theory became the mark by which polytechniciens positioned themselves above mere machinists and soldiers.[54]

Back in 1789 J.-G. Lacuée had proposed mathematics exams as a "non-arbitrary" mechanism for determining admission to high (military) office. Unlike birth or favoritism or fame, exams were "genuinely constitutional" markers of merit, appropriate for the new era in which innate talent would find its reward. At the time, he had urged that preparation for these exams be left in private hands. The state could not bear the cost, and as the corrupt guarantor of the old order, it could not be trusted not to favor the aristocracy. Now, as Napoléon's director of the Ecole Polytechnique, Lacuée presided over an institution that defined those markers of merit, and hence the contours of the new French notability. Like the aristocrats of the ancien régime, this new notability could accommodate capable parvenus. Napoléon proclaimed all careers were open to talent. Yet the Ecole Polytechnique remained an exclusionary institution, with the filter having shifted from ancien régime restrictions based on an amalgam of talent and legal entitlement (birth) to Napoleonic restrictions based on an amalgam of talent and wealth (birth). Mathematics—as an impersonal measure of merit—seemed to open up advancement to democratic competition, even as it limited access to high state office and fostered loyalty to state service. In the statist world of the new French elite, it made young artillery captain Pion des Loches "a man," prepared him for a career, and sheltered him from the political struggles of the 1790s. Mathematics may have offered the captain little help as he directed his cannon across the length and breadth of Europe. But mathematics did help him chart his way through the social turmoil of the post-Revolutionary state.[55]

Apotheosis of a Technocratic Elite

Revolution and war—and war's successes—remade the artillery service in the 1790s. Of course, as befitted wartime, promotion within the artillery now depended far less on mathematical study than on battlefield prowess. So that even though the corps still developed in the pattern fashioned by Gribeauval, the tension between theoretical and practical training continued to exercise the service.

The skills lost to emigration and death had been made up from within the ranks, and with hurriedly trained recruits from the artillery's new Ecole d'Application at Châlons. However, that school hobbled through its five-year existence under the Directory. Instruction was disrupted by administrative changes and the continual siphoning of students into the regiments. Then, during the peace of 1800–1801, senior commanders saw a chance to reestablish order in their service. They worried that the theoretical portion of the curriculum had been slighted. In the past five years, many officers had risen through the ranks on the basis of talent and performance on the field. School-educated engineers complained that these men from the ranks did not show sufficient respect toward the more enlightened officers (*officiers éclairés*), and were behaving with insubordination. Their practical on-the-job training was invaluable, but their abilities would always be specific to their experience. Without the wider view that theoretical training afforded, they imagined that their way of practicing their art was the only way. Yet, paradoxically, each officer had acquired a different way of practicing their art. This was one of the dangers that the Gribeauvalists had associated with the artisanal life. It was essential that all officers receive a common education so that their commands would be uniform, and hence obeyed by their subordinates with utmost precision. Theoretical knowledge would reestablish the social and cognitive hierarchy, as well as serve as a framework for particularist craft knowledge.[56]

A regulation of 1802 set out in excruciating detail both the theoretical knowledge and practical skills expected of each member of the corps, from lieutenant-general to common soldier. This regulation was a monument to the social structure of the new régime. Schooling was made appropriate for every level. Each rank was expected to master the theoretical knowledge—though not necessarily the practical skills—of all subordinate ranks. That way the activities of each level were subsumed as a particular application of the knowledge of the rank above. This allowed for specialization of tasks, while providing clear markers for a hierarchy of merit. In 1802, the Napoleonic administration also transferred the artillery's Ecole d'Application to Metz in tandem with the Ecole du Génie. There, graduates of the Ecole Polytechnique were to be immersed in the Gribeauvalist mix of theory and practice directed toward specific ends.[57]

The products of the school did not always satisfy the nation's first artillerist, however. Visibly frustrated with the mathematical prodigies turned out by his own engineering schools, Napoléon insisted that training be detailed and specific. In a splenetic attack on the school, he rattled off sixteen questions that cut to the core of the artillerists' craft: How often did the cadets shoot target practice? Did they know the dimensions of the carriages? Could they design batteries? Having already passed through the mathematical filter of the Ecole Polytechnique, students now needed grounding in the "cold and fastidious details" of their craft. By 1807, the two-year course included instruction in ballistics, the theory of sieges, and the construction of machines, as well as instruction on how to draw up work plans, how to estimate costs, and how to divide labor.[58] Senior officers continued their education in the regimental schools. Subjects covered at La Fère in 1802–3 included refresher courses in mathematics and technical drawing, as well as seminars on how to assure the perfect sphericity of musket shots, and how to manage an iron foundry. An initiation in the minute details, arcane and pedantic to outsiders, but essential to the daily operation of the service, still constituted the professional identity of the artillery officer.[59]

Yet these technicians no longer lurked on the margins of power. Their hierarchy now reached right up from the lowest citizen-cannoneer to the head of state. As emperor, Napoléon placed the artillery under his direct authority. From their position as a vaguely dishonorable and peripheral branch of the army, the artillerists had moved themselves to the center of the new regime. Was not one of their own emperor of all the French? Were they not trained in educational institutions that validated their merit? Had not glorious French victories (especially after the stunning successes of 1805) been the result of their mobile artillery? Confident in their status and lavishly rewarded by the new régime, the new artillery could justifiably call themselves the ones for whom the Revolution had been made.

What in the meantime had happened to the philosophe's Enlightenment vision of how the cultivation of useful knowledge would lead to a pacific, wealthy, and egalitarian world? To be sure, the Revolution's abolition of formal privilege had gone some way toward fulfilling Condorcet's utopian aspirations for a society of equal citizens, free to develop their talents. But as we have seen, this egalitarian spirit had also made possible a reconstituted hierarchy—to be understood as a stratified order where entry and promotion depended not on birth, but on institutionally validated markers of merit. For its members, this self-organizing system constituted "society," and its code of state service, their professional *raison d'être*. Condorcet had, of course, imagined a society governed by merit too, but he had expected this knowledgeable elite to legislate with the assent of an educated public. The reemergence of a restrictive and anti-democratic hierarchy clearly distressed those Revolutionaries who worried about the formation of an aristocracy within the new régime. But the French state which pursued military campaigns across the Continent clearly thrived on these forms of social organization. And it was to be this hierarchical system which was to govern social and professional life throughout the nineteenth century.

Filling Out the Ranks

For all these social successes and battlefield laurels, however, the role of the state engineers in production remained as controversial as ever. Before returning in the next chapter to gun-making and the creation of the new capitalist order, we need to come to terms with how the post-Revolutionary state managed its relationship with technological innovation. The abolition of the corporations and the retreat of the state from its direct role in the year II left this central question unaddressed. Specifically, the retreat of the polytechniciens from their activist Mongean agenda left unresolved the role of the engineer in the workshop.

Recognizing this, Fourcroy, Grégoire, and the Thermidorean elite fashioned a variety of new technical institutions. I will discuss two of these below: the Ecole des Arts et Métiers, and the Conservatoire des Arts et Métiers. Both had antecedents in the ancien régime. Both were reformulated after Thermidor to define a distinctly "liberal" role for the state in the realm of technology. And both achieved this result by matching a cognitive and social hierarchy. Having given up on Monge's ideal of direct relations between engineers and artisans, the French state created new strata of intermediaries who were to operate between the conception of an artifact and its execution. That is, in the place of democratic technocracy, the state adopted a military model of hierarchy and subordination. Bonaparte called these new technicians "the NCOs of industry" and expected them to serve an analogous position in the productive chain of command. Indeed, they resemble nothing so much as the *garçon-majors* of the old artillery. And again, what legitimated (and bounded) their authority was their position in a cognitive hierarchy as the translators of technical drawings.[60]

The most prominent school for this sort of training was the Ecole Nationale des Arts et Métiers, located first at Compiègne, then at Châlons (the site of the former artillery school). On a school visit in 1802, Napoléon gave a formal address to the students in which he cited technical drawing as one of the linchpins of French industry, tying it directly to the problem of creating a social hierarchy in the world of work.

> In the north I found accomplished foremen (*contremaîtres distingués*), but none capable of drawing a plan (*tracé*). . . . This is a defect in our industry. I wish to fill it here. No more Latin—that will be learned in the lycées which are being organized—[here we will teach] only the practice of the trades, with the theory necessary to ensure their progress. Here we will form excellent foremen for the manufactures.[61]

Unlike the elite engineering schools, here technical drawing was coordinated with a hands-on apprenticeship. This manual labor marked the students as subordinate to the polytechniciens. These gadzarts, as they were called, came from an inferior social strata: between 1806 and 1830 only 21 percent came from the middle and upper reaches of the bourgeoisie, whereas 27 percent came from the lower range of officials, and 29 percent from the popular classes. There was some nervousness over the status of these men. Lucien Bonaparte insisted that slots in the school were reserved for the sons of NCOs, not officers, and that their accommodations be Spartan. Their career prospects were decidedly limited. Having financed their education, Napoléon considered them bound for state service. It is telling, for instance, that in 1808, the imperial administration set the students to work assembling artillery carriages with interchangeable parts—under the authority of artillery officers.[62]

Later in the nineteenth century, many graduates of the school went on to fill the ranks of the *conducteurs*, state employees under the command of the Corps de Ponts et Chaussées (all Polytechnique graduates). They drew up plans, oversaw construction work, and kept accounts. Here too their cognitive role as "translators" between state engineers and subcontractors on the worksite indelibly marked them for an inferior social rank. *Conducteurs* were forcibly excluded from the professional engineering corps until late in the century.[63]

In private industry, however, their position as "translators" of technical drawings eventually gave them greater social mobility and a certain autonomy. Jacques-Eugène Armengaud, who graduated from Châlons in 1825, became a teacher of technical drawing and editor of France's most influential industrial journal. He publicized the use of technical drawing to plan production runs. In the middle of the century, he also set national standards for screw threads. Another graduate was Jules César Houel, who worked his way up from machine-builder to technical director of a successful locomotive firm. In the 1840s, he centralized all decision-making in a bureau d'études, staffed by engineers and draftsmen, thereby usurping the ability of foremen to define the job specifications and breaking their entrepreneurial power to subcontract tasks. Henceforth, technical drawings defined the task, not the foremen. The net effect was to increase the uniformity of the artifacts produced, achieving in some cases a limited degree of interchangeability.[64]

Here we are on the verge of Frederick Taylor's famous system of "scientific management." The purpose of Taylor's system was to increase productivity by breaking "soldiering," the practice by which laborers mutually held down their pace of work. Through an analysis of work efforts, Taylor sought to make the workers' skills explicit, and hence hold down the price of their labor. Among other techniques, his method relied on the extensive use of technical drawings to enable engineers, rather than shop floor workers, to direct the pace and flow of work. Yet Taylor and his European enthusiasts always claimed that they had found an "objective" basis for organizing production, one that should command consensus among both managers and workers. This was a bold attempt to depoliticize one of the central sources of social conflict in early twentieth-century Europe. In the context I have been laying out, however, Taylorism emerges as the logical extension of a hierarchical distribution of technical knowledge. This distribution, reinforced by "rational" markers which clearly delineated the competencies of the various strata, seemed also to depoliticize the question of who held authority in production. Taylorism had wide appeal in early twentieth-century France, in part because French engineering firms had been practicing aspects of it for decades.[65]

The Picture House of Knowledge

Despite its active role in educating technicians, the state's role in generating practical technological innovation remained circumspect. The state funded the Ecoles des Arts et Métiers and the Ecole Polytechnique. And the polytechniciens in their official bureaus authorized the design of roads and bridges, forts and weaponry. But even in these cases, the state always contracted with private capitalists to do the actual construction. State engineers were confined to an admonitory role.

Established under the Thermidorean regime, the Conservatoire National des Arts et Métiers attempted to define an appropriate role for the state in the promotion of French technology, one which did not involve outright ownership of the means of production as had occurred during the Terror. The transformation of this institution can serve as a final example of how the technocrats' public divorce of technology and politics actually served a particular political agenda. In this new climate, the state's role turned on regulating the exchange of technical information. Just as the ancien régime believed the state should police the public marketplace to assure equal access to the exchange of goods, so the Thermidorean state sought to define and police a public marketplace of information about technology.[66] Henri Grégoire and the other founders of this information marketplace did not want a free-for-all forum on the merits of alternative techniques any more than the post-Revolutionary state surrendered the right to supervise exchanges in the marketplace. The technical information available at the Conservatoire was carefully crafted by its administrators to encourage particular forms of industrial organization. This information centered on machine production.

Now famous as a museum, the Conservatoire began as more than a collection of machine models. It also possessed an active machine shop, a drafting office, and a technical school. The Conservatoire inherited machine collections from Vaucanson, the Academy of Sciences, the new patent commission, and the Atelier de Perfectionnement. The machine shop, a direct successor of the Atelier de Perfectionnement, produced textile machinery and metersticks, an excellent example of the Conservatoire's role as the state's replicator of standards. And the drafting office was the direct successor of the office which Prieur de la Côte-d'Or had assembled as part of the Manufacture of Paris. Its draftsmen had illustrated Vandermonde's report on the Klingenthal Manufacture (fig. 6.1), Monge's treatise on cannon production, Hassenfratz's unwritten report on musket-making (figs. 7.6–7.9), and the machines of the Atelier de Perfectionnement (figs. 6.2–6.4).[67] The Conservatoire also inherited the savant-managers of all these institutions. The early governing boards included Vandermonde, Le Roy, Claude-Louis Berthellot, and Claude-Pierre Molard (an assistant of Vandermonde and pupil of Monge).

But it was the collection of drawings and machinery, rather than the machine shop itself, which soon became the focus of the Conservatoire's mission. After Thermidor, the drafting office produced a technical drawing of every machine in the Conservatoire collection (figs. 6.2–6.4). As Grégoire pointed out, such a collection of technical drawings presented several advantages in combination with a *cabinet* of machines. While machines might disappear or break down, drawings would survive. Drawings saved space, an advantage as models proliferated. Above all, drawing would enable the precise exchange of technical information. Berthollet wanted draftsmen to draw the machines so exactly that "a worker could construct the artifact solely from their rendering." Molard insisted that the bureau's drawings be so painstaking that visitors would not need to disassemble the machines to understand their workings or measure their parts. And Grégoire envisaged this technical information flowing from the Conservatoire "to fertilize all of France."[68]

Technical drawings, however, were not merely neutral conduits of information. Like any good rational language—the metric system, *le français national*—technical drawing provided a structure for right thinking. A central tenet of the Enlightenment had been the belief that progress in the mechanical arts depended on finding a uniform and precise language for that subject. Technical drawing would serve as the vernacular of industry in the way mathematics served physics, the chemical nomenclature aided Lavoisier, or the metric nomenclature organized weights and measures. Grégoire echoed Diderot when he asserted that "the language of the mechanical arts is in its infancy" and expressed the hope that technical drawing would organize the mechanical arts. At present, each trade had its own terminology; names differed from region to region, even from manufacture to manufacture. The purpose of the museum was to bring rational order to this state of affairs by classifying tools broadly according to the social needs they served (agriculture, metalworking, textiles), then more narrowly by function (for metalwork: piercing, grinding, engraving, screw-cutting). A drawing was placed alongside each model, with each arranged in the gallery in a "systematic order to facilitate study and demonstrations."[69]

The technocrats hoped that such a standardized language of machinery would standardize technological *practice* as well. In the Vosges they chopped down trees with axes; near Villers Coterets, they used saws. Grégoire saw no reason to doubt that "one of the two methods is incontestably preferable." He was incensed to learn that even the saws came in a variety of types. "Why not indicate which type of tool will enable a man to spend his forces with the most economy and in a manner most advantageous to his health?" And the only way to do this, Grégoire insisted—again echoing Diderot—was to break down the barriers of secrecy that surrounded artisanal practice; yes, even if it enabled foreigners to learn of French methods. Only such a publicly available knowledge would free the artisan from the "rut of routine." But Grégoire was not interested in the diffusion of just any kind of technical information. The Conservatoire was not the place to learn how to knit stockings or weave ribbons, but "how to construct the most accomplished machines." He professed astonishment that some workers still spoke of machines as depriving them of employment. Yet he also boasted that ". . . the use of machines . . . has as its goal: (1) to obtain more work by saving on human strength and the number of individuals needed, and (2) to give the work greater perfection without requiring greater skill from the workers."[70]

This agenda took little cognizance of the possibility that trees in Vosges and Villers Coterets might best be felled with different tools, or that artisans on their *Tour de France* carried tacit know-how among workshops of widely dispersed regions. Implicit in the Conservatoire's classificatory scheme and "geometrized" technical drawings was the assumption that there existed an optimal form of technological development the state should encourage, a style of industrialization which revolved around uniform machine production. The state's role here, as in Grégoire's contemporary advocacy of the metric system and *le français national*, was to bring French practices in line with the dictates of reason and nature—not through direct, coercive intervention, but by promulgating "neutral" standards. This would liberate the laborer (his body and his pocketbook) and the nation.[71]

During the Restoration the Conservatoire also became France's most vibrant center for popular technical education. Under the leadership of Charles Dupin (another acolyte of Monge), the Conservatoire sought to create a labor elite committed to Victorian ideals of progress, much like the British Mechanics Institutes. In Britain, the emphasis was on the principles of political economy; in France, it focused on drawing techniques.[72] The Conservatoire's "Petit Ecole," established in 1806, trained thousands of students in technical drawing skills. Journals and texts published by the teachers of the Petit Ecole—Le Blanc and Armengaud—led French efforts to standardize screw threads and set other industrial norms. Along with the Ecoles des Arts et Métiers, the school helped supply a new layer of mid-level technicians to the burgeoning machine-building industries of the nineteenth century.[73]

Statism and Technocracy

Technocracy is commonly taken to be a monolithic entity whereby politics is reduced to administration, and the social life is subordinated to the demands of the machine. We are familiar with the fallacy that technology is nothing more than applied science. Technocracy is the belief that society is nothing more than applied technology. But to concede this is to concede that there is a single way to configure the relationship between technology and society, or at least, that there is a single group with a monopoly on the knowledge of how this is to be done. In their retreat from the world of active republican politics, the techno-Jacobins (now the techno-Thermidoreans) made just this sort of "concession." That is, they made a strategic retreat to the "apolitical" bureaus and academies in order to pursue a different sort of state politics. As we have seen, the French Revolution precipitated a thorough reevaluation of the relationship between the social and technological order. But in the end the post-Revolutionary régime reverted to a solution largely pioneered in the ancien régime. The tenets of democratic technocracy were set aside in favor of the stratified relations of the bureaus.

Of course, the Thermidoreans did not resolve the debate over the relationship between republicanism and the productive order. The d'Allarde and Chapelier laws of the early 1790s ostensibly established an unambiguous right to engage in production and commerce. Yet as Gail Bossenga has pointed out, in the late 1790s many large-scale merchants used refurbished corporate institutions to reestablish their control over trade. Whatever lip service the central authorities paid to the market principle, many local officials acted in concert with large-scale merchants and workshop masters to limit access to the marketplace. One of their most effective methods was to enforce standards of production.[74]

Alexandre Vandermonde articulated this ambivalence about the state's relationship to the market in his lectures on political economy at the Ecole Normale de l'An III (where he held the world's first academic chair in economics).[75] Vandermonde had supervised the Atelier de Perfectionnement and the founding of the Conservatoire des Art et Métiers. Just before his lectures, he had reported to the Thermidorean Convention on economic conditions in the ravaged town of Lyon. In a politically charged atmosphere, he struck that note of moderation and reconciliation which characterized both the technocratic pose and the Thermidorean *juste milieu*. He deplored the Revolution's disruption of commercial production: the "incoherence" of the *maximum*, the abuse of the requisitions, and the savagery of the political Terror. Yet he also defended a role for government in the management of the economy. Good regulations, government incentives, and protective tariffs were needed to encourage local producers and to restrict the exodus of skilled workers. One must not allow English ideologies to transform French *laissez-faire* into *faire mal*. Against the Le Chapelier law, Vandermonde defended the advantages of voluntary corporatist organizations to protect less-established artisans while they improved their skills. Regulations created a stable product from which innovations might proceed; sales of Piedmont silk had increased after the new regulations had defined standards of quality there. By enforcing similar standards in France, the state could also guarantee the "purity of mercantile morals."[76]

It was as part of this elaboration and refurbishment of the methods of the ancien régime that the artillery service reasserted its authority over the production of guns in Saint-Etienne and elsewhere in France. Their manner of doing so set the tone for French military-industrial policy for much of the nineteenth century. It also doomed the engineering approach to interchangeable parts manufacturing.

Chapter Nine

TECHNOLOGICAL AMNESIA AND THE ENTREPRENEURIAL ORDER

ONE of the most common assumptions about technology is that it "stacks." Whether considered as a form of knowledge, a set of practices, or a collection of hardware, technology is said to "accumulate." This realm, above all others, is said to be one in which we can accomplish more than our predecessors—thanks, in part, to the efforts of our predecessors. One popular version of technological progress implies that technology stacks like a Lego set of modular bricks. Even scholars who doubt that technological progress is a friend of human betterment still assume that technology is cumulative, much like the gung-ho enthusiasts they deplore. And even those historians who place radical disjuncture at the center of technological change agree that new technology builds on old. For instance, in Edward Constant's (Kuhnian) scheme of technological revolutions, innovators must have a thorough grasp of current knowledge (and its limits) to generate new paradigms. Ironically, scholars committed to evolutionary models of technological change come closest to repudiating a simplistic notion of technological accumulation. Human ingenuity, they argue, like nature, is prolific. Much that is spawned, dies. Throughout human society the broken lines of filiation greatly outnumber the successful adaptations. These scholars emphasize the reception of innovation, and hence its social context. Yet this scholarship usually ignores those paths that seem to lead nowhere.[1]

Underlying this book has been a different assumption, and hence a different kind of history. It is time, therefore, to confront one of the more radical implications of the historicist view of technology I have been developing: the possibility that a technology (even a technology accounted "superior") can be rejected, discontinued, and forgotten. In Tokugawa Japan, it should be remembered, all knowledge of firearms was systematically exorcised. Similarly, the ideal of interchangeable parts production pioneered in late eighteenth-century France was repudiated in the nineteenth century. That repudiation was sufficiently thorough that today we know this method of production as the "American system of manufactures." Disbelief and ridicule greeted those machinists who tried to interest the French state in interchangeable gun manufacture in the 1850s. At a time when the British were importing an entire panoply of American machine tools to outfit their Enfield arsenal, the French only half-heartedly integrated a few new machines into their existing armories. Only late in the nineteenth century did the French arms industry become the nexus of an indigenous machine-tool trade (as had been the case in America for several decades). This chronology of discovery and its subsequent fifty-year erasure violates several of our most basic assumptions about the "natural" history of technology—at least as it is supposed to have unfolded in the heroic age of the industrializing West. Why was interchangeable parts manufacturing, with its promise of efficiency and gain, repudiated in France? What can explain this strange "technological amnesia?"

Between 1795 and his death in 1802, Honoré Blanc placed his innovation on a business footing, selling the army interchangeable gunlocks from his Roanne factories. But despite considerable technical success, his factory was caught up in a bureaucratic battle within the post-Revolutionary artillery corps. This intra-service struggle turned on a larger debate over the relationship between the state and capitalist production. In his monumental study of the metallurgical industry in this period, Denis Woronoff describes the tension between the practices of private commerce and the role of the state, both as a client and as a prod to innovation. The relationship between the two was never resolved, Woronoff argues, but varied with the rhythm of war and shifts in the political scene. These tensions were particularly acute in the armaments industry. True, even at the height of the Napoleonic War effort, the armaments industry accounted for only 8 percent of the military budget, and 10 percent of total metallurgical demand. And these percentages were not surpassed in France until the 1860s.[2] European armies of the nineteenth century, like those of the eighteenth, remained armies of men. Yet the mass army's need for guns—Napoléon constantly worried about shortages—strained traditional methods of production. How different groups struggled to resolve these tensions tells us much about the hesitant process by which the modern French polity took shape.

This chapter, therefore, carries through the major theme of this book, arguing that guns are political entities. The debate over the design and fit of nineteenth-century guns mirrored the debate over the structure of French society. The result was French "statism": an uneasy concatenation of liberal capitalism and corporate interest, in which the generation of commercial wealth was superintended by the state, and *raison d'état* was exercised in the service of commercial groups. Gun production was one of those sites on which this amalgam was assembled by the various French elites who contested the inheritance of the Revolution. As Louis Bergeron has argued, the central problem of Napoleonic and nineteenth-century France was to forge a unified elite out of the disparate groups which now claimed to govern France. And governing France meant, in large measure, controlling those unruly forces which had sometimes made the Revolution disruptive to property and men of elite standing. Interchangeable parts production—as the idealization of a particular social program—signaled one set of alliances to carry out this aim. As we will see, a program based on the harmonious functioning of the gun represented another. Interchangeable parts production was repudiated in France not simply because it was a technical and economic "failure," but because it ceased to serve as an *ideal* for organizing the polity. In this sense, the physical qualities of French muskets comprise a record of what was remembered—and what was suppressed.[3]

Entrepreneurial Interchangeability, 1794–1802

Only fifty kilometers north of Saint-Etienne, Roanne was in a different world. The locals spoke French, rather than Franco-Provençal. The town looked northward, rather than inward, and owed its wealth to its position at the navigable head of the Loire River. A thriving town of seven to eight thousand inhabitants during the Revolution, Roanne enjoyed relative calm under a municipal government dominated by wealthy bourgeois of moderate leanings. In 1794 Blanc acquired the spacious Ursuline convent for his factory, plus the Alcock family's water mills four kilometers out of town. There, before his death in 1793, Alcock, the mass producer of buttons, had briefly minted coins. The Roanne site, however, had no indigenous metalworking industry. The price of coal was eight times greater than at Saint-Etienne. And even in Roanne the social and financial chaos of the mid-1790s made it difficult to equip factories and retain workers. Nevertheless, the hope—at least in one faction of the artillery bureaucracy—was that Roanne could draw on some of the human and natural resources of the Forez region without invoking the hostility which uniform production methods provoked in Saint-Etienne. That hope was, in some measure, realized.[4]

In 1795 Blanc promised the government that he would soon be delivering 1,800 gunlocks a year. By September 1797 he had in fact produced roughly 4,000 interchangeable gunlocks and pledged to supply 25,000–30,000 a year—more than Saint-Etienne's output before the Revolution.[5] This promise was never fulfilled, though by 1800 the factory had shipped a total of 11,500 interchangeable gunlocks. These were sent unhardened to other manufactures—Saint-Etienne, Versailles, Liège, Turin—where they were engraved and mounted on muskets. In 1801, with Blanc near death, the entrepreneur Jean-François Cablat bought out the manufacture and contracted to deliver 12,000 *entire* muskets a year. These guns were to be "wholly interchangeable." The inspector at Roanne was randomly to test 5 out every 100 gunlocks for interchangeability. Until the manufacture was closed in 1807, the annual output averaged roughly 10,000 gunlocks and 2,000 muskets. During those same years, by comparison, Eli Whitney was failing to meet his contracts for far fewer muskets, and his gunlocks were not, in fact, interchangeable. By any technical measure, Blanc's was an extraordinary accomplishment.[6]

The Roanne factory achieved this success by amplifying and intensifying the methods used at Vincennes. Although a few parts were still produced on an outwork basis, the centralization of production far surpassed anything previously attempted. According to Inspector Tugny, each milling machine at Roanne performed a specific task, unlike the Vincennes workshop and the Atelier de Perfectionnement. This meant that the accuracy of the machines could be improved, minimizing hand finishing. The number of analytical steps to make the gunlock was 156; the lock plate alone required 32 separate tasks, and the tumbler, 22. Needless to say, each of these tasks was not assigned to a different worker. Yet this relentless division is indicative of the analytical approach to production. After Cablat expanded the manufacture, its 200 workers were divided into 16 basic subtrades, and these were divided more finely by task. Cablat estimated that machines represented about 20–30 percent of the cost of making the lock, with another 10–20 percent for raw materials, and the rest for labor.[7]

There were, however, several limitations to Roanne's achievement worth noting. First, attempts to render the whole gun interchangeable were premature. The Roanne inspector regularly verified that the mounting and trigger, as well as the gunlock, had been rendered fully interchangeable. But lathes capable of turning irregular surfaces, such as the gun stock, though known to the French, did not possess the flexibility or speed of Thomas Blanchard's American lathe of the 1810s. Second, only those locks mounted at Roanne were case hardened and tested on site. Those case hardened at other manufactures might well not be interchangeable. Nor would the army realize one of the professed advantages of interchangeability—ease of repair in the field—until interchangeable gunlocks were designated as such, and accompanied by replacement parts. Third, the Roanne Manufacture continued to employ "adjusters and correctors of the gunlock pieces" who checked the fit of the locks before handing them over to the controllers for inspection. Call them what you will, they functioned as fitters—as the inspector admitted. And fourth, the controllers at Roanne actually had "twenty times" as much work under the new system than the old. The uniformity system obliged them to gauge and measure a large number of parts. Rather than substituting mechanical controls for personal oversight, the system demanded even greater administrative supervision.[8]

All this suggests the central limitation on the Roanne system: artisanal skills had not been supplanted. Blanc always admitted, of course, that his system required highly skilled machinists. Ironically, he recruited these men principally from Saint-Etienne; men such as Jean-Baptiste Lyonnet and Jean-Baptiste Pozon (the two shop chiefs), and Jean-Claude Delahaye, who became the factory's leading technician after Blanc's death. But as for the rest, Blanc had always boasted that he only needed unskilled labor—"workers without talent, what one might well call day laborers"—and he wrangled permission to hire workers with less than the requisite two years' experience in metalworking. In fact, hand filing in jigs continued to be the most time-consuming procedure in the production process, and Blanc had great difficulties attracting this essential class of skilled artisans. As Roanne's inspector acknowledged, the artisans of Saint-Etienne had learned from bitter experience just how precarious government contracts could be, and relied on the private gun market to provide fall-back employment. In Roanne they would have no such security. So the labor shortage was made up with armorer-conscripts; roughly half of the workers at Roanne were young men exempted from the army because of their background in metallurgy.[9]

The failure to realize major savings in labor costs was the principal reason that the factory never turned a clear profit. Blanc acknowledged that high interest rates, the local price of labor, and materials costs prevented him from manufacturing locks as cheaply as those of Saint-Etienne. This made the operation dependent on a state subsidy. Thanks to the intervention of Agoult and Aboville, Blanc had initially received a supplement of 30 *sols* per lock (27%). In return, he was to match the Saint-Etienne price within a year, and undersell the armorers the year after. Two years later, Lazare Carnot (then minister of war), noted that Blanc was still not underselling Saint-Etienne, where prices, admittedly, were dropping. Aboville, however, successfully renewed a smaller subsidy, and by 1802 Roanne was selling gunlocks for 5.6 *francs*, equal in quality and price to those of Charleville—though Cablat complained that at that price he was selling them at a loss.[10] Overall, the evidence indicates that between 1800 and 1804, the Roanne factory was producing roughly ten thousand interchangeable gunlocks a year for a price within 20 percent of the going rate at Saint-Etienne. Certainly, the factory attracted private investors. The most active was Carrier du Réal, brother of the infamous Carrier de Monthieu. Excluded from Saint-Etienne after the Revolution, he finally put 60,000 *livres* of the Carrier family fortune into mechanical production. The Aboville family also invested several thousand *livres*. And Cablat sank 30,000 *livres* into the business[11] Yet the government's protection was essential; the state provided conscript labor, low-interest loans, and subsidies. This made the effort vulnerable to the enmity of a powerful new faction in the artillery bureaucracy led by Jean-Jacques-Basilien Gassendi.

Gassendi rose to power through his personal friendship with Napoléon. He was born in 1748—eighteen years after Aboville and thirty-three years after Gribeauval. He was a distant relation (a nephew, seven generations removed) of the illustrious natural philosopher, and placed great stock in this connection. The artillerist was one of those *roturiers* who entered the artillery school at the moment of Gribeauval's reforms. He graduated at the top of his class, yet he was always ambivalent about the new artillery. This may explain his slow rate of promotion during the ancien régime.

> Trente ans pour mon pays j'ai servi dans les camps,
> Et trente ans mon pays me laisse aux derniers rangs. . . .[12]
>
> For thirty years I served my country in the camps,
> And for thirty years my country let me languish in the ranks. . . .

This changed after he had the good fortune to command young lieutenant Buonaparte [*sic*]. In the last years of the ancien régime, the two men became close. Like Napoléon, Gassendi was ambivalent about the Revolution. He swore allegiance to the Constitution in 1790, but attempted to retire from the artillery shortly thereafter, pleading poor eyesight. (The forty-one-year-old captain had recently contracted a marriage that made him financially independent.) He clearly preferred the bureau to the battlefield. The Jacobins ordered Gassendi out of the army, but, after Thermidor, Napoléon cajoled him into serving on the Central Committee of the Artillery, and then promoted him to general. When the War Office was reorganized along functional, rather than geographic divisions, Gassendi took the powerful position as Director of the Division of Artillery and Engineering, which enabled him to control the liaison between the minister of war and the artillery service. Finally, in 1805, Gassendi succeeded briefly to the office of first inspector-general, and shortly thereafter to the Council of State. From this position of power he was able to crush the Roanne Manufacture.[13]

Why, after decades of commitment to "rational production," did the artillery under Gassendi repudiate interchangeable parts manufacturing? After all, to argue that the method "did not pay" hardly answers the question. During the same period, the United States government—no friend of state expenditures—underwrote gun production with its "armory system" (what in France was known as the *régie* system). And the other French armories also depended on government patronage, with the Entrepreneurs there contracted to buy guns for the state in return for a fixed profit. As for the difference between the cost of gunlocks from Roanne and those from Saint-Etienne, that was never as great as Roanne's competitors liked to claim. Moreover, Aboville made plausible arguments that interchangeable parts would ultimately save the government millions of *livres* in repair costs. Yet Gassendi and his partisans denied that interchangeability was achievable, affordable, or desirable. Why did they do so, and how did they triumph?[14]

Bureaucracy and Business

The bitter bureaucratic battle over interchangeable parts manufacturing in the early nineteenth century was part of a contemporary effort by the Napoleonic elites to redefine the French state after the trauma (as they saw it) of the Revolution. The Revolution had alerted the central authorities to the danger posed by the popular classes and the centrifugal tug of local interests. The Napoleonic solution was to harness local elites and popular classes to the national purpose primarily through the mechanism of the "warfare state." The challenge was to maintain a stable domestic social order while simultaneously pursuing near continuous warfare across the continent. So long as its battlefield successes continued, the French armies were financially self-sustaining.

But warfare itself had a variable effect on domestic industries. Generally speaking, the armaments trade flourished. Yet even amid plenty there remained the crucial question of how to divide the spoils. This depended on the accommodation reached by economic and military elites with *distinct* interests and cultural codes, and distinct relationships to the artisans who actually produced the weapons. Just as the controversy between the Vallièrists and Gribeauvalists had reflected different conceptions of how the ancien régime state should be organized, so the debate between the "Gribeauvalists" (now led by F. M. Aboville) and Gassendi's party reflected divergent views on how post-Revolutionary France should be governed. The guns of the Napoleonic era reflect this struggle.

Even the most basic technological "facts" were contested during this intraservice controversy. The central question was an apparently simple one: could the pieces of Blanc's gunlocks actually be interchanged? To answer it, both sides had recourse to demonstrations along the lines of Blanc's 1790 "proof" in the Hôtel des Invalides. As in the cannon demonstrations of the 1770s, the trappings of scientific experiments were here transposed into a bureaucratic context. And again, witnesses ranked in a formal hierarchy advanced a partisan agenda. At stake was control over the technological memory of the French state.

For his part, Aboville orchestrated several vindications of Blanc's methods. In 1801, as Blanc was on his deathbed, workmen demonstrated to the Central Committee of the Artillery that out of five hundred sample locks, only six had broken parts and only two needed a touch of the file. Armed with this finding, Aboville petitioned the First Consul to take full advantage of Gribeauval's uniformity system. But as much as he admired Gribeauval's legacy, Napoléon wanted the opinion of someone he trusted. Across the top of this petition, he scrawled: "The minister of war will let me know the opinion of the director of the [division of] artillery, [Gassendi], on this subject."[15] Immediately, the opposing faction set to work to discredit the idea of interchangeability. A week later, a test performed under the direction of Gassendi, determined that out of (a different) 492 gunlocks shipped from Roanne, only 152 were approved as serviceable.[16] Control over technological "facts" depended on control over personnel.

So began a series of intricate bureaucratic maneuvers to control the personnel responsible for Roanne. Aboville struck first, sending his eldest son, Augustin-Gabriel, a prominent artillery brigadier, to Roanne as its first official inspector in 1802. In a secret private correspondence that ran in parallel to their official letters, Aboville warned his son that Gassendi masked his animus behind the veil of public disinterestedness. When war called his son away, Aboville sent a hand-picked successor, Inspector Tugny. However, the next year, Aboville's patronage network collapsed. The new director-general of manufactures, Saint-Martin, openly sided with Gassendi. In the shuffle of personnel, Tugny left Roanne. Worst of all, Aboville was reassigned to the Senate at the age of seventy-four, and replaced by the twenty-eight-year-old August-Frédéric-Louis Marmont. There was contentment in the artillery at the replacement of the "vieux papa."[17] A new generation was taking over.

Gassendi, who wanted to ship Blanc's machines to the Conservatoire National des Arts et Métiers as museum pieces, ordered another test. This time only three out of one hundred locks proved interchangeable. Cablat protested, however, and was allowed to demonstrate that far more of the locks were interchangeable than at first appeared.[18] As a compromise Marmont suggested relocating the manufacture to Liège, with its lower labor costs and experienced workforce. (Again, no one dared to suggest transplanting the method to nearby Saint-Etienne). In early 1804, Cablat agreed to the move in return for a sizable advance and a (temporary) supplement of 1 *franc* per gunlock. He obligated himself to produce forty-eight thousand identical gunlocks a year. Within a year, the Liège Manufacture had produced nearly six thousand gunlocks.[19] Then Gassendi succeeded Marmont as first inspector-general. In a final round of "scientific tests," a spot-inspection by one of Gassendi's clients found the gunlocks inadequate; whereas Delahaye, the controller at the Liège factory, found them interchangeable. Shortly thereafter, Director-General Saint-Martin replaced Delahaye with a controller in the patronage network run by Gassendi. The new controller "discovered" many deficiencies in the gunlocks, and for his pains was rewarded with a bonus of 600 *francs*.[20] In a final effort to trump Gassendi, Gaspard Monge, now the Senator for Liège, appealed to Napoléon. Monge denounced Saint-Martin's actions. But Gassendi well understood the ways of bureaucratic infighting.

> One cannot prevent denunciations, [but] they fall into oblivion if you do not respond to them. The Division [of Artillery] does not seek to destroy the manufacture of identical gunlocks, the manufacture itself is self-destructing by not furnishing the aforesaid gunlocks. Right-thinking people have seen this coming; but it was necessary to use proof to destroy the belief of those who had this delusion [regarding interchangeability].[21]

Gassendi then wrote to the minister of war and the emperor giving his reasons for suppressing the manufacture. At that very moment, the Liège factory was presenting fifty gunlocks with interchangeable parts at the 1806 Parisian Exposition of the Products of Industry. Unlike the Crystal Palace Exhibition of 1851 which so stimulated the British public, however, this event passed unnoticed. In short order the gunlock factory was permanently closed.[22] Over the next decade, a few armories experimented with "accelerated methods": state armories at Versailles and at Mutzig (under Aboville's younger son), and private factories under Julien Le Roy (discussed below) and John George Bodmer (who subsequently had an illustrious career in England as the inventor of moveable cranes and a continuous-flow assembly line).[23]

But in the end, the superior clout of Gassendi's party sealed the fate of interchangeable parts production in France. This had lasting repercussions. Gassendi's manipulation of the process of testing and witnessing established the failure of interchangeability as a technological "fact." He then perpetuated these "results" through his *Aide-mémoire*, the "definitive" compendium of information for all artillery officers from 1789 through the Restoration. And Hermann Cotty, who rose under Gassendi to become director-general of the manufactures, noted that Roanne's locks lacked "harmony" in his authoritative book on firearm production (1806) and in his volume on artillery for the *Encyclopédie méthodique* (1822). In that volume, Cotty's entry on "tolerance" pointed out the impossibility of specifying the range of acceptable dimensions for all the complex and interdependent pieces of a musket, and he warned that any such attempt would ruin the action of the gun. For the next fifty years, the consensus was that interchangeable parts manufacturing had been tried, and had failed.[24]

Technological memory, then, at least in the state-led sector, became a struggle to control those institutions with the resources and persistence to sponsor technology. This is not simply because factions in the government directly suppress innovations, but because the uncertainty of these patronage battles makes particular forms of private investment too risky. As A.-G. Aboville, *fils*, complained, private investors would not sink the necessary capital into a manufacture which relied on the support of a small number of government officials.[25] And the converse was true as well: government officials proved loathe to support new technologies when powerful commercial interests opposed them. This was certainly the case for interchangeable parts manufacturing.

To his credit, Aboville always recognized that new methods of production implied new forms of social organization. The old Saint-Etienne armory—with its dispersed artisanal shops coordinated by financier-merchants—had always refused to implement interchangeable production. So in his petition to Napoléon, Aboville acknowledged that he hoped eventually to replace the traditional armories with a single rational manufacture.[26] In his private correspondence with his son, he prophesied that Roanne, "the seed" of that manufacture, would need to be run on the *régie* system.

> [Uniformity production] cannot take place in the armories under the existing Entrepreneurial system—with the exception of Roanne which will always have to fight against the others and will succumb in an instant if it ceases to have the support of the government. The only means to insure its survival is to organize it like our arsenals of construction, with a director-commander and a director-technician, plus the necessary controllers and examiners. Once this first organization is established, it will gradually expand . . . until it sends out a branch and later several, such that the orders from the [traditional] Entrepreneurial armories rise and fall until they are entirely extinct.[27]

Aboville's plan was to supply the core of military need from a state-owned, mechanized factory, while forcing small-time artisanal producers into the fluctuating margins of demand. This is precisely what M. J. Piore says mass production firms try to do to small, flexible production firms. But Aboville was not seeking an efficient production regime, so much as a secure one. That way the state would never again be held hostage by the unruly and rebellious armorers of Saint-Etienne, as had occurred during the Revolution. Precision manufacturing was social discipline, and the "fit" of Roanne's gunlocks demonstrated the artillery's mastery of that discipline. Under Aboville's prodding, the Extraordinary Council of the Matériel of the Artillery recommended that output at Roanne be gradually increased, output elsewhere decreased, and Roanne itself organized as a *régie*. The state would take charge of production. Not that there was any *necessary* link between interchangeability and nationalization. But like the Revolutionary technocrats of the year II, the artillerists of the Aboville school recognized that imposing a technological revolution required the power and purse of the state. As Aboville knew full well, this meant that the technique would arouse powerful opposition from the merchants and artisans of the old armories.[28]

Gassendi and his supporters advocated a very different social strategy for the Napoleonic regime. They argued that assuring an immediate supply of quality firearms to Napoléon's army (the linchpin of the entire imperial system) meant quickly reestablishing the artillery's control over the armories where gun-making expertise resided. To accomplish this, the artillery service needed the acquiescence of the artisanal class and the capital of the wealthiest arms merchants. The artisans were, of course, hostile to those forms of mechanization which transferred control to state inspectors. And merchants—both *fabricants* and *négociants*—were eager to see the state dismantle the *régie* system which had operated in Saint-Etienne since the year II. Gassendi's campaign to close the Roanne factory was part of an ongoing offensive to reverse the Revolutionary nationalization of the armaments industry. He coyly acknowledged as much in a poem.

Les poudres, les canons, les armes portatives
Les voitures à guerre à porter ces objets
Sont les quatre attirails de formes positives
Qu'il faut à l'artilleur qui vole à des succès.
Autrefois on voulait tout mettre à l'entreprise:
En régie aujourd'hui on a tout mis, dit-on;
Autrefois sur un point on n'avait pas raison,
Mais aujourd'hui sur trois on fait une méprise,
Les trois points les voici: poudres, fusils, canons.[29]

Gunpowder, artillery, firearms,
And battle carriages to carry all these

Are the four implements of war
Which every victorious artillerist needs.
Formerly, we wanted everything in private hands.
Today, everything is owned by the state, I hear.
Formerly everyone was wrong on one count.
But today, everyone errs on three.
And these are the three: gunpowder, firearms, artillery.

As poetry this may be execrable, but as policy, it is clear. Only artillery carriages should be built in government-owned workshops; the other three sectors of the French military-industrial complex should be run on the "Entrepreneurial" system, as during the ancien régime. This industrial policy fit into a broader social strategy by which the state would attach itself to private interests. Those interests were defined as capitalist, with the profits to go to the right people. On these grounds, Gassendi rejected a proposal by two small-time entrepreneurs to produce gunlocks by mechanical means: "But an enlightened government must ally itself with [the gain of] individuals, and in the onerous proposal before us, the government finds no such [alliance]."[30]

In this way, Gassendi defined the relationship between the military and commercial elites, the two most prominent groups vying for the spoils of the warfare state. Throughout the eighteenth century it had been the role of the artillery service to mediate between the state and "capitalist" interests. That role was usurped during the Revolution when the Committee of Public Safety put the state's (collective) interest above that of local elites. Between the year II and the year V (1793–97), in Saint-Etienne, as in Paris, the government had either bought its weapons directly from artisans or had itself organized the workshops which made them. Now the Napoleonic state, like its absolutist antecedent, was returning to the system of purchasing its instruments of coercion through local middlemen. This was *statism*—an amalgam of private capitalism and statist direction which governed much of French industrial relations into the twentieth century.

This Napoleonic conception of proper social relations had its correlate in the qualities to be sought in material artifacts. Where Aboville took precision "fit" as the measure of his ability to police the social order, Gassendi's partisans claimed to rule by reference to the ideal of *harmony*. This was not merely a nostalgia for lapsed social forms of the ancien régime. Harmony was also a sign of proper *function*. The gunlock, as the Napoleonic engineers repeatedly stressed, was a machine whose parts had to work together. Gassendi's partisans complained that the advocates of interchangeable parts manufacturing had become so obsessed with precision that they were blind to the operation of the finished mechanism. This, for instance, was how one artillerist deplored the faulty assembly of interchangeable parts guns.

> Harmony is the *sine qua non* of the gunlock; this quality depends on nothing more or less than what can be achieved by ordinary methods, even though this work

> achieves its harmony through continual comparisons; it follows that the manufacture of gunlocks by machines where the manufacture of each piece is total and isolated cannot succeed unless the pieces are rigorously identical, and it is without question that this kind of identity is a chimera.[31]

In practical terms, then, isolated guns parts—like the isolated workers who made them—could not function as a whole. Gassendi's counterdemonstrations, by invoking different criteria of judgment, implied a different organization of production. For his coterie, a harmonious piece of hardware—here, the action of a gunlock—was the sign of their ability to govern the social order.

Saint-Etienne Revisited: The Ancien Regime All Over Again?

The Saint-Etienne armory was a crucial arena for implementing this Napoleonic industrial policy. What happened there set the pattern for military-industrial relations for the rest of the nineteenth century. Military violence was the linchpin of the post-Revolutionary state. Yet the Empire reverted to the ancien régime's system of *purchasing* its instruments of battlefield coercion and using legal sanctions to control the terms of that exchange. In this sense, as in so many others, the artillery service reestablished the pre-1789 order in Saint-Etienne. Tocqueville's analysis applies brilliantly to the French "military-industrial complex"; here too, the Revolution made possible the ancien régime all over again. This is true even in the realm of state-citizen relations, where the Revolution had proclaimed a definitive juridical rupture. In theory, the era of privilege and monopoly was over. At stake now was the extent to which political autonomy would be coterminous with economic autonomy in the crucial realm of armaments production. The Revolution had also disrupted the economics of the arms trade of Saint-Etienne. On the one hand, the net result was to facilitate the creation of a merchant monopoly more potent than any the ancien régime had known. On the other, the mass army's need for guns obliged the elite—both engineer officers and wealthy merchants—to seek some accommodation with local artisans. The process of renegotiating the relationship among these groups was materially substantiated in the guns they produced. Manufacturing tolerance now became a measure of the degree to which artisanal producers would be recognized as citizens, and it demarcated the limits of state authority in the management of private capital. In other words, the fit of a gunlock was a material correlate of the "statist" compromise.

The artillery engineers recognized that harmonious social relations in the armory required a relaxation of the standards of production. Inspector-General Drouot put it this way after his visit to Saint-Etienne in 1806:

> We must have at the head of our armories men who are firm and active and enter into every detail with the sacrifice of all their time; but their firmness must be without severity (*rigeur*); we must not kill the poor workers for faults of proportion

> which do not influence the quality of the arm or its [overall] uniformity. We must gently reconcile three interests which are often opposed, that of the state, the worker and the Entrepreneur. With gentleness, patience and constancy, our guns will always be the best possible.[32]

This state of affairs was not achieved overnight. In chapter 5, we saw how the alliance of artisans and *fabricants*, forged to oppose the engineering vision of production, gave these groups dominance in municipal affairs in the early years of the Revolution (1789–93). In the wake of the Lyonnais occupation (June 1793), control of the municipality began to swing from faction to faction: first to those operating under the Terrorist proconsul Javogues, and then to his declared enemies, the royalist instigators of the "White Terror." Throughout these political gyrations—which included more than ten changes of administration during the four years of the Directory—local civilian control over the armory gave the preeminent faction considerable patronage power.[33] Only with the ascent of Napoléon did the artillery and the wealthy merchant class reestablish their authority. This triumph depended on both market conditions and legal sanctions.

The market for guns had been transformed by the Revolution. Overnight, the trade routes to Africa, the Levant, and central Europe were curtailed. With the coming of war, musket exports were banned, and so too was production for the private market. By 1794 the time of the Terror—the central Parisian authorities had made good on their claim to *all* of Saint-Etienne's output. In effect, the ratio of production in the two sectors reversed itself. Saint-Etienne's repository of gun-making skills followed demand into military production. Whereas private sales had once generated 1.5–1.8 million *francs* per year, they netted only 200,000 *francs* per year around 1800. And under the Empire, 400–700 armorers sold only 3,000–5,000 guns annually on the private market, one-fifteenth the pre-Revolutionary output.[34]

These legal restrictions and market contractions increased the state's leverage in dealing with armorers. So did the post-Revolutionary curtailment on the private ownership of guns. To the Directory's reiterated bans on the private possession of military muskets and concealed weapons, the Napoleonic state added stringent new requirements. After the year XIII (1805), mayors granted gun permits only to those with rural properties (and hence entitled to hunt), and those who either paid their taxes or lived by "honest industry." The cost of these permits eventually rose to 30 *francs*, and the property qualification to fifty hectares. To give these laws bite, the government began to police the sellers as well as buyers. After 1806, retailers had to supply local authorities with lists of all sales. So, after a brief withdrawal during the Revolution, the French state reverted to the ancien régime claim that the possession of guns was an entitlement. Rather than a group privilege, however, the concession now attached to individuals on the basis of their property rights. According to this doctrine, weapons were legitimate only for hunting, which presupposed the possession of rural lands. Even then, the possession of firearms (rather than some other kind of weapon) might be restricted in the interest of public safety. Of course, these bans were widely flouted; even military guns could be purchased illegally from soldiers. Yet these attempts to forestall popular rebellion and retributive justice demonstrate the degree to which guns remained explicitly political artifacts.[35]

These restrictions also made military work the only livelihood available to the armorers of Saint-Etienne—itself temporarily renamed "Armeville." Once the Lyonnais had retreated, the armory was organized as a *régie* under the authority of the proconsul Javogues. The civilian Conseil appropriated *biens nationaux* and built workshops where ironmongers could try their hand at gun production. Meanwhile, established armorers in their own boutiques took on nearly a thousand new apprentices. Over the next 18 months, some 5,000 workers, including novices, women, and children, produced 170,858 muskets.[36] A jury of armorers, selected by the Conseil, was supposed to gauge these weapons, but as was to be expected in wartime, standards were relaxed. Work discipline, on the other hand, was enforced with the social and legal mechanisms of the Terror. As in Paris, production was patriotic. Severe laws which defined the work day, forbade breaks on Sundays or festivals, and prevented artisans from attending to their harvest went hand-in-hand with quasi-democratic consultative institutions, such as those established by Romme in 1793 (and in Paris in 1794).[37]

Then, during the Directory, military demand plummeted. With its territorial expansion, France could now tap the gun-makers of Liège and Turin. Meanwhile, Charleville and Maubeuge had resumed production. With government stockpiles full and inflation rampant, musket production effectively ceased in 1796, leaving the government 150,000 *livres* in arrears to local artisans. For the next two years thousands of armorers were idled without a private market to pick up the slack. This boom-bust cycle meant that when the artillery service returned to Saint-Etienne in 1798, the armorers were desperate for government contracts. At this time, the state reintroduced the Entrepreneur system and output slowly inched upward: from a paltry 5,000 muskets in the year VI (1797–98) to 22,000 in the year VII (1798–99) to 29,000 in the year VIII (1799–1800). (See fig. 9.1.) Inspector Joseph-Benoît de Colomb, who had first come to Saint-Etienne in 1793 as the artillery's representative to the civilian Conseil, now presided over the armory. In 1801 he regained the powers of an ancien régime inspector (though he wielded them with a lighter touch). At that time, the contract between the state and the Entrepreneur was formally reestablished on the basis of the 1777 law. Similarly, the police laws of 1781 and 1782 once again governed the relations between the artillery and armorers. Even the personnel reverted to pre-Revolutionary days. Augustin Merley returned as civilian proof master (with less work to do). And the state even considered reappointing the surviving royal Entrepreneurs (Carrier Du

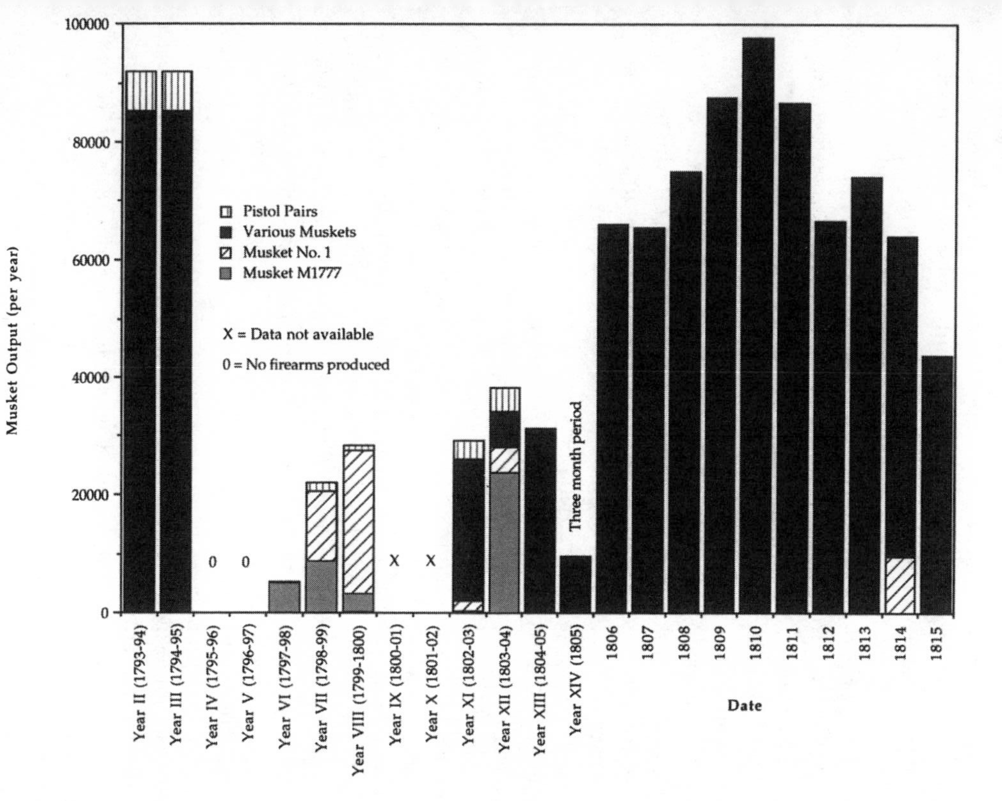

Fig. 9.1. Firearms Production, Manufacture Nationale de Saint-Etienne, 1794–1815. This graph shows the tremendous fluctuations in military musket production at Saint-Etienne during the Revolutionary and Imperial decades: approaching a rate of 100,000 muskets per year in 1794–95, falling to nearly zero in 1796–97, and then gradually rising toward 100,000 again by 1810. Data for years II–VIII are from Guilliaud, *Fabrication d'armes*, [1800]. (I have divided output equally between the years II and III.) Data for the Napoleonic period are from Dubessy, *Manufacture de Saint-Etienne*, 258–59.

Réal and Dubouchet), but settled instead for Jovin and Dubouchet. They soon quarreled, and when Napoléon insisted on a single Entrepreneur, Gassendi saw to it that Jean-Baptiste Jovin was given the contract.[38]

Jovin ran an arms monopoly more effective than any the Carrier family had ever known. His father had made his fortune in coal mining and turning out cheap guns for the slave trade. He had also been a leader in the municipal revolt against the manufacture. Sensing an opportunity in 1792, he had hired the inventor Javelle and converted two water mills and a workshop to the production of military arms. When he was imprisoned as suspect during the year II, dozens of his workers petitioned for his release. His son and heir now had vast powers of patronage; petitions signed by hundreds of arms workers denounced his refusal to hire armorers who had worked for Dubouchet. According to Inspector Tugny the workers were "horribly dependent" on Jovin. His guaranteed 20 percent profit on every gun was enforced by sanctions as formidable as any of the ancien régime.[39]

Despite Revolutionary assertions to have created a free market in labor, the legal status of arms workers remained controversial. Beginning in the year II, armorers of conscript age were exempted from the draft; in return, they were subject to martial law. As Isser Woloch and Alan Forrest have noted, conscription was the central battleground between the state and citizens in the Napoleonic period. Despite popular resistance—the inhabitants of the Forez were notoriously unresponsive to the levies—conscription raised unprecedented levels of manpower. It also gave the state formidable leverage over those who wriggled out of service.[40] Quite a few armorers were draft-dodgers from bourgeois homes. In 1799, the administration decided to weed out shirkers and raise production quality by requiring all arms workers to demonstrate their skills publicly before a panel of three expert armorers. Those who flunked were hustled off to the army; a number fled; the rest were given *cartes de sûreté* to accompany their *livret*. Some five hundred conscripts were employed in the armory at Saint-Etienne in the years of the Empire, 25 percent of the total.[41]

Some prominent artillerists, such as the ill-fated pre-Revolutionary Inspector Lespinasse, wanted to go even farther and completely assimilate armorers to military discipline along the lines of the worker-soldiers. But Inspector Colomb denied that arms workers—aside from the conscripts—could be deprived of their rights simply because they made gun parts for the army. To do so would cause the workers to quit the manufacture. He admitted that the workers had no particular loyalty to the state, and that conscription was a necessary tool of discipline. But rather than use coercion, the state would do better to set *prices* so that work proceeded in proportion to need. The Revolution had taught Colomb that the armorers had options, including rebellion. When the authorities in Paris pushed for uniform rules for all imperial manufactures, Colomb reminded them of the special characteristics of Saint-Etienne: its dispersed workforce, its unique methods of production, the presence of the private trade, its unique division of labor, and its "just-in-time" manufacturing. For instance, gunlocks could not be tested every fifteen days as in Charleville, but had to be gauged as they arrived for fear of disrupting workers whose tasks followed thereafter. Colomb asked that the government restrict itself to specifying *standards* for its gunlocks, and leave the method of production to local arrangements between armorers or merchants.[42]

There remained the question of what *kind* of standards the state would impose. Inspector Pierre Sirodon, in charge of the Tulle armory, warned against setting overly rigid standards. He explained how guns produced in his armory emerged as the outcome of negotiation between artisans and engineers over standards—except that this now involved due deference to arms-workers' skills and engineers' judgments. To be sure, all armorers, controllers, and inspecting engineers should test gun parts with gauges calibrated against a master pattern, itself standardized against national measures. Indeed, one engineer-officer should have sole responsibility for maintaining these standards. But officers must not reproach themselves for slight deviations in the guns they accepted. Many young officers arrived at the manufactures with a habit of submitting to rigid laws, and often succumbed to what Inspector Sirodon called "a fanaticism of measures." They intimidated the controllers and discouraged the workers. The painful truth was that variability permeated the works of man, and perfect identity was a chimera. The dimensions of the lock plate, say, could not be fully specified, and differed from armory to armory. Even the *gauges* differed. What was an officer to do? This was not simply a problem of rule-making, but of practical economics. Officers had to realize that gun-making depended ultimately on the skills of the armorers, and consequently on the *interest* they had in doing the job properly. Certainly, many armorers tried to get away with shoddy work when they could. For certain crucial dimensions (such as the muzzle's caliber) the state should set a well-defined tolerance to eliminate any possible "arbitrariness." But to insist on narrow tolerances for all dimensions would be prohibitively expensive and the workers would quit. Hence, the decision to accept a musket had to be left to the "intelligence of the controller [and] the decision of the inspector." Such a judgment required experience. Controllers must not simply apply gauges mechanically—any apprentice could do that—but judge the *functioning* of the gun. This functioning had many aspects, not all of which could be easily quantified; among them, fit, reliability, durability, and ease of use.[43]

This shift in the engineers' conception of their role can be interpreted as a pragmatic approach to production during wartime. Undoubtedly, this was not the moment for rigor; guns ought to meet minimal standards, and their price ought not be exorbitant, but the state was in no position to bicker over details. However, this shift also signals an expanded conception of manufacturing tolerance as a way to reconcile private interest and the state's interest. The new "play" allowed in standards signaled an adjustment in the relationship of state and private industry. The fit of Napoleonic gunlocks reflected this new political calculation. It also gave the officers a different kind of power over workers. By freeing judgments from objective standards, officers acquired new powers of discretion. Inspector Sirodon instructed his subordinate officers to use this new discretion to punish recalcitrant armorers by rejecting their workpieces and favoring those whose past performance had been prompt and satisfactory. Qualitative judgments of workpieces were recorded, creating a historical record of fidelity and skills. The perfection of any single gun mattered less than the pattern of behavior and personal loyalty.[44]

The artillery's commitment to this new model made for some surprising reversals which shed light on the abandonment of interchangeable parts production. For instance, when the Entrepreneur Jovin introduced "accelerated" methods of machine production he met with protests from artillery engineers. Napoleonic demand for weapons had soared: output at Saint-Etienne went from 34,000 muskets in the year XI (1802–3), to 66,000 in 1806, to 98,000 in 1810 (fig. 9.1). For the first time (the year II excepted), production surpassed the total levels of the ancien régime. Yet the total number of armorers remained roughly 2,000. Equally important, the increase in the number of controllers—from four in 1801 to seven in 1808—did not keep pace with the quintupling of military output.[45]

Within this expansion, the making of gunlocks constituted a critical bottleneck. As before the Revolution, locksmiths remained scattered in the small villages around Saint-Etienne. Of the 584 locksmiths in 1799, only 12 percent worked within the limits of Saint-Etienne proper, whereas 30 percent worked outside of the département de la Loire altogether (map 5.2).[46] What changed was how the locksmith trade was organized. The germ of this reorganization had been laid in the late ancien régime. Recall that after 1763, with Blanc's help, the armory had begun machine milling its tumblers in separate workshops. Now distinct workshops were established to make the lock plate and pan, the cock, and the frizzen. In the first workshop alone—run by Javelle and his son—eighteen different tasks were performed by subcontracted workers. There Javelle, as chef d'atelier, supplied jigs and milling machines analogous to those of Blanc. These prefinished pieces were then distributed to the scattered rural locksmiths, who added their own sears, bridles, springs, and screws to assemble a working lock. All told, some forty-seven different workers made the gunlock, whereas the number in the northern manufactures was only eleven.[47]

In other words, under Jovin's prodding, the armorers had adopted many of Blanc's techniques without any reference to interchangeability. Again, this system should not be equated with mass production. Jovin remained a financial coordinator who rented out space and motive power. The chef d'ateliers acted as subcontractors. And skilled locksmiths in their own ateliers still forged, filed, and fit the pieces to one another. This was merchant capitalism, not industrial capitalism. Nor were such arrangements unusual in the metalworking trades of Saint-Etienne, where the Eustache knife had long been the product of more than thirty hands. This development, moreover, bears out Sabel and Zeitlin's recent admonition that we stop making clear-cut distinctions among manufacturing régimes, as if producers faced an either/or choice between flexible specialization and mass production. Manufacturers often amalgamate technological practices and adapt hardware to new situations. The accumulation of minute innovations—Maxine Berg's "Birmingham" model of technological change—has eluded historians because it is not dramatic, nor easily reified by reference to some "mechanical ideal."[48] In such a situation, technological memory is deceptive. Interchangeable parts manufacturing was repudiated as an "ideal," but many of its practices survived.

On the other hand, ideals matter. Gassendi's partisans worried that Jovin's accelerated methods would disrupt the social equilibrium at Saint-Etienne. Like any properly educated engineer, Inspector Préau (Gassendi's new appointment) acknowledged the advantages of "rational" production: (1) workers doing a single task would perform it more rapidly and hence more cheaply; (2) machines could be designed to assist them and reduce variability in the output; (3) these low-skill workers would become dependent on the Entrepreneur, unable to quit armory, and hence cost less in pay; and (4) the use of raw materials could more easily be monitored. But Préau believed that efficiency, uniformity, deskilling, and surveillance—all goals of the interchangeability system—spelled disaster for the *quality* of Saint-Etienne's locks, already below that of its northern sisters. The reasons for this were social. To begin with, he cited problems with the replication of skills. The new machines cost money and would have to be maintained by highly skilled (and expensive) mechanics who could hold the manufacture hostage. At the same time, the mass of workers would soon lose their lock-making skills, and would no longer serve as a seedbed for future controllers, chef d'ateliers, and mechanics. Both quality and reliability would suffer. Another concern was the problem of coordination. Without a single, central factory, day-to-day monitoring of the intermediate stages of production lay in the hands of the workers in their scattered ateliers. Recriminations had already erupted among the artisans over who was responsible for defective locks. And if the workers *were* to be assembled in a factory, the result would be mutiny and cabals. Either way, the result would be "destruction of all harmony" in the lock, and (implicitly) in the social order. One could not expect "competing workers" who were laboring "mechanically" to make a harmonious gun. And in a final twist, Préau argued that while this division of labor might be desirable in private industry where goods could be priced according to their quality, the state needed guns of uniformly high standards. This could only be assured by "intelligent" artisans with "an eye and an ear to sense and hear his lock."[49]

This ode to tacit knowledge and manual skills marks a stunning reversal for the artillery service. It was not the only such palinode of the period. Many of the Enlightenment projects which the Revolutionaries had labored to realize collapsed in the face of inadequate means and popular resistance. Consider the metric system; heralded as the universal language of the new national market and the patient tutor of the new rational consumer, it was abandoned by Napoléon, who complained that it had thrown local markets into disarray. Or consider the early statistical surveys of France (as described by Marie-Noëlle Bourguet); they too foundered on the incommensurate diversity of provincial customs.[50] The same trajectory might be cited for the reversals in popular education or in the Concordat with the Catholic church. In each case, the ideal of a direct relationship between the state and its citizens proved to be beyond the resources of the state and as yet incompatible with the self-identity of French citizens. Napoléon may have boasted that he had mastered the world of detail, but in fact he ruled through notables who traded their local knowledge for local authority, and hence for some autonomy. In Saint-Etienne, this was the era in which the wealthy merchant elite consolidated their power.

And with their consolidation foundered that preeminent Enlightenment vision of engineering rationality: the dream that the thinking man's conception of the world could be directly translated into materiality. Originally, this was to be accomplished through the mediation of techniques like mechanical drawing and gauges. Human discretion would be minimized and that pernicious enemy of quality—the capitalist who thought only of his own gain—would be sidelined.[51] In the event, the post-Revolutionary engineers were obliged to manage the social order through human agents, and acknowledge that those agents had (pecuniary) interests. Interchangeable parts manufacturing was not compatible with the intermediate rule of these economic agents. Stamping and drop-forging techniques were suppressed in Saint-Etienne by order of the War Office on 29 September 1808.[52]

A parsimonious explanation for the demise of interchangeable parts manufacturing in France would simply note that the system was tried for fifty years but never paid off. As we saw in the case of technological design, however, the criteria for judging technology are themselves subject to debate. Just as the evaluation of cannon performance depended on the kinds of battles anticipated, so did the evaluation of the methods of production depend on the larger purpose of production. The Gribeauvalists had been mainly concerned with uniform production as an instrument of bureaucratic control, field operations, and state security. The successor generation of artillerists—led by Gassendi—increasingly took note of economic considerations. This shift was *not* precipitous. The Gribeauvalists had always paid attention to cost, if only under pressure from their patrons in the War Office and their client merchant-contractors in Saint-Etienne. And the Gassendites still cared about bureaucratic control, if only to assure levels of quality that would satisfy troop commanders. Rather, the new emphasis on private gain indicates a subtle shift in the social settlement of post-Revolutionary France—particularly in the boundary between the state and private capital—a shift which can be measured in the fit of a flintlock.

Technological Inertia: Decoupling Design

All of these changes in military gun-making occurred within the framework of a relatively static design. The flintlock mechanism at the center of these technological struggles differed only in minor details from the mechanism used by soldiers at the end of the reign of Louis XIV. Though aware of the lock's inadequacies, French engineers feared the effect that radical design changes would have on the production process and on the training of troops in the field. Consequently, even the slightest modification in these designs proceeded with the assistance of experienced armorers such as Blanc. The trick was to get design, production, and deployment to transform together. Only then, in the late nineteenth century, would radical new forms of production—like interchangeable parts manufacturing—be possible.

One minor redesign in the flintlock occurred between 1800 and 1804. The process was initiated by a commission convened by Aboville, which included Agoult, Prieur de la Côte-d'Or, Jovin, and the expert-armorer Régnier. And it was completed by Gassendi. In this, as in so many ways, these experts were determined to return to the ancien régime. They repudiated the "bastard" gun, the Republican No. 1, and proposed minor modifications of the M1777.[53] The "new" design did not placate field commanders, however. Complaints about the number of misfires forced the artillerists to experiment with new designs again in 1811. Ten gunlocks from each of the five manufactures were found to average 1 misfire in every 22 attempts. Foreign muskets generally performed better: Switzerland (1/44); Russia (1/28); Austria (1/62.5); England (1/44); and Spain (1/22). This was, of course, a severe embarrassment to Gassendi's artillery service. However, no changes in official design were announced. The experimenters acknowledged that the existing stockpile of guns and parts meant that "any change in dimensions would entail considerable expenses."[54]

One might thereby conclude that the repository of worker skills, patiently built up over the years, acted as a kind of social inertia. Confronting this techno-social inertia seemed particularly inadvisable during wartime. However, an analogous inertia plagued interchangeable parts manufacturing. Blanc's investment in task-specific machine tools had entrenched a particular technological design at Roanne. Indeed, Cablat had been forced to beg for an exemption from the minor modifications of 1800. Unfortunately, so long as Roanne supplied locks to other manufactures, Cablat was obliged to conform.[55] Rather than having the replication of artifacts bound up in the skills of artisans, in the case of interchangeable parts production they were bound up in the capital costs associated with machinery. Here, the socio-*technical* inertia of capitalized machinery supplanted the techno-*social* inertia of artisanal culture. Either way, radical change was constrained.

The great leaps in the effectiveness of nineteenth-century firearms depended on reconfiguring this relationship between design and production.

The first tentative steps in this direction were taken in the final year of the Empire by an armorer-inventor named Julien Le Roy. Le Roy set out to break this pattern of mutual constraint. He had been experimenting with mechanized production at the Charleville Manufacture for several years. Now he promised to make gunlocks with fully interchangeable parts. More to the point, to make his lock *easier* to produce in bulk, he had redesigned the gunlock mechanism and substituted new materials. His new lock was based on the percussion system, first made practical a decade before. It consisted of only nine major pieces (as opposed to twenty-five in the traditional lock). Moreover, its principal pieces could be readily stamped because they were made of copper.[56] Initially, the artillery bureaucracy and the traditional armories "rejected the invention and its inventor." However, in March 1814, the Central Committee of the Artillery tested the gunlock, found it reliable, and secured the emperor's enthusiastic assent to produce seventy thousand locks. Le Roy was given space in the Conservatoire National des Arts et Métiers, and awarded a 1 *franc* supplement per lock. But copper, though easily stamped, wore out sooner than iron. And certain modifications in design demanded by the artillery (perhaps including the return to a flintlock mechanism) made the locks more complex than Le Roy's initial design.[57]

Le Roy's plans were further disrupted by the first return of the Bourbons. Then, during the Hundred Days, the "Organizers of Victory," Lazare Carnot and Prieur de la Côte-d'Or, were once again called to serve in the defense of France: Carnot to sit as minister of the interior, and Prieur to organize Paris as a vast manufacture of arms. Prieur supported Le Roy's project, though he recognized the hostility of rival interests in the capital, and worried about the inevitable delays and abuses that a state-run enterprise would entail.[58] In the final tense days of April 1815, Napoléon pleaded for guns, while railing against Carnot and Prieur for their hesitancies. Why was Le Roy's factory behind schedule? This was no time for a *régie*. Let the manufacturers have their 15–20 percent profit; pay them up front. The emperor, whose bureaucracy had ruthlessly suppressed uniformity production, now demanded that the guns be stamped out *en masse*:

> Are there enough springs and screws? They can be manufactured like buttons. It is the same for all the other pieces, but we must renounce making them in the government workshops.[59]

The emperor's logic here was faulty. The production of uniform gunlocks was an innovation of the state. Private producers had found that it did not pay; capital costs and delays were too great. But Napoléon was out of time. Waterloo followed six weeks later. As for Le Roy, he persisted in seeking state support for his factory. In 1817, he petitioned for "permission to use stamping methods" for the lock plate, and to employ skilled machinists. His request was denied.[60] Interchangeable parts manufacturing was not to be attempted again in France for fifty years.

The Nineteenth Century: Transfer and Triumph

The subsequent triumph of interchangeable parts manufacturing in the United States is well known. In the first half of the nineteenth century the ideal that became known as the "American system of manufactures" served as a rallying cry for the officer-engineers of the Ordnance Department who managed the state-owned armories at Harpers Ferry and Springfield. These American artillerists acted upon the engineering ideals of their French allies. French engineers wrote English-language textbooks and served in the U. S. artillery. The curriculum at West Point was strongly influenced by the Ecole Polytechnique. On a more practical level, the early development of the hollow milling machine and precision gauges can be traced directly to Blanc's example.[61] But there is no reason to rename the interchangeability system the "French system of manufactures." American mechanics created the great bulk of the practical techniques which made interchangeability a success in the early nineteenth century. Technology may travel easily in guise of an ideal, but the practical realization of a transferred technology invariably requires local "naturalization." Fifty years were to pass before the transplanted ideal was made material. In 1851, American gun-makers Robbins and Lawrence astounded the British at the London Crystal Palace Exhibit with their interchangeable parts rifles. And Samuel Colt brilliantly exploited the ideal of interchangeability to publicize his revolvers. But even then it was some time before machine techniques and skills made uniformity sufficiently workable to do away with "fitters."

In France, meanwhile, the emphasis on merchant capital and dispersed artisanal production persisted—and with it, nonindustrial forms of production. Jovin and Sons, still Saint-Etienne's monopoly Entrepreneur, soon became the town's first millionaires. For their part, arms workers were formally divided into two classes. The smaller group registered to work on military weapons were subject to military law and received a pension on retirement. And the group of "free" workers could quit on a month's notice. In 1831 the firm collected a number of armorers under the same roof and supplied them with motive power from a steam engine. Yet even then the armorers hired their own assistants and persisted in handcraft methods. At mid-century, the recruitment of armorers remained remarkably circumscribed. Yves Lequin found that 70 percent followed in their fathers' profession. And there remained considerable resistance to new forms of mechanized production; a round of sabotaging of barrel-turning lathes followed shortly after the coup of 1830. In this atmosphere, Blanc's ideal of interchangeability was anathema; not simply because it required the use of machinery, but because it would have broken the networks of social and economic relations between armorers.[62]

The Restoration had also preserved Napoléon's gunlock design, just as it preserved his state-building design. True, Blanc's role in designing the M1777 musket—once considered the best musket in Europe—was now dismissed as the presumptuous efforts of a "handy workman." Paixhans, the artillery engineer who created the first post-Gribeauval artillery system, argued that Blanc's talents as a gunsmith gave him no license as a gun designer. Only technically trained artillerists could study the gunlock "like any other machine . . . by applying the rules of geometry, mechanics, and reason without worrying about the methods of production." Indeed, some astounding attempts were made in this period to come up with a scientific analysis of the "thick" lock mechanism, including one fifty-page geometric reconstruction of its forces. Yet no improvement in the flintlock was forthcoming, and Paixhans admitted that any design change would cause disarray in the manufactures. Hence, only minor corrections were introduced in 1816 and in 1822 (by men still loyal to Gassendi). Not until 1840 did the French army fully convert to the percussion system. By then, private sporting guns had already relied on the percussion system for twenty years.[63]

Change in France was gradual. When the British decided in 1851 to equip the Enfield armory with American-built machine tools, the news came to the attention of Paris, thanks to the intervention of Frédéric-Guillaume Kreutzberger. Kreutzberger, an Alsatian machinist who had worked for seven years for Remington Arms in Illion, New York, tried to interest the French state in a similar transfer. But he warned that the new production methods would mean reassigning whole categories of workers and redesigning the gun. The artillery refused: the American machines, they said, would perpetuate "too great a revolution in the operation of our arsenals." They did, however, ask Kreutzberger to help French machine-builders—mostly suppliers to the textile industry—adapt their machine tools for the small arms industry.[64] But it was several decades before these machine-builders became the nexus of a thriving metalworking industry, first for armaments, then for sewing machines, and then for bicycles.[65] As Yves Cohen has pointed out, French private industry took up interchangeable parts manufacturing only at the very end of the nineteenth century. He shows—in an anti-teleological stance much like the one I have taken here—that the ideal of interchangeability only gradually shifted from an operational logic (ease of repair on the battlefield) to a logic of market (lower unit costs of production). In the statist armaments sector of the economy, however, this operational logic persisted all through the century, and was brought to full realization—in combination with the assembly line—only during World War I.[66]

The push to bring modern mass production methods to the Saint-Etienne armory began with the introduction of the Chassepot rifle in 1866. At that time, the municipality of Saint-Etienne assisted the Entrepreneur in the construction of a new arms factory, larger than any in France. For the first time since the Napoleonic Wars, output jumped. A further stage in the armory's development occurred in 1874 when the state introduced machinery to produce the Gras rifle. Concurrent efforts to import U.S. machinery from Pratt & Whitney, however, met with frustrations. And a report concluded that manual labor "saved time, as workers make fewer faults than machines."[67] Finally in 1886, with the Lebel rifle, artillery engineer Gustave Ply had Saint-Etienne retooled for true interchangeable parts production. In his manifesto of 1888–89, Ply laid out engineering production in terms which would have been familiar to the Gribeauvalists. He began with the need for a full set of technical drawings: an official copy, and an accessory set for revisions. These drawings also assigned the requisite tolerances (which now included sufficient "play" between pieces to enable them to function together). On this basis, he then analyzed the production process, specifying how automatic machinery would minimize "everything arbitrary resulting from the workers' good or ill will (*plus ou moins bon volonté*), and their degree of skill." Next, he explained how the requisite tolerances would be built into gauges and so regulate the precision of both labor and machine. And finally, he spelled out the division of moral responsibility: the engineer-director's duty to prevent deviations, the controller's duty to apply the gauges to ensure uniformity of output, and the workers' collective duty to machine to tolerance. For all its objective standards, interchangeability remained a moral imperative. Considerations of cost were conspicuous by their absence. In short, the engineering logic of uniformity production (now augmented with powerful new machines) still operated in the military sector with only a passing regard for the logic of the market.[68]

Thereafter annual output was measured in the millions. The late nineteenth century marked a definitive shift in the percentage of the military budget slated for hardware. Breech-loaded rifles, first used with devastating effect against the Russians, became standard with the Chassepot, and were refined with the Gras and Lebel rifles. The concurrent history of the French involvement in Algeria, Crimea, and the war with Prussia mark these stages in design and production. Military historians have asserted that these new gun designs spurred changes in tactics and the character of warfare as experienced on the battlefield. However, the persistence of Napoleonic tactics long after the advent of these new technologies offers yet another confirmation that technology does not drive history (or at least not in the immediate way that implies a monocausal determinism). More to the point, these changes were accompanied by an evolving conception of French colonial power, and of the state's relationship to its citizenry. But that, of course, is another story.[69]

Conclusion

For 'tis the sport to have the enginer
Hoist with his own petar, an't shall go hard
But I will delve one yard below their mines,
And blow them at the moon.
—William Shakespeare, *Hamlet*

In the end, we are left with a paradox that can be stated succinctly. In the fifty years between 1763 and 1815, the French military engineers enjoyed a spectacular political triumph—yet they mostly failed in the realm of technics. Consider first their successes. In the ancien régime these men had lurked on the periphery of power, a corps of technicians from a hodgepodge of social backgrounds, answerable to lordly patrons, hemmed in by rival claimants, and locked in an internecine war. Yet in the final decades of the eighteenth century, and then more dramatically with the coming of the French Revolution, this same group of men moved to the center of state power. They were among those groups who pioneered the new hierarchical and meritocratic social organization by which the French state was regenerated, and they saw their values enshrined in institutions (the Ecole Polytechnique), which sealed their preeminence among the new notability. All this was done in recognition of the engineers' presumed ability to serve the state, an ability grounded in their professional ethic and its epistemological correlate. The engineers claimed this program combined theoretical knowledge and practical know-how in such a way as to address the material (military) needs of the nation. To be sure, various factions of engineers differed (with considerable vehemence) over the most advantageous technological program, the social order which might best realize it, and consequently, the sort of nation France should be. Vallière and Gribeauval, Monge and Laplace, Aboville and Gassendi each asserted the superior effectiveness of their program. Yet in this same period, the technical successes of the engineers were quite limited.

Consider the limits of their success in the realm of military hardware. The mobility of the Gribeauvalist field artillery certainly assisted in the victories of the Revolutionary wars—though by their own admission the accuracy and firepower of their new cannon marked little advance. And while the M1777 may well have made some marginal improvement in the performance of

French muskets, it is debatable whether this firearm was worth the doubled price the state was obliged to pay. More important were the *social* technologies the engineers inaugurated. French weapons outperformed those of her rivals only when operated in tandem with the new mixed tactics (which the artillerists did much to promote), and when handled by self-disciplined Revolutionary officers, cannoneers, and soldiers (for which the artillerists served as a partial model). The engineers thereby showed that a large-scale techno-military system was compatible with republicanism, or at least with national sovereignty of a particular sort.

However, it was in the realm of production that the engineers met their greatest defeat. The Gribeauvalists' attempt to improve the quality of muskets made at Saint-Etienne drove the armory into ruin. And their efforts to manufacture gunlocks with interchangeable parts foundered on the resistance of armorers and merchants, groups who successfully forced the engineers to recognize their economic interests and political rights. Handcraft methods persisted in France well into the nineteenth century, and provincial artisans and commercial elites retained control over production. The result was to define the limits of the French state's involvement in the development of private capital in a realm it considered vital to its interests. Nineteenth-century entrepreneurial capitalism developed in France upon this post-Revolutionary social settlement. Only in the late nineteenth century—at the behest of new calls to enlist the nation-in-arms—did state engineers at last succeed in implementing their revolutionary program for production.

How can we explain this discontinuity, this fifty-year "hiatus" in development of rational production? A practical-minded skeptic might plausibly point to the inadequacy of the technical means available in the late eighteenth century, particularly to the inadequacy of machine tools. Or a skeptical economic historian might argue that, independent of these machines, interchangeable parts manufacturing was abandoned in the early nineteenth century because it simply did not "pay." From either point of view, the engineers' attempt to impose the uniformity system of production appears as utopian as the attempt of their Revolutionary successors to institute patriotic production in 1793–94. Indeed, both programs *were* utopian, if by that one means that they were fundamentally anticapitalist, and hence, not in line with our familiar triumphantalist histories of industrialization. That is, both engineering production and patriotic production asked those groups who actually made these artifacts to subordinate their personal, venal interests to the service of a "higher" cause—in this case, the collective cause of the nation-in-arms. Hence, the failure of the French engineers to develop the machinery needed for uniform production cannot be understood apart from the resistance of the armorers and merchants to mechanical methods, a resistance grounded in a different technological life, with its own social and economic logic. And the failure of the state to persist in this uniformity project cannot be understood apart from the state's *decision* to let private profitability determine what counted as success. By contrast, it was in the United States that interchangeable parts manufacturing was successfully developed over the first half of the nineteenth century. There, mechanization—however resented—was not bitterly opposed, and the state, ironically, proved willing to subsidize the new technology at its arsenals. In short, the failure of interchangeable parts manufacturing in Revolutionary France needs to be understood as part of a *political* struggle. The engineers, despite their success within the state, failed to make uniform production a desirable goal for provincial commercial groups. And the engineers' decision to accommodate these merchants and armorers doomed interchangeable parts production.

This paradox, then, was the engineers' greatest triumph of all. They asserted their right to technocratic rule on the basis of a mastery that they did not possess—and they got away with it. The historical record shows that the social revolution of the engineers long predated their major successes in the realm of technics. Instead, what the eighteenth-century engineers offered was an ideological program. No less than the Revolutionary politicians, the engineers promised a utopian scheme to remake the world in light of present circumstances. Their effort was directed beyond a narrow mastery of material things toward a reordering of the social world. In that world, the designing, making, and using of material things was to be guided by new codes. Values which today go by such names as "efficiency" or "control" were placed in the service of the new sovereignty, and new institutions devised to nurture them. The engineers made those values appear universal—or at least a necessary concomitant of the new national sovereignty.

To be sure, many of the methods developed by the Enlightenment engineers—the tools of rational production, empiricist engineering design—were to have a spectacular future in the latter half of the nineteenth century. And this ultimate triumph has made their eighteenth-century ascent appear inevitable. In this book, I have argued that the history of gunpowder weaponry cannot be read as leading inevitably to this resultant triumph over "external enemies" (nature, say, or the Prussian army). In doing so, I have never denied that engineering ballistics, uniform production, and mobile tactics were techniques which might ultimately prove successful. I simply have argued that the history of these technologies must be recounted from their creators' point of view, creators with nothing more to go on than an *expectation* of how the world *ought* to be. My method has been to blur the line between "external" enemies and "internal" interests, between arguments about how bullets travel through the atmosphere, how troops should be arrayed, and how military promotions should be awarded. Whatever value such analytical divisions have for us, they were transgressed by the participants in these struggles. Blurring this line has made it immediately apparent that the development of these technologies depended also on the advantages and disadvantages they offered various "internal" interests, including bureaucratic factions of engineers and militarists, as well as economic agents, such as artisans, merchants, and laborers. These new technologies emerged as the negotiated outcomes of social conflict.

Above all, I have resisted the easy temptation to read the projections of eighteenth-century men and women as anything more than an attempt to make sense of their past, persuade their contemporaries, and light their own lives a short distance into the future. The men in wigs and powder cannot take up our arguments. It would be churlish, therefore, to sort their beliefs into the bins of "right" and "wrong" by our retrospective lights. Much of this book can be read as applying to techno-science this basic historicist viewpoint. We should not ask what a technology came in time to do for us, but what it did for particular people in their own time. To be sure, I have staked the seriousness of my study on the fact that among the eighteenth-century alternatives I discuss, one particular engineering vision of rational design, uniform production, and mass warfare eventually triumphed in Western Europe and in the rest of the world too. Yet I have tried not to profit overly by this hindsight or assume its success was preordained. After all, its triumph was far from complete within the time frame of this story, whatever the engineers' political success.

Indeed, one of the central lessons of this book concerns the uses and abuses of technological determinism. The invocation of technological determinism is itself an argument in a high-stakes debate over how to organize the social life, and over *who* will provide that technocratic rulership. The Enlightenment artillerist Du Coudray cited four centuries of progressively thinner battle lines as evidence for the determinant impact of firearms on war. This enabled him to plead for a new battle tactics, while painting his cause with the hard veneer of necessity (and patriotism). Yet the controversy surrounding these new tactics, and their rejection in favor of the mixed order, suggests that no such necessity was at work. The Gribeauvalists who tried to reorganize the Manufacture of Arms at Saint-Etienne and the technocrats who organized the Atelier de Perfectionnement both argued that automatic machinery was necessary to enforce work discipline and the division of labor. They thereby hoped to secure support for a particular form of machine-driven industrial development, one which placed them in charge of that process. For his part, Lazare Carnot, in the period after Thermidor, explained how the circumstances of the Revolutionary war had forced him to serve as the reluctant colleague of murderous politicians. And he invoked the division of executive labor and the metaphor of a mechanistic bureaucracy to show his patriotic hands were clean. This socio-technical exculpation directed the attention of his accusers away from the political instruments (including those classed under the name of the Terror) with which he had sought to reshape France's technological life during the previous year. In citing these cases, my goal has been to understand the *purposes* to which engineers such as Du Coudray and Carnot put their versions of history. The argument from technological determinism itself has a history, one intimately connected to the rise of that professional class which insists that it be awarded control of one of the most critical aspects of our social life.

All this epistemological groundwork, however, has been only a means to loosen the bond of technological necessity and reveal the ways in which conflicts over the terms of the technological life were integral to the political conflicts of the French Revolution. These conflicts did not revolve only around icons. Certain material artifacts are also political in a different sense, in that they are shaped by long-term structural changes in knowledge-making, social organization, and economic relations in which the Revolution is but an episode. The gun is clearly one such artifact. The question is: How general is this argument? Does it apply to artifacts other than guns?

It seems uncontroversial to say that artifacts are political in so much as various *state structures* take a close interest in their design, production, and deployment. For centuries, questions relating to weaponry have been intensely debated within all types of states, often becoming part of the explicit agenda of political factions and parties. Cold War America is certainly one place this has happened; consider the controversy over the Strategic Defense Initiative in the 1980s. Of course, as war has become ever more total over the past two centuries, many technologies other than those relating directly to weaponry have attracted partisan attention for military reasons. Military sponsorship of new technology is pervasive, and "spin-offs" have transformed a variety of civilian industries. It has guided technological innovation (the transistor in the 1950s), prompted the creation of new means of production (numerically controlled machine tools), and influenced the development of infrastructure (the layout of the Prussian railway lines). Even outside of military need, state involvement in public works has expanded, and so has the desire of factions to build social policy into many technological projects. This applies certainly to public works, such as Robert Moses's infamous low-level overpasses which prevented New York's inner-city busses from reaching Jones Beach. More recently, environmental legislation has had a tremendous impact on technology (from dams to automobiles). This is *not* to suggest that politics stops at the point where state activity ceases. On the contrary, one is hard-pressed to imagine any important twentieth-century technology whose development has not in some way been shaped by debate in the broader public sphere.[1]

I have also argued that the *design* of an artifact is political in that it involves a conscious choice by specific agents among real alternatives. In eighteenth-century France, gun design lay in the hands of state engineers. Their design decisions did *not* flow mechanically from their techno-scientific knowledge, nor from the demands of their patrons (though both constrained the sorts of designs imaginable). Rather, as practitioners of the art of gunnery, they designed guns in line with their conception of their own proper social role. Against the background of the ancien régime, their engineering ideology based on merit, hierarchy, and instrumentalism looks like a distinct political alternative. And the guns they designed mirrored that new political stance. But what about those artifacts whose design is not the province of state engineers? Artifacts like suspension bridges and the telegraph will also have their associated "sciences" (structural engineering and electrical engineering, respectively) and their associated sponsors (political entities and private corporations). Yet in these cases, too, design will be mediated by social groups like engineers or technicians or craftworkers. Invariably, the design choices of these groups will reflect their training, their understanding of their social role, and their interests vis-à-vis other groups (corporate management, consumer lobbies, political patrons). Gun design may be more explicitly political, but an analogous agenda may well shape design in these other cases too. This does not mean that designers have complete freedom to fashion technology how they wish. My conception of technology does not lead me to claim, as Pinch and Bijker do in their provocative article on the bicycle, that technologies rise and fall *solely* according to their ability to attract powerful interests. Such a claim is only trivially true. Yet an analysis of technology must include room for those who actually conceive of the technology.[2]

I have also argued that the *production* of artifacts is political in that it turns on conflict and negotiation between the different parties to the manufacturing process. At one level, this argument applies to artifacts wherever the state (as above) takes an interest in production, and writes laws that affect economic activity and labor organization. Tariffs, safety standards, union rules, and tax codes are all outcomes of partisan politics, and all shape the organization of production. Again, this seems entirely uncontroversial. In the case of gun production, we also saw how the state (as a "consumer" of military hardware) went so far as to set the standards of production, and how these standards became the site of further negotiation as local economic agents asserted their interests. In this case, gun production proved to be political even though the means of production remained in private hands. Thus, the artisans and *fabricants* of Saint-Etienne acted politically when they allied themselves with the municipality in the early 1780s to seize control over the standards of production. And more explicitly, the workers of the Manufacture of Paris acted politically when they protested wages and working conditions in the years II and III. But can this same logic be applied to other branches of industrial capitalism where the entrepreneur not only owns the means of production, but sets the standards of production as well?

Certainly, the motives of the managers of private capital differ from those of state engineers. Yet here too, defining standards of production can be understood as "political." To be sure, technical drawings, gauges, fixtures, and machinery are material instruments to assist in the making of things. But as we have seen, the manner in which they are deployed affects the distribution of knowledge in the production process. This has real consequences. That is because the same workers' skills which enable them to control the work process also enable them to keep up the price of their labor, even within the factory setting. To the extent, therefore, that standards of production shift control over the work process from workers to managers, they affect the way in which the proceeds of production are divvied up. Control may be here exerted in the service of cost minimization (and profits), not in the name of engineering quality or operational convenience, but the net result may well be the same. In the process, the objective tools of production themselves become subject to conflict and negotiation, and hence are themselves shaped by the politics of production.

This conflict, of course, only becomes *explicitly* political when the battle over control of the productive process depends on social mobilization through formal institutions (corporate guilds, labor unions, manufacturing syndicates) or when it is articulated in a recognizably political language (natural law, republicanism, patriotism). In the period covered in this book, the various roles in the productive process were not clearly differentiated, the familiar nineteenth-century social solidarities did not yet formally exist, and the language of social protest was not yet constituted around worker grievances. Artisans sometimes acted as entrepreneurs, merchants often refused to invest in production, and workers only fleetingly took up the rhetoric of labor rights. Hence, I have deliberately avoided using the vocabulary of class, and restricted myself to noting how alliances gradually formed around particular roles. In this sense, my study can be seen as a contribution to the larger literature on class formation. Only in the middle of the nineteenth century did these class divisions begin to play a crucial role in the overt political life of western Europe. While calling production "political" in this sense may be controversial, it is certainly familiar.

And finally, the way artifacts are *used* is political in that they are the site of cultural conflict. Adjustments in eighteenth-century tactics implied changes in the definition of the citizen-soldier and in the political topography of the nation. In this sense, the artillery engineers acted politically in the 1770s when they advocated a particular type of weapon and a particular kind of offensive tactics to give it meaning. Given the central role that martial codes played in defining national identity during the past two centuries, controversies over weaponry and tactics can properly be called political. Hence, even the "market" for military guns was fashioned through political conflict. Moreover, I have alluded more than once in this book to the ways in which the debate over the *private* possession of guns was politically charged in this period: Was it an entitlement? Was it a right of citizenship? Was it a matter of property? This sense in which guns are political is certainly familiar to Americans in the late twentieth century. But is this argument limited to guns, a technology so directly related to the maintenance (and subversion) of public order? Can we say that textiles or toys or other consumer goods sold on the open market are political, or is the market there an impersonal reflection of aggregate demand?

Although this book does not provide the evidence to answer this question, there appear to have been many times and places where consumer choices took on an explicitly political cast. Wearing the cockade, demanding meat inspections, boycotting grapes were all conscious acts which invested those commodities with political virtues. At times consumption could become explicitly revolutionary. T. H. Breen has shown how the American colonists self-consciously altered their consumption of certain types of cloth, thereby forging a political community on the eve of the American Revolution. Even when consumers do not self-consciously make a political statement with their purchases, their choices are embedded in cultural codes that signal larger allegiances. As Michael Sonenscher has shown, to wear a hat in the ancien régime was to assume a particular social identity. Or as Leora Auslander has pointed out in her study of furniture consumption in nineteenth-century France, "taste" must be cultivated and is a matter of public sanction. Indeed, it is widely acknowledged that consumer choices are conditioned by cultural codes, and that these choices therefore signal social status and reinforce (or subvert) social rankings.[3]

Politics does not lie like a foam upon the deep waters of our material life, as the old Marxist interpretation of the French Revolution assumed. Neither is politics an autonomous force, as the early revisionists implied. On the contrary, the technological life and the political life are mutually constitutive. Nothing in this book should be read as saying that artifacts are *merely* political. We must never turn a blind eye to the materiality of guns, to the reality of a bullet's trajectory, or to the corpses of those killed in battle. And in time, the accuracy, firepower, and mobility of guns did improve. Machine guns, tanks, planes, napalm, and plutonium have all been engineered to escalating effect in the intervening two centuries. And although the truly dramatic changes happened outside the time-frame of this story, those advances were prefigured by the eighteenth-century intellectual and social program of the Enlightenment engineers. That program was contested, however, and its triumph was by no means predetermined. In showing how and why this was so, I have tried to demonstrate that guns are political in the final sense of that term. That is, I have tried to hold out the hope that things might be different.

ABBREVIATIONS

A.A.S.	Archives de l'Académie des Sciences
A.D.L.	Archives Départementales de la Loire
A.D.R.	Archives Départementales du Rhône
A.E.P.C.	Archives de l'Ecole des Ponts et Chaussées
A.M.StE.	Archives Municipales de Saint-Etienne
A.N.	Archives Nationales
A.P.	*Archives parlementaires*
Atel. Perf.	Atelier de Perfectionnement
B.M.R.	Bibliothèque Municipale de Roanne
B.M.StE.	Bibliothèque Municipale de Saint-Etienne
B.N.	Bibliothèque Nationale
Bib.	Bibliothèque
C.A.A.	Commission des Arts et d'Agriculture
C.A.M.N.F.	Commission d'Administration sur la Manufacture Nationale des Fusils
C.I.C.C.A.	Commission Intermédiaire du Comité Central d'Artillerie
C.N.A.M.	Conservatoire National des Arts et Métiers
C.S.P.	Comité de Salut Public
H.H.U.	Houghton Library, Harvard University
Min. Int.	Minister of the Interior
Min. War	Minister of War
P.V.C.I.P.	Guillaume, ed. *Procès-verbaux du Comité de l'Instruction Publique*
R.A.C.S.P.	Aulard, ed. *Recueil des actes du Comité de Salut Public*
S.H.A.T.	Service Historique de l'Armée de Terre

NOTES

Preface

1. Marie-Jean-Antoine-Nicolas de Caritat de Condorcet, *Sketch for a Historical Picture of the Progress of the Human Mind*, trans. June Barraclough (London: Weidenfield and Nicolson, [1795], 1955), 96, quote p. 194. On the *Sketch* and Condorcet's political philosophy, see Keith Michael Baker, *Condorcet: From Natural Philosophy to Social Mathematics* (Chicago: University of Chicago Press, 1975).

2. Langdon Winner, "Do Artifacts Have Politics?" in *The Whale and the Reactor: A Search for Limits in an Age of High Technology* (Chicago: University of Chicago Press, 1986), 19–39.

Introduction
A Revolution of Engineers?

1. Jean-Jacques Rousseau, *Les confessions* (Paris: Gallimard, [1789], 1959), 2: 93–95.

2. Jefferson to John Jay, 30 August 1785, in Thomas Jefferson, *The Papers of Thomas Jefferson*, ed. Julian P. Boyd (Princeton: Princeton University Press, 1950–), 8: 452–56. Dating Jefferson's visit is somewhat problematic. From his Account Book we know that Jefferson paid a guide "1f4" for a tour of Vincennes on 8 July 1785, in ibid., 14: 498. Boyd believes that on this date Jefferson traveled to Vincennes to visit the country home of the César Henri La Luzerne. That excursion, however, occurred in October. See Thomas Jefferson to James Madison, 29 October 1785, in ibid., 8: 683. There remains some uncertainty as to whether Blanc was as yet fully established in his dungeon workshop.

3. Jefferson to Patrick Henry, 24 January 1786, in Jefferson, *Papers*, 9: 212–15.

4. Jefferson to Henry Knox, 12 September 1789, in Jefferson, *Papers*, 15: 421–22. When the package arrived, a number of the pieces had been lost and broken, but a new set was sent in late 1790. Jefferson to William Short, 6 April 1790, Short to Jefferson, 4, 22 August, 27 October 1790, in ibid., 16: 315–16; 17: 317, 412, 642.

5. Robert S. Woodbury, "The Legend of Eli Whitney and Interchangeable Parts," *Technology and Culture* 1 (1959): 235–53. Silvio Bedini, *Thomas Jefferson, Statesman of Science* (New York: Macmillan, 1990), 140–43, 292–93. Merritt Roe Smith, *Harpers Ferry Armory and the New Technology: The Challenge of Change* (Ithaca: Cornell University Press, 1979). See also Felicia Johnson Deyrup, *Arms Makers of the Connecticut Valley: A Regional Study of the Economic Development of the Small Arms Industry, 1798–1870* (Northampton, Mass.: Smith College Press, 1948). I also appreciate the chance to look at the forthcoming study of the Springfield Armory by Carolyn C. Cooper, Robert B. Gordon, Patrick M. Malone, and Michael Raber, *Model Establishment: A History of the Springfield Armory 1794–1918* (New York: Oxford University Press, forthcoming).

6. Nathan Rosenberg, ed. and intro., *Great Britain and the American System of Manufacturing* (Edinburgh: Edinburgh University Press, 1969). See also Edward Ames and Nathan Rosenberg, "The Enfield Armory in Theory and History," *The Economic Journal* 78 (1968): 827–42. Russell I. Fries, "British Response to the American System: The Case of the Small-Arms Industry after 1850," *Technology and Culture* 16 (1975): 377–403. Clive Trebilcock, "'Spin-Off' in British Economic History: Armaments and Industry, 1760–1914," *Economic History Review*, 2d series, 22 (1969): 474–90. David Hounshell, *From the American*

System to Mass Production, 1800–1932: The Development of Manufacturing Technology in the United States (Baltimore: Johns Hopkins University Press, 1983).

7. See the similar anti-teleological approach to late-nineteenth-century production taken by Yves Cohen, "Inventivité organisationnelle et compétitivité: L'interchangeabilité des pièces face à la crise de la machine-outil en France autour de 1900," *Entreprises et histoire* 5 (1994): 53–72. For a similar cautionary warning against the use of twentieth-century terms such as "mass" production and "mass" consumption to describe eighteenth-century developments, see John Styles, "Manufacture, Consumption, and Design in Eighteenth-Century England," in *Consumption and the World of Goods*, eds. John Brewer and Roy Porter (London: Routledge, 1993), 527–52, especially pp. 529–35.

8. Some earlier accounts touch briefly on aspects of the French experiments with interchangeable parts production. The first was Woodbury, "The Legend of Eli Whitney." Edwin A. Battison pointed to some of Blanc's techniques in "Eli Whitney and the Milling Machine," *The Smithsonian Journal of History* 1 (1966): 9–34. Selma Thomas also briefly highlighted Blanc's achievement in her "'La plus grande économie et la précision la plus exacte,' L'oeuvre d'Honoré Blanc," *Le Musée d'Armes, Bulletin trimestriel des Amis du Musée d'Armes de Liège* 7 (1979): 1–4. A nineteenth-century evaluation of Blanc can be found in W. F. Durfee, "The First Systematic Attempt at Interchangeability in Firearms," *Cassier's Magazine* (April 1894): 469–77. For the impact of French ideals in the U.S., see Merritt Roe Smith, "Army Ordnance and the 'American System' of Manufacturing, 1815–1861," in *Military Enterprise and Technological Change*, ed. Merritt Roe Smith (Cambridge: MIT Press, 1985), 44–49; also *Harpers Ferry*, 88–91, 232. For an early attempt to put these questions in a social context, see John E. Sawyer, "The Social Basis of the American System of Manufacturing," *Journal of Economic History* 14 (1954): 361–79; also "The Entrepreneur and the Social Order: France and the United States," in *Men In Business*, eds. A. H. Cole and W. Miller (Cambridge: Harvard University Press, 1952), 7–22. See chapter 6, below.

9. Charles Sabel and Jonathan Zeitlin, "Historical Alternatives to Mass Production: Politics, Markets and Technology in Nineteenth-Century Industrialization," *Past and Present* 108 (1986): 133–76. See chapter 5.

10. John Kasson, *Civilizing the Machine: Technology and Republican Values in America, 1776–1900* (New York: Penguin, 1976), 3–51. Leo Marx, *The Machine in the Garden: Technology and the Pastoral Ideal in America* (New York: Oxford University Press, 1964).

11. E. P. Thompson, *The Making of the English Working Class* (New York: Vintage, 1963). See also E. J. Hobsbawm, *The Age of Revolution, 1789–1848* (Cleveland: World Publishing, 1962). Harold Perkin, *The Origins of Modern English Society* (London: Ark, 1969). For a recent view, see Peter Linebaugh, *The London Hanged: Crime and Civil Society in the Eighteenth Century* (Cambridge: Cambridge University Press, 1992).

12. In fact, Soboul always acknowledged that nineteenth-century class relations could not be imposed on late-eighteenth-century Paris; Albert Soboul, *Les sans-culottes parisiens en l'an II, Histoire politique et sociale des sections de Paris, 2 juin 1793–9 thermidor an II* (La Roches-sur-Yon: Potier, 1958). Alfred Cobban, *The Social Interpretation of the French Revolution* (Cambridge: Cambridge University Press, 1964). George Taylor, "Non-capitalist Wealth and the Origins of the French Revolution," *American Historical Review* 72 (1967): 469–96. François Furet, *Penser la Révolution française* (Paris: Gallimard, 1978). Alexis de Tocqueville, *The Old Régime and the French Revolution*, trans. Stuart Gilbert (Garden City, N.Y.: Doubleday, 1955).

13. A prominent analysis in this vein is Keith Michael Baker, *Inventing the French Revolution: Essays on French Political Culture in the Eighteenth Century* (Cambridge: Cambridge University Press, 1990). See also the comprehensive review of this historiography by William Doyle, *Origins of the French Revolution* (2d ed.; Oxford: Oxford University Press, 1988). Other works will be cited in the text where appropriate.

14. Colin Lucas, "Nobles, Bourgeois and the Origins of the French Revolution," *Past and Present* 60 (1973): 84–126. Lynn Hunt, *Politics, Culture, and Class in the French Revolution* (Berkeley: University of California Press, 1984). Bryant T. Ragan and Elizabeth Williams, eds., *Re-creating Authority in Revolutionary France* (New Brunswick, N.J.: Rutgers University Press, 1992). On this basis, Sarah Maza has predicted a trend back toward a socially grounded history of the Revolution. Sarah Maza, "Politics, Culture, and the Origins of the French Revolution," *Journal of Modern History* 61 (1989): 704–23. Colin Jones, "Bourgeois Revolution Revivified: 1789 and Social Change," in *Rewriting the French Revolution*, ed. **Colin Lucas (Oxford: Clarendon Press, 1991), 69–118; and "The Great Chain of Buying: Medical Advertisement, the Bourgeois Public Sphere, and the French Revolution,"** ***American Historical Review*** **101 (1996): 13–40.**

15. François Crouzet, *Britain Ascendant: Comparative Studies in Franco-British Economic History*, trans. Martin Thom (Cambridge: Cambridge University Press, 1990). Patrick O'Brien and Caglar Keyder, *Economic Growth in Britain and France, 1780–1940: Two Paths to the Twentieth Century* (London: Allen and Unwin, 1978). See chapter 5.

16. William H. Sewell, Jr., *Work and Revolution: The Language of Labor in France from the Old Regime to 1848* (Cambridge: Cambridge University Press, 1981). Michael Sonenscher, *The Hatters of Eighteenth-Century France* (Berkeley: University of California Press, 1987); also *Work and Wages: Natural Law, Politics, and the Eighteenth-Century French Trades* (Cambridge: Cambridge University Press, 1989). William Reddy, *The Rise of Market Culture: The Textile Trade and French Society, 1750–1900* (Cambridge: Cambridge University Press, 1984). Gail Bossenga, "La Révolution française et les corporations: Trois examples lillois," *Annales: Economies, sociétés, civilisations* 43 (1988): 405–26; and "Protecting Merchants: Guilds and Commercial Capitalism in Eighteenth-Century France," *French Historical Studies* 15 (1988): 693–703. See also the essays in the collection by Steven L. Kaplan and Cynthia J. Koepp, eds., *Work in France: Representations, Meaning, Organization, and Practice* (Ithaca: Cornell University Press, 1986).

17. Tessie P. Liu, *The Weaver's Knot: The Contradictions of Class Struggle and Family Solidarity in Western France, 1750–1914* (Ithaca: Cornell University Press, 1994).

18. Thorstein Veblen, *The Engineers and the Price System* (New York: Huebsch, 1921). Edwin Layton, "Veblen and the Engineers," *American Quarterly* 14 (1960): 64–72. David Noble, *America by Design: Science, Technology, and the Rise of Corporate Capitalism* (New York: Knopf, 1977). Edwin Layton, *Revolt of the Engineers: Social Responsibility and the American Engineering Profession* (Baltimore: Johns Hopkins University Press, 1986). One can argue that in this respect the U.S., Germany, and the U.S.S.R. were all enamored of the logic of engineering in the early part of this century. Thomas P. Hughes, *American Genesis: A Century of Invention and Technological Enthusiasm, 1870–1970* (New York: Penguin, 1989), 249–94.

19. Walter Vincenti has brought attention to "design" as the central feature of engineering activity; see *What Engineers Know and How They Know It: Analytical Studies from Aeronautical History* (Baltimore: Johns Hopkins University Press, 1990). And more recently, see the fine book by Louis Bucciarelli, *Designing Engineers* (Cambridge: MIT Press, 1994). See also Herbert A. Simon, *The Sciences of the Artificial* (Cambridge: MIT Press, 1969). My view of engineering need not imply that I agree with Simon that science is somehow "free" from moralizing. Sociologically minded historians have argued that science itself is nothing more than a set of practical rules for connecting instrument readings, a kind of "theoretical technology." For a summary of recent work in this vein, see Andrew Pickering, ed., *Science as Practice and Culture* (Chicago: University of Chicago Press, 1991).

20. M. Norton Wise, "Meditations: Enlightenment Balancing Acts, or the Technologies of Rationalism," in *World Changes: Thomas Kuhn and the Nature of Science*, ed. Paul Horwich (Cambridge: MIT Press, 1993), 207–56. Unfortunately, many cultural critics of an anthro-

pological bent are still seduced by the argument that, between them, the logic of production and the logic of the market define the full range of technological possibilities. Arjun Appadurai, ed., *The Social Life of Things: Commodities in Cultural Perspective* (Cambridge: Cambridge University Press, 1986).

21. François Blondel, *L'art de jetter les bombes* (Paris: L'auteur et Langlois, 1683), preface.

22. For an expansive use of this black-box concept to include even "technological facts," see Bruno Latour, *Science in Action: How to Follow Scientists and Engineers Through Society* (Cambridge: Harvard University Press, 1987). In chapter 3, I use Latour's method of analyzing technological conflict to open the black box of ballistic science.

23. Thomas P. Hughes, "The Evolution of Large Technological Systems," in *The Social Construction of Technological Systems: New Directions in the Sociology and History of Technology*, eds. Wiebe E. Bijker, Thomas P. Hughes, and Trevor Pinch (Cambridge: MIT Press, 1987), 51–82; *Networks of Power: Electrification in Western Society, 1880–1930* (Baltimore: Johns Hopkins University Press, 1983); "The Electrification of America: The Systems Builders," *Technology and Culture* 20 (1979): 124–61. John Law, "Technology and Heterogeneous Engineering: The Case of Portuguese Expansion," in *Social Construction*, eds. Bijker et al., 111–34. For the important role of Enlightenment theoreticians in laying out the ideology of large technological systems, see Rosalind Williams, "Cultural Origins and Environmental Implications of Large Technological Systems," *Science in Context* 6 (1993): 377–403. For content and context in an industrial setting during the Second Industrial Revolution, see W. Bernard Carlson, *Innovation as a Social Process: Elihu Thomson and the Rise of General Electric, 1870–1900* (Cambridge: Cambridge University Press, 1991), 15.

24. Mona Ozouf, *Festivals and the French Revolution, 1789–1799*, trans Alan Sheridan (Cambridge: Harvard University Press, 1988). Maurice Agulhon, *Marianne into Combat: Republican Imagery and Symbolism in France, 1789 to 1880*, trans. Janet Lloyd (Cambridge: Cambridge University Press, 1981). Hunt, *Politics, Class, Culture*.

25. Daniel Roche, *The Culture of Clothing: Dress and Fashion in the "Ancien Régime"*, trans. Jean Birrell (Cambridge: Cambridge University Press, 1994). Cissie Fairchilds, "The Production and Marketing of Populuxe Goods in Eighteenth-Century Paris," in *Consumption*, eds. Brewer and Porter, 228–48. C. Jones, "Great Chain of Buying."

26. Noel Perrin, *Giving Up the Gun: Japan's Reversion to the Sword, 1543–1879* (Boston: Godine, 1979).

27. Donald MacKenzie, *Inventing Accuracy: An Historical Sociology of Nuclear Missile Guidance* (Cambridge: MIT Press, 1990).

28. See the inspiring "double vision" approach in David Joravsky, *The Lysenko Affair* (Cambridge: Harvard University Press, 1970), 271.

Chapter One
The Last Argument of the King

1. "L'histoire de l'artillerie est l'histoire du progrès des sciences, et partant de la civilisation." Louis-Napoléon Bonaparte, *Etudes sur le passé et l'avenir de l'artillerie*, ed. Alphonse Favé (Paris: Dumain, 1846–62), 1: v–xii. Even recent accounts have credited the Gribeauvalist artillery with self-evident superiority and explain the Vallièrist opposition as misguided filial loyalty. Michel de Lombarès, *Histoire de l'artillerie française* (Paris: Charles-Lavauzelle, 1984). The exception proves the rule; it was written by a descendant of one of the principal Vallièrists. A. Guichon de Grandpont, "La querelle de l'artillerie (parti rouge et parti bleu) au XVIIIe siècle," *Bulletin de la Société Académique de Brest*, 2d series, 20 (1894–95): 2–103. An article by Chalmin was the first to break partially with this tradition. Pierre Chalmin, "La querelle des Bleus et des Rouges dans l'artillerie française à la fin du XVIIIe siècle," *Revue d'histoire économique et sociale* 46 (1968): 465–505. Howard Rosen's dissertation offers more detail regarding the Vallièrist opposition, and valuable hints on

engineering "instrumentality" and tactics. Howard Rosen, "The Système Gribeauval: A Study of Technological Development and Institutional Change in Eighteenth-century France," (Ph.D. diss., University of Chicago, 1981). The quarrel is also covered in Nardin's biography, which unfortunately lacks a scholarly apparatus. Pierre Nardin, *Gribeauval, Lieutenant-général des armées du roi, 1715–1789* (Paris: La Fondation pour les Etudes de Défense Nationale, 1982).

2. For a sophisticated version of this story line, which considers the military role of technologies other than weaponry, see Martin van Creveld, *Technology and War: From 2000 B.C. to the Present* (New York: The Free Press, 1989), 143–47. One often encounters this same multistep determinism by which true knowledge, transformed into effective technology, drives history. Consider Alfred Chandler's analogous history of how the nineteenth-century railroad made necessary the staff and line structure of American corporate capitalism, and hence the rise of middle management. A. Chandler, *The Visible Hand: The Managerial Revolution in American Business* (Cambridge: Harvard University Press, 1977).

3. Phillippe-Charles-Jean-Baptiste Tronson Du Coudray, *L'artillerie nouvelle ou examen des changements faits dans l'artillerie français depuis 1765* (Amsterdam: n.p., 1773), 65.

4. The scholarship on the role of war in the formation of the early modern state is vast, though of uneven quality. For the best recent synthesis, see Charles Tilly, *Coercion, Capital, and European States, A.D. 990–1992* (rev. ed.; Cambridge, Mass.: Blackwell, 1992); also "War-Making and State-Making as Organized Crime," in *Bringing the State Back In*, eds. Peter B. Evans, Dietrich Rueschemeyer, and Theda Skocpol (Cambridge: Cambridge University Press, 1985), 169–91. See also William McNeill, *The Pursuit of Power: Technology, Armed Force, and Society since A.D. 1000* (Chicago: University of Chicago Press, 1982). My analysis is also indebted to the argument of Perry Anderson, *Lineages of the Absolutist State* (London: N.L.B., 1975). See also Brian Downing, *The Military Revolution and Political Change: Origins of Democracy and Autocracy in Early Modern Europe* (Princeton: Princeton University Press, 1992).

5. Recent doubts about the Military Revolution thesis, originally offered as the "cause" of the rise of absolutism, have not invalidated the intimate relationship between the new forms of warfare and the increasingly centralized state. The vast literature can be entered through Geoffrey Parker, *The Military Revolution: Military Innovation and the Rise of the West, 1500–1800* (Cambridge: Cambridge University Press, 1988). John A. Lynn, "The *trace italienne* and the Growth of Armies: The French Case," *Journal of Military History* 55 (1991): 297–330, see esp. 324–28.

6. Colin Jones, "The Military Revolution and the Professionalisation of the French Army under the Ancien Régime," in *The Military Revolution and the State, 1500–1800*, ed. Michael Duffy (Exeter: Exeter Studies in History, 1980), 29–48.

7. John A. Lynn, "Recalculating French Army Growth during the Grand Siècle, 1610–1715," *French Historical Studies* 18 (1994): 881–906.

8. S.H.A.T. Bib. #72227 Joseph-Florent de Vallière, *fils*, *Mémoire touchant la supériorité des pièces d'artillerie longues et solides, lu à l'Académie des Sciences, 19 août 1775* (Paris: Imprimerie Royale, 1775), 3–4.

9. Some of the more fanciful weapons can already be seen in the Renaissance "Theaters of Weapons." Francesco di Giorgio Martini, *Trattati di architettura ingegneria e arte militare* (Milan: Polifilo, [1475], 1967).

10. M. A. Basset, "Essais sur l'histoire des fabrications d'armement en France jusqu'au mi-18e siècle," *Mémorial de l'artillerie* 14 (1935): 954. David G. Chandler, *The Art of War in the Age of Marlborough* (London: Batsford, 1976), 177–81. See chapter 4.

11. *Ordonnance royale du 7 octobre 1732*, in Pierre Surirey de Saint-Rémy, ed., *Mémoires d'artillerie, recueillis* (3d ed.; Paris: Rollin, 1745), 3: 450–62, plates included. A parallel attempt was underway to standardize guns in Stuart England. H. C. Tomlinson, *Guns and Government: The Ordnance Department under the Later Stuarts* (London: Royal Historical Society, 1979), 161–62. But see chapter 6, below.

12. See the summary of officers' duties, along with sample forms and checklists, in Saint-Rémy, *Mémoires*, 3: 207–99. D. Chandler, *Marlborough*, 172.

13. Ernest Picard and Louis Jouan, *Artillerie française au XVIIIe siècle* (Paris: Berger-Levrault, 1906), 2–8. M. A. Basset, "Les grands maîtres de l'artillerie: Essai biographique," *Mémorial de l'artillerie* 11 (1932): 865ff. Théodore Le Puillon de Boblaye, *Esquisse historique sur les Ecoles d'Artillerie pour servir à l'histoire de l'Ecole d'Application de l'Artillerie et du Génie* (Metz: Rousseau-Pallez, [1858]), 26–29.

14. Laws of 1703 and 1704 in Le Puillon de Boblaye, *Ecoles d'artillerie*, 29–31. Louis XIV sold posts in the Corps-Royal as hereditary offices on the model of the judicial and financial courts; venal offices in the line army were *not* hereditary. One artillery officer lost his modest fortune and right of patrilineal succession when his compensation depreciated after John Law's system burst. Louis-August Le Pelletier, *Une famille d'artilleurs: Mémoires de Louis-Auguste Le Pelletier, seigneur de Glatigny, Lieutenant général des armées du roi, 1696–1769* (Paris: Hachette, 1896), xiii, 30–34.

15. [Jean-Florent de Vallière, *père*], "Instruction pour les écoles des cinq bataillons," 23 June 1720, in Saint-Rémy, *Mémoires*, 1: 58–59. Le Puillon de Boblaye, *Ecoles d'artillerie*, 35–36, 40–41.

16. D. Chandler, *Marlborough*, 234 shows that major sieges outnumbered land engagements by 167 to 144 in this period.

17. This human and financial cost was reckoned by Vauban himself. Vauban to Racine, 13 September 1696, in Jacques-Anne-Joseph Sébastien Le Prestre de Vauban, *Sa famille, et ses écrits: ses oisivetés et sa correspondance*, ed. Albert de Rochas d'Aiglun (Geneva: Slatkine, 1972), 2: 445–47.

18. Christopher Duffy, *The Fortress in the Age of Vauban and Frederick the Great, 1660–1789* (London: Routledge and Kegan Paul, 1985), 10, 29–31, 71–97. Pierre-Amboise-François Choderlos de Laclos, *Sur l'éloge de Vauban*, in Laclos, *Oeuvres complètes*, ed. Laurent Versini (Paris, Gallimard, [1787], 1979), 569–93.

19. Jacques-Anne-Joseph Sébastien Le Prestre de Vauban, *Traité des sièges et de l'attaque des places*, ed. Augoyat (Paris, [1704], 1829); *Science militaire contenant l'A.B.C. d'un soldat* (La Haye: Moetjens, 1689). Henry Guerlac, "Vauban: The Impact of Science on War," in *Makers of Modern Strategy from Machiavelli to the Nuclear Age*, ed. Peter Paret (Princeton: Princeton University Press, [1943], 1986), 64–90. On the science of fortress-building, see Hélène Vérin, *La gloire des ingénieurs: L'intelligence technique du XVIe au XVIIIe siècle* (Paris: Albin, 1993), 131–80. For a broader overview of the technical sciences as the outgrowth of the state's interest in military matters, see Henry Guerlac, "Science and War in the Old Regime" (Ph.D. diss., Harvard University, 1941).

20. Vauban, "De l'artillerie," in Vauban, *Sa famille*, 1: 299–311.

21. John Wright, "Sieges and Customs of War at the Opening of the Eighteenth Century," *American Historical Review* 39 (1934): 629–44. Even the right to pillage, if it came to that, was formally restricted to a stated number of days. For examples of aristocratic courtesy between antagonists, see Vauban to Le Peletier, 13 October 1693, in Vauban, *Sa famille*, 2: 299.

22. D. Chandler, *Marlborough*, 245–46.

23. This is based on my analysis of the siege data compiled in Lynn, "The *trace italienne*," 324–28.

24. Vauban to Louvois, 17 July 1691, in Vauban, *Sa famille*, 2: 327. Vauban especially regretted the loss of life among his small corps of highly skilled engineers. Vauban to Catinat, 23 October 1688, in Vauban, *Sa famille*, 2: 298. Losses among these men could be extremely high in fortress war. Anne Blanchard, *Les ingénieurs du "roy" de Louis XIV à Louis XVI* (Montpellier: Dehan, 1979), 189–92.

25. Vauban to Racine, 13 September 1696, in Vauban, *Sa famille*, 2: 445–47. Michel Parent and Jacques Verroust, *Vauban* (Paris: Fréal, 1971), 110–13. I have here reversed the inference drawn in Joan DeJean's fine study of the literary meaning of fortress warfare. Joan DeJean, *Literary Fortifications: Rousseau, Laclos, Sade* (Princeton: Princeton University Press, 1984), 20–75. Variations in the conduct of the siege did not alter its preordained outcome, but connoisseurs might appreciate variations introduced by the parties. Eighteenth-century military manuals—like program notes—allowed young noble amateurs to follow the proceedings "with minimal effort." Guillaume Le Blond, *Traité de l'attaque des places* (Paris: Jombert, 1743). The ritualized formality of the siege had become a target for parody in the eighteenth century. Pierre-Amboise-François Choderlos de Laclos, *Les liaisons dangereuses*, trans. P.W.K. Stone (New York: Penguin, 1961), 40.

26. Jacques-Anne-Joseph Sébastien Le Prestre de Vauban, *Le triomphe de la méthode*, eds. Nicholas Faucherre and Philippe Prost (Paris: Gallimard, 1992), 43–44. Gunners tried harder to impress the king than to inflict damage on the enemy. Le Pelletier, *Une famille d'artilleurs*, 91, 102–3.

27. D. Chandler, *Marlborough*, 235. On the contrast between the controlled violence of siege war and the unbridled violence of field war, see John U. Nef, *War and Human Progress: An Essay on the Rise of Industrial Civilization* (Cambridge: Harvard University Press, 1950), 134–40. Siege life had its creature comforts; the kitchen staff could set up for the season.

28. See also David Kaiser, *Politics and War: European Conflict from Philip II to Hitler* (Cambridge: Harvard University Press, 1990), 152–56.

29. Jean-Florent de Vallière, *père*, "Sur les mines et les avantages qu'on en peut tirer pour la défense des places," and Bernard Forest de Bélidor, "Nouvelle manière de faire des épreuves . . . des mines," in Saint-Rémy, *Mémoires*, 3: 49–51, 69–73.

30. Guillaume Le Blond, *L'artillerie raisonnée* (Paris: Jombert, 1761), 393–94.

31. *Ordonnance du roi concernant le Corps Royal de l'Artillerie*, 23 August 1772, p. 47.

32. Le Pelletier, *Une famille d'artilleurs*, 115–19. Picard and Jouan, *Artillerie*, 11–12.

33. Artillerists had long known that on campaign in Germany, they needed to carry more munitions. Saint-Rémy, *Mémoires*, 3: 132–33.

34. Henry Lloyd, *The History of the Late War in Germany* (London: Horsfield, 1766), 94–105. P. Bourcet, *Mémoires historiques sur la guerre que les français ont soutenue en allemagne depuis 1757 jusqu'en 1762* (Paris: Maradan, 1792), 1: 48–57. Also S.H.A.T. A1 3761 Mouy et al., "Mémoire sur différents sujets concernant artillerie," 19 August 1766.

35. Artillerists on both sides of the quarrel warned that these improvised guns would not stand up to lengthy service. Du Coudray, *L'artillerie nouvelle*, 10–11.

36. Nardin, *Gribeauval*, 111–12. Chalmin, "Querelle," 475.

37. Biographical information, here and below, is taken from Nardin, *Gribeauval*, 11–12.

38. S.H.A.T. 9a11 Gribeauval to [Indecipherable], 29 May 1749. The letter mentions that Gribeauval has just lost his father, and thus, the major portion of his income.

39. Nardin, *Gribeauval*, 38–47.

40. Broglie to Louis XV, 1 March 1773, in *Correspondance secrète du comte de Broglie avec Louis XV*, eds. Didier Ozanam and Michel Antoine (Paris: Klincksieck, 1961) 2: 373–74. Frederick II, *Mémoires de Frédéric* (Paris: Plon, 1866), 2: 277. For Gribeauval's biography during this period, see Nardin, *Gribeauval*, 55–100.

41. S.H.A.T. 1a2/2 Dubois, "Mémoire sur l'artillerie," [1762]. Nardin's biography of Gribeauval (pp. 90–92, 108–18) emphasizes the role of Dubois, whose uncle was Entrepreneur of the Manufacture of Charleville, and whose relative was the future Conventionnel, Dubois-Crancé.

42. "Rapport au Ministère," 3 March 1762, in Eugene Hennebert, *Gribeauval, lt.-général des armées du roy* (Paris: Berger-Levrault, 1896), 36.

43. [Guillaume Le Blond], "Artillerie," *Supplément à l'Encyclopédie, ou dictionnaire raisonné des sciences, des arts et des métiers* (Amsterdam: Rey, 1776), 1: 615, 622.

44. *Ordonnance du roi concernant le Corps Royal de l'Artillerie*, 23 August 1772. *Ordonnance du roi, portant règlement pour la fonte, l'épreuve et la réception des bouches à feu*, 15 December 1772. On the Versailles intrigues surrounding Gribeauval, see Honoré Gabriel Riquetti de Mirabeau, *Mémoires du ministère du duc d'Aiguillon* (3d edition; Paris: Buisson, 1792), 86–96.

45. See the twenty-one separate publications on the subject, some containing as many as eleven separate memoirs, listed in Heinrich Othon von Scheel, *Mémoires d'artillerie contenant l'artillerie nouvelle, ou les changements faits dans l'artillerie française depuis 1765* (Copenhagen: Philibert, 1777), 205–12. On the emerging function of public opinion, see Keith Michael Baker, "Public Opinion as Political Invention," in *Inventing the French Revolution*, 167–99. Also Sarah Maza, *Private Lives and Public Affairs: The Causes Célèbres of Prerevolutionary France* (Berkeley: University of California Press, 1994).

46. Voltaire to Saint-Auban, 10 October 1774, 15 February 1778, in Guichon de Grandpont, "Querelle," 73–75. "Eloge de Jean-Florent de Vallière, [*père*]," *Histoire de l'Académie des Sciences* (1759): 249–58. "Eloge de Joseph-Florent de Vallière, [*fils*]," *Histoire de l'Académie des Sciences* (1776): 53–64. For a pro-Vallièrist speech by an academician, see S.H.A.T. Bib. #72359 Tressan, "Mémoire sur l'artillerie," 23 August 1775, in Jacques Antoine Baratier de Saint-Auban, *Mémoires sur les nouveaux systèmes d'artillerie* (n.p., 1775). The Vallièrist party even achieved some recognition from Frederick the Great; in 1780, Frederick ordered his Academy to elect Saint-Auban. Frederick II to Saint-Auban, 2 August 1779, 15 June 1780, 14 March 1783, in Guichon de Grandpont, "Querelle," 72–73. [Guillaume Le Blond], "Affût," "Artillerie," and "Canon de campagne," *Supplément à l'Encyclopédie*, 1: 190–92, 603–24; 2: 202–9. These articles were signed "A. A., January 1773," but Scheel (*Mémoires d'artillerie*, 205–12) identifies the author as Le Blond. The editors of the *Supplément à l'Encyclopédie* seem to have been caught flat-footed by the gyrations in official policy. By the time the article appeared in 1776, the Gribeauvalists were back in power. Diderot was no longer the editor at this time, and the *Supplément* had shed much of his sharper social criticism. See "Avertissement," *Supplément à l'Encyclopédie*, 1: ii–iii.

47. Chalmin, "Querelle," 494–96. Gribeauval (b. 1715) was two years older than Vallière, *fils*, and three years younger than Saint-Auban, the Vallièrists' chief publicist.

48. S.H.A.T. Bib. #72379 [Jacques Antoine Baratier de Saint-Auban], *Supplément aux considérations sur la réforme des armes* (n.p., [1775]).

49. S.H.A.T. MR1739 Richelieu, Contades, Soubise, and Broglie, "Procès-verbal," 26 March 1774. See the debates published in S.H.A.T. Bib. #72359 [Jean-Baptiste de Gribeauval and Joseph Florent de Vallière, *fils*], *Collection des mémoires authentiques qui ont été présentés à MM. les Maréchaux de France* (Aléthopolis: Neumann, 1774). For the reaction of the salons, see [Louis Petit de Bachaumont], *Mémoires secrets pour servir à l'histoire de la république des lettres en France* (London: Adamson, 1780–89), 20 March 1774, vol. 27.

50. The Maritz family, father and son, had operated the Lyon foundry since the 1730s. Within two decades they controlled all the kingdom's foundries: training Du Pont at Rochefort, Dartein at Strasbourg, and forming a family alliance with Bérenger at Douai. The younger Maritz became a close confederate of Gribeauval and helped him with numerous metallurgical problems, including those related to the manufacture of small arms. On the Maritz family and their methods, see the contemporary descriptions collected in Carel de Beer, ed. *The Art of Gunfounding: The Casting of Cannon in the Late 18th Century* (Rotherfield, U.K.: Boudriot, 1991), especially 153–54 on the advantages of horizontal boring. Basset, "Fabrications d'armement," 1033–65.

51. MacKenzie, *Inventing Accuracy*. Chalmin ("Querelle") and Rosen ("Gribeauval") both note the importance of mobility to the Gribeauvalists; the challenge is to explain the social logic behind this goal.

52. For tables giving the cannon's respective weights, see [Le Blond], "Canon," 2: 203. In fact, because the new carriages substituted iron for many wooden parts, they actually weighed *more* than the old carriages. But as the Gribeauvalists pointed out, weight is not the only factor in mobility; the new carriages had bigger wheels, a more even distribution of weight during transport, et cetera.

53. The best summary of the Gribeauval system is Rosen, "Gribeauval," 30–50.

54. Picard and Jouan, *Artillerie*, 66. James Pritchard, *Louis XV's Navy, 1748–1762: A Study of Organization and Administration* (Kingston, Ont.: McGill-Queen's University Press, 1987), 151–53. Maritz's boring method spread across Europe in the eighteenth century, despite French efforts to keep it secret: Switzerland, 1716; France, 1730s-60s; Holland, 1740s; England, 1770s; and Germany, in the early nineteenth century. Beer, *Gunfounding*, 3–12.

55. S.H.A.T. 1a2/2 Gribeauval, "Mémoire," 1764. Du Coudray, *L'artillerie nouvelle*, 72–75. For a step-by-step description of actions of each of the fifteen men who serviced a piece, and each of its several modes of operation, see [Le Blond], "Artillerie," 1: 622–24.

56. Jean Du Teil, *De l'usage de l'artillerie nouvelle dans la guerre de campagne, connaissance nécessaire aux destinés de commander toutes les armées* (Paris: Charles-Lavauzelle, [1778], 1924), 1–2, 36–37, 39, 43, 49, 53, quote on p. 50. For the influence of this doctrine among the Gribeauvalists, see S.H.A.T. MR1744 Ecole de Besançon, "Etat des objets qui ont été traités tous les samedis à la salle de conferences depuis septembre 1788 jusqu'au septembre 1789," [1789]. Joseph Du Teil, *Napoléon Bonaparte et les Généraux Du Teil, 1788–1794: L'Ecole d'Artillerie d'Auxonne et le siège de Toulon* (Paris: Picard, 1897). Arthur Chuquet, *La jeunesse de Napoléon* (Paris: Colin, 1897–99), 1: 350–55.

57. Throw-weight is here defined as each cannon's shot-pounds multiplied by the number of pieces. It does not take into account the impact velocity of the ball. Nardin, *Gribeauval*, 145, 150.

58. Jacques-Antoine-Hippolyte de Guibert, *Essai général de tactique*, vols. 1 and 2, in *Oeuvres militaires* (Paris: Magimel, year XII [1803]), 2: 194–234. Nardin, *Gribeauval*, 145.

59. Jean Du Teil, *L'usage*, ii–v, 1–2, 118–19. R. R. Palmer, "Frederick the Great, Guibert, Bülow: From Dynastic to National War," in *Makers of Modern Strategy from Machiavelli to the Nuclear Age*, ed. Peter Paret (Princeton: Princeton University Press, [1943], 1986), 91–119, see p. 57.

60. S.H.A.T. Yb668-Yb670 *Registres*, 1763–1789. Between 1765–69, the corps accepted one hundred new officers a year; after which the number reverted to the steady-state of thirty to forty annual admissions.

61. S.H.A.T. MR1739 "Dépenses annuelles de l'artillerie," [1788]. Rosen, "Gribeauval," 165. These seem to be the first budgets which provide a detailed breakdown of expenditures. Annual construction costs fluctuated around 1.5 million *livres*.

62. Claude C. Sturgill, "The French Army's Budget in the Eighteenth Century: A Retreat from Loyalty," in *The French Revolution in Culture and Society*, eds. David G. Troyansky, Alfred Cismaru, and Norwood Andrews, Jr. (New York: Greenwood Press, 1991), 123–34.

63. Claude-Louis de Saint-Germain, *Mémoires de M. le comte de Saint-Germain* (Switzerland: Libraires Associés, 1779), 47. See also Léon Mention, *Le comte de Saint-Germain et ses réformes* (Paris: Clavel, 1884).

64. This view is expressed most completely in Robert S. Quimby, *The Background of Napoleonic Warfare: The Theory of Military Tactics in Eighteenth-Century France* (New York: Columbia University Press, 1957).

65. Charles de Secondat de Montesquieu, *Esprit des lois* (Paris: Flammarion, 1979), 1: 270–71, 273–76. Edmond Silberner, *La guerre dans la pensée économique du XVIe au XVIIIe siècle* (Paris: Sirey, 1939), 186–220. Nef, *War and Human Progress*, 258–70. Hans Speier, "Militarism in the Eighteenth Century," *Social Research* 3 (1936): 304–36.

66. Jacques-Antoine-Hippolyte de Guibert, *Observations sur la constitution des armées de sa majesté prussienne* (Amsterdam: n.p., 1778), 119–22. Frederick II, *Essai sur la grande guerre* (London: Compagnie, 1761), 137–47. On the image of the soldier at the end of the ancien régime, see Jean Chagniot, *Paris et l'armée au XVIIIe siècle: Etude politique et sociale* (Paris: Economica, 1985), 611–45. On the extent to which elite public opinion shared in this revanchist spirit, see Bailey Stone, *The Genesis of the French Revolution: A Global-Historical Interpretation* (Cambridge: Cambridge University Press, 1994), 57–63, 140–44. Also R. R. Palmer, "The National Idea in France before the Revolution," *Journal of the History of Ideas* 1 (1940): 95–111.

67. For an assessment of the French conduct during the war, see Lee Kennett, *The French Armies in the Seven Years' War: A Study in Military Organization and Administration* (Durham, N.C.: Duke University Press, 1967), 117–18.

68. Samuel Scott, "The French Revolution and Professionalization of the French Officer Corps: 1789–93," in *On Military Ideology*, eds. Morris Janowitz and Jacques van Doorn (Rotterdam: Rotterdam University Press, 1971), 5–52. On the surprisingly similar English case, see John Brewer, *The Sinews of Power: War, Money, and the English State, 1688–1783* (New York: Knopf, 1989), 55–63. The classic account of military professionalism insists that this process was only completed in the late nineteenth century. Samuel Huntington, *The Soldier and the State: The Theory and Politics of Civil-Military Relations* (Cambridge, Mass.: Belknap Press, 1957).

69. Saint-Germain, *Mémoires*, 39–42, 59–61, 234–35.

70. Guibert, *Essai général*, 1: 15–16. Guibert's later retracted some of the more bellicose sentiments of his *Essai général*. Palmer, "Frederick the Great." Julie de Lespinasse and Germaine de Staël have left fascinating portraits of Guibert; see Jacques-Antoine-Hippolyte de Guibert, *Ecrits militaires, 1772–1790* (Paris: Copernic, 1977), 37–41.

71. For the most detailed call for a citizen army in the late ancien régime, see [Joseph Servan], *Le citoyen soldat, ou vue patriotique sur la manière la plus avantageuse de pourvoir à la défense du royaume* ("Dans la pays de la liberté," 1780). Servan became minister of war during the Revolution.

72. On the controversy over the division, see Saint-Germain, *Mémoires*, 39–47. Steven T. Ross, "The Development of the Combat Division in Eighteenth-Century French Armies," *French Historical Studies* 4 (1965–66): 84–94.

73. Jay Michael Smith, *The Culture of Merit: Nobility, Royal Service and the Making of Absolute Monarchy in France, 1600–1789* (Ann Arbor: University of Michigan Press, 1996). "Que chacun fût fils de ses oeuvres et de ses mérites: toute justice serait accomplie et l'Etat serait mieux servi," quoted in Guy Chaussinand-Nogaret et al., *Histoire des élites en France du XVIe au XXe siècle: L'honneur, le mérite, l'argent* (Paris: Tallandier, 1991), 232–33. Saint-Germain, *Mémoires*, 39–40. On the nobility's justification of its preeminence more generally, see William H. Sewell, Jr., "Etats, Corps, and Ordre: Some Notes on the Social Vocabulary of the French Old Regime," in *Sozialgeschichte Heute: Festschrift für Hans Rosenberg zum 70. Geburtstag*, ed. Hans-Ulrich Wehler (Göttingen: Vandenhoeck und Reprecht, 1974), 49–68.

74. Guibert, *Eloge de Catinat*, [1775], in Guibert, *Ecrits militaires*, epigraph. Guibert's dedication to "Ma Patrie" was only partially retracted by calling the king the nation's father. Guibert, *Essai général*, 1: 1. See the similar collapse of king and nation in Phillipe-Auguste de Saint-Foix d'Arcq, *La noblesse militaire, ou le patriote français* (3d ed.; n.p., 1756), iv.

75. Mark Motley, *Becoming a French Aristocrat: The Education of the Court Nobility, 1580–1715* (Princeton: Princeton University Press, 1990). This complex mix of skills explains why selection based on the older type of "birth" merit had in fact enabled some brilliant officers to rise to the top.

76. S.H.A.T. Bib. MS174 Du Châtelet, "Observations sur le règlement concernant les

appointments," Comité Militaire, 1782, 2: fol. p. 105. These sources were first brought to light in a brilliant article by David Bien, "The Army in the French Enlightenment: Reform, Reaction, and Revolution," *Past and Present* 85 (1979): 68–98. See also David Bien, "La réaction aristocratique avant 1789: L'example de l'armée," *Annales: Economies, sociétés, civilisations* 29 (1974): 23–48, 505–34.

77. Jacques-Antoine-Hippolyte de Guibert, *Mémoire adressé au public et à l'armée sur les opérations du Conseil de Guerre* (n.p., [1789]), in *Oeuvres*, 5: 191–320. See Guibert's biography in Matti Lauerma, *Jacques-Antoine-Hippolyte de Guibert, 1743–1790*, Annales Academiae Scientiarum Fennicae, Series B, vol. 229 (Helsinki: Suomalainen Tiedeakatemia, 1989), 183–211.

78. S.H.A.T. Bib. MS174 Comité Militaire, "Réponse," 17 April 1782, 2: fol. pp. 253–67. The Comité Militaire was responding to an earlier version of this proposal floated by Min. War Ségur on 8 April 1782.

79. S.H.A.T. MR1790 [Conseil de Guerre], "Projet," [1788]. Anon., "Proposition au roi relatif au Conseil de Guerre," [1788].

80. This was the opinion of contemporaries, see Charles François du Périer Dumouriez, *Mémoires* (Paris: Didot, [1794], 1848), 2: 86. For accounts by historians, see Lauerma, *Guibert*, 209–13. Jean Egret, *The French Prerevolution*, trans. Wesley D. Camp (Chicago: University of Chicago Press, 1977), 47–54.

81. Jacques-Antoine-Hippolyte de Guibert, *Précis de ce qui s'est passé à mon égard à l'Assemblée du Berry, 25 mars 1789* (n.p., [1789]). Anon., *Lettre à M. le comte de Guibert sur le précis de ce qui est arrivée à Berry* (n.p., [1789]).

82. Jacques-Antoine-Hippolyte de Guibert, *De la force publique considérée dans tous ses rapports* (Paris: Didot l'aîné, 1790), "Avant propos." Jacques-Antoine-Hippolyte de Guibert, *Défense du système de guerre moderne*, vols. 3 and 4, in *Oeuvres*, 3: 73–74.

83. James C. Riley, *The Seven Years' War and the Old Regime in France: The Economic and Financial Toll* (Princeton: Princeton University Press, 1986). J. F. Bosher, *French Finances, 1770–1795: From Business to Bureaucracy* (Cambridge: Cambridge University Press, 1970). The claim is not that the French debt was so great (as a percentage of national wealth it was less than that of Britain), but that the unequal mechanisms for servicing that debt were perceived as burdensome.

84. Paul Kennedy, *The Rise and Fall of the Great Powers: Economic Change and Military Conflict from 1500–2000* (New York: Random House, 1987), 109–39. McNeill, *Pursuit of Power*. Kaiser, *Politics and War*. Stone, *Genesis*.

85. Guglielmo Ferrero, *Aventure, Bonaparte en Italie, 1796–97* (Paris: Plon, 1936), 114. In this context, Chalmin, "Querelle," 486, has also pointed out that Guibert was an Encyclopédiste.

86. Theda Skocpol, *States and Social Revolution: A Comparative Analysis of France, Russia, and China* (Cambridge: Cambridge University Press, 1979). Also Theda Skocpol and Meyer Kestnbaum, "Mars Unshackled: The French Revolution in World-Historical Perspective," in *The French Revolution and the Birth of Modernity*, ed. Ferenc Fehér (Berkeley: University of California Press, 1990), 13–29. Samuel Scott, "Professionalization of the French Officer Corps"; *The Response of the Royal Army to the French Revolution* (Oxford: Oxford University Press, 1978). Sewell wants to claim a kind of autonomy for ideology. But in the case of the army and in the case of the reform of weights and measures (which he cites), the impetus behind these universalist Enlightenment projects needs to be understood with reference to the specific social and economic agenda of the Revolutionaries in charge of the central state. William H. Sewell, Jr., "Ideologies and Social Revolutions: Reflections on the French Case," *Journal of Modern History* 57 (1985): 57–85.

87. The classic work that bridges the "generation" gap is Daniel Mornet, *Les origines*

intellectuelles de la Révolution française (Paris: Colin, 1933). For two different approaches to the "underground" currents of radical critique in the eighteenth century, see Robert Darnton, *The Literary Underground of the Old Regime* (Cambridge: Harvard University Press, 1982), 1–40; and Margaret Jacob, *The Radical Enlightenment: Pantheists, Freemasons and Republicans* (London: Allen and Unwin, 1981). For a more fully social treatment of late-eighteenth-century cultural tensions, see Sarah Maza, *Private Lives and Public Affairs*.

88. Tocqueville, *Old Régime*. The classic extension of Tocqueville's thesis is C.B.A. Behrens, *Society, Government, and the Enlightenment: The Experiences of Eighteenth-century France and Prussia* (New York: Harper and Row, 1985). For a fine extension of Tocqueville's analysis to the European military in this period, see Kaiser, *Politics and War*, 203–37.

89. Voltaire to Richelieu, 18 June 1757, in [François-Marie Arouet] Voltaire, *Voltaire's Correspondence*, ed. Theodore Besterman (Geneva: Institut et Musée Voltaire, 1953–65), letter D7293. Although Voltaire was mocking himself, he persisted with his chariot project for fifteen years, even taking it to Empress Catherine of Russia.

90. Voltaire, "La tactique," in *Oeuvres complètes* (Paris: Garnier, [1773], 1877), 10: 187–94.

91. These verses caused a spat between that Enlightenment odd couple, Voltaire and Frederick. Later, Voltaire added notes to the poem which praised Frederick's acts of kindness on the battlefield. Christiane Mervaud, *Voltaire et Frédéric II: Une dramaturgie des lumières, 1736–1778*, in *Studies on Voltaire and the Eighteenth Century*, 234 (Oxford: The Voltaire Foundation, 1985), 467–71.

92. Bien, "The Army," 98.

Chapter Two
A Social Epistemology of Enlightenment Engineering

1. Steven Shapin and Simon Schaffer, *Leviathan and the Air-Pump: Hobbes, Boyle, and the Experimental Life* (Princeton: Princeton University Press, 1985). Much recent work in the history of science has explored a social history of objectivity from different perspectives. For three examples, see Steven Shapin, *A Social History of Truth: Civility and Science in Seventeenth-Century England* (Chicago: University of Chicago Press, 1994). Theodore Porter, *Trust in Numbers: Objectivity in Science and Public Life* (Princeton: Princeton University Press, 1995). Lorraine J. Daston, "Objectivity and the Escape from Perspective," *Social Studies of Science* 22 (1992): 597–618. For an epistemological treatise which makes a comparable case, see Helen E. Longino, *Science as Social Knowledge: Values and Objectivity in Scientific Inquiry* (Princeton: Princeton University Press, 1990). On the moral codes of corporate bodies, including the professions, see Emile Durkheim, *Professional Ethics and Civic Morals*, trans. Cornelia Brookfield (London: Routledge and Kegan Paul, [1898], 1957), 4–8.

2. Académie Française, "Ingénieur," *Le dictionnaire de l'Académie* (1st ed.; Paris: Coignard, 1694), 1: 596. [Le Blond], "Ingénieur," *Encyclopédie*, 8: 741–43. For a full etymology, see Vérin, *Gloire des ingénieurs*, 19–42, 58–62.

3. Temporary schools existed in all the services, including the artillery (1679); Le Puillon de Boblaye, *Ecoles d'Artillerie*, 35–36. For the Ecole du Génie at Mézières, see Blanchard, *Ingénieurs*, 188–212. The Ponts et Chaussées were not military engineers, though a military ethos informed their corporate identity, and military need inspired many of their civilian projects; see Antoine Picon, *L'invention de l'ingénieur moderne: L'Ecole des Ponts et Chaussées, 1747–1851* (Paris: Presses de l'Ecole Nationale des Ponts et Chaussées, 1992). For an overview of engineering schools, see Peter Lundgreen, "Engineering Education in Europe and the U.S.A., 1750–1930: The Rise to Dominance of School Culture and the Engineering Profession," *Annals of Science* 47 (1990): 33–75. Even in England the term "engineer" referred to a military man in the eighteenth century; those engaged in invention

were known as "practical philosophers" or "mechanics," depending on their social station. The term "civil" engineer came to be used to distinguish these men from military engineers.

4. The artillery schools are considered the forerunner of all European technical education in the best survey of the subject. Frederick B. Artz, *The Development of Technical Education in France, 1500–1850* (Cambridge: MIT Press, 1966), 98. "Les hommes n'ayant pas accoutumé de former le mérite, mais seulement le récompenser où ils le trouvent formé," in Blaise Pascal, *Pensées*, ed. Louis Lafuma (Paris: Editions du Luxembourg, 1951), 1: 496 (frag. 935). Max Weber, "Science as a Vocation," and "Politics as a Vocation," in *From Max Weber*, ed. and trans. H. H. Gerth and C. Wright Mills (New York: Oxford University Press, 1946), 77–156. Michel Foucault, *Discipline and Punish: The Birth of the Prison*, trans. Alan Sheridan (New York: Random House, 1979).

5. J. M. Smith, *Culture of Merit*. For a general discussion of merit in the artillery schools of Piedmont, see Vincenzo Ferrone, "Les mécanismes de formation des élites de la maison de Savoie: Recrutement et sélection dans les écoles militaires du Piémont au XVIIIe siècle," *Paedagogica Historica* 30 (1994): 341–70. See also the other articles in that volume, edited by Dominique Julia. For the academicians' solution to the tension between innate capacity and traditional rank, see Daniel Roche, *Le siècle des lumières en province: Académies et académiciens provinciaux, 1680–1789* (Paris: Mouton, 1978), 233–55. For the Masonic lodges, see Margaret Jacob, *Living the Enlightenment: Free Masonry and Politics in Eighteenth-Century Europe* (New York: Oxford University Press, 1991), 5–8. For the salons, see Deena Goodman, "Governing the Republic of Letters: The Politics of Culture in the French Enlightenment," *History of European Ideas* 13 (1991): 183–99. The standard cultural history on this topic does not explain the specific mechanisms which marked out the new elite as meritorious; Chaussinand-Nogaret et al., *Elites*. For a study that places "meritocratic individualism" at the center of contention in France in the 1780s, though without exploring its social basis, see Patrice Higonnet, "Cultural Upheaval and Class Formation During the French Revolution," in *Birth of Modernity*, ed. Fehér, 69–102. For an excellent study of the development of the I.Q. test in the context of a mass army, see John Carson, "Army Alpha, Army Brass, and the Search for Army Intelligence," *Isis* 84 (1993): 278–309.

6. Edwin Layton, "Mirror-Image Twins: The Communities of Science and Technology in Nineteenth-Century America," *Technology and Culture* 12 (1971): 562–80.

7. Eda Kranakis, "The Social Determinants of Engineering Practice: A Comparative View of France and America in the Nineteenth Century," *Social Studies of Science* 19 (1989): 5–70. Picon, *Ingénieur moderne*, introduction, 127. T. Porter, *Trust in Numbers*, 114–47.

8. For Columbia, see Frank Safford, *The Ideal of the Practical: Columbia's Struggle to Form a Technical Elite* (Austin: University of Texas Press, 1976).

9. Each level of the nineteenth-century hierarchy has been separately studied. Terry Shinn, *L'Ecole Polytechnique, 1794–1914* (Paris: Presses de la Fondation Nationale des Sciences Politiques, 1980). John Hubbel Weiss, *The Making of Technological Man: The Social Origins of French Engineering Education* (Cambridge: MIT Press, 1982). Charles R. Day, *Education for the Industrial World: The Ecoles d'Arts et Métiers and the Rise of French Industrial Engineering* (Cambridge: MIT Press, 1987). The engineering hierarchy came under pressure near the end of the nineteenth century when highly capitalized private firms began to lure the "top" engineers from the state services.

10. John Hubbel Weiss, "Bridges and Barriers: Narrowing Access and Changing Structure in the French Engineering Profession, 1800–1850," in *Professions and the French State*, ed. Gerald Geison (Philadelphia: University of Pennsylvania Press, 1984), 15–65.

11. On the Napoleonic reforms and the structure of French science, see Robert Fox and George Weisz, "Introduction," in *The Organization of Science and Technology in France, 1808–1914*, eds. Robert Fox and George Weisz (Cambridge: Cambridge University Press, 1980), 1–28. Roger Hahn, "Le rôle de Laplace à l'Ecole Polytechnique," in *La formation*

polytechnicienne, 1794–1994, eds. Bruno Belhoste, Amy Dahan Dalmedico, and Antoine Picon (Paris: Dunod, 1994), 47.

12. A similar mediating role was played by the technical corps of the American military in the nineteenth century. Terrence J. Gough, "Isolation and Professionalization of the Army Officer Corps: A Post-revisionist View of *The Soldier and the State*," *Social Science Quarterly* 73 (1992): 420–36.

13. [Philippe-Charles-Jean-Baptiste Tronson Du Coudray], *L'ordre profond et l'ordre mince, considérés par rapport aux effets de l'artillerie* (Metz: Ruault, 1776), 80–82.

14. Francis Bacon, *New Organon*, in *Advancement of Learning* (Chicago: Encyclopedia Britannica, 1952), I. 39–44, pp. 109–10.

15. Bacon, *New Organon*, I. 64, p. 114; quotes from I. 69, pp. 116–17, and I. 99, pp. 126–27.

16. Francis Bacon, *New Organon*, I. 95, p. 126. On Bacon, see Robert Proctor, *Value-Free Science: Purity and Power in Modern Knowledge* (Cambridge: Harvard University Press, 1991), 25–38. Some historians now believe the Scientific Revolution was fueled by the activities of these practical philosophers; see J. A. Bennet, "The Mechanics' Philosophy and the Mechanical Philosophy," *History of Science* 24 (1986): 1–28. Alexander Keller, "Mathematics, Mechanics and the Origins of the Culture of Mechanical Invention," *Minerva* 23 (1985): 348–61. Many of these themes can be found in Paolo Rossi, *Philosophy, Technology and the Arts in the Early Modern Era*, trans. Salvator Attanasio (New York: Harper and Row, 1970); and Edgar Zilsel, "The Sociological Roots of Science," *The American Journal of Sociology* 47 (1942): 544–62.

17. Francis Bacon, *New Atlantis*, in *Advancement of Learning*. On Bacon and the armaments industry, see Anthony F. C. Wallace, *The Social Context of Innovation: Bureaucrats, Families, and Heroes in the Early Industrial Revolution, as Foreseen in Bacon's "New Atlantis"* (Princeton: Princeton University Press, 1982), 11–61. On the tension between the utopian promise of the *New Atlantis* and the destabilizing tendencies of innovative technology, see J. C. Davis, "Science and Utopia: The History of a Dilemma," in *Nineteen Eighty-Four: Science Between Utopia and Dystopia*, eds. Everett Mendelsohn and Helga Nowotny (Dordrecht: Reidel, 1984), 21–48.

18. d'Alembert, "Discours préliminaire," *Encyclopédie*, 1: xxiv, vi–vii. Also, see "Théorie," "Pratique," *Encyclopédie*, 16: 253; quote on 13: 264. Also, Denis Diderot, *Pensées sur l'interprétation de la nature* (Paris: Vrin, [1754], 1983), 2, 11–12. The anti-Cartesianism of mid-century is a familiar theme of the Enlightenment. For a recent view of this antipathy and its ambiguities, see Jessica Riskin, "The Quarrel over Method in Natural Science and Politics during the Late Enlightenment," (Ph.D. diss., University of California at Berkeley, 1995). In fact, the rationalist legacy of Cartesianism informed the claims of the empiricists more than they cared to admit; and Descartes himself had always been an advocate of a useful and practical science. See René Descartes, *Discours de la méthode* (Paris: Fayard, [1637], 1987), sixième partie, pp. 54–68. See also Jean Le Rond d'Alembert, *Nouvelles expériences sur la résistance des fluides* (Paris: David, 1777), preface. For Diderot's attack on the guilds, see chapter 4.

19. Diderot, "Art," "Eclectisme," in *Encyclopédie*, 1: 716; 5: 284–85. For his attacks on secrecy, see Diderot, "Encyclopédie," ibid., 5: 647.

20. Georgius Agricola, *De Re Metallica*, trans. Herbert Clark Hoover and Lou Henry Hoover (New York: Dover, [1556], 1950), xxx–xxxi, quote on p. 4. Rossi, *Philosophy*, 46–56. Pamela O. Long, "The Openness of Knowledge: An Ideal and Its Context in Sixteenth-Century Writings on Mining and Metallurgy," *Technology and Culture* 32 (1991): 318–55.

21. Le Pelletier, *Une famille d'artilleurs*, 27, 38–39. Le Pelletier had been trained by his artillerist father at the age of eleven. Only in his twenties did he attend the La Fère regimental school.

22. S.H.A.T. MR1739 "Dépenses annuelles de l'artillerie pendant 6 années dont 2 en

guerre et 4 en paix," [1788]. School costs included the salary of professors and rent for the terrain of the practice polygon.

23. Saint-Rémy, *Mémoires*, 1: 54–72. Le Puillon de Boblaye, *Ecoles d'artillerie*, 53, 56–61.

24. *Ordonnance* of 1720, quoted in Artz, *Technical Education*, 96–97.

25. Bernard Forest de Bélidor, *Architecture hydraulique, ou l'art de conduire, d'élever et de ménager les eaux* (Paris: Jombert, 1737–53), 1: ii. Bélidor, *La science des ingénieurs dans la conduite des travaux de fortification et d'architecture civile* (Paris: Jombert, 1729), 1: 2. See also S.H.A.T. 2b5/1 [Du Puget], "Mémoire relatif au travail de l'officier et des deux professeurs," 1765.

26. "Instruction que son Altesse Royale," 23 June 1720 in Saint-Rémy, *Mémoires d'artillerie*, 1: 71–72.

27. *Ordonnance*, 5 July 1729. S.H.A.T. 1a2/2 Note of Gribeauval,in Dubois, "Mémoire sur l'artillerie," [1762].

28. Bélidor, *La science des ingénieurs*, 1: 3–4. Bernard Forest de Bélidor, *Oeuvres diverses* (Amsterdam: Jombert, 1764), xx. Bélidor, *Architecture hydraulique*, iv.

29. S.H.A.T. 2b5/1 "Circulaire à MM. de Mouy, de Gribeauval," November 1765. S.H.A.T. 2b5/1 [Du Puget], "Travail de l'officier," 1765.

30. Amédée-François Frézier, *La théorie et la pratique de la coupe des pierres et des bois . . . ou traité de stéréotomie* (Strasbourg: Doulsseker, 1737–39), 1: i–vii. Also Frézier, *Elémens de stéréotomie à l'usage de l'architecture pour la coupe des pierres* (Paris: Jombert, 1760).

31. Bélidor, *Architecture hydraulique*, ii.

32. S.H.A.T. Yb668-Yb670 *Registres*, 1763–1789. According to Minister of War Saint-Germain, out of three hundred graduates of the Ecole Militaire, some forty entered the technical arms. Roger Hahn, "L'enseignement scientifique aux écoles militaires et d'artillerie," in *Ecoles techniques et militaires au XVIIIe siècle*, eds. Roger Hahn and René Taton (Paris: Hermann, 1986), 513–45; see p. 525.

33. Bien expresses a certain agnosticism about the practical uses of mathematics, even for the technical officers. "Rightly or wrongly, it was assumed that aiming and firing cannon on land, as well as at sea, required a knowledge of geometry." Bien, "Military Education in 18th-Century France: Technical and Non-Technical Determinants," in *Science, Technology and Warfare: Third Military History Symposium*, eds. Monte Wright and L. Paszek (Washington, D.C.: General Publishing Office, 1969), 53.

34. Foucault, *Discipline and Punish*, 156–57. His source is E. Gerspach, *La Manufacture de Gobelins* (Paris, 1892).

35. J. M. Smith offers a telling Foucauldian analysis of how the monarch's universal gaze "democratized" merit in the late ancien régime. Smith, *Culture of Merit*.

36. S.H.A.T. 2a59 Le Pelletier, "Règlement," 1749.

37. S.H.A.T. 2a59 "Ecole d'artillerie de Metz," October 1767. The central cadet school at La Fère was closed by Vallière, *fils*, and when Gribeauval returned to power he confirmed the decision. S.H.A.T. 2a60/4 Anon., "Mémoire sur la formation d'une école," [1780s]. S.H.A.T. 2a59 Joseph Du Teil, "Ecole d'Auxonne, ordre du 1 juin 1786, portant règlement pour la salle de dessin," 1786.

38. S.H.A.T. 2a59 Anon. to Gribeauval, 20 September 1788. S.H.A.T. Xd260 "Inspection de 1773." Variation in teaching methods became an argument for changing the Camus textbook, since many of the professors were adding improvised material to their courses. S.H.A.T. 2a59 "Ecole de Metz," October 1767.

39. For the desirability of mathematics as a uniform curriculum, see Paris de Meyzieu, "Ecole Militaire," in *Encyclopédie*, 5: 307–8. Also, H.H.U. MS Eng601.65 Anon., *General Establishment of the Officers Military and Civil . . . belonging to the Ecole Royale Militaire de Paris* (n.p., 1770). For the perceived dangers of rhetoric, see Quentin Skinner, *Reason and Rhetoric in the Philosophy of Hobbes* (Cambridge: Cambridge University Press, 1996).

40. Le Pelletier, *Une famille d'artilleurs*, 27, 38–39. S.H.A.T. Xd260 "Inspection de 1721."

41. For evaluations at various periods, see S.H.A.T. 2a60/4 "Examen des officiers, Besançon," January 1748; S.H.A.T. Xd248 Camus, "Mémoire," 3 March 1762; S.H.A.T. Xd249 Bézout, "Etat des élèves," June 1768; and Xd249 Laplace to [Indecipherable], 18 August 1789. "Ordonnances," 21 December 1761, 17 September 1763, in Le Puillon de Boblaye, *Ecoles d'artillerie*, 56–57. For other evaluations, see Rosen, "Gribeauval," 78–80.

42. S.H.A.T. Xd249 Gribeauval, "Les officiers surnuméraires," [1779].

43. S.H.A.T. 2b5/1 Guibert [to Dubois], "Projet d'un nouveau cours de mathématiques pour les écoles du Corps Royal de l'Artillerie," 25 November 1761.

44. S.H.A.T. 2a59 "Projet de règlement concernant l'instruction du Corps de l'Artillerie," 19 thermidor, year IX [7 August 1801]. The need for a uniform code became even more urgent after the Revolution when the social origins of officers varied widely, but it made itself felt in the pre-Revolutionary period as well. See chapter 8.

45. T. Porter, *Trust in Numbers*. On England, see John Gascoigne, "Mathematics and Meritocracy: The Emergence of the Cambridge Mathematical Tripos," *Social Studies of Science* 14 (1984): 547–84. On the examiners, see Prosper Jules Charbonnier, *Essais sur l'histoire de la ballistique* (Paris: Société d'Editions Géographiques Maritimes et Coloniales, 1928), 244. On the limits of mathematics, see S.H.A.T. 2a60/4 Gomer, "Mémoire sur la marche de l'instruction," 28 February 1783; Anon., "Remarques critiques," 1 February 1784.

46. S.H.A.T. 2b5/1 Guibert [to Dubois], "Projet d'un nouveau cours de mathématiques," 25 November 1761. Also see S.H.A.T. 2b5/1 Bron, "Mémoires sur l'instruction," May 1766. S.H.A.T. 2a59 "Ecole d'artillerie de Metz," October 1767.

47. S.H.A.T. 2b5/1 "Circulaire à MM. de Mouy, Gribeauval," November 1765.

48. Etienne Bézout, *Cours de mathématiques à l'usage du Corps Royal de l'Artillerie* (Paris: Pierres, [1772], 1781). S.H.A.T. 2b5/1 [Du Puget], "Tavail de l'officier," 1765.

49. On various approaches to mathematics in engineering education in this period, see Antoine Picon, "Les ingénieurs et la mathématisation: L'example du génie civil et de la construction," *Revue d'histoire des sciences* 52 (1989): 155–71.

50. The standard view of eighteenth-century physics as a gloss on Newton is reaffirmed in Kuhn's model, which implies that eighteenth-century "normal" science was devoted to solving practical problems within the Newtonian paradigm. More properly, the form of rational mechanics labeled "Newtonian" science was largely a product of the eighteenth-century Continental analysts. John L. Greenberg, "Mathematical Physics in Eighteenth-Century France," *Isis* 77 (1986): 59–78. Clifford Truesdell, *Essays in the History of Mechanics* (Berlin: Springer-Verlag, 1968), 85–183. Also see the historiographical articles in Roy Porter, ed., *The Ferment of Knowledge: Studies in the Historiography of Eighteenth-Century Science* (Cambridge: Cambridge University Press, 1980), especially, Simon Schaffer, "Natural Philosophy," 55–91; H.J.M. Bos, "Mathematics and Rational Mechanics," 327–55; and J. L. Heilbron, "Experimental Natural Philosophy," 357–87.

51. Bonaiuto Lorini, *Delle fortificazioni*, in Rossi, *Philosophy*, 61–62.

52. On mixed mathematics in the eighteenth century, see Jean Etienne Montucla, *Histoire des mathématiques* (Paris: Blanchard, [1799], 1960).

53. On descriptionism in eighteenth-century physics, see J. L. Heilbron, *Weighing Imponderables and Other Quantitative Science around 1800*, Historical Studies in the Physical and Biological Sciences, Supplement to vol. 24 (Berkeley: University of California Press, 1993), 141–46. On guns and water wheels, see chapter 3.

54. d'Alembert, *Résistance des fluides*, x, xxviii–xxix, xli–xlii. For general complaints about the superabundance of calculations, see Heilbron, *Weighing Imponderables*, 29–31.

55. Vincenti, *What Engineers Know*, 138–39, 160–62.

56. For the growing emphasis on analysis in instruction, see L.W.B. Brockliss, *French Higher Education in the Seventeenth and Eighteenth Centuries: A Cultural History* (Oxford: Clarendon Press, 1987), 337–90. On Laplace, see Denis I. Duveen and Roger Hahn, "La-

place's Succession to Bézout's Post of Examinateur des Elèves de l'Artillerie," *Isis* 48 (1957): 423. On Laplace's use of his position as artillery examiner in the post-Revolutionary period, see chapter 8. Legrendre and Lacroix, two prominent analysts, also served as artillery examiners; see Charbonnier, *Ballistique*, 244.

57. Euler in Jean-Louis Lombard, ed. and trans., *Nouveaux principes d'artillerie, commentés par Léonard Euler* by Benjamin Robins (Dijon: Frantin, 1783), 85. On Condillac, see Robin Rider, "Measure of Ideas, Rules of Language: Mathematics and Language in the 18th Century," in *The Quantifying Spirit in the Eighteenth Century*, eds. Tore Frängsmyr, J. L. Heilbron, and R. Rider (Berkeley: University of California Press, 1990), 113–40. Bos notes our dearth of knowledge about mathematical training in the engineering schools in Bos, "Rational Mechanics."

58. S.H.A.T. 2b5/1 "Circulaire à MM. de Mouy, Gribeauval," November 1765. Du Puget, letter, 21 September 1765. Carnot, "Réflexions sur la métaphysique du calcul infinitésimal," 1797, in Lazare Carnot, *Révolution et mathématique*, ed. Jean-Paul Charnay (Paris: L'Herne, 1985), 2: 457–58. Ironically, his work of 1785 on machines returned to a geometric mode of explanation; see Charles C. Gillispie and A. P. Youschklevitch, *Lazare Carnot, Savant: A Monograph Treating Carnot's Scientific Work* (Princeton: Princeton University Press, 1971), 121–23, 139–43. Picon, *Ingénieur moderne*, 302–7, 622. Norton Wise sees analysis as essentially static, but he too highlights the importance of the subject to Enlightenment thinking. Wise, "Meditations," 227–37, 239–40.

59. Bélidor, *La science des ingénieurs*, 1: 3–4. On optimization in the theory of machines (such as water wheels), see Jean-Pierre Séris, *Machine et communication: Du théâtre des machines à la mécanique industrielle* (Paris: Vrin, 1987), 283–341, 456–57. This form of economic analysis was tentatively performed by Ponts et Chaussées engineers in the eighteenth century; see François Etner, *Histoire du calcul économique en France* (Paris: Economica, 1987), 47–66. Porter shows how engineers deployed economic analysis in different cultural contexts in the nineteenth and twentieth centuries; see T. Porter, *Trust in Numbers*, 114–46.

60. S.H.A.T. 2a59 Le Pelletier, "Instructions qui seront données sur le dessein à l'Ecole d'Artillerie de Metz," 1749. "Ecole de l'artillerie de Metz," October 1767. *Ordonnance du roi concernant le Corps Royal de l'Artillerie*, 3 October 1774, pp. 109–10. Even the new geometry was most successful when coupled with analysis, as in Monge's descriptive geometry. Lorraine J. Daston, "The Physicalist Tradition in Early Nineteenth-Century French Geometry," *Studies in the History and Philosophy of Science* 17 (1986): 269–95.

61. Architects were concurrently taking this route to professional differentiation, also using a stratified system of drawing education. Jacques-François Blondel, *Discours sur la nécessité de l'étude de l'architecture, cinquième cours public* (Paris: Jombert, 1754), 15, 79–82, 98–99. For a historian's (dis)analogies between architects and engineers in this period, see Antoine Picon, *Architectes et ingénieurs au siècle des lumières* (Marseilles: Parenthèses, 1988).

62. S.H.A.T. 2a59 Joseph Du Teil, "Salle de dessin," 1786. For a text on shadows used by students at the military engineering school at Mézières in the late ancien régime, see [Monge?], "Traité des ombres dans le dessin géométral," [1774], in Théodore Olivier, ed., *Applications de la géométrie descriptive* (Paris: Carillian-Goeury, 1847), 6–8. For a text from the Revolutionary period, see Gaspard Monge, "Stéréotomie," *Journal de l'Ecole Polytechnique* 1 (year III [1795]): 9. Today, manuals still warn students of the difficulties of "reading" technical drawings. W. Abbott, *Technical Drawing* (London: Blackie, 1962), 14. For a general history of technical drawing and its cognitive bearing on engineering, see Eugene Ferguson, *Engineering and the Mind's Eye* (Cambridge: MIT Press, 1992), 87–96. Auslander points out that while artisans produced shaded drawings for potential customers, they did not shade the private drawings they used in production. Leora Auslander, "The Creation of

Value and the Production of Good Taste: The Social Life of the Furniture Trade in Paris, 1860–1914," (Ph.D. diss., Brown University, 1990), 385–90.

63. S.H.A.T. 2a59 Saint-Auban, "Instruction," 25 October 1765.

64. Quote in *Ordonnance du roi concernant le Corps Royal de l'Artillerie*, 3 October 1774, pp. 111–12. S.H.A.T. 2a59 "Ecole d'artillerie de Metz," October 1767.

65. S.H.A.T. MR1744 Ecole de Besançon, "Etat des objets qui ont été traités," 1789. See also Hahn, "Ecoles militaires," 531, 539–41. On the chemistry lab, see Joseph Du Teil, "Mémoire," 19 September 1781, in Joseph Du Teil, *Napoléon*, 220–21, and S.H.A.T. 2a59 "Règlement concernant le Service du Corps de l'Artillerie dans les écoles," April 1792. Woronoff doubts that the French metals industry saw significant improvements in this period. Denis Woronoff, *L'industrie sidérurgique en France pendant la Révolution et l'Empire* (Paris: Editions de l'Ecole des Hautes Etudes en Sciences Sociales, 1984).

66. [Bachaumont], *Mémoires secrets*, 23 October 1769, vol. 4; 1 December 1769, 20 November 1770, vol. 5.

67. See the connections between Jars and Wendel and the artillery corps in Charles C. Gillispie, *Science and Polity at the End of the Old Regime* (Princeton: Princeton University Press, 1980), 435–38. Laurent Versini, "Un maître de forges au service des Lumières et de la Révolution américaine: François Ignace de Wendel, 1741–1795," *Studies on Voltaire and the Eighteenth Century* 155 (1976): 2137–2206. For a general survey of the state's role in research, see Patrice Bret, "Les origines de l'institutionnalisation de la recherche militaire en France (1775–1825)," in Colloque International d'histoire militaire, *L'influence de la Révolution française sur les armées en France, en Europe, et dans le monde*, Actes no. 15 (Vincennes: Commission Française d'Histoire Militaire, 1991), 345–62.

68. Gilbert Bodinier, *Les officiers combattants de la Guerre d'Indépendance des Etats Unis de Yorktown à l'an II* (Vincennes: S.H.A.T., 1983), 97–99, 104.

69. Du Coudray, "Mémoire" to U.S. Congress, quoted in André Lasseray, *Les français sous les treize étoiles, 1775–1783* (Mâcon: Protat Frères, 1935), 86–107, 444–54. For the tempestuous career of Du Coudray, which came to an end in the depths of the Schuylkill River near Philadelphia, see Ken Alder, "Forging the New Order: French Mass Production and the Language of the Machine Age, 1763–1815" (Ph.D. diss., Harvard University, 1991), 132–34.

70. S.H.A.T. Yb668-Yb670 *Registres*, 1763–1789. For the general army, see Bien, "Réaction aristocratique," 30–35. The Ségur Law of 1781 reduced the number of commoners entering the artillery to 5 percent (the law still permitted entry into the artillery for anyone with a brother, father, or grandfather in the corps). No commoners were admitted into the line army at this time.

71. S.H.A.T. Yb668-Yb670 *Registres*, 1763–1789. The scribe only listed the occupations of half the fathers. However, my numbers are corroborated by those found by Bodinier for the American expedition, where 29 percent of artillerists had noble fathers in civilian offices. Bodinier, *Officiers combattants*, 98–99.

72. When one historian speculated that "half" the artillerists came from "the bourgeoisie," this is the group he must have had in mind. Albert Babeau, *La vie militaire sous l'ancien régime: Les officiers* (Paris: Firmin-Didot, 1890), 2: 76–77, 88. See Louis Bergeron and Guy Chaussinand-Nogaret, *Les "Masses de granit:" Cent mille notables du Premier Empire* (Paris: Editions de l'Ecole des Hautes Etudes en Sciences Sociales, 1979), 33–36, 63–64. The classic account of flux in social categories in this period is Cobban, *Social Interpretation*. See also Sarah Maza, "Luxury, Morality, and Social Change: Why There was No Middle Class Consciousness in Prerevolutionary France," unpublished typescript, 1995.

73. S.H.A.T. MR1744 [Grenoble], "Mémoire," [1777]. Lucas, "Nobles, Bourgeois." The same pattern of recruitment can be found in the Corps du Génie: before 1747, roughly half were commoners; the number fell to 43 percent in 1748–1777; and to 25 percent in 1778–

1791. The decrease was filled by young men of recent nobility. Blanchard, *Ingénieurs*, 229–46. See also Roger Chartier, "Un recruitment scolaire au XVIIIe siècle, l'Ecole Royale du Génie à Mézières," *Revue d'histoire moderne et contemporaine* 20 (1973): 353–75.

74. S.H.A.T. Yb668-Yb670 *Registres*, 1763–1789. My data are corroborated by Bodinier, who found that 11 percent of the artillery officers in the French expeditionary force of 1781 had artillerist fathers. Bodinier, *Officiers combattants*, 98. Some 20 percent of the Génie also had a father in that service. Blanchard, *Ingénieurs*, 238–43.

75. S.H.A.T. MR1790 Anon., "Projet d'une nouvelle constitution du Corps de l'Artillerie," [1788]. Bien, "Military Education." Chaussinand-Nogaret et al., *Elites*. Motley, *Becoming a French Aristocrat*.

76. On promotions, see Rosen, "Gribeauval," 61–62; Le Pelletier, *Une famille d'artilleurs*, 57–58, 118, 135. For two evaluations of the pupil Schoderlos [*sic*] de Laclos, see S.H.A.T. Xd248 Camus, "Etat des étudiants," 15 January 1760; and "Mémoire," 31 January 1761. Laclos was forced out of the service in 1788, not because of *Les liaisons dangereuses*, but because he questioned Vauban's contribution to the security of France. Georges Poisson, *Choderlos de Laclos, ou l'obstination* (Paris: Grasset, 1985), 174–86, 206.

77. *Ordonnance du roi concernant le Corps Royal de l'Artillerie*, 3 October 1774, p. 15. This amplified the laws of 1763. Such a system had operated for artillery NCOs since 1759. Picard and Jouan, *Artillerie*, 18–19, 35–37. The regular army adopted a modified version of this method for its own NCOs in 1764. Jean Chagniot, in *Histoire militaire de la France*, ed. André Corvisier, Anne Blanchard et al. (Paris: Presses Universitaires de France, 1992–94), 2: 44.

78. None of the leading historians of the late-eighteenth-century French army—Bien, Bertaud, Forrest, Scott, or Lynn—mention these electoral institutions of the ancien régime artillery. The Revolutionary army of 1793–95 briefly allowed troops to nominate three of their comrades for a vacancy in the NCOs, from whom officers of superior rank made a final selection. Superior officers were selected by those still higher in the chain of command. The elections in the army were curtailed by the Thermidoreans, and abolished altogether by Napoléon in 1805. See Scott, *The Response of the Royal Army*, 7–8; Jean-Paul Bertaud, *The Army of the French Revolution: From Citizen-Soldiers to Instrument of Power*, trans. R. R. Palmer (Princeton: Princeton University Press, 1988), 88–89, 171–75; and chapter 3 below.

79. Scheel, *Mémoires d'artillerie*, 438.

80. Max Weber has remarked on this "guild-like" closure of officialdom in such cases. Weber, "Bureaucracy," in *From Max Weber*, 200. Durkheim, *Professional Ethics*, actually suggests this corporate model be reimposed on otherwise "unorganized" economic activities. On credentialing as a form of social monopolization, see Eliot Freidson, *Professional Powers: A Study of the Institutionalization of Formal Knowledge* (Chicago: University of Chicago Press, 1986), 63–91. In certain respects this system resembles the tenure and promotion methods still used in academia.

81. S.H.A.T. 2a60/4 Gomer, "Constitution du Corps Royal," 1761.

82. Gilbert Bodinier, "Les officiers de l'armée royale et la Révolution," in *Le métier militaire en France aux époques de grandes transformations sociales*, ed. André Corvisier (Vincennes: S.H.A.T., 1980), 61.

83. Scheel, *Mémoires d'artillerie*, 429–30. Saint-Auban, *Nouveaux systèmes d'artillerie*, 101–2.

84. *Ordonnance du roi concernant le Corps Royal de l'Artillerie*, 3 October 1774.

85. *Instructions que le roi a fait expédier aux Inspecteurs-Généraux du Corps Royal de l'Artillerie*, 8 August 1771.

86. Scheel, *Mémoires d'artillerie*, 439. Saint-Auban had been born a *roturier* and had been ennobled for his military service only in 1739. S.H.A.T. 1a2/2 Gribeauval, "Réponses aux objections," May 1765.

87. S.H.A.T. 2a59 Le Pelletier, No title, August 1763.

88. S.H.A.T. 2a59 Saint-Auban, "Instruction," 25 October 1765.

89. S.H.A.T. 2b5/1 "Circulaire à MM. Mouy, Gribeauval," November 1765.

90. S.H.A.T. 1a2/2 Gribeauval, "Importance du chef de brigade," [1765]. The duties of the rank had already been formulated by War Office bureau chief Dubois; see S.H.A.T. 1a2/2 Dubois, "Mémoire sur l'artillerie," [1762]. Saint-Auban objected to the new rank on the grounds that it produced favoritism, see Saint-Auban, *Nouveaux systémes d'artillerie*, 106. Similarly, the routinization of industrial work in the late nineteenth century went hand in hand with the rationalization of managerial work. Shoshana Zuboff, *In the Age of the Smart Machine: The Future of Work and Power* (New York: Basic Books, 1988).

91. This is the earliest secular-bureaucratic usage of "hiérarchie" I have found. S.H.A.T. 2a60/4 Anon., "Remarques critiques," 1 February 1784. In the late 1780s the term became more widely used by military men. See S.H.A.T. Ya269 M. Thomé, "Mémoire," 18 August 1788. For the classic use of the term by a social historian, see Roland Mousnier, *Les hiérarchies sociales de 1450 à nos jours* (Paris: Presses Universitaires de France, 1969).

92. [Conseil de Guerre], *Ordonnance du roi portant règlement sur la hiérarchie de tous les emplois militaires*, 17 March 1788 (Paris: Imprimerie Royale, 1788). On the reaction of the salons, see [Servan], "Hiérarchie," in *Art militaire*, in *Encyclopédie méthodique* (Paris, 1797), vol. 4. All the standard French-usage dictionaries locate the first military usage in the post-Napoleonic era. On this, and Necker's usage, see Necker, "Administration financière," [1791], in Le Robert, "Hiérarchie," *Dictionnaire de la langue française*.

93. Lacuée, "Cadet," *Art militaire*, in *Encyclopédie méthodique*, 4: 97–99. Emphasis his. Lacuée, "Examen," *Art militaire*, in *Encyclopédie méthodique*, 4: 315–20.

94. Harold Perkin, *The Third Revolution: Professional Elites in the Modern World* (London: Routledge, 1996), chap. 4. Ezra Suleiman, *Politics, Power and Bureaucracy in France: The Administrative Elite* (Princeton: Princeton University Press, 1974).

95. C. Jones, "Bourgeois Revolution," 95–111.

96. The scholarship on the "revolutionary" role of the ancien régime legal profession—especially in the aftermath of the Maupeou crisis—is reviewed in Sarah Maza, *Private Lives and Public Affairs*, 86–97, quote by François Chavray de Boissy, p. 95.

97. Matti Lauerma, *L'artillerie de campagne française pendant les guerres de la Révolution: Evolution de l'organisation et de la tactique*, Annales Academiae Scientiarum Fennicae, Series B, vol. 96 (Helsinki: 1956), 93, 104–5. Despite the desertion of several Gribeauvalist reformers to the émigré camp, artillerists were less likely than their counterparts in the infantry or cavalry to serve there: 31 percent versus 54 percent. Bodinier, *Officiers combattants*, 500. On the retention of artillerists, and the comparable rates among the Corps du Génie and the Géographes, see S.H.A.T. 1a5/9 "Etat des officiers d'artillerie, Extraits de l'annuaire militaire," [1792]. Also, Bodinier, *Officiers combattants*, 500, 510, 513, 516; and Bodinier, "Officiers de l'armée," 64, 68, 71. While John Lynn discounts the attachment of the artillerists to the new order, their *relative* loyalty seems telling. John A. Lynn, "En avant! The Origins of the Revolutionary Attack," in *Tools of War: Instruments, Ideas and Institutions of Warfare, 1445–1871*, ed. John A. Lynn (Urbana: University of Illinois Press, 1990), 173–74; *The Bayonets of the Republic: Motivation and Tactics in the Army of Revolutionary France, 1791–94* (Urbana: University of Illinois Press, 1984), 207–12.

98. Officers of fortune in the artillery were more likely to stay than comparable officers in the regular army: 63 percent versus 48 percent. Bodinier, *Officiers combattants*, 506. The artillery troops themselves were largely in favor of the Revolution—according to Samuel Scott because of their peasant origins, better pay, extended drill, and hence, their greater esprit de corps. Samuel Scott, "The Regeneration of the Line Army during the French Revolution," *Journal of Modern History* 42 (1970): 310, 317–18, 320, 324, 328. A.N. AF III 153 "Etat par ordre d'ancienneté," nivôse, year V [January 1791]. The names here have checked against S.H.A.T. Yb668-Yb670 *Registres*, 1763–1789; and A.N. AF II 293c *Re-*

gistres, year III [1795]. For comparison with the line army, see Bertaud, *Army of the French Revolution*, 184.

99. After the defection of Dumouriez, Aboville swore a special oath to the Republic. *Moniteur*, 21 April 1793, 16: 17–18. Bertaud, *Army of the French Revolution*, 187. See chapters 8 and 9.

100. On the broader pattern of military professionalism across the modern period, see André Corvisier, "Rapport de synthèse," *Le métier militaire*, ed. Corvisier, 8. For a study of military professionalization in the post-Revolutionary period, see Howard G. Brown, "Politics, Professionalism, and the Fate of the Army Generals after Thermidor," *French Historical Studies* 19 (1995): 133–52.

101. Antoine-Augustin-Flavien Pion des Loches, *Mes campagnes, 1792–1815: Notes et correspondance du colonel d'artillerie* (Paris: Firmin-Didot, 1889), 31.

102. Pion des Loches, *Mes campagnes*, 32–34.

103. Jean-Louis Lombard, *Traité du mouvement des projectiles* (Dijon: Frantin, year V [1796–97]), xii–xiii.

Chapter Three
Design and Deployment

1. For a thoughtful analysis along similar lines, see Bucciarelli, *Designing Engineers*. He calls the applied-science fallacy the savant's "object world," and the derived-society fallacy, the utilitarian's "marketplace."

2. Vincenti, *What Engineers Know*, 6–12. Also Walter G. Vincenti, "Engineering Knowledge, Type of Design, and Level of Hierarchy: Further Thoughts about *What Engineers Know*," in *Technological Development and Science in the Industrial Age*, eds. Peter Kroes and Martijn Bakker (Dordrecht: Kluwer, 1992), 17–34.

3. Even today, these two sciences may be said to differ less "in kind" than "in degree." For ballistics, see MacKenzie, *Inventing Accuracy*. As for military science, poor as its predictive power may be, it also purports to offer game-theoretical calculations of outcomes.

4. On the theme of extending the laboratory to the world, see Bruno Latour, *The Pasteurization of France*, trans. Alan Sheridan and John Law (Cambridge: Harvard University Press, 1988). On the idea of normalizing a part of experience to make it more amenable to quantification, see T. Porter, *Trust in Numbers*.

5. [Le Blond], "Artillerie," 1: 617. Gaston Bachelard, *Le nouvel esprit scientifique* (Paris: Presses Universitaires de France, 1949), 12–14.

6. Niccolò Tartaglia, *La nova scientia* (Venice, 1537). This theory of ballistics operated within an Aristotelian metaphysics, though its mechanics owed something to the impetus theory of the Renaissance. See Alexandre Koyré, "La dynamique de Niccolò Tartaglia," *La science au seizième siècle, Colloque international de Royaumont* (Paris: Hermann, 1960), 91–116. In his later works, Tartaglia claimed the ball curved continuously toward the earth, though this deflection might initially be undetectable. This new theory became the basis of the "point blank" method of aiming guns. Niccolò Tartaglia, *Three bookes of colloquies concerning the arte of shooting*, trans. Cyprian Lucar (London: Harrison, 1588), 11–14. Other artillery theorists elaborated on this scheme over the next century, deriving range tables. Diego Ufano, *Artillerie, c'est à dire vraie instruction de l'artillerie*, trans. Théodore de Brye ([Zutphen]: Aelst, 1621), 54–55. Francis Malthus, *Pratique de la guerre concernant l'usage de l'artillerie, bombes, et mortiers* (Paris: Guillemont, 1646).

7. Galileo Galilei, *Two New Sciences*, trans. Stillman Drake (Madison: University of Wisconsin Press, 1974).

8. A. Rupert Hall, *Ballistics in the Seventeenth Century: A Study in the Relations of Science and War with Reference Principally to England* (Cambridge: Cambridge University Press,

1952); and "Gunnery, Science, and the Royal Society," in *The Uses of Science in the Age of Newton*, ed. John G. Burke (Berkeley: University of California Press, 1983), 111–41. Hall's interpretation has been recently underscored again by Robert Westfall, "Science and Technology during the Scientific Revolution: An Empirical Approach," *Renaissance and Revolution: Humanists, Scholars, Craftsmen, and Natural Philosophers in Early Modern Europe*, eds. J. V. Field and Frank A.J.L. James (Cambridge: Cambridge University Press, 1993), 63–72.

9. See Hall, *Ballistics*, 158–65. Hall hints at a broader understanding of the "science" of early modern artillery when he discusses the rational basis of craft skill (p. 58). Boris Hessen, *The Social and Economic Roots of Newton's "Principia"* (New York: Howard Fertig, 1971). Robert K. Merton, *Science, Technology and Society in Seventeenth-Century England* (New York: Howard Fertig, [1938], 1970). For evidence of a practical "science" among seventeenth-century British artillerists, see Frances Willmouth, "Mathematical Sciences and Military Technology: The Ordnance Office in the Reign of Charles II," in *Renaissance and Revolution*, eds. Field and James, 117–31.

10. Brett D. Steele, "Muskets and Pendulums: Benjamin Robins, Leonard Euler, and the Ballistics Revolution," *Technology and Culture* 35 (1994): 348–82. A central theme of Steele's tale of Robins's neglected genius is that his discoveries were ignored for a hundred years. Steele uses "admittedly anachronistic" terms, and explicitly evokes the "lag" model. "Muskets," 381, 365. See also Steele, "The Ballistics Revolution: Military and Scientific Change from Robins to Napoléon" (Ph.D. diss., University of Minnesota, 1994).

11. Ronald Kline, "Constructing 'Technology' as 'Applied Science': Public Rhetoric of Scientists and Engineers in the U.S., 1880–1945," *Isis* 86 (1995): 194–221. Stuart W. Leslie, *The Cold War and American Science: The Military-Industrial-Academic Complex at MIT and Stanford* (New York: Columbia University Press, 1993).

12. François Blondel, *L'art de jetter les bombes* (La Haye: Arnout, 1685), 510–12.

13. Deschiens de Ressons, "Méthode pour tirer les bombes avec succès," *Mémoires de l'Académie des Sciences* (1716): 79–86. Ressons's procedures hardly differed from those of the sixteenth century. John Francis Guilmartin, *Gunpowder and Galley: Changing Technology and Mediterranean Warfare at Sea in the Sixteenth Century* (Cambridge: Cambridge University Press, 1974), 157–75.

14. Saint-Rémy, *Mémoires*, 2: 250–51; 3: 154–55.

15. Malthus, *Pratique*, 144–45.

16. Saint-Rémy, *Mémoires*, 1:115–16. Le Blond, *Artillerie raisonnée*, 107–8. If the breech and muzzle did not have the same diameter, or the lie of the bore was not "true," artillerists built a brass rod, called a "dispart" to compensate. [Henri Gauthier], *Instruction pour les gens de guerre* (Paris: Cognard, 1692), 27–28. Robert Norton, *The Gunner, Shewing the Whole Practice of Artillerie* ([London]: Robinson, 1628), 80–84.

17. [Le Blond], "Artillerie," 1: 617. Malthus, *Pratique*, 72–77. More complex algorithms can be found in Ufano, *Artillerie*, 133–35; F. Blondel, *Art de jetter* (1685), 31–38; and Thomas Binning, *A Light to the Art of Gunnery* (London: Darby, 1676), 114–22. On comparable English practices, see David McConnell, *British Smooth-Bore Artillery: A Technological Study* (Ottawa: Minister of the Environment, 1988), 375.

18. Although many gunners believed accuracy at such a range was illusory, handbooks purported to show artillery men how to hit targets *à tout volonté* (or what the English called, "at random"). Saint-Rémy, *Mémoires*, 1: 115. For a hybrid Aristotelian-Galilean physics coupled to an experimental program, see Robert Anderson, *To Hit a Mark* (London: Morden, 1690).

19. For Tartaglia's twelve-point scale, see Tartaglia, *Three*, 1–5. The sinusoidal scale meant that doubling the measure from mark two to mark four would theoretically double the range of the shot. F. Blondel, *Art de jetter* (1685), 96–112, 265–69. Saint-Rémy, *Mémoires*, 3: 58–60.

20. Vallière, *fils*, in Scheel, *Mémoires d'artillerie*, 19–20, 331, 336.

21. S.H.A.T. 2a60/4 [Joseph Du Teil], "Projet d'instruction pour l'Ecole d'Auxonne," 1 April 1780. One British artillery expert argued that the quadrant was of no use even to train the eye, since all adjustments are made by guesswork. John Muller, *A Treatise of Artillery* (London: Millan, 1768), 147–56.

22. Frederick II, *Instruction secrètte* [*sic*], trans. Le Prince de Ligne (Brussels: Hayez, 1787), 120–22. Guibert, *Essai général*, 2: 244–46. Edme Jean Antoine Du Puget, *Essai sur l'usage de l'artillerie* (Amsterdam: Arkstee et Merkus, 1771), 47, 53.

23. *Ordonnance* of 16 October 1732, in Saint-Rémy, *Mémoires*, 3: 450–62. Scheel, *Mémoires d'artillerie*, 251–52.

24. Du Puget, *Usage de l'artillerie*, 29–30. S.H.A.T. Bib. #72227 Anon., *Procès-verbal des épreuves faites aux Ecoles d'Artillerie de Douai* (Amsterdam: Arkstee et Merkus, 1772), 25–26. [Le Blond], "Artillery" and "Canon," 1: 617; 2: 204–6. See plates in "Art Militaire," *Supplément à l'Encyclopédie*, pl. 1, no. 2.

25. See table in [Le Blond], "Canon," 2: 203. Du Coudray, *L'artillerie nouvelle*, 15–16.

26. Philippe-Charles-Jean-Baptiste Tronson Du Coudray, *L'état actuel de la querelle sur l'artillerie* (n.p., 1774), 60. Scheel, *Mémoires d'artillerie*, 331–47. B. P. Hughes emphasizes these developments. Basil Perronet Hughes, *Firepower: Weapons Effectiveness on the Battlefield, 1630–1850* (New York: Scribner's, 1974), 18. The British did not begin to introduce these devices until around 1780, and not definitively until the late 1790s. McConnell, *British Smooth-Bore*, 213, 376–80.

27. Du Coudray, *L'artillerie nouvelle*, 45–50, 209–11. A.N. T591(6) Anon., "Dissertation sur la hausse," n.d. For those who assume the *hausse* required the use of mathematics, see Edme Jean Antoine Du Puget, *Réflexions sur la pratique raisonnée du pointement des pièces de canon* (Amsterdam: Arkstee et Merkus, 1771), 46–49; and Rosen, "Gribeauval," 35, 80–82.

28. On tacit knowledge, the classic works are Michael Polanyi, *Personal Knowledge* (London: Routledge and Kegan Paul, 1958); and *The Tacit Dimension* (London: Routledge and Kegan Paul, 1967). More recently, see Donald MacKenzie and Graham Spinardi, "Tacit Knowledge, Weapons Design and the Uninvention of Nuclear Weapons," *American Journal of Sociology* 101 (1995): 44–99. For more on artisanal knowledge, see chapters 4 and 5, below.

29. Louis Tousard, *American Artillerist's Companion, or Elements of Artillery* (Philadelphia: Concord, 1809), 2: x.

30. Bernard Forest de Bélidor, *Nouveau cours de mathématique à l'usage de l'artillerie et du génie* (Paris: Jombert, 1725), 381–87. See also Bélidor, "Mémoire sur la longueur que doivent avoir les cannons par rapport à leur calibre," in *Oeuvres divers*, 74, 79. For his tables, see Bélidor, *Le bombardier français, ou nouvelle méthode de jeter les bombes avec précision* (Paris: Imprimerie Royale, 1731). On seventeenth-century tests by the English, see Hall, "Gunnery." The innovative seventeenth-century French artillerist Frézeau de la Frézelière was later accused of proceeding without "reasoning" by [Gauthier], *Instruction*, 99–100, 121–29. On ancient catapult design, see Barton Hacker, "Greek Catapults and Catapult Technology: Science, Technology and War in the Ancient World," *Technology and Culture* 9 (1968): 34–54.

31. Jean-Florent de Vallière, *père*, *Mémoire sur les charges et les portées des bouches à feu* (Paris: Imprimerie Royale, 1741). Anon., "Preface," in Bélidor, *Oeuvres divers*. "Eloge de Bélidor," in *Histoire de l'Académie des Sciences* (1761): 167–81.

32. My analysis has benefited greatly from discussions of contemporary water-wheel engineering. See Terry Reynolds, *Stronger than a Thousand Men: A History of the Vertical Water Wheel* (Baltimore: Johns Hopkins University Press, 1983). Edwin Layton, "Escape from the Jail of Shape: Dimensionality and Engineering Science," in *Technological Development*, eds. Kroes and Bakker, 35–68; "Millwrights and Engineers: Science, Social Roles, and the Evo-

lution of the Turbine in America," in *The Dynamics of Science and Technology: Social Values, Technical Norms and Scientific Criteria in the Development of Knowledge*, eds. Wolfgang Krohn, Edwin Layton, and Peter Weingart (Dordrecht: Reidel, 1978), 61–87. Svante Lindqvist, "Labs in the Woods: The Quantification of Technology During the Late Enlightenment," in *Quantifying Spirit*, eds. Frängsmyr, Heilbron, and Rider, 291–314. And Bruno Belhoste et al., *Le moteur hydraulique en France au XIXe siècle: Concepteurs, inventeurs et constructeurs* (Nantes: Université de Nantes, 1990).

33. There is a large scholarly literature on the "problem" of experiment. For technological tests, the best work is MacKenzie, *Inventing Accuracy*, 340–81. On the social work necessary to reproduce scientific results in the early modern period, see Shapin and Schaffer, *Leviathan*. On the gentleman's laboratory, see Steven Shapin, "The House of Experiment in Seventeenth-Century England," *Isis* 79 (1988): 373–404. On gentlemanly natural philosophers and their invisible technical assistants, see Steven Shapin, *Social History of Truth*. For the importance of public experimentation in the eighteenth century, see Jan Golinski, *Science as a Public Culture: Chemistry and Enlightenment in Britain, 1760–1820* (Cambridge: Cambridge University Press, 1992). For science more generally, the most searching critical work has been done by H. M. Collins, "The Seven Sexes: A Study in the Sociology of a Phenomenon, or the Replication of Experiments in Physics," *Sociology* 9 (1975): 205–24.

34. In France, a public audience for science was first cultivated by Fontenelle. For the political undercurrents of the public debate over a controversial science on the eve of the French Revolution, see Robert Darnton, *Mesmerism and the End of the Enlightenment in France* (Cambridge: Harvard University Press, 1968).

35. Choiseul to Gribeauval, 29 April 1764, in Nardin, *Gribeauval*, 125.

36. S.H.A.T. 4b25(b) "Epreuves de 1764." For a summary of the Strasbourg results, see S.H.A.T. A1 3761 Mouy, Gribeauval et al., "Mémoire sur différens sujets concernant l'artillerie," 19 August 1766. For Gribeauval's private account of events, see S.H.A.T. 9a11 Gribeauval to "Cher ami," 18 July 1764.

37. Du Coudray, *L'artillerie nouvelle*, 12–13.

38. Saint-Auban, quoted in Thomas du Morey and Chev. de Frazan, "Rapport," *Journal de physique* 16 (1780): 128.

39. Anon., *Epreuves de Douai*. [Le Blond], "Cannon," 1: 205. Guichon de Grandpont, "Querelle," 27, 54.

40. [Gribeauval and Vallière, *fils*], *Mémoires authentiques*, vi–xii, 4–7; quote by Saint-Auban on p. xi. Du Coudray, *L'artillerie nouvelle*, 105–6. Tests for the American Strategic Defense Initiative were similarly rigged in the 1980s. "Lies and Rigged Star Wars Test Fooled the Kremlin and Congress," *The New York Times*, 18 August 1993, p. 1.

41. Guibert, *Essai général*, 1: 447–49.

42. S.H.A.T. 2a60/4 Anon., "Mémoire sur les expériences projetées," 1766. Nardin, *Gribeauval*, 172–73.

43. Jean-Florent de Vallière, *père*, "Mémoire sur les charges et les portées des bouches à feu," in *Artillerie polémique de 1772*, ed. Joseph-Florent de Vallière, *fils* (Aléthopolis: Neumann, 1772), 116–17.

44. Vallière, *père*, quoted in [Le Blond], "Cannon," 2: 204. For a similar argument, see Du Puget, *Pratique raisonnée*, 34–42.

45. Vallière, *fils*, in [Gribeauval and Vallière, *fils*], *Mémoires authentiques*, 103. Vallière, *père*, *Sur les charges et les portées*, 41.

46. For the manuscript translations of Robins undertaken by Jean-Baptiste Le Roy and Lombard around 1750, see Lombard, *Nouveaux principes*, quote on p. iii. On the work of Patrick d'Arcy, see below. Leonard Euler, "Recherches sur la véritable courbe que décrivent les corps jetés dans l'air," *Mémoires de l'Académie de Berlin* (1753): 321–55. Jean-Charles de Borda, "Sur la courbe décrite par les boulets," *Mémoire de l'Académie des Sciences* (1769): 247–71. Benjamin Robins, *Nouveaux principes d'artillerie*, trans. Du Puy (Grenoble: Durand,

1771). In the 1770s, the publisher Panckoucke stood ready to print two hundred copies of any new translation, see Condorcet to Turgot, July, 17 August 1774; and Turgot to Euler, n.d., in Jean-Antoine-Nicolas de Caritat de Condorcet, *Correspondance inédite de Condorcet à Turgot*, ed. Charles Henry (Paris: Perrin, n.d.), 178–80, 193–94, 245–47. The usual explanation for Robins's obscurity—Euler's "refutation" of his work—does not hold water. Euler disagreed with Robins on the reasons why rifling worked, but he seconded Robins's basic approach.

47. Vallière, *fils*, *La supériorité des pièces*, 13. Du Coudray, *L'artillerie nouvelle*, 103–4. Steele, "Ballistics Revolution," 209–10.

48. Benjamin Robins, *New Principles*, in Robins, *Mathematical Tracts*, ed. James Wilson (London: Nourse, [1742], 1761), 1: 38–52.

49. Robins, "Practical Maxims," in *Mathematical Tracts*, 1: 271.

50. Ibid., 1: 269.

51. Robins, *New Principles*, 1: 74–83, 91–101, 145–51.

52. Robins, "A Comparison of the Experimental Ranges of Cannon and Mortars with the Theory Contained in the Preceding Pages," in *Mathematical Tracts*, 1: 230–44. The promise was made in *New Principles*, 1: 53–54. The results of Hutton's experiments were appended to Lombard, *Nouveaux principes*, 508–31.

53. Euler kindly hypothesized that Robins's faulty assumptions canceled out. Euler, in Lombard, *Nouveaux principes*, 62–68, 92–102, quote on p. 107. For doubts about the practical usefulness of the new curves, see Du Puget, *Pratique raisonnée*, 34–35.

54. Robins always acknowledged the practical problem of using range as a parameter for gunners in the field. Robins, "Practical Maxims," 1: 276–77.

55. Like Bélidor and Robins, d'Arcy sought the optimal ratio of charge to barrel length, but did so by measuring the gun's recoil. Patrick d'Arcy, "Sur les effets de la poudre sur l'artillerie," *Histoire de l'Académie des Sciences* (1751): 1–14. Saint-Auban accused d'Arcy of being a stalking horse for Moor and Stark. S.H.A.T. Bib. #72215 Jacques Antoine Baratier de Saint-Auban, *Observations et expériences sur l'artillerie* (Aléthopolis: Neumann, [1750s]), 147–71. Jacques Antoine Baratier de Saint-Auban, "Lettres," *Mercure* (October 1752), 59; (June 1752), 6, 21–22, 45.

56. Patrick d'Arcy, "Lettres," *Mercure* (April 1752), 95; (June 1752), 49.

57. Patrick d'Arcy, *Essai d'une théorie d'artillerie* (Dresden: Walther, 1766), 8–10, 80–82. Some of his early tests in front of colleagues proved disastrous. *Histoire de l'Académie des Sciences* (1753): 70–72. On Robins's troubles with funding, see Steele, "Ballistics Revolution," 118.

58. Jacques François Louis Grobert, *Mémoire pour mesurer la vitesse initiale des mobiles de différens calibres* (Paris: n.p., 1804).

59. On "rules of practice," see Lombard, *Nouveaux principes*, i–ii. Tables can be found in Jean-Louis Lombard, *Tables du tir des canons et des obusiers* (Auxonne: n.p., 1787); for confessions of their inadequacy, see pp. 30–31. Subsequent experiments through the Revolutionary period did little to alleviate the limits of Lombard's approach. P. L. Villantroys, intro. and trans., *Nouvelles expériences d'artillerie* by Charles Hutton (Paris: Magimel, year X [1801–02]), x–xii.

60. Demands for accessible tables go back to S.H.A.T. 2b5/1 [Du Puget], "Travail de l'officier," 1765. Lombard, *Tables*, vi. The evidence suggests Napoleonic gunners seldom performed mathematical calculations in the field; and that the practical effect of the new technology was simply to increase the *rate* of fire. Hughes, *Firepower*, 126, 167–68.

61. T. Porter, *Trust in Numbers*.

62. [Le Blond], "Artillerie," 1: 606–7, 613.

63. Azar Gat, *The Origins of Military Thought from the Enlightenment to Clausewitz* (Oxford: Clarendon, 1989), 25.

64. Jacques-François de Chastenet de Puységur, *Art de la guerre par principes et par règles* (Paris: Jombert, 1749), 1: 3, 5–6.

65. Melissa Scott, "The Victory of the Ancients: Tactics, Technology, and the Use of Classical Precedent" (Ph.D. diss., Brandeis University, 1992). McNeill, *Pursuit of Power*, 128–33. For the classics as mental training for officers, see [Marc-René de Montalembert], *Essai sur l'intérêt des nations en général et de l'homme en particulier* (n.p., 1749).

66. Jean-Charles de Folard, *Nouvelles découvertes sur la guerre* (2d ed.; Brussels: Foppens, 1724). Folard's works were much republished in the eighteenth century.

67. Maurice de Saxe, *Mes rêveries, ou mémoires sur l'art de la guerre* (Amsterdam: Arkstee et Merkus, 1757), 1: 41–43.

68. François-Jean de Graindorge d'Orgeville de Mesnil-Durand, *Projet d'un ordre français en tactique* (Paris: Boudet, 1755), xiv, 104–5, 121. Brent Nosworthy, *The Anatomy of Victory: Battle Tactics, 1689–1763* (New York: Hippocrene, 1990), 329–41.

69. Guibert, *Observations*, 123–27. For a fascinating recent attempt to calculate the relative efficiencies of eighteenth-century instruments of war, see B. P. Hughes, *Firepower*. Data on men admitted to the Invalides in the first half of the eighteenth century suggest that 80 percent had been wounded by firearms. André Corvisier, *L'armée française de la fin du XVIIe siècle au ministère de Choiseul* (Paris: Presses Universitaires de France, 1964), 2: 674.

70. [Le Blond], "Artillerie," 1: 610, 619. Du Puget, *Usage de l'artillerie*, v–xxxviii. Saint-Auban had embraced Folard's claim that gunpowder was of little importance during his polemic with d'Arcy in the 1750s. Saint-Auban, "Lettre," *Mercure de France* (October 1752), 45–46.

71. Jacques Lambert Alphonse Colin, *L'infanterie au XVIIIe siècle: La tactique* (Paris: Berger-Levrault, 1907). Others who follow his view include Quimby, *Napoleonic Warfare*.

72. [Du Coudray], *L'ordre profond*, 11–27, 93.

73. On Fontenoy, see [Gribeauval and Vallière, *fils*], *Mémoires authentiques*, 120. François-Jean de Graindorge d'Orgeville de Mesnil-Durand, *Réponse à la brochure intitulée "L'ordre profond"* (Amsterdam: Cellot and Jombert, 1776), 28–32. Mesnil-Durand, *Projet d'un ordre français*, xiv.

74. Colin, *Infanterie*, 242. S.H.A.T. MR1819 Félix Wimpfen, "Rélation du camp de Bayeux," 1778.

75. Quimby, *Napoleonic Warfare*, 231–48.

76. S.H.A.T. MR1715 [Broglie], *Extrait du projet d'instruction pour . . . l'ordre français* (Paris: Imprimerie Royale, 1778).

77. Guibert, *Système de guerre*, 3: 208–10. [Bachaumont], *Mémoires secrets*, 3, 19 October 1778, 12: 136, 150–51.

78. Guibert, *Système de guerre*, 3: 208–10.

79. S.H.A.T. MR1819 Broglie to Louis XVI, 13 October 1778.

80. S.H.A.T. MR1819 Wimpfen, "Rélation du camp de Bayeux," 1778.

81. Guibert, *Système de guerre*, 3: 309–10.

82. S.H.A.T. MR1819 Wimpfen, "Rélation du camp de Bayeux," 1778.

83. François-Jean de Graindorge d'Orgeville de Mesnil-Durand, *Observations sur le canon par rapport à l'infanterie en générale* (Paris: Jombert, 1772), 13–14, 23–32.

84. [Du Coudray], *L'ordre profond*, 13, 35–67, 73–85.

85. Mesnil-Durand, *Réponse à la brochure*, 132.

86. [Du Coudray], *L'ordre profond*, 90–93.

87. Guibert, *Système de guerre*, 3: 303.

88. S.H.A.T. MR1715 Anon., "Sur l'ordre profond et l'ordre mince, tirer du no. XX du *Journal des sciences et beaux arts*," [1778]. The Moderns marshaled elite public opinion behind them using the established mechanisms of scientific publicity. For the favorable reaction of the salons to the Moderns, see [Bachaumont], *Mémoires secrets*, 3, 19 October 1778, 12: 136, 150–51.

89. Mesnil-Durand, quoted in Guibert, *Système de guerre*, 3: 216–17.

90. [Douazac], *Dissertation sur la subordination* (2d ed.; Avignon: Dépense de la Compagnie, 1753).

91. S.H.A.T. MR1715 Anon., "Mémoire sur la constitution militaire," [1780s]. Colin, *Infanterie*, 177. Egret, *Prerevolution*, 50.

92. Guibert, *Système de guerre*, 3: 220–21. Guibert believed that the widespread introduction of firearms had produced a general uniformity in the military constitutions of all the European powers; not that he found this admirable or optimal. Guibert, *Essai général*, 1: 146–47.

93. S.H.A.T. MR1715 Castries, "Observations," 30 September 1778. Colin, *Infanterie*, 233–40.

94. This new order was designed to be put to use immediately for instructing troops. Colin, *Infanterie*, 254–59.

95. For an early programmatic statement, see Jean-Paul Bertaud, "Voies nouvelles pour l'histoire militaire de la Révolution," *Voies nouvelles pour l'histoire de la Révolution française* (Paris: B.N., 1978), 185–205. For the follow-up, see Bertaud, *Army of the French Revolution*. Scott, *Response of the Royal Army*. Alan Forrest, *The Soldiers of the French Revolution* (Durham, N.C.: Duke University Press, 1990). Peter Wetzler, *War and Subsistence: The Sambre and Meuse Army in 1794* (New York: Lang, 1985). Lynn, *Bayonets of the Republic*.

96. John A. Lynn, "French Opinion and the Military Resurrection of the Pike, 1792–1794," *Military Affairs* 41 (1977): 1–7.

97. Soboul, *Sans-culottes*, 653–54. Collot-d'Herbois, *Moniteur*, 22 September 1793, 17: 708–9.

98. Robespierre, 10 February 1792, in *Oeuvres complètes* (Paris: Leroux, 1912), 8: 160–61.

99. Municipalité de Paris, "Arrêté relatif aux piques," *Moniteur*, 14 February 1792, 11: 370–71. [Municipalité de Paris], *Arrêté concernant l'incorporation des citoyens armés de piques*, 11 July 1792. On the tensions between active and passive citizens in Paris, see Florence Devenne, "La Garde Nationale: Création et évolution, 1789-août 1792," *Annales historiques de la Révolution française* 283 (1991): 49–66, on pikes, 64.

100. *A.P.*, 31 July 1792, 47: 332. R. B. Rose, *The Enragés: Socialists of the French Revolution?* (Sydney: Sydney University Press, 1965), 57–58.

101. Carnot, "Rapport et projet de décret sur une distribution de piques," 25 July 1792, in *Révolution et mathématique*, 2:103–7. For an early call to deploy pikes in the army, see *A.P.*, 11 April 1792, 41: 490–97.

102. Joseph Servan, *Plan d'organisation pour les bataillons de piquiers* (Paris, 1792), quoted in Lynn, *Bayonets of the Republic*, 190.

103. Poisson, *Choderlos de Laclos*, 323–24.

104. Collot d'Herbois, 6 September 1793, but echoing a speech from a year before, in Bertaud, *Army of the French Revolution*, 155–56.

105. Alexandre Tuetey, ed., *Répertoire général des sources manuscrites de l'histoire de Paris pendant la Révolution française* (Paris: Imprimerie Nouvelle, 1890–1914), 26 May 1793, 8: 329–30. Pike production was definitively suspended on 31 August 1793. Carnot, *A.P.*, 1 August 1792, 47: 361–66. Carnot to C.S.P., 9 October 1793, in *Révolution et mathématique*, 2: 136–37. Bertaud, *Army of the French Revolution*, 154–55. On the evolution of Carnot's military strategy, see Marcel Reinhard, *Le grand Carnot, de l'ingénieur au conventionnel, 1753–1792*, vol. 1 (Paris: Hachette, 1950), 117–36; *Le grand Carnot, l'organisateur de la victoire, 1792–1823*, vol. 2 (Paris: Hachette, 1952), 45–54.

106. On the co-optation of the *armées révolutionnaires*, see Richard Cobb, *The People's Armies: The armées révolutionnaires, Instrument of the Terror in the Departments, April 1793 to Floréal, year II*, trans. Marianne Elliott (New Haven: Yale University Press, 1987), 50–51, 523, 600–601. On the justice system, see Colin Lucas, "Revolutionary Violence, the People and the Terror," in *The Terror*, vol. 4, *The French Revolution and the Creation of Modern*

Political Culture, ed. Keith Michael Baker (Oxford: Elsevier, 1994), 57–79. On the *levée en masse*, see Bertaud, *Army of the French Revolution*, 102–5.

107. Prieur, *R.A.C.S.P.*, 5 March 1794, 11: 553–54. Lynn notes that the attack column of Guibert was tried and modified after near disaster. Lynn, "En avant!" 161–62, 168–71. See also Colin, *Infanterie*, 263–73.

108. David G. Chandler, *The Campaigns of Napoléon* (New York: Macmillan, 1966), 133–201. Peter Paret, "Napoléon," in *Makers of Modern Strategy from Machiavelli to the Nuclear Age*, ed. Peter Paret (Princeton: Princeton University Press, [1943], 1986), 123–42.

109. Lynn, *Bayonets of the Republic*, 97–118.

110. Bertaud, *Army of the French Revolution*, 81–90, 150–53, 157–71.

111. Laclos to Mde. Laclos, in Laclos, *Oeuvres*, 903.

112. Elting E. Morison, "Gunfire at Sea: A Case Study of Innovation," *Men, Machines, and Modern Times* (Cambridge: MIT Press, 1966), 17–44.

113. Laclos to Mde. Laclos, 16, 18, 23 May, 13, 25 June 1800, in Laclos, *Oeuvres*, 893, 894, 896, 904, 909. Laclos, *Eloge de Vauban*, in *Oeuvres*, 569–93. DeJean, *Literary Fortifications*, 191–201. For a record of the interminable itinerary of one artillerist, see Louis-Joseph Bricard, *Journal du cannonier Bricard, 1792–1802* (Paris: Delagrave, 1891). Also see the sixteen-page densely written itinerary of Pion des Loches, *Mes campagnes*, 483–99. For a prescient Vallièrist warning against this phenomenon, see [Le Blond], "Artillerie," 1: 614.

114. Johann Wolfgang von Goethe, *The Campaign in France, 1792*, trans. Robert R. Heitner (New York: Surkamp, 1987), 19 September 1792, 5: 651–52. On Valmy, see Lauerma, *L'artillerie de campagne*, 160–67. Emmanuel Hublot, *Valmy, ou la défense de la nation par les armes* (Paris: Fondation pour les Etudes de la Défense Nationale, 1987), 293–99.

115. Lauerma, *L'artillerie de campagne*, 41–42, 120–28, Custine quote on p. 160.

Chapter Four
The Tools of Practical Reason

1. David Landes, *The Unbound Prometheus: Technological Change and Industrial Development in Western Europe from 1750 to the Present* (Cambridge: Cambridge University Press, 1969). Joel Mokyr, *The Lever of Riches: Technological Creativity and Economic Progress* (New York: Oxford University Press, 1990). R. H. Coase, "The Nature of the Firm," in *The Firm, the Market and the Law* (Chicago: University of Chicago Press, [1937], 1988), 33–55. A. Chandler, *Visible Hand*. For one influential theoretical statement of proto-industrialization, see Peter Kriedte, Hans Medick, and Jürgen Schlumbohn, *Industrialization before Industrialization: Rural Industry in the Genesis of Capitalism* (Cambridge: Cambridge University Press, 1981).

2. Landes and Mokyr deserve considerable credit for putting technology back at the center of the history of industrialization. But as Mokyr himself has noted, the essential process of technological imitation and innovation can occur only when others are persuaded that the new methods are better; yet resistance to new technology can be considerable and new technologies themselves are almost invariably fragile. Joel Mokyr, "Technological Inertia in Economic History," *Journal of Economic History* 52 (1992): 325–38. Business historians have assumed the existence of the entrepreneur; whereas considerable empirical evidence suggests that a well-defined entrepreneurial role was slow to develop in much of Western Europe. Reddy, *Rise of Market Culture*. See also the debate between Stephen Marglin, "What Do Bosses Do?: Part I," *The Review of Radical Political Economy* 6 (1974): 60–112; and David Landes, "What Do Bosses Really Do?" *The Journal of Economic History* 46 (1986): 585–623. Recently, Tessie Liu, Maxine Berg, and others have noted that proto-industrialization theory takes for granted that industrializing capitalists will triumph over artisanal producers. They

provide evidence, particularly from domestic industries, that this is by no means always the case. Liu, *The Weaver's Knot*, 22–44. Maxine Berg, *The Age of Manufactures: Industry, Innovation, and Work in Britain, 1700–1820* (Totowa, N.J.: Barnes and Noble Books, 1985), 77–86.

3. Sabel and Zeitlin, "Historical Alternatives." David Hull, "In Defense of Presentism," *History and Theory* 18 (1979): 1–15.

4. For a recent programmatic statement on technological representation and the distribution of power in the workplace, see Steve Lubar, "Representation and Power," *Technology and Culture* 36 (1995): S54-S81.

5. Werner Sombart, *Krieg und Kapitalismus* (New York: Arno Press, [1913], 1975), 74–116. Lewis Mumford, *Technics and Civilization* (New York: Harcourt, 1934), 89–98. John Nef denies Sombart's assertion that war has contributed to economic progress, but he does not address the claim that war *preparation* has contributed to the pattern of European industrialization. Nef, *War and Human Progress*, 65–88. For military "spin-off" before the Industrial Revolution, see Trebilcock, "'Spin-Off' in British Economic History." For doubts that military spin-off mattered much before the twentieth century, see Mokyr, *Lever of Riches*, 183–86. Max Weber, *Economy and Society: An Outline of Interpretive Sociology* (Berkeley: University of California Press, 1968), 2: 1155–56. David Noble, "Command Performance: A Perspective on the Social and Economic Consequences of Military Enterprise," *Military Enterprise*, ed. M. R. Smith, 329–46.

6. For a nuanced Marxist account, which treats local circumstances in all their complexity, but which still suffers from this penchant for seeing social classes as stand-ins for innate ideas, see David Noble, *Forces of Production: A Social History of Industrial Automation* (New York: Knopf, 1984). See the critique in Charles Sabel, *Work and Politics: The Division of Labor in Industry* (Cambridge: Cambridge University Press, 1982), 59–70.

7. Woodbury, "Legend of Eli Whitney." For a discussion of standards in modern mass production and their dependence on standards as embodied in drawings, gauges, and machine tools, see Earle Buckingham, *Principles of Interchangeable Manufacturing* (2d ed.; New York: The Industrial Press, 1941), 1–17.

8. There is a large literature on trust in economic relations; see Allan Fox, *Beyond Contract: Work, Power and Trust Relations* (London: Faber and Faber, 1974).

9. My thinking here depends essentially on a scholarly literature devoted to the social studies of science and technology, which argues that agreement among scientists on standards represents the (temporary) closure of controversy. See principally, T. Porter, *Trust in Numbers*.

10. From the point of view of the shop floor, engineers and managers invariably come to production as outsiders, imposing standards, novel work procedures, and new production schedules. Even within a modern firm, relations between employer and employees are in some sense a highly controlled exchange. Michael Buroway, *Manufacturing Consent: Changes in the Labor Process under Monopoly Capitalism* (Chicago: University of Chicago Press, 1979), 5–56.

11. Julien Offray de La Mettrie, *L'homme machine*, ed. Paul-Laurent Assoun (Paris: Renoel, 1981). Sewell, *Work and Revolution*. For the rise of a "geometric" physiognomy of the body, see Barbara Stafford, *Body Criticism: Imaging the Unseen in Enlightenment Art and Medicine* (Cambridge: MIT Press, 1991), 107–13, 148–52. For a brilliant reexamination of automata as a demonstration of managerial control, see Simon Schaffer, "Enlightenment and the Automaton," in *The Sciences in Enlightened Europe*, eds. William Clark, Jan Golinski, and Simon Schaffer (Chicago: Chicago University Press, forthcoming).

12. These data are surveyed in Bélidor, *Architecture hydraulique*, 1: 40–45. Eugene Ferguson, "The Measurement of the 'Man-Day,'" *Scientific American* 225 (1971): 96–103.

13. Charles Augustin Coulomb, "Résultats de plusieurs expériences destinées à détermi-

ner la quantité d'action que les hommes peuvent fournir," *Théorie des machines simples* (Paris: Bachelier, 1821), 255–96. Coulomb had been presenting this material to the Academy of Sciences since 1778. See C. Stewart Gillmor, *Coulomb and the Evolution of Physics and Engineering in Eighteenth-Century France* (Princeton: Princeton University Press, 1971), 18–25, 77–78.

14. To explain why the porter had not himself worked at this efficiency, Coulomb resorted to psychology; the laborer did not want to appear too weak to carry the heavier load. This ergonomic program was to bear great fruit in the nineteenth century, in part under the direction of the French artillery service. A. J. Morin did tests at the Strasbourg foundry in the 1830s. I. Grattan-Guinness, "Work for the Workers: Advances in Engineering Mechanics and Instruction in France, 1800–1830," *Annals of Science* 41 (1984): 1–33, p. 23. More famously, it became the background of the Taylorist time-and-motion studies. Anson Rabinbach, *The Human Motor: Energy, Fatigue and the Origins of Modernity* (New York: Basic Books, 1990).

15. Sonenscher, *Hatters*, 1–11, 70–71, 164–65.

16. Sewell, *Work and Revolution*, 16–39. There is a certain tension between Sewell and Sonenscher's understanding of the relationship between language and work practices which I gloss over here. Michael Polanyi, *Personal Knowledge*. Still valuable on the corporations is Emile Coonaert, *Les corporations en France avant 1789* (Paris: Gallimard, 1941). For the somewhat different situation in England, see John Rule, "The Property of Skill in the Period of Manufactures," *The Historical Meanings of Work*, ed. Patrick Joyce (Cambridge: Cambridge University Press, 1987), 99–118. On secrecy in the gun trade, see Cesar Fiosconi and Jordam Gusero, *Espiarda Perfeyta, or The Perfect Gun*, trans. and ed. Rainier Daehnhardt and W. Keith Neil (London: Sotheby, [1718], 1974), ix–xi, 9.

17. On Diderot's attack on secrecy, see his "Encyclopédie," *Encyclopédie*, 5: 647. Though Cynthia Koepp deplores Diderot's vision of progress and Stephen Werner celebrates it, both agree that Diderot shifts attention away from human effort to that of the machine. Koepp, "The Alphabetical Order: Work in Diderot's *Encyclopédie*," in *Work in France*, eds. Kaplan and Koepp, 229–57. Stephen Werner, *Blueprint: A Study of Diderot and the Encyclopédie Plates* (Birmingham, Ala.: Summa, 1993).

18. Diderot, "Art," *Encyclopédie*, 1: 716–18. d'Alembert, "Discours préliminaire," *Encyclopédie*, 1: xiii. A.E.P.C. #2385 Jean-Rodolphe Perronet, "Description de la façon dont on fabrique les épingles," 7 January 1740. On the rationalization of work as represented in the *Encyclopédie*, see Antoine Picon, "Gestes ouvriers, opération et processus technique: La vision de travail des encyclopédistes," *Recherches sur Diderot et sur l'Encyclopédie* 13 (1992): 131–47. Also, Koepp, "The Alphabetical Order."

19. Diderot, "Art" and "Encyclopédie," *Encyclopédie*, 1: 716; 5: 636, 645–46. Compare this with Diderot, *Interprétation de la nature*. For a suggestive hint about the "natural history" of trades in this period, see Charles C. Gillispie, "The Natural History of Industry," *Isis* 48 (1957): 398–407, esp. pp. 405–7.

20. On the inadequacy of verbal description, see Diderot, "Bas de métier," *Encyclopédie*, 2: 98. Also Diderot, "Prospectus," in Diderot, *Oeuvres complètes* (Paris: Garnier, 1876), 13: 141–43, quote on p. 143. Diderot acknowledged his debt to the *Theatrum* and boasted of besting them in Diderot, "Encyclopédie," *Encyclopédie*, 5: 645. William H. Sewell, Jr., "Visions of Labor: Illustrations of the Mechanical Arts before, in, and after Diderot's *Encyclopédie*," in *Work in France*, eds. Kaplan and Koepp, 258–86. For a bibliography of the vast literature on the plates, see Werner, *Blueprint*.

21. Diderot, *Plan of a University for the Russian Government*, in *French Liberalism and Education in the Eighteenth Century*, ed. François de La Fontainerie (New York: McGraw-Hill, 1932), 230. René Descartes, *Dioptrique*, in *Discours*, 71–208.

22. On the artisanal drawing schools, see Arthur Birembaut, "Les écoles gratuites de

dessin," in *Ecoles techniques*, eds. Hahn and Taton, 441–76. Also Antoine Léon, *La Révolution française et l'education technique* (Paris: Société des Etudes Robespierristes, 1968), 60, 70. On the Ponts et Chaussées, see Picon, *Ingénieur moderne*. On the Corps du Génie, see Bruno Belhoste, Antoine Picon, and Joël Sakarovitch, "Les exercices dans les écoles d'ingénieurs sous l'ancien régime et la révolution," *Histoire de l'education* 46 (1990): 53–109.

23. Peter Geoffrey Booker, *A History of Engineering Drawing* (London: Northgate, 1979), 128–49.

24. Yves Deforge, *Le graphisme technique: Son histoire et son enseignement* (Seyssel: Vallon, 1981). See also Ferguson, *Mind's Eye*.

25. Monge, "Stéréotomie," 1. Monge, *Géométrie descriptive, Leçons de l'an III* (Paris, year VII [1799]), xvi. On the methods of masons, see Lon Shelby, "The Geometrical Knowledge of the Medieval Master Masons," *Speculum* 47 (1972): 395–421. Desargues still defined his axes in terms of the object represented, whereas Monge's axes were entirely independent of the object. For an excellent discussion of different forms of technical drawing, see Booker, *Engineering Drawing*.

26. One of Monge's exercises was to show how the descriptive geometry can be used to construct perspective views. The opposite transformation, however, is not possible. The classic essay on the new Renaissance "grammar" of visualization is W. M. Ivins, *On the Rationalization of Sight* (New York: Metropolitan Museum, 1938). On the translatability of Renaissance perspective, see Samuel Y. Edgerton, *The Renaissance Discovery of Linear Perspective* (New York: Harper and Row, 1976). A vast literature explodes the truism about photography. Bosse quoted in Mark Schneider, "Gerard Desargues, The Architectural and Perspective Geometry: A Study in the Rationalization of Figure" (Ph.D. diss., Virginia Polytechnic Institute, 1984), 142. For a history of how these debates over representation played themselves out in the institutional life of the French arts, with the "realists" Bosse and Desargues trying to recruit the artisans, and the "fine arts" academicians lining up for royal patronage, see Nicolas Mizroeff, "Pictorial Sign and Social Order: L'Académie Royale de Peinture et Sculpture, 1638–1752" (Ph.D. diss., University of Warwick, 1990).

27. Daston, "Objectivity." Daston and Galison map out some of the features of the "objective" representation of scientific phenomena without examining the methods of pictorial representation that seek to ensure that objectivity. Lorraine J. Daston and Peter Galison, "The Image of Objectivity," *Representations* 40 (1992): 81–128. Frézier, *Elémens de stéréotomie*, 1: ix–x.

28. Ken Baynes and Francis Pugh, *The Art of the Engineer* (Woodstock, N.Y.: Overlook Press, 1981), 36–37, 60–69. Jacques Payen, *Capital et machine à vapeur au XVIIIe siècle: Les frères Périer et l'introduction en France de la machine à vapeur de Watt* (Paris: Mouton, 1969).

29. Bruno Latour, "Drawing Things Together," in *Representations in Scientific Practice*, eds. Michael Lynch and Steve Woolgar (Cambridge: MIT Press, 1990), 20–69.

30. Gaspard Monge, "Développements sur l'enseignement adopté pour l'Ecole Centrale des Travaux Publics," 21 ventôse, year II [11 March 1794], in Janis Langins, *La République avait besoin des savants; Les débuts de l'Ecole Polytechnique: L'Ecole Centrale des Travaux Publics et les cours Révolutionnaires de l'an III* (Paris: Belin, 1987), 245.

31. Michael Lynch, "Discipline and the Material Form of Images: An Analysis of Scientific Visibility," *Social Studies of Science* 15 (1985): 37–65.

32. Belofsky notes that the descriptive geometry comes in two "dialects," which are the product of separate and historically contingent cultural developments. Harold Belofsky, "Engineering Drawing: A Universal Language in Two Dialects," *Technology and Culture* 32 (1991): 32–34.

33. S.H.A.T. 2a59 Le Pelletier, "Instructions qui seront données sur le dessein à l'Ecole d'Artillerie de Metz," 1749. Joseph Du Teil, "Salle de dessin," 1786. "Ecole d'Artillerie de Metz," October 1767. Saint-Auban, "Instruction," 25 October 1765.

34. Desargues quoted in Mark Schneider, "Desargues," 100. Booker, *Engineering Drawing*, 185–97. On computers and their role in representing work, see Zuboff, *Age of the Smart Machine*.

35. Jean-Baptiste Vaquette de Gribeauval, *Tables de construction des principaux attirails de l'artillerie proposées ou approuvées depuis 1764 jusqu'en 1789*, ed. Jacques Charles Mason (Paris, 1792).

36. Michael E. Gorman and W. Bernard Carlson, "Interpreting Invention as a Cognitive Process: The Case of Alexander Graham Bell, Thomas Edison, and the Telephone," *Science, Technology and Human Values* 15 (1990): 131–64.

37. J.-F. Blondel, *Etude de l'architecture*, 15. Auslander, "Creation of Value," 149–57. See also Leora Auslander, "Perceptions of Beauty and the Problem of Consciousness: Parisian Furniture Makers," in *Rethinking Labor History: Essays on Discourse and Class Analysis*, ed. Lenard R. Berlandstein (Urbana: University of Illinois Press, 1993), 149–81.

38. Jean-Jacques Rousseau, *Emile, or on Education*, trans. Allan Bloom (New York: Basic Books, 1979), 143–46, 195–203.

39. [Jean-Jacques Bachelier], *Détails sur l'origine et l'administration de l'Ecole Royale Gratuite de Dessin* (Paris, 1768). Enrollment is calculated from A.N. F17 2499 Bachelier to Min. Interior, 19 December 1792. See also Dominique Julia et al., *Atlas de la Révolution française* (Paris: Editions de l'Ecole des Hautes Etudes en Sciences Sociales, 1988), 2: 65. The best account of ancien régime education is Harvey Chisick, *The Limits of Reform in the Enlightenment: Attitudes toward the Education of the Lower Classes in Eighteenth-Century France* (Princeton: Princeton University Press, 1981).

40. Louis-Sébastien Mercier, *Tableau de Paris* (new ed.; Amsterdam: n.p., 1782–88), 10: 99. B.N. V23842(9) Jean-Jacques Bachelier, *Discours sur l'utilité des écoles élémentaires en faveur des arts mécaniques*, 10 September 1766.

41. Bachelier, *Discours sur l'utilité*, 7–8. Bachelier, *Collection des discours* (Paris: Imprimerie Royale, 1790), 39. On artful design and eighteenth-century markets, see Styles, "Manufacturing, Consumption, Design," 543–47.

42. *Lettres patentes du roi portant établissement d'une Ecole Gratuite de Dessin à Paris*, 20 October 1767. Bachelier, *Collection des discours*, [1771], 19. The school's patrons included much of court society. On the virtues of neoclassicism, see Stafford, *Body Criticism*, 11–15. Jean Starobinsky, *1789: The Emblems of Reason*, trans. Barbara Bray (Charlottesville: University Press of Virginia, 1982). Madelyn Gutwirth, *The Twilight of the Goddesses: Women and Representation in the French Revolutionary Era* (New Brunswick: Rutgers University Press, 1992).

43. B.N. V23849 "Extrait du procès-verbal de l'Assemblée Nationale, 4 September 1790, in [Bachelier], *Mémoire sur l'origine, les progrès, et la situation de l'Ecole Royale Gratuite de Dessin* (Paris: Imprimerie Royale, 1790), 1–7. A.N. F17 2449 [Bachelier], *Pétition des souscripteurs de l'Ecole Gratuite de Dessin* (Paris: Imprimerie Royale, 1792). Suddenly in 1789 Bachelier began to speak with approbation of the suppression of the corporations in 1776 because it had given everyone, including women, the right to join the professions. He now wanted girls to be given the same kind of drawing-based technical education given to boys. Bachelier, *Mémoire sur l'education des filles* (Paris: Imprimerie Royale, 1789), 10, 20–21. On the uncertainties aroused by the abolition of the guilds, see Michael Sonenscher, "The Sans-culottes of the Year II: Rethinking the Language of Labor in Revolutionary France," *Social History* 9 (1984): 301–28. Also, Steven L. Kaplan, "Réflexions sur la police du monde du travail, 1700–1815," *Revue historique* 261 (1979): 17–77.

44. As Shapin and Schaffer note, savants had to visit Boyle's laboratory to see his pump in operation before they could build a replica. The dispute diminished only after Pupin developed a "commercially available" pump. But might not the same problem of reproducibility plague differences in production *within* a workshop as *between* rival workshops? Shapin and Schaffer, *Leviathan*, 225–82.

45. For a suggestive hint on gauges, see Lubar, "Representation and Power."

46. On gauges in early eighteenth-century gun-making, see Fiosconi and Gusero, *Espiarda Perfeyta*, 47, 51–55, 195.

47. Robert B. Gordon, "Who Turned the Mechanical Ideal into Reality?" *Technology and Culture* 29 (1988): 744–78. Robert B. Gordon and Patrick Malone, *The Texture of Industry: An Archeological View of the Industrialization of North America* (New York: Oxford University Press, 1994), 373–80, 386–88.

48. Historians have previously located the first use of the term and concept in the mid-nineteenth century; see Booker, *Engineering Drawing*, 187–89. Gordon and Malone, *Texture of Industry*, 374.

49. T. Porter, *Trust in Numbers*.

50. E. P. Thompson, "Time, Work-Discipline, and Industrial Capitalism," *Past and Present* 38 (1967): 56–97. Linebaugh, *London Hanged*, 371–401. An analogous transition also occurred when the French state obliged producers to switch from the anthropomorphic work-based measures of the ancien régime to the new rational measures of the metric system. Ken Alder, "A Revolution to Measure: The Political Economy of the Metric System in France," in *The Values of Precision*, ed. M. Norton Wise (Princeton: Princeton University Press, 1995), 37–71.

51. Fred Colvin, *Gages and their Use in Inspection* (New York: McGraw-Hill, 1942).

52. Scheel, *Mémoires d'artillerie*, 143–46. S.H.A.T. 9a11 Gribeauval to Manson, 7 March 1765. Du Coudray, *L'artillerie nouvelle*, 58.

53. Charles M. S. Dartein, *Traité élémentaire sur les procédés en usage dans les fonderies pour la fabrication des bouches à feu d'artillerie* (Strasbourg: Levrault, 1810), 260.

54. S.H.A.T. 9a11 Gribeauval to Manson, 7 March 1765. S.H.A.T. 4d4 Choiseul to Chateaufer, 31 March 1765.

55. Du Coudray, *L'artillerie nouvelle*, 80.

56. Jean Du Teil, *L'usage*, 16.

57. S.H.A.T. 4d4 Choiseul to Chateaufer, 12 January 1765; Maritz to Choiseul, 24 January 1765. Du Coudray, *L'artillerie nouvelle*, 66–67. F. M. Aboville, *père*, to Min. War, 5 prairial, year IV [24 May 1796], in M. R. Cain, "Historique de la création de l'Atelier de Précision de l'Artillerie," *Mémorial de l'artillerie française* 28 (1954): 381. Aboville claimed he proposed the cylinders in 1760, and they were initially rejected by Maritz, who later turned around and endorsed them.

58. S.H.A.T. 4d4 Choiseul to Chateaufer, 31 March 1765. Du Coudray, *L'artillerie nouvelle*, 66.

59. Du Coudray, *L'artillerie nouvelle*, 58, 62. Alder, "Revolution to Measure."

60. S.H.A.T. 9a11 Gribeauval to Manson, 7 March 1765. M. Norton Wise, ed. and intro., *The Values of Precision* (Princeton: Princeton University Press, 1995).

61. Lombard, *Projectiles*, xiv.

62. All of these rational tools were gradually being introduced in ship-building. For an overview of the changing regulations governing French ship-building and the training of naval engineers, see Martine Acerra, "Les constructeurs de la Marine (XVIIe–XVIIIe siècle)," *Revue historique* 273 (1985): 283–304. For contemporary geometric construction, work organization, and inspection, see Jean Boudriot, ed. *The Seventy-four Gun Ship: A Practical Treatise on the Art of Naval Architecture*, trans. David Roberts (Beccles, U.K.: Naval Institute Press, 1986). For a recent synthesis, see Vérin, *Gloire des ingénieurs*, 210–20.

63. S.H.A.T. 9a11 Gribeauval to "Monsieur et cher ami," 18 July 1764.

64. S.H.A.T. 9a11 Gribeauval to Manson, 7 March 1765. In fact, the new carriages weighed *more* than the old, primarily because they substituted iron axles for wooden ones and added other reinforcing iron parts.

65. S.H.A.T. 9a11 Gribeauval to Manson, 7 March 1765. Jean Du Teil, *L'usage*, 13–15. Du Coudray, *L'artillerie nouvelle*, 86–87.

66. S.H.A.T. 4c3/2 Jollain, 16 September 1740.

67. A set of *Tables de construction* for carriages made at Douai had existed since 1754, though these were far from complete. S.H.A.T. 4c3/2 Belle-Isle to [indecipherable], 30 September 1760. Calls for standardized carriages had been heard since the seventeenth century. Basset, "Fabrications d'armement," 975, 986. S.H.A.T. 1a2/2 Dubois, "Mémoire sur l'artillerie," [1762].

68. Le Duc, "Mémoire concernant la connaissance, le détail et l'usage des principaux attirails de l'artillerie," 1750, cited in Rosen, "Gribeauval," 135–37.

69. S.H.A.T. 4c3/2 Gribeauval, "Mémoire," 26 November 1781.

70. S.H.A.T. 4c3/2 Anon., "Proportion des ferrures pour le canon . . . avec leur poids, longueur, largeur, épaisseur," [1770]. Du Coudray, *L'artillerie nouvelle*, 86. Gribeauval, *Tables de construction*.

71. Du Coudray, *L'artillerie nouvelle*, 81.

72. Jean Du Teil, *L'usage*, 23.

73. Picard and Jouan, *Artillerie*, 25, 28–29, 32. On methods of payment and recruitment, see S.H.A.T. 1a2/2 Mouy, "Article 22," March 1764. Register lists of worker-soldiers can be found in S.H.A.T. Xd75-Xd79.

74. *Ordonnance du roi concernant le Corps Royal de l'Artillerie*, 3 October 1774, pp. 25, 28–29. The same pay differential applied to corporals, appointees, second-class workers, and apprentices. The pay of children over ten was added to their father's wages. Workers also received a small bonus pay during wartime. Le Puillon de Boblaye, *Ecoles d'artillerie*, 66. A.N. AD VI 40 "Règlement concernant le service des Arsenaux de Construction," 1 April 1792. Nardin, *Gribeauval*, 326–27. S.H.A.T. 9a3 Le Duc, 18 May 1774.

75. S.H.A.T. 1a2/2 Gribeauval, "Réponse aux objections," May 1765. Quote in S.H.A.T. 1a2/2 Mouy, "Article 22," March 1764. S.H.A.T. 9a3 Le Duc, 18 May 1778. A.N. AD VI 40 "Règlement concernant le service des Arsenaux de Construction," 1 April 1792.

76. Jean Du Teil, *L'usage*, 16. Du Coudray, *L'artillerie nouvelle*, 83. On the variation in measures in ancien régime France, see Alder, "Revolution to Measure."

77. A.N. AD VI 40 "Règlement concernant le service des Arsenaux de Construction," 1 April 1792. Nardin, *Gribeauval*, 138–39. The *livret* had antecedents in the ancien régime and Revolutionary period. Kaplan, "Police du monde du travail." Du Coudray, *L'artillerie nouvelle*, 83.

78. S.H.A.T. 9a11 Gribeauval to Manson, 7 March 1765.

79. Du Coudray, *L'artillerie nouvelle*, 86.

80. Ibid., 84. Though Du Coudray's book was partisan, there seems little reason to doubt his account. The ethic of the Gribeauvalist party was to create a potent, mobile artillery—cost was deliberately deemphasized. He also took care to point out all the exceptions to these savings. For example, the new caissons required more complex ironwork than the old, and were consequently more expensive.

81. Du Coudray, *L'artillerie nouvelle*, 84, 210–11. Fries, "British Response," 383–85.

82. Saint-Auban, in Scheel, *Mémoires d'artillerie*, 252–55. Du Coudray denied that the new carriages were more fragile—the iron axles broke far less often—and even if they did, the worker-soldiers still could make simple field repairs as they always had. Du Coudray, *L'artillerie nouvelle*, 220–22.

83. S.H.A.T. 9a11 Gribeauval, 29 December 1775. The delay in publication of the *Tables de constructions* was also due to military secrecy. 4c3/2 Gribeauval, "Additions et corrections proposées aux *Tables de construction*," September 1767. 9a11 Gribeauval to [Chev. de H.], 21 August 1767. S.H.A.T. 9a3 Le Duc, 18 May 1778. Nardin, *Gribeauval*, 340–41.

84. Nardin, *Gribeauval*, 150. S.H.A.T. 9a11 Gribeauval to Thiboutot, 1 July 1782. S.H.A.T. MR1739 "Dépenses annuelles de l'artillerie," [1788]. *Ordonnance du roi concernant le Corps Royal de l'Artillerie*, 3 October 1774, pp. 76, 86. However, the pay allotted for the companies of worker-soldiers stayed the same at 240,000 *livres* per annum.

85. Scheel, *Mémoires d'artillerie*, 150–55; and Heinrich Othon von Scheel, *Treatise of Artillery Containing a New System or Alterations Made in the French Artillery since 1765*, trans. Jonathan Williams (Philadelphia: Fenno, 1800), 110–22.

86. [Du Coudray], *L'ordre profond*, 92.

Chapter Five
The Saint-Etienne Armory: Musket-Making and the End of the Ancien Régime

1. Jean-Marie Roland de la Platière, *Lettres écrites de Suisse, d'Italie, de Sicile et de Malthe, Par M.**, avocat en Parlement à Mlle.** à Paris en 1776, 1777 et 1778* (Amsterdam: n.p., 1780), 6: 460–62. Roland's impression of a population explosion at Saint-Etienne is borne out by the data. Louis Messance, *Nouvelles recherches sur la population de la France avec quelques remarques importantes sur divers objets d'administration* (Lyon: Frères Perisse, 1788), 68–69, 119. F. Tomas, "Problèmes de démographie historique: Le Forez au XVIIIe siècle," *Cahiers d'histoire* 13 (1968): 381–99.

2. Roland, *Lettres*, 6: 461.

3. S.H.A.T. 4f3 Gribeauval to Ségur (Min. War), 5 June 1785. Emphasis by Gribeauval.

4. The War Office claimed it had 140,000 muskets in stock, of which only 75,000 were new. Assessments of war readiness were highly politicized, and hence suspect, however. *A.P.* 26 July, 16 September 1792, 47: 158–59; 50: 62.

5. A.D.L. L945 Roland to Administration du département du Rhône et Loire, 29 August 1792.

6. A.D.L. L945 Ravel and Pourret to [Roland], 3 September 1792. Carrier de Thuillerie and Dubouchet to [Ravel and Pourret], 5 September 1792. Ravel and Pourret, letters, 4 September–8 October 1792.

7. M. J. Piore, "Dualism as a Response to Flux and Uncertainty," and "The Technological Foundations of Dualism and Discontinuity," in *Dualism and Discontinuity in Industrial Societies*, eds. S. Berger and M. J. Piore (Cambridge: Cambridge University Press, 1980), 13–81. Sabel and Zeitlin, "Historical Alternatives." For examples of flexible specialization in America, see Philip Scranton, "Diversity in Diversity: Flexible Production and American Industrialization, 1880–1930," *Business History Review* 65 (1991): 27–90. An example of the "free rider" problem would be a firm's reluctance to train skilled workers for fear that other firms will poach its technicians. This may be less of a problem if municipal taxes fund worker training. Pierre-Joseph Proudhon, *Système des contradictions économiques*, vol. 1 in *Oeuvres complètes* (Geneva: Slatkine, 1982).

8. Sabel and Zeitlin, "Alternatives to Mass Production," 175.

9. Ibid., "Alternatives to Mass Production," 147–50.

10. Michael P. Hanagan, *The Logic of Solidarity: Artisans and Industrial Workers in Three French Towns, 1871–1914* (Urbana: University of Illinois Press, 1980). Michael P. Hanagan, *Nascent Proletarians: Class Formation in Post-Revolutionary France* (Oxford: Basil Blackwell, 1989). For a contrasting view of nineteenth-century family life in the Forez, which also operates within the class-formation mode, see Elinor Accampo, *Industrialization, Family Life, and Class Relations: Saint Chamond, 1815–1914* (Berkeley: University of California Press, 1989).

11. Sabel and Zeitlin are here quoting approvingly from P. Pic and J. Godart, eds., *Le mouvement économique et social dans la région lyonnaise en 1901* (Paris, 1902), 87–88; see "Alternatives to Mass Production," 155. Most recent accounts of nineteenth-century Saint-Etienne suggest that its stormy political struggles can be traced to antagonism within a fractured bourgeoisie that pitted newcomer industrialists against an established merchant elite, itself only recently able to dominate master artisans. David M. Gordon, *Merchants and Capi-*

talists: Industrialization and Provincial Politics in Mid-Nineteenth-Century France (Tuscaloosa, Ala.: University of Alabama Press, 1985). Also, David McKinney Pemberton, "Industrialization and the Bourgeoisie in Nineteenth-Century France: The Experience of Saint-Etienne, 1820–1872" (Ph.D. diss., Rutgers State University, 1984). For the attempt of various republican radicals to organize a coherent working-class political movement among Saint-Etienne artisans, see Ronald Aminzade, *Ballots and Barricades: Class Formation and Republican Politics, 1830–1871* (Princeton: Princeton University Press, 1993), 139–73, 223–32.

12. Landes, *Prometheus*. On the travails of the family proprietor, see David Landes, "Religion and Enterprise: The Case of the French Textile Industry," in *Enterprise and Entrepreneurs in Nineteenth- and Twentieth-Century France*, eds. Edward C. Carter II, Robert Foster, and John N. Moody (Baltimore: Johns Hopkins University Press, 1976), 41–86. O'Brien and Keyder, *Economic Growth*. Richard Roehl, "French Industrialization: A Reconsideration," *Explorations in Economic History* 13 (1976): 233–81. Robert Aldrich, "Late Comer or Early Starter? New Views on French Economic History," *Journal of European Economic History* 16 (1987): 89–100.

13. The number of works on textiles is too vast to cite. For the metals industry and the Birmingham model, see chapter 6 below. Maxine Berg has been at the forefront of expanding the study of eighteenth-century technological change beyond the textile industry. Berg, *Age of Manufactures*, 19–20, 40–43. The sole work on gun production in France is François Bonnefoy, *Les armes de guerre portatives en France du début du règne de Louis XIV à la veille de la Révolution (1660–1789): De l'indépendance à la primauté* (Paris: Libraire de l'Inde, 1991). Unfortunately, while this monograph attempts to be comprehensive, it fails to provide any analysis of social relations in the industry. Some general investigations into the relationship between the state and the metals industry can be found in Harold T. Parker, *The Bureau of Commerce in 1781 and Its Policies with Respect to French Industry* (Durham: North Carolina Academic Press, 1979); Bertrand Gille, *Les origines de la grande industrie métallurgique en France* (Paris: Domat, 1947); Charles Ballot, *Introduction du machinisme dans l'industrie française* (Geneva: Slatkine, [1923], 1978); and especially, Woronoff, *L'industrie sidérurgique*.

14. Jean-Marie Roland de la Platière, *Manufactures, arts et métiers*, in *Encyclopédie méthodique* (Paris: Panckoucke, 1790), vol. 2, part 2: 46–47. On the structure of the Saint-Etienne metals trade, see Louis-Joseph Gras, *Essai sur l'histoire de la quincaillerie à Saint-Etienne* (Saint-Etienne: Théolier, 1904).

15. On water power, see R. Dubessy, *Historique de la Manufacture de Saint-Etienne* (Saint-Etienne: n.p., 1900), 10–11. Messance, *Nouvelles recherches*, 121. A.N. F12 1315B John Holker, "Suite des observations sur Saint-Etienne en Forez," 1756. On the concept of "second nature," and the role of standards in transforming "endowments" into "resources," see William Cronon, *Nature's Metropolis: Chicago and the Great West* (New York: Norton, 1991).

16. Louis-Joseph Gras, *Le Forez et le Jarez navigables* (Saint-Etienne: Théolier, 1930).

17. My discussion of coal is based on: Roland, *Lettres*, 6: 451–57; E. Leseure, *Historique des mines de houille du département de la Loire* (Saint-Etienne: Théolier, 1901), 21–63; and Jean Merley, "La situation de bassin houiller stéphanois en 1783," *Bulletin du Centre d'Histoire Régionale de Saint-Etienne* 2 (1977): 55–69.

18. Paul Tézenas du Montcel, *Etude sur les assemblées provinciales: L'Assemblée du département de Saint-Etienne* (Saint-Etienne: Théolier, 1903), May 1788, 2: 258–69. "Cahier de Montaud" and "Cahier de Saint-Genest-Lerpt," in *Cahiers de doléances de la province de Forez*, Etienne Fournial and Jean-Pierre Gutton, eds. (Saint-Etienne: Centre d'Etudes Foréziennes, 1974), 1: 217; 2: 335–37. Jean-Baptiste Galley, *Saint-Etienne et son district pendant la Révolution* (Saint-Etienne: Imprimerie de la Loire Républicaine, 1903–09), 1: 58–60. The local sub-délégué strenuously lobbied his superiors to protect the Reserve. Messance, *Nouvelles recherches*, 124–30.

19. For the proof rules and their violation, see A.N. F12 1309 Gaudin, 3 October 1765. Also, see below.

20. Of the sixty thousand guns per year made in the 1780s, 45 percent went to Bordeaux (for southwestern France and the Atlantic trade), 27 percent to Aix (for southern France, the Levant, and North Africa), 15 percent to Paris and Orléans, and 14 percent to Lyon and Dijon (for eastern France and markets in Switzerland and Germany). In 1787, the merchants of Saint-Etienne also shipped out nearly five thousand gun barrels. In northern France, these guns competed with the muskets of Liège. Paul Maguin, *Les armes de Saint-Etienne* (Saint-Etienne: Maguin, 1990), 51. Roland, *Manufactures*, 2, part 2: 46–48. H.H.U. fms Typ 432.1 Auguste-Denis Fougeroux de Bondaroy, "Art du cannonier, armurier à Saint-Etienne," 1763. Louis-Joseph Gras, *Histoire de l'armurerie stéphanoise* (Saint-Etienne: Théolier, 1905), 100. On French gun trade in the New World, see M. L. Brown, *Firearms in Colonial America: The Impact of History and Technology, 1492–1792* (Washington, D.C.: The Smithsonian Institution Press, 1980), 178–88. On British and Dutch guns in the Atlantic trade, see W. A. Richards, "The Importance of Firearms in West Africa in the Eighteenth Century," *Journal of African History* 21 (1980): 43–59.

21. Honoré Blanc, *Mémoire important sur les fabrications des armes de guerre* (Paris: Cellot, 1790), 7.

22. For an inspiring treatment of the cultural meaning of hats, see Sonenscher, *Hatters*, 12–17. François Billaçois, *The Duel: Its Rise and Fall in Early Modern France*, trans. Trita Selous (New Haven: Yale University Press, 1990), 174–81.

23. The ancien régime prohibited the manufacture or possession of concealed weapons, such as pocket pistols (laws of 18 December 1660 and 23 March 1728). On the ways in which privilege structured economic activity, see Liliane Hilaire-Pérez, "Invention and the State in Eighteenth-Century France," *Technology and Culture* 32 (1991): 911–31. On the growing influence of liberal thinking, see Elizabeth Fox-Genovese, *The Origins of Physiocracy: Economic Revolution and Social Order in Eighteenth-Century France* (Ithaca: Cornell University Press, 1976), 61–62, 100–103, 304–6. Also, Simone Meyssonnier, *La balance et l'horloge: La genèse de la pensée libérale en France au XVIIIe siècle* (Montreuil: Editions de la Passion, 1989).

24. A.N. F12 1309 "Mémoire: réponse des négociants aux objections des Entrepreneurs," [1765]. In 1783, of the 185 merchants, 85 shipped fewer than 10 muskets, and 42 only 1 musket. Maguin, *Armes*, 48.

25. S.H.A.T. 4f5 Danzel de Rouvroy, "Etat et dénombrement général des ouvriers," 1 January 1782.

26. On Charleville and Maubeuge, see Bonnefoy, *Armes de guerre*, 237–68. The navy ran a manufacture of firearms at Tulle. The army also had a bayonet and sword manufacture at Klingenthal; see chapter 6. On Birmingham, see chapter 6.

27. Reddy, *Market Culture*, 19–47.

28. This is what David Landes calls "fission within a trade," rather than an analytical division of labor. David Landes, *Revolution in Time: Clocks and the Making of the Modern World* (Cambridge: Harvard University Press, 1983), 204. On the number and distribution of the armorers, see S.H.A.T. 4f5 Danzel, "Etat et dénombrement," 1 January 1782. To identify locales, see J. E. Dufour, *Dictionnaire topographique du Forez et des paroisses du Lyonnais et du Beaujolais formant le département de la Loire* (Mâcon: Protat Frères, 1946). On Saint-Etienne's social geography, see Jean-Baptiste Galley, *L'élection de Saint-Etienne à la fin de l'ancien régime* (Saint-Etienne: Ménard, 1903), 248–51.

29. Pierre Jarlier, *Répertoire d'arquebusiers et de fourbisseurs français* (Saint-Julian-du-Sault: Clobies, 1976). A.D.L. L945 "Dénombrement de tous les platineurs de la municipalité de la Tour [en Fouillouse]," [1793–95]. Maguin, *Armes*, 45.

30. S.H.A.T. 4f5 Danzel, "Etat et dénombrement," 1 January 1782. A.D.L. L930(2)

Dubouchet, "Dénombrement des ouvriers," and "Tableau des ouvriers," 18 nivôse, year II [7 January 1794].

31. Denis Descreux, *Notices biographiques stéphanoises* (Saint-Etienne: Constantin, 1868), 191–93, 259–62. A.M.StE. 1D1 Conseil Municipal, *Registres*, 18 September 1766, p. 8.

32. Roland, *Lettres*, 6: 462–63. In the year II the Revolutionary government learned of the difficulty of transforming even experienced metalworkers into skilled gun-makers; see chapter 7.

33. S.H.A.T. 4f22 Sirodon, "Mémoire sur les proportion dans les armes," year XIII [1804–5].

34. S.H.A.T. 4f12 [Agoult], "Mémoire sur des épreuves de batterie de platines," January 1769.

35. Montbeillard, "Mémoire," in Galley, *Election*, 390.

36. Pierre Gardette, *Atlas linguistique et ethnographique du Lyonnais* (Paris: C.N.R.S., 1956–76). Pierre Gardette, *Etudes de géographie linguistique* (Strasbourg: Société de Linguistique Romane, 1983). The inhabitants of Saint-Etienne considered the citizens of Roanne, only fifty kilometers away, to be "foreign."

37. Jean Chapelon, *Oeuvres complètes*, ed. Annie Elsass (Saint-Etienne: Centre d'Etudes Foréziennes, 1985), 350. For an analysis of dialect poetry which seeks to dispel the moralizing tone of elite investigations of working-class life, see William Reddy, "The Moral Sense of Farce: The Patois Literature of Lille Factory Laborers, 1848–70," in *Work in France*, eds. Kaplan and Koepp, 364–92.

38. Chapelon, *Oeuvres*, 350, 356. "Au fargerons de se leva matïn/Et de sugir toujours lou meillour vïn."

39. A.N. F12 1309 Messance to Intendant at Lyon, [1774].

40. Montbeillard, "Mémoire," in Galley, *Election*, 390.

41. Georges Straka, ed., *Poèmes du XVIIIe siècle en dialecte de Saint-Etienne (Loire), Edition avec commentaires philologiques et linguistiques* (Paris: Société d'Edition les Belles Lettres, 1964), 45–46, 477–535.

42. Straka dates the poem to 1782–83, the height of the troubles in the Royal Manufacture. Straka, *Poèmes*, 539–45.

43. Ibid., 553–55.

44. Bonnefoy, *Armes de guerre*, 283–84. On the contemporary increase in worker discontent in Lyon, see Cynthia Truant, "Independent and Insolent: Journeymen and Their 'Rites' in the Old Regime Workplace," in *Work in France*, eds. Kaplan and Koepp, 131–75.

45. Bonnefoy, *Armes de guerre*, 237–68.

46. A.D.L. L943 Jovin to [Deselles?], 10 February 1782.

47. For this "sweated" outcome, see Liu, *Weaver's Knot*. The Saint-Etienne ribbon trade did fall into this pattern; see James R. Lehning, *The Peasants of Marles: Economic Development and Family Organization in Nineteenth-Century France* (Chapel Hill: University of North Carolina Press, 1980). Also, Henri Guitton, *Industrie des rubans de soie en France* (Paris: Sirey, 1928), 38. Accampo, *Family Life*, 28–33. Roland, *Manufactures*, vol. 2, part 2: 47–48.

48. On the spectacular rise of Titon, see François Bonnefoy, "Maximilien Titon, directeur-général des magasins d'armes de Louis XIV, et le développement des armes portatives en France," *Histoire, économie et société* 5 (1980): 353–80. To possess a "monopsony" is to be the sole person able (or entitled) to buy a good from various sellers; a "monopoly" is the sole right (or ability) to produce or sell a given good.

49. Fabricants de la ville de Saint-Etienne to the Controller-General, c. 1700, in *Correspondance des contrôleurs-généraux des finances avec les intendants de provinces, 1685–1699*, A. M. de Boislisle, ed. (Paris: Imprimerie Nationale, 1874–97), 2: 20. At the same time,

excluded merchants undermined Titon's empire by selling arms to regiments exempted from his monopsony, as well as to those officially forbidden. For a poetic description of the crisis, see Chapelon, *Oeuvres*, 350.

50. S.H.A.T. 4f7/1 Conseil de Guerre, "Mémoire," 12 May 1716. S.H.A.T. 4f2/3 "Adjudication," 1 September 1716. "Ordonnance," 25 October 1716. "Règlement," 4 January 1717. See the various contracts in S.H.A.T. 4f2/2. S.H.A.T. 4f3/1 "Mémoire des Entrepreneurs de Saint-Etienne," 6 June 1760.

51. S.H.A.T. 4f7/1 Vallière, *père*, to Min. War, 10 April 1749. Carrier de Monthieu, "Mémoire," 1765. For the attitude of the nobility to commercial enterprises, see d'Arcq, *Noblesse militaire*.

52. For the Carrier family's convoluted genealogical tree, see Dubessy, *Manufacture de Saint-Etienne*, 8, 115–21. In 1746–54 the artisanal class bought 264,000 *livres*' worth of property and sold 225,000 *livres*, for a net gain of 37,000 *livres*. This, however, was dwarfed by the merchants' net gain of 449,000 *livres*. Of the merchants' loans to artisans, 51 percent were for sums between 10–99 *livres*, and 40 percent were for sums between 100–499 *livres*. Josette Garnier, *Bourgeoisie et propriété immobilière en Forez aux XVIIe et XVIIIe siècles* (Saint-Etienne: Centre d'Etudes Foréziennes, 1982), 106–25, 145–51, 252–54, 476–77, 479–83. Compare with Harpers Ferry's ruling "Junto" in M. R. Smith, *Harpers Ferry*, 140–83.

53. Etienne Fournial et al., *Saint-Etienne: Histoire de la ville et de ses habitants* (Roanne: Horvath, 1976), 138. Galley, *Election*, 303–4. See the satire of these local "bourgeois gentilshommes" in Chapelon, *Oeuvres*, 294.

54. S.H.A.T. 4f7 Jean Maritz, "Mémoire résumé," 1763. The price, including a bayonet, went from 13 *livres* 10 *sous* to 22 *livres*. Bonnefoy, *Armes de guerre*, 492. S.H.A.T. 4f12 Montbeillard, "Mémoire sur les platines," 6 May 1764.

55. S.H.A.T. 4f2/2 "Actes," 5 February, 3 March 1765.

56. S.H.A.T. 4f7 [Bellegarde], "Observations sur la situation actuelle à Saint-Etienne," 8 September 1765. Montbeillard lost his position because he and Buffon crossed swords with the Gribeauvalists on metallurgical questions relating to the artillery quarrel.

57. S.H.A.T. 4f7 Bellegarde, "Mémoire," [1769]. Montbeillard, "Mémoire," 18 April, 12 September 1764.

58. *Lettres patentes pour l'establissement d'une Manufacture Royale d'Armes à Feu à Saint-Etienne en Forez*, 5 August 1769.

59. *Jugement du Conseil de Guerre tenu par ordre du roi, à l'Hôtel des Invalides du octobre 12, 1773* (Paris: Imprimerie Royale, 1773). Bellegarde was sentenced to twenty years. Paradoxically, the trial enabled the Gribeauvalists to paint themselves as the victims of "ministerial despotism." For a full treatment of the trial, see Alder, "Forging the New Order," 44–58; Nardin, *Gribeauval*, 219–52; and S.H.A.T. 5a9/1.

60. B.M.StE. #542 [H.-M. Bellegarde], "Mémoire à consulter," in [Carrier de Monthieu et al.], *Mémoires à consulter pour le sieur de Monthieu, Entrepreneur de la Manufacture Royale des Armes à Saint-Etienne* (Paris: Valleyre, 1773). Gribeauval later condemned all the guns produced since 1746 as contemptible, and judged those produced between 1763 and 1766 as mediocre, "since the workers were only just then emerging from barbarism." S.H.A.T. 5a9/1 Gribeauval to Montbarey (Min. War), 2 April 1776.

61. Some historians claim that Choiseul always intended the guns for the American rebels, but the inspections began in the late 1760s when an imminent revolt in America hardly seemed likely. Philibert-Jérome Gaucher Passac, "Précis sur M. de Gribeauval," [1816], in *Revue d'artillerie* 34 (1889): 98–109. See Alder, "Forging the New Order," 54–58, for the picaresque story of how these defective guns eventually made their way to the American Continental Army, thanks to the cupidity of Carrier de Monthieu, the intrigues of Pierre-Augustin Caron de Beaumarchais, and the treason of Silas Deane.

62. Le Duc, "Réponse aux notes demandées par M. de Saint-Auban," 14 January 1772; Linguet, "Consultation," 6 October 1773; and L.-A. Bellegarde, "Mémoire," in [Carrier de Monthieu et al.], *Mémoire à consulter*.

63. S.H.A.T. MR1741 Blanc, "Mémoire historique sur les progrès de la fabrication des armes pour les troupes du roi," 27 April 1777. For objections to the Vallièrist gun, see S.H.A.T. 4f22/1 Biron to [indecipherable], 20 June 1774. Colonels who no longer purchased weapons for their troops had less incentive to see that their soldiers took proper care of them. S.H.A.T. 4f57 "Mémoire d'artillerie," 1775.

64. S.H.A.T. 4f22/1 Gribeauval, Montbarey, Du Châtelet, No title, 24 January–7 February 1776. The Régiment du Roi was one of those exempted from using regulation guns. S.H.A.T. 4f3 Blanc, "Etat de dépenses," 18 March 1777. Patrick d'Arcy complained bitterly that Gribeauval had rejected his design out of hand because of d'Arcy's support for the Vallièrists. [Bachaumont], *Mémoires secrets*, 12 May 1777, 10: 171.

65. For an overview of changing gunlock designs, see Jean Boudriot, "L'évolution de la platine de l'arme d'infanterie française," *Gazette des armes* 68 (1979): 35. These 1788–90 tests still used range as an index of the superiority of one design over another. A.N. T591(3) Givry et al., "Journal des épreuves des armes de guerre," 14 October 1788; "Tableau contenant le journal d'épreuves," 1790.

66. Lombard, *Projectiles*, 186–201. See also Guibert, *Essai général*, 1: 223–51.

67. S.H.A.T. MR1741 Honoré Blanc, "Mémoire historique sur les progrès," 27 April 1777. This program of "experimentation" was already underway in the 1760s. S.H.A.T. 4f12 [Agoult], "Epreuve sur la batterie," January 1769.

68. Some evidence on reliability can be found in the reduction in the number of soldiers admitted to the Hôtel des Invalides after having been maimed by their own musket: 8 in 1715–16 and only 2 in 1762–63. Corvisier, *L'armée française*, 2: 676–80. On gun design, see Maurice Bottet, *Monographies de l'arme blanche (1789–1870) et de l'arme à feu portative (1718–1900) des armées françaises de terre et de mer* (Paris: Haussmann, 1959), 135–62. Torsten Lenk, *The Flintlock: Its Origin and Development*, trans. G. A. Urquatt (London: Holland Press, 1965), 109–20. Arthur Norris Kennard, *French Pistols and Sporting Guns* (Hamlyn, N.Y.: Country Life Collectors' Guides, 1972), 38–43.

69. S.H.A.T. 4f22/1 Gribeauval, Montbarey, Du Châtelet, No title, 24 January–7 February 1776. [Gribeauval], "Règlement des proportion que doit avoir le fusil," 26 February 1777. Gribeauval to [Indecipherable], 26 February 1777.

70. S.H.A.T. 4f3/1 [Entrepreneurs de Saint-Etienne], "Extrait d'un mémoire donné au ministre de la guerre," 6 June 1760. A.M.StE. HH12 Anon., "Concernant les nouveaux règlements établis dans la Manufacture de Saint-Etienne," [1759]. Anon., "Mémoire concernant les nouveaux arrangements établis dans la Manufacture de Saint-Etienne," [1759]. At the time, the War Office drew up a list of the prices paid to workers over the past fifty years. S.H.A.T. 4f3/1 "Etat concernant les prix payés aux différents ouvriers, 1711 à 1760," [1760.]

71. S.H.A.T. 4f3 "Tableau du temps employé à la fabrication des pièces d'armes dans les différentes Manufactures," [1770s]. For Perronet, see chapter 4. Compare with A. Chandler, *Visible Hand*.

72. S.H.A.T. 5a9/2 Carrier Du Réal to Gribeauval, 12 October 1785. S.H.A.T. 4f3 Gribeauval to Ségur, 5 June 1785. Dubessy, *Manufacture de Saint-Etienne*, 116–21 and appendix 6.

73. Gras, *Armurerie*, 42–48. A recent discussion of Turgot's ban can be found in Steven L. Kaplan, "Social Classification and Representation in the Corporate World of Eighteenth-Century France: Turgot's 'Carnival,'" in *Work in France*, eds. Kaplan and Koepp, 176–228.

74. "Règlement de 26 février 1777," in Gras, *Armurerie*, 47.

75. Kaplan, "Police du monde du travail." On the arquebusier corporations outside of Saint-Etienne, see René de Lespinasse, *Les métiers et corporations de la ville de Paris* (Paris: Imprimerie Nationale, 1892), 2: 341–56.

76. S.H.A.T. 4f5 [Gribeauval], "Mémoire," 29 October 1781.

77. S.H.A.T. MR1739 Agoult, "Mémoire remis à M. de Saint-Sernin," 14 November 1782. S.H.A.T. 4f7 Danzel, "Mémoire sur les causes de la décadence de la Manufacture de Saint-Etienne," 19 September 1784. The total budget for firearms, after briefly rising from 500,000 *livres* in 1782 to 1,200,000 *livres* in 1783, fell to 600,000 *livres* in 1784 and thence to 100,000 *livres* in 1785–87. As the king was approximately one to three years in arrears, this suggests a sizable drop in purchases sometime in the early 1780s. S.H.A.T. MR1739 "Dépenses annuelles de l'artillerie," [1788].

78. Montbeillard, "Mémoire," in Galley, *Election*, 390. S.H.A.T. 4f12 Montbeillard, "Mémoire," 16 August 1763; "Mémoire sur les platines," 6 May 1764. Dubessy, *Manufacture de Saint-Etienne*, 18. There were three Jean Reys and two Lalliers listed as proof masters in this period.

79. S.H.A.T. 9a10 Vallière, *père*, to Chamillart, 10 November 1728. S.H.A.T. 4f2 "Mémoires des entrepreneurs de Saint-Etienne," 26 February 1729. S.H.A.T. 4f7/1 Vallière, *père*, to Pierrebrune, 10 April 1749. The condemnation of Bonnand comes from S.H.A.T. 4f7/1 [Bellegarde], "Mémoire," 1765.

80. Descreux, *Notices biographiques*, 63–64. S.H.A.T. 4f3 Vallière, *fils*, No title, 15 January 1773. S.H.A.T. 4f3 Anon., "Examen des raisons qui ont rendu susceptible d'augmentations plusieurs pièces du nouveau modèle du fusil de soldat," [1777?]. S.H.A.T. 4f7 [Bellegarde], "Mémoire," 1765.

81. Montbeillard, "Mémoire," in Galley, *Election*, 386–90.

82. S.H.A.T. 4f7 Montbeillard, "Mémoire," August 1763. S.H.A.T. 4f3 Vallière, *fils*, 15 January 1773. Blanc, vaguely implicated in the Bellegarde scandal, returned to Saint-Etienne only when the Gribeauvalists reestablished their authority. S.H.A.T. 6c5 Monteynard (Min. War) to Inspector Jaunay, 8 January 1774. Blanc to Gribeauval, 12 April 1782.

83. S.H.A.T. 5a9/1 [Agoult?], "Mémoire," 1777.

84. These drawings have unfortunately been stolen from the military archives, and I wish to thank Louis Viau for supplying me with photocopies of them. S.H.A.T. 4f22 [Département de la Guerre], *Règlement fixant les principales dimensions des armes portatives, 1 vendémiaire, year XIII* [23 September 1804] (Paris: Imprimerie Nationale, XIII [1804]). For the debates over specifying gun dimensions in 1804, see Sirodon, "Mémoires sur les proportions dans les armes," year XIII [1804–5]; Tuffet Saint-Martin, "Observations sur le règlement," year XIII [1804–5]; C. G. Dufort, "Observations sur le règlement," year XIII [1804–5]. S.H.A.T. 4f7 [Blanc et al.], "Mémoire sur la fabrication des armes à feu à la Manufacture de Saint-Etienne," 1777.

85. On gauges in U.S., see M. R. Smith, *Harpers Ferry*, 102, 109–10. On gauges in France, see Cohen, "Inventivité organisationnelle." On the use of gauges in the manufactures, see S.H.A.T. 4f3 [Gribeauval], "Circulaire à Montbeillard, Bellegarde, et Minard, inspecteurs des trois manufactures," March 1766. On the local replication of master gauges, see S.H.A.T. 4f6/3 Agoult, "Projet d'un règlement," [1770s].

86. S.H.A.T. 4f10 Blanc, "Observations sur 4 cannons," 24 June 1781. S.H.A.T. 6c5 Danzel, "Aperçu des travaux," 16 May 1783.

87. S.H.A.T. 4f12 [Agoult], "Mémoire," January 1769. Montbeillard, "Mémoire," in Galley, *Election*, 386–90. Battison, "Eli Whitney," 11–16. For more information on the hollow milling machine, see chapter 5. On the gauges and jigs, see Montbeillard, "Mémoire," in Galley, *Election*, 386–90. S.H.A.T. 4f6/3 "Projet de règlement pour la Manufacture de Saint-Etienne," 21 July 1773.

88. S.H.A.T. MR1741 Blanc, "Mémoire historique sur les progrès," 27 April 1777. S.H.A.T. 4f1 [Agoult], "Règlement provisoire pour la Manufacture de Saint-Etienne," 26 February 1777.

89. S.H.A.T. 4f3 [Blanc] to Gribeauval, 1 February 1778.

90. S.H.A.T. 4f7 Agoult, "Mémoire," 1785. See also S.H.A.T. MR1741 Blanc, "Mémoire historique sur les progrès," 27 April 1777. S.H.A.T. 4f1 Agoult, "Règlement provisoire," 26 February 1777.

91. S.H.A.T. 4f3 Anon., "Tableau du temps," [1770s]. My discussion of barrel-making draws on the following sources: S.H.A.T. MR1741 Honoré Blanc, "Mémoire historique sur les progrès," 27 April 1777; S.H.A.T. 4f7 [Blanc et al.], "Mémoire sur la fabrication des armes," 1777; H.H.U. fms Typ 432.1 Fougeroux de Bondaroy, "Art du cannonier," 1763; and Dubessy, *Manufacture de Saint-Etienne*, 78–81, 223–39. On Blanc's efforts, see S.H.A.T. 4f10 Blanc, "Observations sur 4 cannons," 24 June 1781. Interestingly, the same regional variation in production methods can be found in the contemporary United States: at Harpers Ferry, the armorers also did not overlap the two edges, while at the northern Springfield armory, they did. M. R. Smith, *Harpers Ferry*, 95–96.

92. A.N. AD VI 39 Gillier-Renard and Siauve, *Les députés extraordinaires de la commune de Saint-Etienne aux députés de la Convention Nationale* (Paris: Girouard, [1793]). Claude Journet to Javogues, [1793–94], and Pierrotin to Assemblée Nationale, 30 December 1789, in Galley, *Révolution*, 1: 71–72. A.D.L. L835 [Guilliaud], *La fabrication d'armes de guerre à Saint-Etienne* (n.p., [1800]). M. Thiroux, *Instruction théorique et pratique d'artillerie* (3d ed.; Paris: Dumaine, 1849), 170. Compare with M. R. Smith, *Harpers Ferry*, 113–16.

93. S.H.A.T. 4f7 Danzel, "Décadence de la Manufacture," 19 September 1784. S.H.A.T. 4f10 Blanc, "Observations sur 4 canons," 24 June 1781. The same working conditions prevailed for bayonet grinders at the Manufacture of Klingenthal. S.H.A.T. Xd234 Eblé, "Inspection de Klingenthal," 1808. Aboville, "Inspection de Klingenthal," 20 November 1810.

94. For a description of the lathes, see S.H.A.T. 4f8 C. P. Griffet de la Beaume, "Rapport sur les trois machines inventées par le Sr. Javelle," 18 January 1787. A. Paulin-Desormeaux, *Manuel de l'armurier, du fourbisseur et de l'arquebusier* (Paris: Roret, 1832), 270–73. On Javelle, see Descreux, *Notices biographiques*, 205–6. On lathes in the north, see S.H.A.T. 4f6/2 Agoult, "Notes pouvant servir à la formation d'un règlement d'armes," [1777]. On American lathes, see M. R. Smith, *Harpers Ferry*, 117–24.

95. This calculation of cost savings did not take into account capital or development costs. S.H.A.T. 4f7 (16) Agoult, No title, [1790–91]. A.D.L. L834 Javelle, "Brevet: Machines propre à achever et polir extérieurement les canons de fusils, sans meules," 13 February 1792. On conflict over the lathes in the early 1800s, see S.H.A.T. 6c5 A. G. Aboville, *fils*, to F. M. Aboville, *père*, 5 germinal, year X [26 March 1802]. Lathes finally came into use in the 1840s. Dubessy, *Manufacture de Saint-Etienne*, 399–404. Thiroux, *Artillerie*, 172.

96. Steven L. Kaplan, *Provisioning Paris: Merchants and Millers in the Grain and Flour Trade during the Eighteenth Century* (Ithaca: Cornell University Press, 1984), 25–33.

97. This incident is described in partisan terms by a number of letters and memoirs in A.N. F12 1309. On new barrel design, see Kennard, *French Pistols*, 42.

98. A.N. F12 1309 "Ordonnance de par le roi portant règlement sur l'épreuve et l'expédition des armes de commerce qui se fabriquent à Saint-Etienne," 10 February 1780. Augustin Merley to Necker, 17 September 1780.

99. Vauban, quoted in Bonnefoy, *Armes de guerre*, 131–32. A.N. F12 1317 Maire et échevins et les principaux arquebusiers de la ville de Saint-Etienne to Bertin, [1763]; Choiseul to Echevins, 25 April, 28 May 1763; Bertin to Trudaine, 31 July 1763. S.H.A.T. 4f7/1 "Délibérations des échevins de Saint-Etienne," 15 October 1764 (with comments by Montbeillard). See the summary in A.N. F12 1309 [Fleselle], 22 January 1781. On the

efficacy of the double test, see H.H.U. fms Typ 432.1 Fougeroux de Bondaroy, "Art du cannonier," 1763.

100. S.H.A.T. 4f7/1 "Mémoire pour la fabrication des fusils," [1720]. S.H.A.T. 4f22/1 "Règlement du 26 février 1777, des proportions que doit avoir un fusil d'infanterie."

101. S.H.A.T. 4f3 Gribeauval to Ségur, 5 June 1785. A.N. F12 1309 Lullier to Necker, 13 April 1781.

102. A.N. F12 1309 Jean Lullier to Necker, 13 April 1781.

103. A.N. F12 1309 "Etat de quelques ouvriers mis en prison par order de M. d'Agoult," [1781]. Jean Lullier, "Mémoire," 15 June 1781. If the witticism was in French, then it was meant to be understood.

104. A.N. F12 1309 "Ouvriers mis en prison par Agoult," [1781]. Linebaugh, *London Hanged.*

105. A.M.StE. 1D1 Conseil Municipal, *Registres*, 2 October 1781. A.N. F12 1309 Neyron et al. to Contrôleurs-Généraux, 29 December 1780.

106. A.N. F12 1309 Fleselle to [Necker], 22 January 1781. Agoult to Necker, 28 July 1780. Anon., "Mémoire," 5 June 1781.

107. "Arrêté du Conseil d'Etat du Roi, portant règlement pour la police de la Manufacture d'Armes établie à Saint-Etienne," 17 January 1782, in Roland, *Manufactures*, vol. 2, part 2: 49–50. On the retreat of the War Office, see S.H.A.T. 9a11 Ségur to Gribeauval, 18 January 1782. Gribeauval to Joly de Fleury, 23 February 1782.

108. A.M.StE. 1D1 Conseil Municipal, *Registres*, 2 October 1781, p. 142. A.D.L. L948 Conseil Municipal, "Elections," 1786–89.

109. William H. Sewell, Jr., *A Rhetoric of Bourgeois Revolution: The Abbé Sieyes and "What is the Third Estate?"* (Durham, N.C.: Duke University Press, 1994). It is worth noting that the *fabricants* preferred a laissez-faire state, while Sieyes wanted the state deeply involved in economic life.

110. S.H.A.T. 9a11 Gribeauval to Ségur, 8 January 1782.

111. S.H.A.T. 9a11 Danzel, "Mémoire," 3 June 1782. Louis XVI, "De par le roi," 15 June 1782.

112. "Ordonnance du roi, portant règlement sur le bon ordre que Sa Majesté veut être observé dans la Manufacture d'Armes à Feu établie à Saint-Etienne," 7 July 1783, in Roland de la Platière, *Manufactures*, vol. 2, part 2: 50–53.

113. S.H.A.T. 4f7 Danzel, "Décadence de la Manufacture," 19 September 1784. The proposals of Carrier de la Thuillerie and Dubouchet (in Saint-Etienne) and Ignace de Wendel (in Charleville) are summarized in S.H.A.T. 4f3 Gribeauval to Ségur, 5 June 1785.

114. S.H.A.T. 4f3 Gribeauval to Ségur, 5 June 1785. Also S.H.A.T. 4f3 Gribeauval, "Mémoire," 1785.

115. S.H.A.T. 4f3 Gribeauval to Ségur, 5 June 1785.

116. On iron prices, see A.D.L. L948 Messance to [Intendant], 11 May 1788. Messance to Terray, 9 January 1789. On the eve of the Revolution, Inspector Lespinasse claimed that 360 gunsmiths were registered. S.H.A.T. 5a9/2 Lespinasse et al., "Observations des contrôleurs et reviseurs de la Manufacture de Saint-Etienne," 14 February 1789. On the new contract, see Dubessy, *Manufacture de Saint-Etienne*, 63.

117. Lynn Hunt, "Committees and Communes: Local Politics and National Revolution in 1789," *Comparative Studies in Society and History* 18 (1976): 321–46, quote on 321. See also Lynn Hunt, *Revolution and Urban Politics in Provincial France: Troyes and Reims, 1786–1790* (Stanford: Stanford University Press, 1978).

118. Hunt, "Committees and Communes," 346.

119. Taylor, "Non-capitalist."

120. The rising price of foodstuffs touched off two outbursts, one in April 1789 and one in August 1790. A.M.StE. HH9 Conseil Municipal, *Registres*, 8–18 April 1789. Jean Merley,

"Réflexions sur le cours des grains à Saint-Etienne et au Puy aux XVIIe et XVIIIe siècles (1660–1793)," *Bulletin du Centre d'Histoire Régionale, Université de Saint-Etienne* 1 (1981): 39–70.

121. Randall points out that Luddism was most intense where the concentration of putting-out capital was greatest, and presumably, social mobility was least. Adrian Randall, "Work and Resistance in the West of England," in *Regions and Industries: A Perspective on the Industrial Revolution in Britain*, ed. Pat Hudson (Cambridge: Cambridge University Press, 1989), 175–98.

122. Sauvade quoted in Galley, *Révolution*, 1: 74–76. Descreux, *Notices biographiques*, 317–18. Gras, *Quincaillerie*, 98–99. For other instances of machine-breaking by artisans in the early days of the Revolution, see Gay Gullickson, *Spinners and Weavers of Auffay: Rural Industry and the Sexual Division of Labor in a French Village, 1750–1850* (Cambridge: Cambridge University Press, 1986), 89–90.

123. A.D.L. L1039 Lespinasse to Tour du Pin (Min. War), 13 August, 28 October 1789; Lespinasse to Agoult, 18 August 1789. Galley, *Révolution*, 1: 60–67.

124. A.D.L. L1039 Lespinasse to Tour du Pin (Min. War), 27, 28 November 1789; Lespinasse to Agoult, 11, 12, 13, 19 November 1789.

125. Fournial, *Saint-Etienne*, 159–64.

126. A.D.L. L1039 Lespinasse to Tour du Pin (Min. War), 27, 28 November 1789. On the Council, see Galley, *Révolution*, 1: 81–83.

127. A.D.L. L945 "Délibération prise par les officiers municipaux de la ville de Saint-Etienne," 23 September 1790. The mayor had been away on September 23, and the next day the municipality moderated its stand somewhat, denying that it had called for the complete suppression of the manufacture.

128. *A.P.*, 19 December 1791, 36: 249.

129. Corps Municipal to Servan (Min. War), 25 May 1792, in Dubessy, *Manufacture de Saint-Etienne*, 130. See also *A.P.*, 28 May 1792, 44: 177.

130. On the Commission, see *A.P.*, 11, 16 June, 4 July 1792, 45: 99–100, 270–71; 46: 108. A.D.L. L945 Min. War to Lespinasse, 21 July 1792. The Assembly authorized 3 million *livres* for these purchases. The proposal to relax standards was opposed by Jean-Gérard Lacuée who sought to reestablish the authority of the artillery over the manufactures. *A.P.*, 17 July 1792, 42: 552-56.

131. *A.P.*, 3 December 1791, 35: 558.

132. On the new legal monopoly, see A.D.L. L945 "Arrêté," 29 September 1792. A.N. AD VI 40 "Projet de décret proposé à la Convention Nationale," 9 October 1792. *A.P.*, 13 October 1792, 52: 487. On Romme's Council, see A.D.L. L945 Romme et al. to Municipalités (Saint-Genest-Lerpt, Vilard, Fouillouse, etc.), 23 September 1792. "Extrait de registre de la municipalité de Roche la Molière," 25 September 1792. A.M.StE. 2H2 Romme, Minutes of meetings, 7, 9 October 1792. For complaints about patronage and favoritism, see A.M.StE. 2H2 Robert et al. to Citoyens Administrateurs, 9 April 1793. *A.P.*, 10 September, 2, 30 October 1792, 49: 548; 52: 276; 53: 63. Galley, *Révolution*, 1: 366–71.

133. *A.P.*, 9 October 1792, 52: 415. A.D.L. L945 Crouzat and Coignet, "Journal," 18 December 1793. The town's merchants submitted 16,742 guns, but over a quarter had to be returned for repairs, and a third had not yet been verified. Lespinasse to Administrateurs du District, 16 December 1792. A.D.R. 42L171* (58) "Certificats," December 1793.

134. A.D.R. 42K171* Bouillet to Min. War, 10, 14 May, 23 July 1793. *A.P.*, 10 February, 20 March, 2 April 1793, 58: 439; 60: 349; 61: 107. A.D.L. L945 "Etat de la déclaration des armes et munitions des citoyens des municipalités de Valbenoîte, Fauet, La Valette et Le Métar," 6 September 1792.

135. *A.P.*, 3 April, 30 May, 20 June 1793, 61: 280; 65: 616; 67: 22. A.D.R. 42K171* Letters of Bouillet to Min. War, May–August 1793. Colin Lucas, *The Structure of the Terror: The Example of Javogues and the Loire* (Oxford: Oxford University Press, 1973), 35–42.

Chapter Six
Inventing Interchangeability:
Mechanical Ideals, Political Realities

1. William A. Johnson, ed. and trans., *Christopher Polhem: The Father of Swedish Technology* (Hartford: Trinity College Press, 1963).

2. S.H.A.T. 6c5 Ressons to Son Altesse Sérénissime, 21 August 1723. "Etat des épreuves de Deschamps," [1727]. Breteuil (Min. War), "Mémoire," 29 September 1723. S.H.A.T. 6c5 6203(8/9) Anon., No title, [1727–28]. S.H.A.T. 4f12 3385 Anon., No title, 28 September 1725. The sole published reference to Deschamps's work is the scathing post-Revolutionary critique by Gassendi, engaged at the time in suppressing Blanc's interchangeable parts manufacture; see chapter 9. Jean-Jacques-Basilien Gassendi, *Aide-mémoire à l'usage des officiers d'artillerie de France* (5th ed.; Paris: Magimel, 1819), 591.

3. S.H.A.T. 6c5 Guérin, "Mémoire sur la fabrication," [1720s]. Ressons, three letters of 21 August 1723, 21 February 1724, [1724]. S.H.A.T. 4f7 Vallière, *père*, "Mémoire," 1727. S.H.A.T. 4f5 "Arrêté du roi," 4 February 1727. Deschamps's confederate Bicot was appointed controller at Saint-Etienne during the 1720s, and earned 4,500 *livres* a year, a salary comparable to that of the inspector. S.H.A.T. 4f5 [Vallière, *père*], "Mémoire," 7 December 1736.

4. Gordon, "Mechanical Ideal."

5. S.H.A.T. 4f12 Montbeillard, "Rapport sur la fabrication des platines," 6 May 1764. On Blanc's personal history and character, see S.H.A.T. 6c5 Du Châtelet to Ségur, 27 July 1786. S.H.A.T. 4f7 Bellegarde, "Observations sur la situation," 8 September 1765. See Blanc's own résumé of his career in his *Mémoire important*, 5–7. A letter of 1798, in his own hand, indicates that Blanc was literate, though without higher schooling. S.H.A.T. 6c5 Blanc to [F. M. Aboville, *père*], 26 germinal, year VI [15 April 1798]. Blanc's efforts have been briefly alluded to by various students of the French military and French industrialization. Nardin, *Gribeauval*, 344–46; Galley, *Election*, 399–400; and Ballot, *Machinisme*, 501.

6. For the best study of this social type in France, see James M. Edmonson, *From Mécanicien to Ingénieur: Technical Education and the Machine Building Industry in Nineteenth-Century France* (New York: Garland, 1987).

7. As a local controller at Saint-Etienne, Blanc was implicated in the Bellegarde affair and dispatched to Charleville in 1774 on the orders of Vallière, *fils*. S.H.A.T. 6c5 Monteynard (Min. War) to Inspector Jaunay, 8 January 1774. On Blanc's duties, see S.H.A.T. 4f5 Gribeauval, "Mémoire," 21 October 1781. For his shop, see Danzel, "Aperçu des travaux," 16 May 1783. S.H.A.T. 4f12 Blanc, "Inventoire des outils pour le travail des platines," 5 November 1784.

8. S.H.A.T. 4f3 Gribeauval to Ségur (Min. War), 5 June 1785.

9. S.H.A.T. 6c5 Gribeauval to Agoult, 31 July 1785.

10. S.H.A.T. MR1739 "Dépenses annuelles de l'artillerie," [1788]. On Guibert's hopes for cost-saving from Blanc's techniques, see Nardin, *Gribeauval*, 355. The War Office seemed more pleased with the accounting work of Rolland de Bellebrune (Gribeauval's secretary) than with any of Blanc's accomplishments. S.H.A.T. 6c5 Ségur (Min. War) to Gribeauval, 22 July 1785, 7 July 1786, 25 May 1787. For Gribeauval's defense of the program, see S.H.A.T. 6c5 Note by Gribeauval, in Danzel, "Aperçu des travaux," 16 May 1783. For the accounts themselves, see S.H.A.T. 6c5 Rolland de Bellebrune, "Recapitulation des dépenses," 12 June 1790; S.H.A.T. MR1739 Rolland de Bellebrune, "Mémoire concernant la comptabilité," 28 September 1790.

11. S.H.A.T. 4f3 Gribeauval to Ségur (Min. War), 5 June 1785.

12. S.H.A.T. 6c5 Danzel, "Aperçu des travaux," 16 May 1783.

13. S.H.A.T. MR1739 Rolland de Bellebrune, "Mémoire," 28 September 1790. Bonnefoy, *Armes de guerre*, 393–411.

14. S.H.A.T. 6c5 Rolland de Bellebrune, "Mémoire concernant la fabrication," [late 1791]. F. M. Aboville, *père*, to Narbonne (Min. War), 27 January 1792. Locks were shipped untempered to other manufactures, where they were engraved with the name of that manufacture. S.H.A.T. 6c5 Régnier et al., "Expérience faite . . . sur cent platines," 15 germinal, year XII [5 April 1804].

15. S.H.A.T. 4f7 Danzel, "Décadence de la Manufacture," 19 September 1784. S.H.A.T. MR1743 Dubois d'Escordal, "Mémoire sur la constitution et manufacture d'armes à feu," [1788–90].

16. S.H.A.T. 6c5 [Givry et al.], "Précis des motifs qui ont fait autoriser le Sr. Blanc," January 1792.

17. S.H.A.T. 6c5 Blanc to Gribeauval, 12 April 1782. Blanc, *Mémoire important*, 12, 18.

18. Quote in S.H.A.T. 6c5 [Givry et al.], "Précis," January 1792. On the transfer to Vincennes, see Ségur (Min. War) to Gribeauval, 22 July 1785. Du Châtelet to Ségur, 27 July 1786.

19. Blanc, *Mémoire important*, 16–17.

20. S.H.A.T. 6c5 Givry et al., "Résultats," [November 1791]. [Givry et al.], "Précis," January 1792.

21. S.H.A.T. 9a11 Gribeauval to Ségur (Min. War), 16 April 1783. Givry's personal papers are available in A.N. T591 (1–6). For a general description of the Klingenthal armory, see Bonnefoy, *Armes de guerre*, 287–334.

22. For complaints about the rings in Saint-Etienne, see S.H.A.T. T591(1/2) Givry to Fyard, 12 February 1783.

23. For the artisans' original complaint, see A.N. T591(1/2) Jean Schmidt et al., "Aujourd'hui," 29 January 1783. For the government correspondence on this matter, see A.N. T591(1/2) Givry to Ségur, 22 October 1782; Givry to Gribeauval, 23 January 1783; Givry, "Mémoire," 3 February 1783; Gribeauval to Ségur, 29 January 1783; Ségur to Gribeauval, 11 March 1783.

24. S.H.A.T. 9a11 Villeneuve to Gribeauval, 4 April 1783.

25. Ibid. S.H.A.T. T591(1/2) Givry to Agoult, 15 March 1783.

26. S.H.A.T. Bib. MS175 Comité Militaire, 2 April 1783, 3: fol. pp. 194–206.

27. S.H.A.T. 9a11 Villeneuve to Gribeauval, 4 April 1783.

28. S.H.A.T. Bib. MS175 Comité Militaire, 2 April 1783, 3: fol. pp. 194–206. S.H.A.T. 9a11 Givry to Ségur, 24 April 1783.

29. S.H.A.T. Bib. MS175 Comité Militaire, 2, 9, 16, 23 April 1783, 3: fol. pp. 194–206, 222, 234–44.

30. A.N. T591(1/2) Givry to Villeneuve, 22 April and 31 July 1783. A.N. T591(3) Givry and Bisch, "Notes de ce que l'on peut tolérer sur les dimensions de la baïonnettes du modèle 1777," 30 January 1784. "Ordonnance portant règlement," 1 April 1784; Givry to Gau, *fils*, 21, 23 August, 22 September 1784; Givry to Ségur, 25 August, 17 September 1784; Givry to Gribeauval, 1 September 1784; Givry to Gomer, 22 September 1784.

31. A.N. T591(3) Givry to Bisch, 31 July 1783. A.N. T591(6) [Veimarq] to Givry, 17 October 1784. Givry's library is catalogued in A.N. T591(1/2), as is his personal correspondence and daily diary-accounts.

32. A.N. T591(6) Givry to Ségur, 3 October 1784; Perrier to Givry, 25 March 1785. *Lettres patentes du roi . . . entreprise de la Manufacture d'Armes Blanches de Klingenthal* (Strasbourg: Levrault, 1786). Alexandre-Théophile Vandermonde, *Procédés de la fabrication des armes blanches* (Paris: Imprimerie du Département de la Guerre, year II [1793–94]), 5–6.

33. S.H.A.T. 6c5 Danzel, "Aperçu des travaux," 16 May 1783.

34. Berg, *Age of Manufactures*, 264–314. On Boulton and the general trend toward broadly based consumption, see Neil McKendrick, "Commercialization and the Economy," in *The Birth of a Consumer Society: The Commercialization of Eighteenth-Century England*, eds.

Neil McKendrick, John Brewer, and J. H. Plumb (Bloomington: Indiana University Press, 1982), 9–194, esp. pp. 66–77. For the earlier history of Birmingham, see Marie B. Rowlands, *Masters and Men in the West Midland Metalware Trades before the Industrial Revolution* (Manchester: Manchester University Press, 1975).

35. For an analysis of Saint-Etienne's hardware trades, see A.N. F12 1315B Michael Alcock, "Mémoire," [1750s]; also [John] Holker, "Suite des observations sur Saint-Etienne en Forez," 1756. On changes in Saint-Etienne's cutlery trade, see Auguste-Denis Fougeroux de Bondaroy, "Art du coutelier en ouvrages communs," in *Descriptions des arts et métiers, faites ou approuvées par Messieurs de l'Académie Royale des Sciences*, ed. Jean Elie Bertrand (Neuchâtel: Société Typographique, 1780), 14: 365–68. Also A.D.L. L930 Guilliaud, *Mémoire sur la mise en oeuvre de tous les metaux dans le Département de la Loire* (Lyon: Louis Cutty, [1795–96]). For the history of Alcock and Trudaine's negotiation, see A.N. F12 1315B Alcock to Trudaine, 16 September 1763. On the paradoxes of Trudaine's policies, see Hilaire-Pérez, "Invention and the State," and more generally, Fox-Genovese, *Physiocracy*. On the inadequate models of the classical political economists when confronted with the complexities of the Birmingham metals trade, see Berg, *Age of Manufactures*, 69–77, 264–314. The topic of technology transfer is vast. For the eighteenth century see Peter Mathias, "Skills and the Diffusion of Innovations from Britain in the Eighteenth Century," in *The Transformation of England* (New York: Columbia University Press, 1979), 21–44. John Harris has used the Alcock example to illustrate the phenomenon of technology transfer, without noting its influence on the development of interchangeable parts production. John Harris, "Michael Alcock and the Transfer of Birmingham Technology to France before the Revolution," in *Essays in Industry and Technology in the Eighteenth Century* (Hampshire, U.K.: Variorum, 1992), 113–63.

36. On Alcock's early years in France, see Bernard Guineau, *La Manufacture Royale de Quincaillerie, Taillanderie, et Bijouterie de toutes sortes de métaux façon d'Angleterre de La Charité-sur-Loire, 1756–1809* (La Charité-sur-Loire: Association des Amis de La Charité-sur-Loire, 1985), 27–28. On the move to Roanne, see A.N. F12 1315A Alcock to Trudaine, 29 March 1763. B.M.R. 2C4 "Manufacture de boutons d'Alcock," 1777. On the Roanne operation, see B.M.R. Robert Bouiller, "La fabrique de boutons Alcock à Roanne, au XVIIIe siècle," *La famille Alcock: 200 ans d'histoire en Roannais* (Roanne: Centre Forézien d'Ethnologie, 1985), 35. Bouiller had access to the papers of the Alcock family. The firm faced competition from French imitators; see Ballot, *Machinisme*, 482–88. On Alcock's state contracts, including government contracts to mint money, see A.N. F12 1315B Trudaine to Alcock, 30 January 1777; Du Châtelet to Trudaine, 1 March 1777. B.M.R. 6F55 "Mémoire," 21 September 1791, 26 May 1792. Francisque Pothier, *Roanne pendant la Révolution, 1789–1796* (Roanne: Durand, 1868), 155–56.

37. On the transfer of the property to Blanc, see Corenfustier, "Rapport," *A.P.*, 25 October, 1 November 1793, 77: 524–25; 78: 134–35.

38. "Règlement," 12 June 1758, in Guineau, *Quincaillerie*, 30–31. A.N. F12 1305B Baillison, "Mémoire," 11 April 1765.

39. A.N. F12 2232 Joseph Alcock, 10 February 1779, 28 March 1780. A.N. F12 1315B Brisson (inspector at Lyon), 28 September 1779. When Joseph Alcock went to spy out the latest Birmingham methods in 1779, he brought back the first report of Watt's new steam engine; see Harris, "Alcock," 143–44.

40. A.N. T591(4/5) Givry and Wendel, "Mémoire sur l'exploitation des mines," 27 March 1784. Spying was a two-way business—with the English pilfering French secrets as well. Albert Edward Musson and Eric Robinson, *Science and Technology in the Industrial Revolution* (Manchester: Manchester University Press, 1969), 216–30.

41. Some evidence on British stamping of gun parts can be found in Rowlands, *Master and Men*, 156. Quote in W. O. Henderson, *Industrial Britain Under the Regency: The Diaries*

of Escher, Bodmer, May, and de Gallois, 1814–1818 (London: Cass, 1968), 81–87. On the respective machine-tool industries, see L.T.C. Rolt, *A Short History of Machine Tools* (Cambridge: MIT Press, 1965). J. W. Roe, *English and American Tool Builders* (New Haven: Yale University Press, 1916). Ballot, *Machinisme*, 522–25. André Garanger, "L'expansion du mécanisme," in *Histoire générale des techniques*, ed. Maurice Daumas (Paris: Presses Universitaires de France, 1962–79), 3: 116–19, 129–31. Robert S. Woodbury, "The History of the Lathe to 1850," in *Studies in the History of Machine Tools* (Cambridge: MIT Press, 1972).

42. Information is sparse on the Birmingham gun trade in the eighteenth century. W.H.B. Court, *The Rise of the Midland Industries* (London: Oxford University Press, 1938), 142–47. De Witt Bailey and Douglas A. Nie, *English Gunmakers: The Birmingham and Provincial Gun Trade in the Eighteenth and Nineteenth Centuries* (New York: Arco, 1978). For the nineteenth century, see John D. Goodman, "The Birmingham Gun Trade," in *Birmingham and the Midland Hardware District*, ed. Samuel Timms (London: Harwicke, 1866), 412–14. Howard Blackmore, *British Military Firearms, 1650–1850* (London: Jenkins, 1961), 49–57. On the international gun trade, see Richards, "Africa."

43. Fries, "British Response." Rosenberg, "Introduction," in *American System*, 30–43.

44. Carolyn C. Cooper, "The Portsmouth System of Manufacture," *Technology and Culture* 25 (1984): 182–225. On the context of Samuel and Jeremy Bentham's work in the shipyards, see Linebaugh, *London Hanged*, 371–401.

45. François Crouzet, *Britain Ascendant*, 139–42.

46. Gribeauval to Choiseul (defunct), 3 April 1789, in Nardin, *Gribeauval*, 373.

47. S.H.A.T. 6c5 F. M. Aboville, *père*, to Narbonne (Min. War), 27 January 1792. Le Duc to Gimel, 17 May 1789, in Nardin, *Gribeauval*, 373.

48. These threats and insinuations against Blanc are countered in S.H.A.T. MR1739 Rolland de Bellebrune, "Mémoire concernant la comptabilité," 28 September 1790. S.H.A.T. 6c5 [Givry et al.], "Précis," January 1792. A draft version of Blanc's *Mémoire important* (differing in minor details) exists in the A.D.L. entitled "Mémoire à Nosseigneurs de l'Assemblée Nationale pour le Sr. Honoré Blanc," [1790]. The earliest mention of the *Mémoire* is 3 November 1790, at a hearings of the C.A.A. Fernand Gerbaux and Charles Schmidt, eds., *Procès-verbaux des Comités d'Agriculture et de Commerce* (Paris: Imprimerie Nationale, 1906–10), 1: 623. On Blanc's output, see S.H.A.T. 6c5 Blanc to Min. War, 3 jour comp., year V [19 September 1797].

49. Blanc, *Mémoire important*, 12, 19–21, quote 20–21.

50. There is no evidence that Roanne ever employed blind or deaf workers. In passing this boast on to Bonaparte, Aboville admitted that it would "shock many prejudices," but he wanted to underscore the degree of deskilling possible. S.H.A.T. 6c5 F. M. Aboville, *père*, to Bonaparte, 16 pluviôse, year X [5 February 1802].

51. S.H.A.T. 6c5 [War Office], "Instructions pour les commissaires chargés de procéder à l'examen des opérations faites par le Sr. Blanc," [Summer 1791]. A.A.S. *Procès-verbaux*, 10 November 1790, 109: 232–33. Gillmor, *Coulomb*, 15–16. Duveen and Hahn, "Laplace's Succession."

52. A.A.S. File: "Le Roy" Le Roy, "Mémoire," 18 August 1751. Landes, *Time*, 162–67, 221–25. Le Roy had met Jefferson before Franklin's departure on 11 July 1785—before Jefferson's encounter with Blanc. The two became better acquainted later. Le Roy to Jefferson, 28 September 1786, in Jefferson, *Papers*, 10: 410–11. Charpentier built portable presses for Jefferson over the next few years. Ferdinand Le Grand to Jefferson, 10 August 1786, in ibid., 10: 217–18.

53. A.A.S. *Procès-verbaux*, 10 November 1790, 16, 19 March 1791, 109: 232–33, 300–309. The "Le Roy" file contains a draft version of this report with a few minor corrections in Le Roy's hand. Le Roy, Laplace, Coulomb, Borda, *Rapport fait à l'Académie Royale des*

Sciences, le samedi 19 mars 1791, d'un Mémoire important de M. *Blanc* (Paris: Moutard, 1791). Present at the assembly were all the members of the jury, as well as academicians such as Vandermonde and Monge, who later promoted interchangeable parts manufacturing.

54. S.H.A.T. 6c5 [War Office], "Instructions pour les commissaires," [Summer 1791]. S.H.A.T. 6c5 Givry et al., "Examen des travaux du Sr. Blanc fait par MM. les officiers du Corps Roîal de l'artillerie nommés commissaires à cet effet," [September–October 1791].

55. Sources on production methods are not as rich as we might like. The commission's surviving reports provide a clocked summary of Blanc's procedures. The most detailed description is still Le Roy's *Rapport*, which has been partially translated by Edwin Battison, "Eli Whitney," 13–16. Also see Edwin A. Battison, "A New Look at the 'Whitney' Milling Machine," *Technology and Culture* 14 (1973): 592–98. For the Parisian Atelier de Perfectionnement (chap. 7) and Blanc's Manufacture of Roanne (chap. 9) we have manuscript evidence and drawings. For the Versailles manufacture, see Hermann Cotty, *Mémoire sur la fabrication des armes portatives de guerre* (Paris: Magimel, 1806), 63–75.

56. *R.A.C.S.P.*, 5, 17 March 1794, 11: 553; 12: 16. A.N. F12 1311 Vandermonde and Hassenfratz to Périer, 2 messidor, year II [20 June 1794]. A.N. F12 1312A Antelmy to Hassenfratz, 20 frimaire, year III [10 December 1794]. A.N. F12 1313 Mégnié, "Extrait sommaire du travail de mécanique," 17 nivôse, year III [6 January 1795]. On Hall, see M. R. Smith, *Harpers Ferry*, 232–34.

57. S.H.A.T. 4f6/4 Tugny, "Mémoire sur la Manufacture d'Armes de Roanne," messidor, year XI [June–July 1803]. S.H.A.T. 4f4 Tugny and Cablat, "Devis des prix courants des différentes opérations," 29 fructidor, year XI [16 September 1803].

58. This confirms Battison's conjecture that Eli Whitney's sketch of a "Tumbler Mill" depicts a double milling machine, and was invented by Blanc, not Whitney. As Battison remarks, this machine also resembles the much larger device described by Gaspard Monge for cutting the cylindrical trunnions of artillery cannon. Battison, "Eli Whitney," 29–30. Gaspard Monge, *Description de l'art de fabriquer les canons* (Paris: Imprimerie du C.S.P., year II [1794]), 191–94, plates 50–52. For the procedures at the Atelier de Perfectionnement, see A.N. F12 1313 Mégnié, "Travail de mécanique," 17 nivôse, year III [6 January 1795].

59. S.H.A.T. 6c5 Givry et al., "Résultats," [November 1791]. Le Roy et al., *Rapport.*

60. S.H.A.T. 6c5 [Givry et al.], "Précis," January 1792. S.H.A.T. 4f6/4 Tugny, "Mémoire sur la Manufacture d'Armes de Roanne," messidor, year XI [June–July 1803]. S.H.A.T. 6c5 Min. War to F. M. Aboville, *père*, 29 nivôse, year X [19 January 1802].

61. Charles Sabel and Jonathan Zeitlin, "Stories, Strategies, Structures: Rethinking Alternatives to Mass Production," intro. to *Worlds of Possibility: Flexibility and Mass Production in Western Industrialization*, forthcoming; typescript published (New York: Center for Law and Economic Studies Columbia School of Law, 1993).

62. Unfinished pieces were locked in a safe overnight to prevent chicanery. Blanc was permitted to substitute new pieces for any he considered to be substandard, but (1) this had to occur *before* the examination for interchangeability, (2) the commissioners were to record this fact, and (3) the extra time needed to construct the substitute piece was to be included in the total. The commissioners broke open some of the pieces to test their solidity and hardness; they were judged the equal of ordinary gunlocks. S.H.A.T. 6c5 Givry et al., "Résultats," [November 1791]. For a brilliant use of industrial archeology, see Gordon, "Mechanical Ideal."

63. S.H.A.T. 4f3 Anon., "Tableau du temps," [1770s]. S.H.A.T. 6c5 Givry et al., "Résultats," [November 1791]. S.H.A.T. 6c5 [Givry et al.], "Précis," January 1792. S.H.A.T. 4f4 Tugny and Cablat, "Devis des prix," 29 fructidor, year XI [16 September 1803].

64. Gribeauval hoped to find skilled machinists in Paris, see S.H.A.T. 6c5 Gribeauval to Agoult, 31 July 1785. On the effect of the Revolution, see S.H.A.T. 6c5 [Givry et al.], "Précis," January 1792.

65. The pamphlet which took note of the arms factory was B.N. Lb39 9753 Anon., *Détail de la grande révolution arrivée au Château de Vincennes et au faubourg Saint-Antoine* (Paris: La Barre, [1791]). Jean Paul Marat, *L'ami du peuple*, 3 March 1791, in Philippe-Joseph-Benjamin Buchez and Prosper-Charles Roux, eds., *Histoire parlementaire de la Révolution française, ou journal des assemblées nationales, depuis 1789 jusqu'en 1815* (Paris: Pauline, 1834–38), 9: 115–16. Bailly (Mayor), "Proclamation relative au donjon de Vincennes, du 28 février 1791," in *Moniteur*, 2 March 1791, 7: 506. On the same day a band of black-cloaked aristocrats rushed to the Tuileries to rescue Louis XVI from his popular "protectors," in what has become known as the "affair of the daggers." See *Chronique de Paris*, 2 March 1791, 61: 243–44; and *Moniteur*, 1, 6 March 1791, 7: 504, 541. [Georges Lefebvre], "L'affaire du donjon de Vincennes," *Annales historiques de la Révolution française* 22 (1950): 263–65.

66. S.H.A.T. 6c5 [War Office], "Instructions pour les commissaires," [Summer 1791]. Blanc to Duportail (Min. War), January 1792.

67. S.H.A.T. 6c5 Givry et al., "Résultats," [November 1791].

68. Ibid. S.H.A.T. 6c5 [Givry et al.], "Précis," January 1792. Blanc, "Rapport" to C.I.C.C.A., 26 prairial, year VII [14 June 1799]. *A.P.*, 25 October, 1 November 1793, 77: 524–25; 78: 134–35. Prieur, *R.A.C.S.P.*, 8 March 1794, 11: 590–91.

69. For a claim that gun-making motivated Renaissance machine design, see Vernard Foley et al., "Leonardo, the Wheel Lock, and the Milling Process," *Technology and Culture* 24 (1983): 399–427. For a claim that the watch industry (and hence the private sector) spurred interchangeable parts manufacturing, see Donald R. Hoke, *Ingenious Yankees: The Rise of the American System in the Private Sector* (New York: Columbia University Press, 1990), 20–99. On the wider impact of the armaments industry on early industrial development, see Trebilcock, "'Spin-Off' in British Economic History."

70. Landes, *Time*, 292–93, 308–20. Lindy Biggs, *Rational Production* (Baltimore: Johns Hopkins University Press, 1996). Stephen Meyer, *The Five Dollar Day: Labor Management and Social Control in the Ford Motor Company, 1908–1921* (Albany: State University of New York Press, 1981). Jonathan Zeitlin, "Flexibility and Mass Production at War: Aircraft Manufacture in Britain, the United States, and Germany, 1939–1945," *Technology and Culture* 36 (1995): 46–79.

71. Coase, "Firm." A. Chandler, *Visible Hand*. Joseph A. Schumpeter, *The Theory of Capitalist Development* (Cambridge: Harvard University Press, 1934).

Chapter Seven
The Machine in the Revolution

1. A.N. W77 plaq. 1, pièce 15 Section de l'Indivisibilité, 22 ventôse, year II [12 March 1794]. For the investigation, see A.N. W77 plaq. 5 "Extrait du registre du C.S.P.," 12 pluviôse year II [31 January 1794]. The placard announced that the top salary of locksmiths would be raised by 2 *livres* a day. *R.A.C.S.P.*, 9 March 1794, 11: 607.

2. *A.P.*, 20 March 1794, 86: 726–29.

3. *A.P.*, 26 March 1794, 87: 384.

4. *A.P.*, 20 April 1794, 89: 95. Georges Lefebvre, *The French Revolution: From 1793 to 1799*, trans. J. H. Stewart and J. Friguglietti (New York: Columbia University Press, 1964), 89.

5. *R.A.C.S.P.*, 8 November 1793, 8: 291.

6. Barère, *A.P.*, 23 August 1793, 72: 675. Although the *levée en masse* speech was delivered by Barère, the draft version was in Carnot's hand. See *A.P.*, 23 August 1793, 72: 688–89.

7. Barère quoted Guibert in *A.P.*, 1 February 1794, 84: 172–83. On early calls for mass

armies, see Brissot, *A.P.*, 29 December 1791, 36: 607. On the domestic implications of the call for war, see T.C.W. Blanning, *The Origins of the French Revolutionary Wars* (London: Longman, 1986).

8. *A.P.*, 3 July 1793, 68: 209–11.

9. François Furet and Mona Ozouf, eds., *A Critical Dictionary of the French Revolution*, trans. Arthur Goldhammer (Cambridge, Mass.: Belknap, 1989).

10. Soboul, *Sans-culottes*. Kåre D. Tønnesson, *La défaite des sans-culottes: Mouvement populaire et réaction bourgeoise de l'an III* (Drammen, Norway: Presses Universitaires d'Oslo, 1959). George Rudé, *The Crowd in the French Revolution* (Oxford: Clarendon Press, 1959). For a synthesis in this mold, see R. B. Rose, *The Making of the Sans-Culottes: Democratic Ideas and Institutions in Paris, 1789–1792* (Manchester: Manchester University Press, 1983).

11. Andrews notes that the sans-culottes made up a tiny fraction of the population, and were preeminently master artisans, large-scale employers, and the sons of well-placed professionals. Richard Mowery Andrews, "Social Structures, Political Elites and Ideology in Revolutionary Paris, 1792–94: A Critical Examination of Albert Soboul's *Les sans-culottes parisiens en l'an II*," *Journal of Social History* 19 (1985): 71–112. Sonenscher argues that rather than expressing the empathy of master artisans for their employees, the myth of the sans-culotte described the journeyman's version of the "good master." Sonenscher, "The Sans-culottes of the Year II." See also his elaboration in *Work and Wages*, 328–62. I owe the tone of this analysis to William H. Sewell, Jr., "The Sans-culotte Rhetoric of Subsistence," in *Terror*, ed. Baker, 4: 249–69.

12. Haim Burstin, "Problems of Work during the Terror," in *Terror*, ed. Baker, 4: 271–93; quote on p. 284. On the changing juridical standing of work, see Sewell, *Work and Revolution*, 92–142.

13. *R.A.C.S.P.*, 8 November 1793, 8: 291. Laborers employed on public construction projects also asserted that their work was patriotic; see Allan Potofsky, "Work and Citizenship: Crafting Images of Revolutionary Builders, 1789–1791," in *The French Revolution and the Meaning of Citizenship*, eds. Renée Waldinger, Philip Dawson, and Isser Woloch (Westport, Conn.: Greenwood Press, 1993), 185–99.

14. Camille Richard, *Le Comité de Salut Public et les fabrications de guerre sous la Terreur* (Paris: Rieder, 1921), 714–15. See the enthusiastic review by Albert Mathiez, "Compte rendue," *Annales Révolutionnaires* 14 (1922): 166–68. For the influence of World War I on Mathiez, see Albert Mathiez, *La victoire en l'an II: Esquisses sur la défense nationale* (Paris: Félix Alcan, 1916). S.H.A.T. Bib. #84889 Françoise Gaugelin, "Les ouvriers de la Manufacture Extraordinaire d'Armes de Paris, 23 août 1793 au 1 pluviôse, an III," (Mémoire de maîtrise, Université de Paris I, 1975–76), written under the direction of Albert Soboul.

15. Peuchet, *Moniteur*, 9 July 1791, 9: 70.

16. *A.P.*, 31 May 1793, 65: 652. Tuetey, *Sources manuscrites*, 2 June 1793, 8: 455. Jacques-René Hébert, *Le Père Duchesne* (Paris: E.D.H.I.S., 1969), n.d., no. 274, p. 6.

17. Billaud-Varenne, *A.P.*, 29 September 1793, 74: 535.

18. *R.A.C.S.P.*, 30 January 1794, 10: 536.

19. For the arrests and counterarrests, see A.N. AF II 214A dos. 1834 Prieur, 13 brumaire, year II [3 November 1793]. On the exemption from the draft granted armorers, see *A.P.*, 18 September, 8 November 1793, 74: 362–63; 78: 598–600. Also, *R.A.C.S.P.*, 29 September 1793, 25 February 1794, 7: 109–11; 11: 385. On the limitations placed on nonmilitary gun production, see *A.P.*, 14 December 1793, 26 January 1794, 81: 442–44; 83: 677–78.

20. On the charity ateliers, see A.N. F7 3688(1) Bailly (Mayor) to Tour du Pin (Min. War), 12 April 1790. Alan Forrest, *The French Revolution and the Poor* (New York: St. Martin's Press, 1981). Carnot, "Rapport sur la Manufacture," in *A.P.*, 3 November 1794, 78: 207–21. For a police report citing gaiety over the armament ateliers, see Tuetey, *Sources manuscrites*, 17 September 1793, 9: 405. Also, "Charmont," 11 February 1794, in Pierre

Caron, ed., *Paris pendant la Terreur: Rapports des agents secrets du Ministre de l'Intérieur* (Paris: Alphonse Picard, 1910–78), 4: 43.

21. *R.A.C.S.P.*, 14 April 1794, 12: 584.

22. *R.A.C.S.P.*, 14 April 1794, 12: 584–85. On Bergerac, see Alder, "Forging the New Order," 608–16.

23. Carnot, "Rapport sur la Manufacture," in *A.P.*, 3 November 1794, 78: 207–21. Carnot, *A.P.*, 14 December 1793, 81: 442–44. For the army's complaints, see *R.A.C.S.P.*, 28 March 1794, 12: 247. Also, Bertaud, *Army of the French Revolution*, 242–45. Output can be calculated from reports given every ten days (*décade*). A.N. AD VI 40 "Situation des Fabrications Nationales des Armes, 10 frimaire, an II-20 nivôse, an III," [20 December 1793–9 January 1795].

24. *R.A.C.S.P.*, 9 November 1793, 27 April 1794, 8: 305; 13: 96. I am assuming five persons per household.

25. Saint-Just, "Discours non-prononcé du 9 thermidor, an II," in Richard, *Fabrications de guerre*, 660. Georges Bouchard, *Un organisateur de la victoire: Prieur de la Côte-d'Or, Membre du Comité de Salut Public* (Paris: Clavreuil, 1946).

26. On the reduction of the War Office's authority, see *R.A.C.S.P.*, 12 November, 24 December 1793, 15 February 1794, 8: 359–62; 9: 627; 11: 160–61. For the ins and outs of the administration of the Manufacture and the resulting tensions with Jacobin War Minister Bouchotte, see Richard, *Fabrications de guerre*, 5, 82–88, 631–55. Later the Committee ordered all artillery officers at the armories to subordinate themselves to the civilian commissioners. *R.A.C.S.P.*, 10 September 1793, 14 April 1794, 6: 399–400; 12: 581. On the centralization of the Parisian Manufacture, see *R.A.C.S.P.*, 13 February 1794, 11: 113–14. For the final triumph of the Committee over the War Office, see Carnot, *A.P.*, 1 April 1794, 87: 694–99.

27. Orders of 4, 6, ventôse, year II [22, 24 February 1794], in Richard, *Fabrications de guerre*, 638–39.

28. The staff of the bureaucracy is listed in A.N. AD VI 40 *Organisation du Commission Administrative de la Manufacture Nationale des Fusils établie à Paris*, 3 floréal, year II [22 April 1794] (Paris: Imprimerie du C.S.P., year II [1794]). The desirable qualities of these administrators were outlined in C.N.A.M. 1° 170 C.S.P. to Société du Bureau de Consultation, 23 brumaire, year II [13 November 1793]. For Vandermonde's appointment, see *R.A.C.S.P.*, 8 December 1793, 9: 225.

29. For workers from Maubeuge and Liège, see *R.A.C.S.P.*, 27 August 1793, 9 January 1794, 6: 127; 10: 142. On reports of the Maubeuge workers' discontent, see "Rolin" and "Bacon," 12 January 1794, in Caron, *Paris pendant la Terreur*, 2: 314, 318. Also, Barère, *A.P.*, 1 February 1794, 84: 172–83. On Miramiones, see *R.A.C.S.P.*, 23 November 1793, 8: 647. On purges of "Liègeois," see *R.A.C.S.P.*, 19 December 1793, 9: 515–16.

30. The lists can be found in S.H.A.T. Xd233 "Tableau des ouvriers en fer enregistrés dans la Manufacture, Section d'Arcis, Champs Elysées, Beaurepaire, Montagne, Cité, Bonconseil, Amis de la Patrie, Marat, Mutius Scaevola, Bonne Nouvelle, Poissonière, Homme Armé," August 1793. Frédéric Braesch, "Essai de statistique de la population ouvrière de Paris vers 1791," *La Révolution française* 63 (1912): 289–321. Sonenscher, "The Sans-culottes of the Year II." For the work requirements, see Barère, *A.P.*, 8 November 1793, 78: 598; *R.A.C.S.P.*, 11 December 1793, 9: 322–23.

31. S.H.A.T. Xd233 "Ouvriers des usines de Paris; Ateliers Républicain, Carmagnol, Sans-Culotte," [1794–95].

32. Carnot, "Rapport sur la Manufacture," *A.P.*, 3 November 1794, 78: 207–21. Richard, *Fabrications de guerre*, 66. Barère, *A.P.*, 1 February 1794, 84: 172–83.

33. On Ciraud, see A.N. T1164(2) "Extrait du registre de délibération du Comité de la

Section de la Place Vendôme," 3 May 1791. See the *décade*-by-*décade* lists of output and workers in [Ciraud], "Etat des récépissés," ventôse, year II [February 1794] to nivôse, year III [January 1795]. A.N. T1164(1) [Ciraud], "Etat des ouvriers," 3ième *décade*, ventôse, year II [March 1794]; and 21 thermidor, year II [8 August 1794]. On the attempts to weed out the nonproducers, see Deschassaux to Dauffe, 1 thermidor, year II [19 July 1794]; Administration Générale des Armes Portatives to Ciraud, 24 germinal, year II [13 April 1794].

34. Hassenfratz was commissioned to write a description of the procedures employed in the production of muskets, but he never got around to it. The students in the Ecole des Armes were sent to do the drawings. *R.A.C.S.P.*, 20, 24 January 1794, 10: 331–32, 416. For rejects in the Quinze-vingt Atelier, see output lists by *décade* in A.N. T1164(1). On workers' complaints about production standards and corruption in the barrel proof, see A.N. AF II* 129 Guillemardet to Conseil d'Administration, 17 fructidor, year II [3 September 1794]. S.H.A.T. Xd233 Anon., "Deschamps," [1795]. S.H.A.T. 4f7/1 "Instruction pour les agents chargés de recevoir les baïonnettes," 4 frimaire, year II [24 November 1794]. In the year III, pattern locks were made "to the final degree of perfection." S.H.A.T. Xd233 Commission des Armes et Poudres to the Agence d'Atelier de l'Observatoire, 13 frimaire, year III [3 December 1795].

35. Sabel, *Work and Politics*.

36. For examples of insurrections, see below; also Rudé, *Crowd*, 134–35. For the administrators' responses, see *R.A.C.S.P.*, 12 December 1793, 9: 347–50. A total of fifty-five metalworkers were executed, eight of them arms workers. Donald Greer, *The Incidence of the Terror during the French Revolution: A Statistical Interpretation* (Gloucester, Mass.: Peter Smith, 1966), 156–57.

37. For Prieur's use of the language of artisanal self-sufficiency, see *R.A.C.S.P.*, 6 April 1794, 12: 425. Workers had been warned that wage-work was temporary in *R.A.C.S.P.*, 12 December 1793, 9: 347–50. The complaints of workers regarding compensation at the Camp sous Paris of 1791 prefigure in many ways the problems at the Manufacture of Paris; see Frédéric Braesch, *La Commune du dix août 1792: Etude sur l'histoire de Paris du 20 juin au 2 décembre 1792* (Paris: Hachette, 1911), 756–803, especially pp. 776–78.

38. A.N. F12 1309 J. H. Hassenfratz et al., "Assemblée des commissaires des sections de Paris et de ceux nommés par le Ministre de la Guerre pour la fixation du prix du travail," [15 October 1793]. On the changing wage schemes, see *R.A.C.S.P.*, 20 September, 26 October, 11 December 1793, 6: 577–78; 8: 21–22; 9: 322–23. Even the daily wage was soon made to vary between 3 and 5 *livres* a day—at the discretion of the *directeur*—making it too dependent on productivity.

39. Pension law of 19 August 1792. For injury claims, see A.N. T1164(2) "Extrait des registres de C.A.M.N.F.," 28 floréal, year II [17 May 1794]. Gaugelin, "Ouvriers," 108–12. Richard, *Fabrications de guerre*, 705–9.

40. [Hassenfratz] to Kobiersky, 15 frimaire, year II [5 December 1793], in Lazare Carnot, *Correspondance générale de Carnot, publiée avec des notes historiques et biographiques, août 1792–mars 1795*, ed. Etienne Chavaray (Paris: Imprimerie Nationale, 1892–1907), 4: 229.

41. For Capucins, see *R.A.C.S.P.*, 4, 19 December 1793, 9: 147, 515–16. For Hassenfratz's appointment, see *A.P.*, 14 December 1793, 81: 442–44. On trouble in the ateliers, see "Prevost," 24 December 1793, in Caron, *Paris pendant la Terreur*, 1: 384. For the Maché-aux-Poissons atelier, see *R.A.C.S.P.*, 12, 16 December 1793, 9: 236, 439. Richard, *Fabrications de guerre*, 700.

42. *R.A.C.S.P.*, 11, 12 December 1793, 9: 322–23, 347–50.

43. For agitation that winter, see "Monic," 23 December 1793, 2 January 1794; "Pourvoyeur" 29 January 1794; "Mercier," 4 February 1794; and "Bacon," 19 February 1794, in Caron, *Paris pendant la Terreur*, 1: 366; 2: 145; 4: 199; 5: 325; 5: 219. On the patriotic

speeches of the armorers, see *A.P.*, 20 March 1794, 86: 726–29. For the declarations in favor of Hébert, see A.N. W174 "Bacon," 1, 5, 8 germinal, year II [21, 25, 28 March 1794].

44. Worker quoted in Rudé, *Crowd*, 133. Wage rates were reassessed in *R.A.C.S.P.*, 4 April 1794, 12: 383–84. A.N. AF II 215A dos. 1844 C.S.P. to Ouvriers, Instructeurs, Sous-chefs, 19 germinal, year II [8 April 1794]. A.N. AD VI 40 "Organisation du C.A.M.N.F.," 3 floréal, year II [22 April 1794]. The Quinze-vingt Atelier sent its representatives in late April, A.N. T1164(2) C.A.M.N.F. to Ciraud, 8 floréal, year II [27 April 1794].

45. *R.A.C.S.P.*, 27 April 1794, 13: 98–100. A.N. T1164(2) "Extrait du registre des procès-verbaux de l'Assemblée," 10 germinal, year II [30 March 1794].

46. Reddy, *Market Culture*, 64–66.

47. *R.A.C.S.P.*, 29 September, 9 December 1793, 7: 109–11; 9: 281. Richard, *Fabrications de guerre*, 21. Gaugelin, "Ouvriers," 67.

48. *R.A.C.S.P.*, 16 April 1794, 12: 618–19. The Commission modified the formula in the upcoming months, but its bias remained.

49. Carnot, "Rapport sur la Manufacture," *A.P.*, 3 November 1794, 78: 207–21. On the juries for inventions, see A.N. AD VI 40 "Décret de la Convention Nationale," 2 April 1793. *R.A.C.S.P.*, 28 January 1794, 10: 492; 9 March 1794, 11: 607–8.

50. *R.A.C.S.P.*, 9, 27 April 1794, 12: 485; 13: 98–100. At the same time, another atelier was dedicated to making steel springs and another to making screws; see *R.A.C.S.P.*, 24 March, 4 May 1794, 12: 146; 13: 267. A.N. T1164(2) "Extrait des registres de l'Assemblée," 17 ventôse, year II [7 March 1794].

51. On Jouvet and Sueur, and Blanc's delays, see A.N. F12 1311 Régnier et al., "Rapport de l'Administration Générale des Armes Portatives," [May 1794]. On Huet, see A.N. F12 2197 Berthier, "Rapport," 12 pluviôse, year III [31 January 1795]; and C.A.A. to Commission des Armes, 13 ventôse, year III [3 March 1795].

52. *R.A.C.S.P.*, 27 April 1794, 13: 98–100. See also S.H.A.T. B12* 48 Administration Générale des Armes, 1 floréal, year II [20 April 1794]. Planning for the Atelier had begun somewhat earlier. A.N. F12 1311 Antelmy to Conseil d'Administration des Armes, 29 germinal, year II [18 April 1794]. For the sole published discussion of the Atelier, see René Tresse, "Un atelier pilote de la Révolution française: L'Atelier de Perfectionnement des Armes Portatives, mai-décembre 1794," *Techniques et civilisations* 5 (1956): 54–64.

53. Quote in A.N. F12 1311 Régnier et al., "Rapport de l'Administration Générale des Armes Portatives," [May 1794]. Tresse attributes these remarks to Vandermonde and Hassenfratz, "Un atelier pilote," 54–55. For appointments, see *R.A.C.S.P.*, 3 May 1794, 13: 240.

54. A.N. F12 1310 Anon., "Rapport sur l'Atel. Perf.," [year II]. For the test of the locksmiths, see A.N. F12 1311 Conseil de la Manufacture de Paris, 26 floréal, year II [15 May 1794]. Antelmy, 7 prairial, year II [26 May 1794]. For requests for docile workers, see Antelmy to Conseil de l'Administration, 29 germinal, year II [18 April 1794]. Hassenfratz and Vandermonde, 4 prairial, year II [23 May 1794]. For attempts to make tools, see A.N. F12 1313 "Atelier des platineurs," [December 1794]. See also orders for outside tools, dies, punches, and pattern gunlocks in A.N. F12* 233 Agence des Armes Portatives, 29 floréal, 27, 28 prairial, 4 messidor, year II [18 May, 15, 16, 22 June 1794]. For the evaluation of the test, see A.N. F12 1313 [Mégnié], "Travaux de l'an III," [1794–95].

55. A.N. F12 1313 Mégnié, "Travail de mécanique," 17 nivôse, year III [6 January 1795].

56. Prieur, *R.A.C.S.P.*, 27 April 1794, 13: 98–100. A.N. F12 1310 [Antelmy], "Rapport sur l'Atel. Perf.," year II [1794].

57. On the press forges, see A.N. F12 1313 Mégnié, "Travail de mécanique," 17 nivôse, year III [6 January 1795]. On Mégnié, see Maurice Daumas, *Les instruments scientifiques aux XVIIe et XVIIIe siècles* (Paris: Presses Universitaires de France, 1953), 344, 359–63. On the quarrel, see A.N. F12 1310 Antelmy to C.A.A., 15, 18, thermidor, 25 fructidor, year III [2,

5 August, 11 September 1795]. On Charpentier's lathe, see *R.A.C.S.P.*, 7 December 1793, 9: 236–37.

58. On Glaësner, see *R.A.C.S.P.*, 10 February 1794, 9: 44. On the other artisans, see A.N. F12 1310 Mégnié, "Travail de mécanique," 17 nivôse, year III [6 January 1795]. On Beguinot, see A.N. F12 1311 Labolle to Hassenfratz, [1794].

59. C.N.A.M. 10° 438 Vandermonde, "Rapports sur l'Atel. Perf.," [1794–95].

60. On the drop presses, see *R.A.C.S.P.*, 5, 17 March 1794, 11: 553; 12: 16. A.N. F12 1311 Vandermonde and Hassenfratz to Périer, 2 messidor, year II [20 June 1794]. For procedures for lockplates and tumblers, see A.N. F12 1313 Mégnié, "Travail de mécanique," 17 nivôse, year III [6 January 1795]. "Atelier de Savart," [late 1794]. "Atelier des platineurs," [December 1794]. Compare with M. R. Smith, *Harpers Ferry*, 85–139.

61. The worker's complaint is reported in A.N. F12 1310 C.A.A. to C.S.P., 24 ventôse, year III [14 March 1795]. Chefs d'ouvriers de l'Atel. Perf. to C.A.A., 21 messidor, year III [9 July 1795]. Vandermonde to C.A.A., 25 germinal, year III [14 April 1795].

62. A.N. F12 1310 Mégnié to C.A.A., [1794–95]. On conflicts in the shop, see A.N. F12 1311 Antelmy to Min. Int., 9 nivôse, year V [29 December 1796]. A.N. F12 1310 Mégnié to C.A.A., [1794–95]. A.N. F12 1312A Antelmy to Bénézech, 21 pluviôse, year III [9 February 1795]. Time sheets bear out the complaints of the administration. A.N. F12 1310 "Extrait de registre de C.A.A.," prairial, year III [May–June 1795]. On Antelmy's corruption charges, see A.N. F12 1311 Dubois, "Rapport au Min. Int.," 11 thermidor, year V [29 July 1797].

63. Prieur, "Rapport sur les moyens préparés pour établir l'uniformité des poids et mesures," in *P.V.C.I.P.*, 24 fructidor, year III [10 September 1795], 6: 665.

64. *R.A.C.S.P.*, 2 January 1795, 19: 226–27. C.N.A.M. 10° 438 Vandermonde, "Rapports sur l'Atel. Perf.," [1794–95].

65. A.N. F12 1556 Anon., "Rapport de l'Agence des Arts et Manufactures à la C.A.A.," [1795]. C.N.A.M. 10° 438 Vandermonde, "Rapports sur l'Atel. Perf.," [1794–95].

66. B.N. Lb41 1120 A.-L. Frécine, *Mort aux tyrans! Au nom du peuple français*, 23 prairial, year II [11 June 1794].

67. S.H.A.T. Reg. A/8 "Rapport de l'Agence des Armes à la Commission des Armes et Poudres," 12 messidor, year II [30 June 1794].

68. For output, see A.N. T1164(2) [Ciraud], "Etat des récépissés," ventôse, year II [February 1794] to nivôse, year III [January 1795]. On the new incentive wages, see A.N. F9 58 *Procès-verbaux des séances pour la fixation des prix de fabrication de fusils*, 1–5 prairial, year II [20–24 May 1794] (Paris: Vatar, year II [1794]), 17. This time, with the Paris Commune under the control of the Committee, the wage cap would be enforced for *all* workers in the capital. *R.A.C.S.P.*, 5 July 1794, 14: 736–37.

69. Barère, *A.P.*, 10 June 1794, 91: 489. A.N. AF II* 129 Guillemardet to Conseil d'Administration des Armes et Poudres, 26 messidor, year II [14 July 1794].

70. Rudé, *Crowd*, 136–37. A.N. AF II 48, plaq. 374, fol. 10 Lescot-Levriot (Mayor) to Commandant-Général de la Force Armée, 9 thermidor, year II [27 July 1794].

71. For the retrospective denunciation, see *Moniteur*, 19 November 1794, 22: 529. On the military policies of the new Committee, see *R.A.C.S.P.*, 2, 24 August 1794, 15: 600; 16: 310–13. On the new *maximum*, see *R.A.C.S.P.*, 11 August 1794, 16: 26. Rudé, *Crowd*, 145. On the police laws, A.N. AF II* 129 Guillemardet to Conseil d'Administration des Armes et Poudres, 12 thermidor, year II [30 July 1794]. *R.A.C.S.P.*, 16 August 1794, 16: 54–55.

72. For the denunciations, see *R.A.C.S.P.*, 6 December 1794, 18: 546–47. On the cost differences, see *Journal des débats*, 22 frimaire, year III [12 December 1794], no. 812, pp. 1213–14.

73. For rumors and speeches, see *A.P.*, 31 August 1794, 96: 129–30, 146–47; Lakanal, 31

August 1794, 96: 143; Treilhard, 1 September 1794, 96: 174–76. On the Unité fire, see *A.P.*, 20 August 1794, 95: 321–23. For Carrier's accusations, see François-Alphonse Aulard, ed., *La Société des Jacobins, Recueil des documents du Club des Jacobins de Paris* (Paris: Jouast, 1889–97), 1, 3 September 1794, 6: 404–17.

74. Police reports and rumors can be found in *Journal de Perlet*, 1, 2 September 1794; "Gillet et Le Camus," 2 September 1794; and "Viard et Ollivier," 3, 10 September 1794, in Aulard, *Paris thermidorienne*, 1: 74–77, 79, 90.

75. *R.A.C.S.P.*, 1 September 1794, 16: 447–50. Richard, *Fabrications de guerre*, 770–74.

76. S.H.A.T. Xd233 Commission des Armes et Poudres, "Ouvriers des usines de Paris," [1794].

77. *Journal des débats*, 27–30 brumaire, year III [17–20 November 1794], no. 785, pp. 804–5; no. 786, pp. 814; no. 787, pp. 838–39; no. 788, pp. 858–59.

78. *Journal des débats*, 28 brumaire, year III [18 November 1794], no. 786, pp. 813–15.

79. Daniel Guérin, *La lutte de classes sous la Première République: Bourgeois et "bras nues," 1793–97* (3d ed.; Paris: Gallimard, 1946), 155. For commentary on Guérin, see Rudé, *Crowd*, 134.

80. *R.A.C.S.P.*, 14, 18 November 1794, 18: 140, 210–11. A.N. T1164(2) "Extrait des registres de C.A.M.N.F.," 28 brumaire, year III [18 November 1794].

81. Aulard, *Paris thermidorienne*, 17 November 1794, 1: 251. A.N. T1164(2) "Extrait des registres de C.A.M.N.F.," 24 brumaire, year III [14 November 1794]. S.H.A.T. Xd233 Boyle et al., 28 brumaire, year III [18 November 1794]. For police reports, see A.N. F7 2524 plaqs. 18, 19, Anon., no title, 2, 3 frimaire, year III [22, 23 November 1794].

82. Montmayon, *Journal des débats*, 4 frimaire, year III, [24 November 1794], no. 792, pp. 911–15, quote p. 915.

83. S.H.A.T. Xd233 Agence des Armes Portatives to Commission des Armes et Poudres, 22 fructidor, year II [8 September 1794]. *R.A.C.S.P.*, 12 August 1794, 16: 50. "Alletz et Babille," 18 November 1794, in Aulard, *Paris thermidorienne*, 1: 254.

84. "Babille," 11 December 1794, in Aulard, *Paris thermidorienne*, 1: 304–5. On the movement, see Tønnesson, *La défaite des sans-culottes*, 22–29.

85. *Journal des débats*, 23, 24 frimaire, year III [13, 14 December 1794], no. 811, pp. 1197–99; no. 812, pp. 1213–17. Administrators had been ordered to restore calm in A.N. T1164(2) Commission des Armes et Poudres to Ciraud, 30 brumaire, year II [20 November 1793].

86. Capon, Bénézech, in *R.A.C.S.P.*, 11 December 1794, 18: 628–29. On Maubeuge, see *R.A.C.S.P.*, 14 December 1794, 18: 706. On the *journée* of germinal, see Gaugelin, "Ouvriers," 159. *R.A.C.S.P.*, 29 March 1795, 21: 376.

87. Louis-Bernard Guyton-Morveau, *Sur l'état de situation des arsenaux*, 14 pluviôse, year III [2 February 1795] (Paris: Imprimerie Nationale, year III [1795]), 20–21. Other claims for the output are greater; thus Paris produced 219,259 muskets and was capable of turning out 27,000 per month. C.N.A.M. 1°100 Anon., "Etat de tous les établissements," [1795].

88. Law of 18 floréal, year III [7 May 1795], in Pierre Nardin, "Le Comité de l'Artillerie et la réalisation des matériels d'armement, dès origines à 1870," *Mémorial de l'artillerie française* 50 (1976): 471. S.H.A.T. MR2129 Prieur to Commission des Armes et Poudres, 8 floréal, year II [27 April 1794]; Capon, "Rapport au C.S.P.," 25 prairial, year II [13 June 1794].

89. S.H.A.T. 4f7/1 Agoult, "Observations sur l'instruction," 28 germinal, year VI [17 April 1798]. "Commission des Armes à Feu," 2 fructidor, year VIII [20 August 1800]. The commissioners also included Prieur de la Côte-d'Or and Régnier.

90. Gassendi, *Aide-mémoire* (3d ed.; Paris: Magimel, 1801), 2: 548–49.

91. Kasson, *Civilizing the Machine*, 3–106.

Chapter Eight
Terror, Technocracy, Thermidor

1. On the technological work of the Academy, see Roger Hahn, *Anatomy of a Scientific Institution: The Paris Academy of Sciences, 1666–1803* (Berkeley: University of California Press, 1971), 65–72, 116–26.

2. On the polity of international science during the Enlightenment, see Lorraine J. Daston, "The Ideal and the Reality of the Republic of Letters in the Enlightenment," *Science in Context* 4 (1991): 367–86. For a survey of the political involvement of the savants, see Roger Hahn, "The Triumph of Scientific Activity: From Louis XIV to Napoléon," *Proceedings of the Annual Meeting of the Western Society for French History* 19 (1989): 204–11. Never since then have individuals with scientific training held posts of such political responsibility in any European nation (the one possible exception being France's Third Republic). Even so, only 12 percent of all legislators were medical professionals, academicians, or military men; whereas roughly half were lawyers. See Alison Patrick, *The Men of the First French Republic: Political Alignments in the National Convention* (Baltimore: Johns Hopkins University Press, 1972), 260. Quote from Monge to Lacroix, 28 June 1791, in P.-V. Aubry, *Monge, le savant ami de Napoléon Bonaparte, 1746–1818* (Paris: Gauthier-Villars, 1964), 76–77. On Condorcet, see Baker, *Condorcet.*

3. Hahn, *Anatomy*. Hahn has partially retracted his claim that a modern profession of science emerged from the institutional shake-up of the Revolution. Roger Hahn, "Scientific Research as an Occupation in Eighteenth-Century Paris," *Minerva* 13 (1975): 501–13. Dorinda Outram, "The Ordeal of Vocation: The Paris Academy of Sciences and the Terror, 1793–95," *History of Science* 21 (1983): 251–73.

4. Gabrielle Hecht has pointed out that more than one kind of technocratic ideal has been operative in twentieth-century France in "Rebels and Pioneers: Technocratic Ideologies and Social Identities in the French Nuclear Workplace, 1955–1969," *Social Studies of Science* (August 1996), forthcoming.

5. On nineteenth-century Germany, see Keith Anderton, "The Limits of Science: A Social, Political, and Moral Agenda for Epistemology in Nineteenth-Century Germany" (Ph.D. diss., Harvard University, 1993). On the Weimar period and the Vienna Circle, see Richard Beyler, "From Positivism to Organicism: Pascual Jordan's Interpretations of Modern Physics in Cultural Context" (Ph.D. diss., Harvard University, 1994).

6. Monge, *Canons*. Alexandre-Théophile Vandermonde, Gaspard Monge, and Claude-Louis Berthollet, *Avis aux ouvriers en fer sur la fabrication de l'acier* (Paris: Imprimerie du Département de la Guerre, year II [1793–94]). On Meudon, see *R.A.C.S.P.*, 20 April 1794, 12: 706. Richard, *Fabrications de guerre*, 612–32. Prieur and Robespierre prevented the Convention from discussing Meudon openly, a secrecy which flew in the face of avowed republican and scientific virtues. *R.A.C.S.P.*, 28 October 1793, 8: 57, 78. Prieur, *A.P.*, 26 September 1794, 98:79–82.

7. For an account of Jacobin hostility to science, see Charles C. Gillispie, "The *Encyclopédie* and the Jacobin Philosophy of Science: A Study of Ideas and Consequences," in *Critical Problems in the History of Science*, ed. Marshall Clagett (Madison: University of Wisconsin Press, 1959), 255–89. On Marat's hostility to establishment science, see Darnton, *Mesmerism*, 92–94; also, Joseph Dauben, "Marat: His Science and the French Revolution," *Archives internationales d'histoire des sciences* 22 (1969): 235–61. For a survey of these themes, see Martin Staum, "Science and Government in the French Revolution," in *Science, Technology and Culture in Historical Perspective*, ed. Louis A. Knafla et al. (Calgary: University of Calgary Press, 1976), 105–26.

8. Lavoisier, *Oeuvres*, 4: 618, quoted in Hahn, *Anatomy*, 234. Even reforms proposed by

Lavoisier, Talleyrand, and Condorcet to save the Academy of Sciences as a "National Society" were deemed too elitist. Hahn, *Anatomy*, 195–225, 247–51.

9. Carnot, "Rapport sur la Manufacture," in *A.P.*, 3 November 1794, 78: 207–21. On the rescue efforts, see *R.A.C.S.P.*, 16 December 1793, 9: 439; and Aulard, *Jacobins*, 22 July 1793, 5: 311. Savants in the life sciences could still justify their activities by pointing to their usefulness at the Jardin des Plantes, now known as the Muséum d'Histoire Naturelle; see Michael Osborne, "Applied Natural History and Utilitarian Ideals: 'Jacobin Science' at the Muséum d'Histoire Naturelle, 1789–1870," in *Re-creating Authority*, eds. Ragan and Williams, 125–43.

10. For this argument, see Henry Guerlac, "Some Aspects of Science during the French Revolution," in *Essays and Papers in the History of Modern Science* (Baltimore: Johns Hopkins University Press, 1977), 477. Outram, in "Ordeal of Vocation," argues that the savants experienced a disproportionate burden of repression. But her prosopographical method does not allow for a comparison with other elites, or give a sufficiently detailed sense of *why* individual savants were persecuted.

11. *P.V.C.I.P.*, 15 nivôse, year II [4 January 1794], 3: 239. Jean-Baptiste-Joseph Delambre, *Grandeur et figure de la terre*, ed. Guillaume Bigourdan (Paris: Gauthier-Villars, [1827], 1912), 213.

12. To take just the example of Hassenfratz: A.N. F7 4739 Hassenfratz to C.S.P., 23 thermidor, year II [10 August 1794]. Hassenfratz to Comité de Sûreté Générale, [July–August 1794]. C.S.P. to Comité de Sûreté Générale, 23 thermidor, year II [10 August 1794]. Aulard, *Jacobins*, 24 February, 12 April, 31 December 1793, 5: 39, 134, 587–88. See his exculpation in Gustave Laurent, "Un mémoire historique du chimiste Hassenfratz," *Annales historiques de la Révolution française* 1 (1924): 163–64. Outram, "Ordeal of Vocation," 266–68.

13. For Lalande and a balanced approach to this question, which has exercised many historians of science, see Roger Hahn, "Fourcroy, Advocate of Lavoisier?" *Archives internationales d'histoire des sciences* 12 (1959): 285–88. See Guerlac, "Aspects of Science," 442.

14. Although François Furet may wish to set a terminal date to this debate, the controversy over the meaning of the French Revolution continues—a fact to which his own celebrity attests. Furet, *Penser*.

15. Pierre Larousse, "Carnot, Lazare-Nicolas-Marguerite," *Le grand dictionnaire universel du XIXe siècle* (Geneva: Slatkine, [1867], 1982), 3: 427. For a collection of judgments on Carnot's career, see *Révolution et mathématique*, 1: 15–77.

16. Jules Michelet, *Histoire de la Révolution française* (Paris: Chamerot, 1847–53), vol. 6, book 13, chap. 2, pp. 235–39, and "Judgement final." François-Alphonse Aulard, *Etudes et leçons sur la Révolution française* (Paris: Alcan, 1893), 189–211. Albert Mathiez, *The French Revolution*, trans. C. A. Phillips (New York: Russell and Russell, [1922–27], 1962), 373–74. Albert Mathiez, "Les divisions dans les comités de gouvernement à la veille du 9 thermidor d'après quelques documents inédits," *Revue historique* 118 (1915): 70–87. Jean Jaurès, *Histoire socialiste de la Révolution française* (Paris: Editions Sociales, 1952), 6: 202–3. Albert Sorel, *L'Europe et la Révolution française* (Paris, 1891), 3: 513–14. Hippolyte-Adolphe Taine, *La Révolution* (Paris: Hachette, 1878–82), vol. 3, chap. 2, sec. 3, pp. 233–39.

17. Reinhard, *Le grand Carnot*. Gillispie and Youschklevitch, *Carnot, Savant*, 21–22.

18. *A.P.*, 22 September 1794, 97: 350–51.

19. B.N. Le38 1127 Antoine-François Fourcroy, *Sur les arts qui ont servis à la défense de la République, séance du 14 nivôse, an III*, [3 January 1795] (Paris: Imprimerie Nationale, year III [1794]). See also, W. A. Smeaton, *Fourcroy, Chemist and Revolutionary, 1755–1809* (Cambridge, U.K.: Heffer, 1962).

20. Grégoire, *A.P.*, 31 August 1794, 96: 154. For the Dumas-Lavoisier story, see Grégoire, "Troisième rapport sur le vandalisme," 24 frimaire, year III [14 December 1794],

p. 2. Fourcroy explicitly seconded Grégoire's pursuit of the "Vandals" in *A.P.*, 31 August 1794, 96: 145. Outram, "Ordeal," 258–59. On the reconstitution of scientific life, see Hahn, *Anatomy*, 286–312. James Guillaume, "Un mot légendaire: 'La république n'a pas besoin de savants,'" *La Révolution française* 38 (1900): 385–99.

21. Anon., *Leurs têtes branlent*, [1795], in Bouchard, *Prieur*, 321.

22. Carnot and Prieur, *Moniteur*, 7 germinal, year III [27 March 1795], 24: 49–53. Carnot's repudiation of his role in Danton's execution can be found in "Réponse au rapport fait sur la conjuration du 18 fructidor," 8 floréal, year VI [27 April 1798], in *Révolution et mathématique*, 1: 286–87. On the anonymous shout of 20 May 1795, see Bouchard, *Prieur*, 333–34, 464–66.

23. Carnot, *Moniteur*, 7 germinal, year III [27 March 1795], 24: 50. Emphasis by Carnot.

24. On Prieur's orders, see Bouchard, *Prieur*, 318. George Armstrong Kelly, *Victims, Authority, and Terror: The Parallel Deaths of d'Orléans, Custine, Bailly, and Malesherbes* (Chapel Hill: University of North Carolina Press, 1982), 199.

25. On *raison d'état*, see Carnot, "Rapport à la Convention sur la réunion de Monaco," 14 February 1793, in *Révolution et mathématique*, 2: 310. Bronislaw Baczko, *Ending the Terror: The French Revolution after Robespierre*, trans. Michel Petheram (Cambridge: Cambridge University Press, 1994). This "necessitarian" thesis is one of the three core interpretations of the Terror which have dominated interpretation ever since. Mona Ozouf, "The Terror after the Terror? An Immediate History," in *Terror*, ed. Baker, 4: 3–18.

26. Hunt, *Politics, Culture, and Class*, 3.

27. Gillispie, "Jacobin." Joseph Fayet, *La Révolution française et la science, 1789–1795* (Paris: Marcel Rivière, 1960). Kline, "Constructing 'Technology.'"

28. For the claim that science flourished in this period, see Jean-Baptiste Biot, *Essai sur l'histoire générale des sciences pendant la Révolution française* (Paris: Duprat, 1803). Also, Georges Pouchet, *Les sciences pendant la Terreur d'après les documents du temps et les pièces des Archives Nationales*, ed. James Guillaume (Paris: Société de l'Histoire de la Révolution Française, 1896).

29. François-Alphonse Aulard, *The French Revolution: A Political History*, trans. Bernard Maill (New York: Russell and Russell, [1910], 1965), 2: 248–49. On the activities of Robespierre in favor of the war effort, see Richard, *Fabrications de guerre*, 674–81. For a more recent narrative of the activities of the savants during the Revolution, which essentially tows the post-Jacobin line, see Jean and Nicole Dhombres, *Naissance d'un nouveau pouvoir: Science et savants en France, 1793–1824* (Paris: Payot, 1989).

30. Guérin, *Lutte des classes*. François Furet and Denis Richet, *La Révolution française* (Paris: Fayard, 1965–66), 1: 219. Jean-Paul Charnay, "Ruptures, répétitions et contradictions dans la fortune politique et l'assise sociale de Lazare Carnot," in *Lazare Carnot, ou le savant-citoyen*, ed. Jean-Paul Charnay (Paris: Université de Sorbonne, 1990), 31–45. See also the other articles in this collection.

31. Cuvier, *Recueil des éloges historiques*, cited in Outram, "Ordeal of Vocation," 262.

32. For a view that much of the politics of science in this period turns on education policy, see L. Pierce Williams, "The Politics of Science in the French Revolution," in *Critical Problems*, ed. Clagett, 291–308. Woloch shows these schemes, though ephemeral, marked out a new state role in supplying universal popular education. Isser Woloch, "The Right to Primary Education in the French Revolution: From Theory to Practice," in *Meaning of Citizenship*, eds. Waldinger, Dawson, and Woloch, 137–52. Fourcroy, *Sur les arts*.

33. The plans for the school can be found in *R.A.C.S.P.*, 12, 17 February, 20 March 1794, 11: 89, 222; 12: 70–72. Louis-Bernard Guyton-Morveau et al., *Programmes des cours révolutionnaires sur la fabrication des salpêtres, des poudres et des canons, 1, 11, 21 ventôse, 5 germinal, an II* [19 February, 1, 11, 25 March 1794] (Paris: Imprimerie du C.S.P., year II

[1794]). For quotes, see Barère, *A.P.*, 18 February 1794, 85: 208–10. For a history of the course in the context of revolutionary changes in education methods, see Janis Langins, "Words and Institutions during the French Revolution: The Case of 'Revolutionary' Scientific and Technical Education," in *The Social History of Language*, eds. Peter Burke and Roy Porter (Cambridge: Cambridge University Press, 1987), 136–60. Also see the extensive description in Richard, *Fabrications de guerre*, 479–86.

34. C.N.A.M. Bib. 295 [C.S.P.], *Programmes des cours révolutionnaires sur l'art militaire, 5 fructidor, an II-13 vendémiaire, an III* [22 August–4 October 1794] (Paris: Imprimerie du C.S.P., year III [1794]). Barère, *A.P.*, 14 July 1794, 93: 148. Guillaume, "Introduction," *P.V.C.I.P.*, 4: xxi. *R.A.C.S.P.*, 16 December 1794, 18: 754.

35. Calls for a single engineering school had been heard since 1793. Lecointe-Puyraveau, in *P.V.C.I.P.*, September 1793, 5: 630–32. This call was taken up by the C.S.P.; see Barère, in *P.V.C.I.P.*, 21 ventôse, year II [11 March 1794], 5: 628.

36. Bruno Belhoste, "Les origines de l'Ecole Polytechnique," *Histoire de l'éducation* 42 (1989): 13–53. Picon, *Ingénieur moderne*. Janis Langins, "The Ecole Polytechnique (1794–1804): From Encyclopedic School to Military Institution" (Ph.D. diss., University of Toronto, 1979). On Monge, see René Taton, *L'oeuvre scientifique de Gaspard Monge* (Paris: Presses Universitaires de France, 1951), 50–100.

37. Monge, "Géométrie descriptive," *Séances [et leçons] des Ecoles Normales, recueillies par les sténographes, et revues par les professeurs* (Paris: Reyner, year III [1795]), 4: 88. Also, Monge, "Objet des études dans les écoles pour les artistes et les ouvriers," [August 1793], in *Ecole Normale de l'An III: Leçons de mathématiques*, ed. Jean Dhombres (Paris: Dunod, 1992), with an "introduction" by B. Belhoste and R. Taton, pp. 269–94, 574–78. Hassenfratz and Monge had been the driving force behind the request of the city of Paris in 1793 for a technical secondary school based on the descriptive geometry. *P.V.C.I.P.*, 15 September 1793, 2: 408–29, esp. p. 414.

38. Monge, *Géométrie descriptive*, xvi.

39. Ibid., xv. Booker, *Engineering Drawing*, 48–67, 86–113.

40. Sylvestre-François Lacroix, *Discours sur l'instruction publique, prononcé à la distribution des prix des Ecoles Centrales, le 29 thermidor, an VIII* [17 August 1800] (Paris: Duprat, year IX [1800]), 20. Catherine Mérot, "Le recruitment des Ecoles Centrales sous la Révolution," *Revue historique* 274 (1985): 357–85. Also, Julia et al., *Atlas*, 2: 40–45. Even this drawing-based curriculum was a stripped-down version of the original proposal for technical education. Compare Lakanal's plans to include drawing, agriculture, and the other mechanical arts, in *P.V.C.I.P.*, 26 frimaire, year III [16 December 1794], 5: 299–309, with the final law voted in with the Institut in *P.V.C.I.P.*, 23 vendémiaire, year IV [15 October 1795], 6: 794–95. This transition is sketched in Dominique Julia, *Les trois couleurs du tableau noir: La Révolution* (Paris: Belin, 1981), 257–82, 290–92.

41. Antoine Louis Claude Destutt de Tracy, *Observations sur le système actuel d'instruction publique* (Paris: Panckoucke, year IX [1800–1801]), 2. For the role of grammar as a foundational subject in this period, see Patrice Higonnet, "The Politics of Linguistic Terrorism and Grammatical Hegemony during the French Revolution," *Social History* 5 (1980): 41–69. For the new type of fine arts drawing, see [Hautes Alpes], *Procès-verbal de l'inauguration de l'Ecole Centrale* (Gap: Allier, year V [1796–97]).

42. The law announcing the school makes this clear; see Fourcroy and Prieur, "Projet," in *P.V.C.I.P.*, 22 September 1794, 5: 79–81. Langins, "Ecole Polytechnique," 55–102. Belhoste, "Origines de l'Ecole Polytechnique." See also the discussion in James Guillaume, "Documents nouveaux sur la création de l'Ecole Centrale des Travaux Publics," in *P.V.C.I.P.*, 5: 627–53.

43. Monge, "Développements," in Langins, *Savants*, 26–29, 227–47. For a summary of the descriptive geometry at the Ecole in its early years, see Belhoste et al., "Exercices,"

91–108. Also Joël Sakarovitch, "La géométrie descriptive, Une reine déchue," in *Formation polytechnicienne*, eds. Belhoste, Dalmedico, and Picon, 77–93.

44. Monge, "Géométrie descriptive," in *Leçons*, 1: 285; 3: 63–64. In keeping with the compression of effort required by the Revolution, the first class of students were also expected to cram three years' worth of learning into one.

45. Monge, "Extrait des procès-verbaux des séances du Conseil d'Administration," 20 pluviôse, year III [8 February 1795], in Langins, *Savants*, 116–19.

46. [Prieur de la Côte-d'Or], "Avant-propos," *Journal de l'Ecole Polytechnique* 1 (year III [1795]): v. See also Belhoste et al., "Exercices," 98–101.

47. For a denouncement of the students' political activities and Monge's defense of them, see *P.V.C.I.P.*, 6, 16 vendémiaire, year IV [28 September, 8 October 1795], 6: 724, 746, also p. xxiv. Fourcroy, "Rapport," 7 vendémiaire, year III [28 September 1794], in Langins, *Savants*, 200–226. On the government's efforts to restrain the political activities of the Ecole's students, see Shinn, *Ecole Polytechnique*, 19–21. On the school's governance and finance, see Amboise Fourcy, *Histoire de l'Ecole Polytechnique*, ed. Jean Dhombres (Paris: Belin, 1987), 71, 79.

48. For the name change, see Prieur, "Rapport," in *P.V.C.I.P.*, 12 fructidor, year III [29 August 1795], 6: 601–4. Prieur, *Mémoire sur l'Ecole Centrale des Travaux Publics*, 30 prairial, year III [18 June 1795] (Paris: Imprimerie de la République, messidor, year III [1795]).

49. On Prieur's evasiveness at this time, see Bouchard, *Prieur*, 336–38. Laurent, "Hassenfratz." Fourcroy, "Rapport," in *P.V.C.I.P.*, 30 vendémiaire, year IV [22 October 1795], 6: 839–50. One can trace this transformation in the laws reprinted in *Journal de l'Ecole Polytechnique* 4 (year V [1796–97]): ix–xxviii.

50. Langins, "Ecole Polytechnique," 122–51.

51. Guyton, "Discours," *Journal de l'Ecole Polytechnique* 7 (year VII [1798–99]): 211–12.

52. Margaret Bradley, "Scientific Education versus Military Training: The Influence of Napoléon Bonaparte on the Ecole Polytechnique," *Annals of Science* 32 (1975): 415–49; Napoléon to Lacuée, 2 germinal, year XIII [23 March 1805], p. 446. Fourcy, *Ecole Polytechnique*, 177, 252.

53. Janis Langins, "Sur l'enseignement et les examens à l'Ecole Polytechnique sous le Directoire: A propos d'une lettre inédite de Laplace," *Revue d'histoire des sciences* 40 (1987): 145–77; see p. 166 for a remark on the oddity of a theoretical engineering school. On the artisanal assistants, see Langins, "Ecole Polytechnique," 229. On curriculum, see Fourcy, *Ecole Polytechnique*, 376–77.

54. This has been a topic of considerable debate. While Langins and Glas emphasize the hostility between Monge and Laplace, Hahn sees their conflict as transitory. Eduard Glas, "On the Dynamics of Mathematical Change in the Case of the Monge and the French Revolution," *Studies in the History and Philosophy of Science* 17 (1986): 249–68. Hahn, "Rôle de Laplace." Grattan-Guinness concedes that Monge's school declined, but rather than see his dispute with Laplace as a debate between practice and theory, he points instead to the diverging interests of engineer-savants and mathematical physicists, both of whom married mathematics and rational mechanics, though to different ends. I. Grattan-Guinness, "The *ingénieur-savant*, 1800–1830: A Neglected Figure in the History of French Mathematics and Science," *Science in Context* 6 (1993): 405–33.

55. Pion des Loches, *Mes campagnes*, 31–34.

56. On Châlons, see Le Puillon de Boblaye, *Ecoles d'artillerie*, 77–136. S.H.A.T. 2a59 Anon. "Instruction du Corps de l'Artillerie," 19 thermidor, year IX [7 August 1801].

57. S.H.A.T. 2a59 Berthier (Min. War) "Règlement sur l'instruction des troupes d'artillerie," *Extrait des registres*, 3 thermidor, year XI [22 July 1803] (Paris: Imprimerie de la République, year XI [1803]), 3, 10.

58. Bonaparte, "Décision," 16 September 1807, *Correspondance* (Paris: Imprimerie

Impériale, 1858–69), 16: 45–46. Augustin Lespinasse, *Essai sur l'organisation de l'arme de l'artillerie* (Paris: Magimel, year VIII [1799–1800]), 40, 71–76. S.H.A.T. 2a59 *Programmes de la Commission Mixte d'Officiers d'Artillerie et du Génie* (Metz: Collignon, 1807).

59. S.H.A.T. 2a59 Demarçay, "Projet d'une nouvelle composition des écoles régimentaires d'artillerie," 21 thermidor, year XI [9 August 1803]. S.H.A.T. Xd260 Chanteclaire, "Rapport concernant l'Ecole de la Fère," 1 germinal, year XI [22 March 1803].

60. The history of the school has been treated by Day, *Industrial World*, quote by Napoléon, p. 71.

61. Léon, *Education technique*, 250, also pp. 77–82, 177–84, 247–56.

62. I am using Day's data as reclassified by Weiss, *Technological Man*, 83–84. The remaining 23 percent had fathers in the military, so their social origins cannot be determined. Napoléon to Clark (Min. War), 8 March 1808, *Correspondance*, 18: 379.

63. The *conducteurs* increasingly protested their exclusion. Kranakis, "Social Determinants," 15–16. At the very end of the Restoration, another intermediary engineering school was established—the Ecole Centrale des Arts et Manufactures—this time to supply engineers for private industry; see Weiss, *Technological Man*.

64. Edmonson, *From Mécanicien*, 306–20. Day, *Industrial World*, 215–17. On technical drawing as a tool for social advancement, see Patrice Bourdelais, "Employés de la grande industrie: Les dessinateurs du Creusot, Formations et carrières (1850–1914)," *Histoire, économie et société* 8 (1989): 437–46.

65. For Taylorism in Europe, see Rabinbach, *Human Motor*.

66. Kaplan, *Provisioning Paris*, 25–33.

67. René Tresse, "Les dessinateurs du C.S.P.," *Techniques et civilisations* 5 (1956): 1–10.

68. Bénézech (Min. Int.), "Règlement," 1 August 1796, in [C.N.A.M.], *Recueil des lois, décrets, ordonnances, arrêtés, décisions et rapports relatifs à l'origine . . . du Conservatoire* (Paris: Imprimerie Nationale, 1889), 19–23. Grégoire, *A.P.*, 29 September 1794, 98: 148–53, quote on p. 152. A.N. F12 1556 Berthollet, "Rapport au C.A.A.," [1794–95]. C.N.A.M. 10° 325 Molard, [No title], [1790s].

69. Grégoire, *A.P.*, 29 September 1794, 98: 150–51.

70. Grégoire, "Rapport," 15 May 1798, in [C.N.A.M.], *Recueil*, 32–33.

71. On the metric system, see Alder, "Revolution to Measure." On patois, see Higonnet, "Linguistic Terrorism."

72. Charles Dupin, *Discours et leçons sur l'industrie* (Paris: Bachelier, 1825), 2: 153–391. On the role of the Conservatoire in the early nineteenth century, see Robert Fox, "Education for a New Age, The Conservatoire des Arts et Métiers," in *Artisan to Graduate*, ed. D.S.L. Cardwell (Manchester: Manchester University Press, 1974), 23–38. On the Mechanics Institutes, see Maxine Berg, *The Machinery Question and the Making of Political Economy* (Cambridge: Cambridge University Press, 1980).

73. Day, *Industrial World*, 19. Edmonson, *From Mécanicien*, 181–200. The Conservatoire's rich collection of technical drawings from 1800 to 1850 have been reviewed by Deforge, *Graphisme technique*, 100–106.

74. On the new property regime, see Sewell, *Work and Revolution*. Bossenga, "Les corporations." The Directory similarly established Bureaux des Poids et Mesures to reregulate exchanges in the marketplace, even as they attempted to transform France into a single national market. Alder, "Revolution to Measure."

75. Vandermonde, "Economie politique," *Séances [et leçons] des Ecoles Normales, recueillies par les sténographes, et revues par les professeurs* (Paris: Reyner, year III [1795]), 2: 233–45, 290–302, 447–63; 3: 145–61, 437–45; 4: 168–80, 452–71; 5: 89–109. Jacqueline Hecht, "Un example de multidisiplinarité: Alexandre Vandermonde (1735–1796)," *Population* 26 (1971): 641–76.

76. B.N. Vp 5301 Vandermonde, "Rapport sur les fabriques et le commerce de Lyon, 15 brumaire, an III," [5 November 1794], *Journal des arts et manufactures* 1 (year III [1794]): 1–48, quote p. 21. Vandermonde, *Séances*, 2: 244–45; 4: 170–72, 460–61.

Chapter Nine
Technological Amnesia and the Entrepreneurial Order

1. Edward W. Constant II, "A Model for Technological Change Applied to the Turbojet Revolution," *Technology and Culture* 14 (1973): 553–72. For a technological pessimist who believes in cumulative technology, see the works of Jacques Ellul, *The Technological Society*, trans. John Wilkinson (New York: Knopf, 1964). For a fertile approach to evolutionary models of technological change, see Mokyr, *Lever of Riches*. On the realization by biologists that many lines of descent fail, see Stephen J. Gould, *Wonderful Life: The Burgess Shale and the Nature of History* (New York: Norton, 1989).

2. Woronoff, *L'industrie sidérurgique*, 369–420; for data on Napoleonic France in 1806, see pp. 379, 410. François Crouzet, "Recherches sur la production d'armements en France (1815–1913)," *Revue d'histoire* 509 (1974): 45–84; 510 (1975): 409–22.

3. Louis Bergeron, *France Under Napoléon*, trans. R. R. Palmer (Princeton: Princeton University Press, 1981).

4. Serge Dontenwill, "Roanne au dernier siècle de l'ancien régime: Aspects démographiques et sociaux," *Etudes foréziennes* 4 (1971): 49–73. See also Pothier, *Roanne*. Lucas, *Structure of the Terror*, 33–34, 56–57, 363–69. On the sale of the *biens nationaux*, see B.M.R. 2Q3 Actours and Brissat, "Estimation des biens nationaux à Roanne," 7 August 1790.

5. S.H.A.T. B12* 64 Reg. 1 Chaulet to Commission d'Artillerie, "Correspondance de la Commission d'Artillerie avec la Commission des Armes et Poudres," 5 thermidor, year II [23 July 1794]. S.H.A.T. 6c5 Blanc to Min. War, 3 jour comp., year V [19 September 1797]. Blanc to C.I.C.C.A., 26 prairial, year VIII [15 June 1800].

6. The daily accounts of the Roanne factory under Cablat have survived; see B.M.R. 3R2 "Journal général de la Manufacture Nationale d'Armes de Roanne, Commencé le 7 nivôse, an X," [28 December 1801]. S.H.A.T. 4f2 Cablat and Min. War, "Traité pour l'entreprise," 30 fructidor, year X [17 September 1802]. S.H.A.T. 4f6/4 Tugny, "Mémoire sur la Manufacture," messidor, year XI [June–July 1803]. On Whitney, see Woodbury, "The Legend of Eli Whitney," and Gordon, "Mechanical Ideal."

7. S.H.A.T. 4f6/4 Tugny, "Mémoire sur la Manufacture," messidor, year XI [June–July 1803]. S.H.A.T. 4f4 Tugny and Cablat, "Devis des prix," 29 fructidor, year XI [16 September 1803].

8. S.H.A.T. 4f6/4 Tugny, "Mémoire sur la Manufacture," messidor, year XI [June–July 1803]. S.H.A.T. 6c5 A. G. Aboville, *fils*, to F. M. Aboville, *père*, 7 ventôse, year X [26 February 1802]. Régnier et al., "Expérience faite . . . sur cent platines," 15 germinal, year XII [5 April 1804]. S.H.A.T. 4f4 Tugny and Cablat, "Devis des prix," 29 fructidor, year XI [16 September 1803].

9. S.H.A.T. 4f6 A. G. Aboville, *fils*, "Observations sur le mémoire présenté au premier consul," germinal, year X [March–April 1802]. For Blanc's boast, see S.H.A.T. B12* 64 Reg. 1 Chaulet to Commission d'Artillerie, "Correspondance," 5 thermidor, year II [23 July 1794]. For a list of conscripts, see A.N. AF III 173 Blanc, "Manufacture de platines," 21 pluviôse, year IV [10 February 1796].

10. S.H.A.T. 6c5 Blanc, "Prix que les Entrepreneurs," vendémiaire, year VI [September–October 1797]. For the history of Blanc's subsidy, see S.H.A.T. 6c5 Agoult, "Historique des travaux du Sr. Blanc," 13 vendémiaire, year VI [4 October 1797]. Carnot (Min. War) to C.I.C.C.A., 14 prairial, year VIII [3 June 1800]. C.I.C.C.A. to Min. War, "Rapport," 26

messidor, year VIII [15 July 1800]. Cablat complained, however, that he was selling those at a loss. S.H.A.T. 6c5 Tugny, "Mémoire présenté au Conseil d'Etat," 16 vendémiaire, year XII [9 October 1803].

11. S.H.A.T. 6c5 A. G. Aboville, *fils*, to F. M. Aboville, *père*, 5 germinal, year X [26 March 1802]. Marmont to Min. War, 14 frimaire, year XII [6 December 1803]. B.M.R. 3R2 "Journal général de . . . Roanne." Blanc died with 14,000 *livres* of personal debt.

12. Jean-Jacques-Basilien Gassendi, *Mes loisirs* (Dijon: Frantin, 1820).

13. Paul Gaffarel, "Le Général de Gassendi," *Mémoire de la Société Bourguignonne de Géographie et d'Histoire* 19 (1903): 387–459. On some occasions, Napoléon praised Gassendi, but in his *Journal* he wrote: "One can see that this man has not been to war, he's contemptible, a dunce (*un pleutre, un ignare*)." *Le Mémorial de Saint-Hélène* (Paris: Gallimard, 1956), 1041–42.

14. Gassendi regularly asserted that genuinely identical gunlocks would cost four times as much as ordinary ones. S.H.A.T. 6c5 Gassendi to Min. War, 27 November 1806. F. M. Aboville, *père*, to Min. War, 19 messidor, year IX [8 July 1801].

15. S.H.A.T. 6c5 F. M. Aboville, *père*, et al., "Procès-verbal de remontage de 500 platines," 17 vendémiaire, year X [9 October 1801]. C.I.C.C.A. to Min. War, 22 frimaire, year X [13 December 1801]. F. M. Aboville, *père*, to Bonaparte, 16 pluviôse, year X [5 February 1802]. See the note by Bonaparte in the upper left-hand corner.

16. S.H.A.T. 6c5 Lamogère et al., "Procès-verbaux sur 492 platines," 25 pluviôse, year X [14 February 1802].

17. One can follow these intrigues in S.H.A.T. Xd415 F. M. Aboville, *père*, to Gassendi, 16 pluviôse, year X [5 February 1802]. S.H.A.T. 6c5 F. M. Aboville, *père*, to A. G. Aboville, *fils*, 13 ventôse, 4 floréal, year X [4 March, 24 April 1802]. For one young artillerist's view of the elder Aboville, see Pion des Loches, *Mes campagnes*, 124–25.

18. S.H.A.T. 6c5 Jacques et al., "Epreuve faite . . . sur cent fusils," 12 February 1804. Cablat to Min. War, 8 nivôse, year XII [30 December 1803]. Régnier et al., "Expérience faite . . . sur cent platines," 15 germinal, year XII [5 April 1804].

19. S.H.A.T. 6c5 Marmont to Min. War, 14 frimaire, year XII [6 December 1803]. Cablat and Min. War, 15 ventôse, year XII [6 March 1804]. Claude Gaier, "Note sur la fabrication des 'platines identiques' et sur la Manufacture Impériale de Platines de Liège," *Le Musée d'Armes, Bulletin trimestriel des Amis du Musée d'Armes de Liège* 7 (1979): 7.

20. For the test, countertest polemic, see S.H.A.T. 6c5 Lefebvre, "Examen fait de cinq platines," 10 prairial, year XIII [30 May 1805]. Delahaye, "Manufacture Impériale de Platines établie à Liège," 10 prairial, year XIII [30 May 1805]. Boucher et al., "Manufacture de Platines de Liège," 12 fructidor, year XIII [30 August 1805]. On the punishment of Delahaye, see S.H.A.T. Xd414 Saint-Martin to Gassendi, 9 November 1806. S.H.A.T. 6c5 Evain to Min. War, July 1806.

21. S.H.A.T. Xd414 Gassendi, note on Saint-Martin to Gassendi, 8 November 1806.

22. S.H.A.T. 6c5 Gassendi to Min. War, 27 November 1806. [Gassendi] to Bonaparte, 9 December 1806. Gaier, "'Platines identiques," 9. However, I have found no evidence of the exhibit in A.N. F12 985 "Etat général des objets admis à l'Exposition des Produits de l'Industrie," 1806. The vestigial musket-making operation in Roanne was closed down at roughly the same time. S.H.A.T. 6c5 Gassendi to Min. War, 26 June 1807.

23. For a comparison of production methods at Versailles and Roanne, see Cotty, *Fabrication des armes*, 63–75. S.H.A.T. 4f6/4 A. M. Aboville, *cadet*, "Rapport sur Mutzig," 20 November 1810. On Bodmer, see Henderson, *Industrial Britain*. Roe, *Tool Builders*, 75–76.

24. Gassendi, *Aide-mémoire* (4th ed., Paris: Magimel, 1809), 2: 589–90; also (5th ed., 1819), 2: 591–92. Cotty acknowledged certain advantages of "accelerated" methods, but denied that identity was possible. Cotty, *Fabrications des armes*, 69–75; *Dictionnaire de l'artillerie*, vol. 23 of *Encyclopédie méthodique* (Paris: Agasse, 1822), 185, 337–43, 456.

25. S.H.A.T. 4f6 A. G. Aboville, *fils*, "Observations sur le mémoire présenté au Premier Consul," germinal, year X [March–April 1802].

26. S.H.A.T. 6c5 F. M. Aboville, *père*, to Bonaparte, 16 pluviôse, year X [5 February 1802].

27. S.H.A.T. 6c5 F. M. Aboville, *père*, to A. G. Aboville, *fils*, 13 ventôse, year X [4 March 1802].

28. S.H.A.T. 6c5 F. M. Aboville, *père*, "Conseil Extraordinaire de l'Artillerie, Séance du 24 fructidor, an X," [11 September 1802].

29. Gassendi, "Prédiction trompée sur ****," *Mes Loisirs*.

30. S.H.A.T. 6c5 Gassendi, "Rapport de la C.I.C.C.A. sur la proposition que font citoyens Poterat et Cabanel," 16 fructidor, year VII [2 September 1799].

31. S.H.A.T. Xd234 Charles Lucio, "Liège: Rapport d'inspection de 1808," 6 July 1808. Lucio, however, admitted that he had never seen Blanc's mechanical methods in actual operation.

32. Quoted in S.H.A.T. Bib. #8402 Jean Rousseau, "La vie quotidienne dans les manufacture d'armes à l'époque de la Révolution et de l'Empire," (Mémoire de maîtrise, IVe Section de l'Ecole Pratique des Hautes Etudes, 1969), 58. See also S.H.A.T. Xd235 Drouot, "Rapport sur la Manufacture," 1808.

33. A.N. AF III 153 Reverchon to Carnot (Dir. Exéc.), 29 ventôse, year IV [19 February 1796]. [Guilliaud], *Fabrication d'armes*, [1800]. Lucas, *Structure of the Terror*.

34. For gun production during the Directory and Consulate, see [Guilliaud], *Fabrication d'armes*, [1800]. A.N. F12 1955 "Rapport au Min. Int.," 27 messidor, year IX [14 September 1801]. For private gun sales during the Empire, see A.D.R. 2F16 Anon., "Mémoire," 15 ventôse, year XIII [6 March 1805]; "Armes de commerce," 1811–1815.

35. A.N. F7 3016 Min. de la Guerre, "Extrait des registres du Directoire Exécutif," 8 ventôse, year IV [27 February 1796]. Comité de Sûreté Générale, 11 brumaire, year IV [2 November 1795]. "Extrait de registres du Département de l'Averion," 20 vendémiaire, year XIII [12 October 1804]. Dorgant to Préfet de Police des Basses Pyrenées, 28 January 1806. "Décret impérial," 11 July 1810. Dépt. de Cher, "Chasse de port d'armes," 20 August 1813. Ministère de la Police Générale, "Circulaire aux préfets," 14 November 1813.

36. *R.A.C.S.P.*, 17 October, 10 December 1793, 7: 464; 9: 303–4. For nearly nine hundred new apprentices, see A.D.L. L935-L941 "Brevet d'apprentissages," [1794–95]. For output figures, see [Guilliaud], *Fabrication d'armes*, [1800]. The definitive work on the relationship between the central authorities and local institutions in the Forez during the Terror is Lucas, *Structure of the Terror*.

37. A.D.L. L933 Bonnand et al., "Extrait des registres de . . . Commission des Ateliers d'Armes," 25 germinal, year II [14 April 1794]. Two armorers were executed in the year II as counter-revolutionaries, along with the royal Entrepreneur Carrier de Thuillerie. A.D.L. L1039 "Tableau des contrerévolutionnaires," year II [1794]. In September 1794, eighteen rural armorers were denounced as rebels because they insisted on taking in their harvest rather than working in the ateliers. A.D.L. L933 Antoine Bruel, "Dénonciations," 16 fructidor, year II [2 September 1794]. Boyer to Administrateur du District, 18 nivôse, year II [28 December 1793].

38. On stockpiles, see A.N. AF III 153 Anon., "Etat général des fusils," 1 fructidor, year IV [18 August 1796]. On production, see [Guilliaud], *Fabrication d'armes*, [1800]. On the reestablishment of the statutes of the ancien régime in Saint-Etienne, see S.H.A.T. 4f2/2 [Agoult], "Observations sur la soumission," year V [1796–97]. S.H.A.T. 4f6/1 Truffet Saint-Martin, "Observations sur les différents régimes," year X, [1801–2]. Colomb, "Observations sur le règlement de police," 27 January 1806. A.D.R. 2H2 *Règlement pour la police* (Saint-Etienne: Boyer, 1810). B.M.StE. F609(6) Dubouchet, *Mémoire*, 28 thermidor, year IX [16 August 1801] (Saint-Etienne: Imprimerie Expéditive et Economique, [1801]).

39. A.D.L. L936 Javelle, "Etat des ouvriers travaillant dans la manufacture de Jovin," 8 nivôse, year II [28 December 1793]. A.D.R. 4LL182* F. Jovin to Cit. Rep., 27 vendémiaire, year II [18 October 1793]. S.H.A.T. 4f6/3 Tugny to Songis, 30 vendémiaire, year XIII [22 October 1804]. "Pétition," 5 thermidor, year VIII [24 July 1800], in Dubouchet, *Mémoire*, 36–38.

40. Isser Woloch, "Napoleonic Conscription: State Power and Civil Society," *Past and Present* 111 (1986): 101–29. Also, Alan Forrest, *Conscripts and Deserters: The Army and French Society during the Revolution and Empire* (New York, Oxford University Press, 1989), 53–56. Philippe Gonnard, "Moeurs administrative du Premier Empire: La conscription dans la Loire," *Revue d'histoire de Lyon* 7 (1908): 81–110.

41. S.H.A.T. 6c5 Colomb, "Direction générale des Manufactures d'Armes," 5 messidor, year XII [24 June 1804]. A.D.L. L934 Lardon et al., "Registre pour servir," 26 vendémiaire, year VIII [18 October 1799]. On conscripts, see S.H.A.T. Xd235 Drouot, "Rapport," 1808. For a rare surviving *livret*, see J. Rousseau, "Vie quotidienne," 94–98, 108.

42. Lespinasse, *Organisation de l'artillerie*, 56–57, 107–8. S.H.A.T. 4f6/1 Colomb, "Police des Manufactures d'Armes," 27 January 1806.

43. S.H.A.T. 4f7 Sirodon, "Mémoire sur les proportions dans les armes," year XIII [1804–5].

44. Ibid. Equally novel was the bureaucratic process by which the central authorities consulted the various inspectors regarding the gun's specifications. See S.H.A.T. 4f22 Tuffet Saint-Martin, "Observations sur le règlement," year XIII [1804–5]; C. G. Dufort, "Observations sur le règlement," year XIII [1804–5]; Sirodon, "Sur la platine," year XIII [1804–5].

45. Of the 1892 total workers in 1808, 605 were locksmiths, 483 were conscripts, and 8 were chefs d'atelier. S.H.A.T. Xd235 Drouot, "Rapport," 1808.

46. A.D.L. L942 Berardier-Merley et al., "Etat nominatif de tous les réquisitionnaires," 15 floréal, year VIII [5 May 1800].

47. S.H.A.T. 4f12 Préau, "Précis de la fabrication," 25 April 1812. Gassendi, *Aide-mémoire* (1819), 553–54. Dubessy, *Manufacture d'armes*, 410–11. According to Colomb, this bottleneck in gunlock production had also temporarily emerged during the year II. S.H.A.T. 46f/3 Colomb, "Mémoire," 1 prairial, year XI [21 May 1803].

48. Sabel and Zeitlin, "Stories and Strategies." Berg, *Age of Manufactures*.

49. S.H.A.T. 4f12 Préau, "Précis de la fabrication," 25 April 1812.

50. Alder, "Revolution to Measure." Marie-Noëlle Bourguet, *Déchiffrer la France: La statistique départementale à l'époque napoléonienne* (Paris: Edition des Archives Contemporaines, 1988).

51. In his report to the emperor of 1810, Gassendi repeated Gribeauval's claim that the Entrepreneur had no say over the methods of production. S.H.A.T. Xd415 Gassendi to Bonaparte, 5 June 1810. In the Restoration, Cotty reiterated the advantages of Gribeauval's use of Entrepreneur-merchants, allowing that the state might intervene directly only if no suitable merchant presented himself. Cotty, *Artillerie*, 163–66.

52. Dubessy, *Manufacture de Saint-Etienne*, 205.

53. S.H.A.T. 4f7/1 F. M. Aboville, *père*, et al., "Commission des Armes à Feu," 2 fructidor, year VIII-11 ventôse, year IX [20 August 1800–2 March 1801]. S.H.A.T. 4f22 [Département de la Guerre], *Règlement fixant les principales dimensions des armes portatives* (Paris: Imprimerie Nationale, [1804]). Jean Boudriot, "Système an IX, XIII," *Armes à feu françaises: Modèles d'ordonnance* (Paris, 1961).

54. S.H.A.T. MR1981 Cotty et al., "Rapport sur les expériences," 16 March 1811.

55. S.H.A.T. 6c5 Min. War to F. M. Aboville, *père*, 29 nivôse, year X [19 January 1802]. Cablat to Min. War, 7 nivôse, year XII [29 December 1803].

56. S.H.A.T. 6c5 Anon., "Notes sur les procédés méchaniques," June 1810. S.H.A.T. 6c2/1 Gassendi to Min. War, 30 January 1814. I have not been able to determine whether Le Roy was kin to Jean-Baptiste Le Roy or the rest of the clockmaker family.

57. S.H.A.T. 6c2/1 Evain, "Rapport fait au Ministre," 10 March 1814. S.H.A.T. Xd414 Division de l'Artillerie to Bonaparte, 12 March 1814. A.N. F12 1565 Commission sur la Perfectionnement des Armes, "Rapport sur le fusil modèle de M. Julien Le Roy," [1815]. S.H.A.T. 6c2/1 Cotty, "Rapport sur le fusil," 4 December 1815.

58. S.H.A.T. 6c2/1 Cotty et al., "Rapport au Ministre de la Guerre," 4 June 1815. A.N. F12 1565 Prieur to Carnot, [1815].

59. A.N. F12 1565 Bonaparte, "Notes sur les moyens employées pour la fabrication des fusils dans la capitale," 26 April 1815.

60. S.H.A.T. 6c2/1 Le Roy, "Rapport," 4 July 1817.

61. Peter Michael Molloy, "Technical Education and the Young Republic: West Point as America's Ecole Polytechnique, 1802–1833" (Ph.D. diss., Brown University, 1975). Americans adapted French gun designs as well, modifying them to make them easier to mass produce. M. R. Smith, *Harpers Ferry*, 280.

62. Pemberton, "Industrialization," 76–90, 264. Cotty, *Artillerie*, 226, 312. Arms-workers in the surrounding countryside began to take up permanent residence in the city; see Yves Lequin, *Ouvriers de la région lyonnaise* (Lyon: Presse Universitaire de Lyon, 1977), 224. André Pauze, "L'émeute des ouvriers canonniers en 1831," *Bulletin des Amis du Vieux Saint-Etienne* (1991): 59–62.

63. S.H.A.T. 4f12 Paixhans, "Rapport fait au C.I.C.C.A.," [1816]. Anon., "Essai sur la construction de la platine," n.d. The French army was the first European power to experiment with percussion locks in some of its muskets, but it did not adopt them until the 1840s. Boudriot, "L'évolution de la platine," 30–35. Kennard, *French Pistols*, 49–55. Bottet, *Monographies de l'arme*, 163–64.

64. S.H.A.T. 4f8/1 Remington, 8 September 1855. Min. War to Président du Comité d'Artillerie, 10 January 1856. Président du Comité d'Artillerie, 19 January 1856. Kreutzberger, "Rapport addressé à M. le Président du Comité d'Artillerie," 1856. Comité d'Artillerie, "Avis sur une proposition de Kreutzberger," 14 March 1856. Kreutzberger to Min. War, 26 September 1858. See also Edmonson, *From Mécanicien*, 423–29.

65. Edmonson, *From Mécanicien*, 434–38. For the sewing machine, see Monique Peyrière, "Recherches sur la machine à coudre en France, 1830–1889" (Mémoire de maîtrise, Ecole des Hautes Etudes en Sciences Sociales, 1990). For the U.S., see Hounshell, *American System*.

66. Cohen, "Inventivité organisationnelle," 71. Other works emphasize the role of the military and the war economy in the further development of interchangeability. Patrick Fridenson, "The Coming of the Assembly Line to Europe," in *Dynamics of Science and Technology*, eds. Krohn, Layton, and Weingart, 159–75. Aimée Moutet, "Introduction de la production à la chaine en France du début du XXe siècle à la grande crise de 1930," *Histoire, économie et société* 6 (1983): 63–82.

67. The Saint-Etienne gun-makers thereafter kept tabs on progress at Enfield; see M. C. Jalambert, *Rapport à l'Exposition de Londres en 1862: Pour l'arquebusiers* (Saint-Etienne: Robins, 1862). For an overview of nineteenth-century developments in the arms industry, see Christiane Lacombe, "Introduction du machinisme dans les fabrications d'armement en France au XIXe siècle," *Revue internationale d'histoire militaire* 41 (1979): 37–69, quote on p. 40. For the troubled introduction of Pratt & Whitney machines, see S.H.A.T. 4f8/1 Coleman to Min. War, 18 April 1878.

68. Ply conceded that some of these automatic machines did not exist yet, not even in the advanced machine-tool industries of America and Britain. Gustave Ply, "Etude sur l'organisation du service technique dans les manufactures d'armes," *Revue d'artillerie* 32 (1888): 344–90; 33 (1888–89): 5–47, 101–42, 211–43, 297–332.

69. On French military production, see Crouzet, "Recherches sur la production." A good review of European changes in armaments production in this period can be found in Joseph Bradley, *Guns for the Tsar: American Technology and the Small Arms Industry in Nineteenth-*

Century Russia (Dekalb, Ill.: Northern Illinois University Press, 1990), 3–42. Also Daniel Headrick, *The Tools of Empire: Technology and European Imperialism in the Nineteenth Century* (New York: Oxford University Press, 1981), 83–126. For a "lag-theory" history linking armaments design and tactics, see Steven T. Ross, *From Flintlock to Rifle: Infantry Tactics, 1740–1866* (London: Associated University Press, 1979).

Conclusion

1. For a synthesis of how weapons research became central to Cold War science, see Leslie, *Cold War and American Science*. On military sponsorship of transistors, see Thomas Misa, "Military Needs, Commercial Realities, and the Development of the Transistor, 1948–1958," in *Military Entrepreneurship*, ed. M. R. Smith, 253–87. For the Air Force sponsorship of machine tools, see Noble, *Forces of Production*. For a politico-military comparison of Prussian and American railroads, see Colleen Dunlavy, *Politics and Industrialization: Early Railroads in the United States and Prussia* (Princeton: Princeton University Press, 1994). On the involvement of the American state in a variety of technological public works, some military, some not, see T. P. Hughes, *American Genesis*. On the pervasive role of America's military in the postwar society, see Michael Sherry, *In the Shadow of War: The United States Since the 1930s* (New Haven: Yale University Press, 1995). The Moses example comes from Winner, "Do Artifacts."

2. The subject of design is only now receiving full attention. On telegraphy, see Paul Israel, *From Machine Shop to Industrial Laboratory: Telegraphy and the Changing Context of American Invention, 1830–1920* (Baltimore: Johns Hopkins University Press, 1992). For bridges, see Kranakis, "Social Determinants." For bicycles, see Trevor J. Pinch and Wiebe Bijker, "The Social Construction of Technological Facts and Artifacts: Or How the Sociology of Science and the Sociology of Technology Might Benefit Each Other," in *Social Construction*, eds. Bijker, Hughes, and Pinch, 17–50.

3. T. H. Breen, "Narrative of Commercial Life: Consumption, Ideology, and Community on the Eve of the American Revolution," *William and Mary Quarterly* 50 (1993), 471–501; also, "The Meaning of Things: Interpreting the Consumer Economy in the Eighteenth Century," in *Consumption*, eds. Brewer and Porter, 249–60. Sonenscher, *Hatters*. Leora Auslander, *Taste and Power: Furnishing Modern France* (Berkeley: University of California Press, 1996). Pierre Bourdieu, *Distinction: A Social Critique of the Judgement of Taste*, trans. Richard Nice (Cambridge: Harvard University Press, 1984).

BIBLIOGRAPHY

Archival sources are listed below, with particular attention to those which proved most germane to armaments and engineers. The materials housed in the current Manufacture d'Armes de Saint-Etienne are extremely limited for the period before 1814 when the Austrians burned the Manufacture. Surviving records can be found in Dubessy, *Manufacture de Saint-Etienne*. In 1995 the artillery collection at S.H.A.T. was being renumbered; I have retained the old numeration which will correspond on a one-to-one basis with the new numeration. For Revolutionary Saint-Etienne, I also consulted the Archives of the Département du Rhône in Lyon, which until 1793 was the seat of the Département du Rhône et de la Loire. Ephemeral pamphlets found in the archives which I cite only once have been fully referenced in the endnotes and are not reproduced in the list of primary sources.

Archival Material

Archives de l'Académie des Sciences, Paris (A.A.S.)
  Files on various academicians: Le Roy, Coulomb
  Procès-verbaux of the Academy of Sciences
Archives Départementales de la Loire, Saint-Etienne (A.D.L.)
  Revolutionary series: L
Archives Départementales du Rhône, Lyon (A.D.R.)
  Bouillet's correspondence in 1793: 42K171*
  Series on denunciations in the Loire: 4LL
Archives of the Ecole des Ponts et Chaussées (A.E.P.C.)
  Perronet's analysis of the pin factory: #2385
Archives Municipales de Saint-Etienne (A.M.StE.)
  Local petitions: HH
  Meetings of the Municipal Council
Archives Nationales, Paris (A.N.)
  Cartons on the Atelier de Perfectionnement: F^{12} 1310–1314
  Cartons, "Ciraud" (Quinze-vingt Atelier, Manufacture of Paris): T1164(1–2)
  Cartons deposited by Givry (artillery engineer, inspector): T591(1–6)
  Government series during the Revolution: AD VI 39, AD VI 40, AF* II, AF IV
  Series on industry and the state: F^{12}
  Series on military matters: F^{9}
  Series on public works: F^{13}
  Other series: F^{4}, F^{7}, F^{14}
Bibliothèque Municipale de Roanne, Roanne (B.M.R.)
  Journal général of the Roanne factory: 3R2
  Various manuscripts, pamphlets, and local monographs
Bibliothèque Municipale de Saint-Etienne, Saint-Etienne (B.M.StE.)
  Various local pamphlets and monographs
Bibliothèque Nationale, Paris (B.N.)
  Various pamphlets and rare books

The Conservatoire National des Arts et Métiers (C.N.A.M.)
Archives of C.N.A.M.
Bibliothèque of C.N.A.M.
Collection of technical drawings, especially series #13571
Houghton Library, Harvard University, Cambridge, Mass. (H.H.U.)
Papers of Fougeroux de Bondaroy
Service Historique de l'Armée de Terre, Archives, Vincennes (S.H.A.T.)
Carton on Honoré Blanc: 6c5
Cartons on the Manufacture of Paris: B[12]*48, B[12]*64, Xd233–Xd235
Cartons on Roanne Manufacture: Xd414–Xd415
Correspondence series: 9a
Personnel registers for the artillery: Yb668–Yb672
Series on armory finances: 5a, 5b
Series on the cannon foundries: 4c
Series of military memoirs: MR
Series on the royal, national, and imperial armories: 4f
Various series on the artillery service and its schools: 1a, 2a, 2b, Xd249–Xd251
Service Historique de l'Armée de Terre, Bibliothèque, Vincennes (S.H.A.T. Bib.)
Meetings of the Comité Militaire: MS173–MS176
Military ordonnances
Rare books on the military
Theses on the armaments industry

Primary Sources

Académie Française, *Le dictionnaire de l'Académie*. 1st ed. Paris: Coignard, 1694.

Agricola, Georgius. *De Re Metallica*. Translated by Herbert Clark Hoover and Lou Henry Hoover. New York: Dover, [1556], 1950.

Alembert, Jean Le Rond d'. *Nouvelles expériences sur la résistance des fluides*. Paris: David, 1777.

Anderson, Robert. *To Hit a Mark*. London: Morden, 1690.

Anon., *Lettre à M. le comte de Guibert sur le précis de ce qui est arrivée à Berry*. N.p., [1789].

Anon., *Procès-verbal des épreuves faites aux Ecoles d'Artillerie de Douai*. Amsterdam: Arkstee et Merkus, 1772. (S.H.A.T. Bib. #72227)

Archives parlementaires de 1787 à 1860: Recueil complet des débats législatifs et politiques des chambres françaises. Paris: Imprimé par ordre du Sénat et de la Chambre des Députés, 1879–.

Arcq, Phillipe-Auguste de Saint-Foix d'. *La noblesse militaire, ou le patriote français*. 3d ed. N.p., 1756.

Arcy, Patrick d'. *Essai d'une théorie d'artillerie*. Dresden: Walther, 1766.

Aulard, François-Alphonse, ed. *La Société des Jacobins, Recueil des documents du Club des Jacobins de Paris*. Paris: Jouast, 1889–97.

———. *Paris pendant la réaction thermidorienne et sous le Directoire*. Paris: Cerf, 1898–1902.

———. *Recueil des actes du Comité de Salut Public*. Paris: Imprimerie Nationale, 1889–1951.

[Bachaumont, Louis Petit de]. *Mémoires secrets pour servir à l'histoire de la république des lettres en France*. London: Adamson, 1780–89.

Bachelier, Jean-Jacques. *Collection des discours*. Paris: Imprimerie Royale, 1790.

———. *Discours sur l'utilité des écoles élémentaires en faveur des arts mécaniques*, 10 September 1766. (B.N. V23842(9))

———. *Mémoire sur l'origine, les progrès, et la situation de l'Ecole Royale Gratuite de Dessin*. Paris: Imprimerie Royale, 1790. (B.N. V23849)

Bacon, Francis. *Advancement of Learning*. Chicago: Encyclopedia Britannica, 1952.

Bélidor, Bernard Forest de. *Architecture hydraulique*. Paris: Jombert, 1737–53.

———. *La science des ingénieurs dans la conduite des travaux de fortification et d'architecture civile*. Paris: Jombert, 1729.

———. *Le bombardier français, ou nouvelle méthode de jeter les bombes avec précision*. Paris: Imprimerie Royale, 1731.

———. *Nouveau cours de mathématiques à l'usage de l'artillerie et du génie*. Paris: Jombert, 1725.

———. *Oeuvres diverses*. Amsterdam: Jombert, 1764.

Bézout, Etienne. *Cours de mathématiques à l'usage du Corps Royal de l'Artillerie*. Paris: Pierres, [1772], 1781.

Binning, Thomas. *A Light to the Art of Gunnery*. London: Darby, 1676.

Blanc, Honoré. *Mémoire important sur les fabrications des armes de guerre*. Paris: Cellot, 1790.

Blondel, François. *L'art de jetter les bombes*. Paris: L'auteur et Langlois, 1683.

———. *L'art de jetter les bombes*. La Haye: Arnout, 1685.

Blondel, Jacques-François. *Discours sur la nécessité de l'étude de l'architecture, cinquième cours public*. Paris: Jombert, 1754.

Boislisle, A. M. de, ed. *Correspondance des contrôleurs-généraux des finances avec les intendants de provinces, 1685–1699*. Paris: Imprimerie Nationale, 1874–97.

Bonaparte, Napoléon. *Correspondance*. Paris: Imprimerie Impériale, 1858–69.

———. *Le Mémorial de Saint-Hélène*. Paris: Gallimard, 1956.

Boudriot, Jean, ed. *The Seventy-four Gun Ship: A Practical Treatise on the Art of Naval Architecture*. Translated by David Roberts. Beccles, U.K.: Naval Institute Press, 1986.

Bricard, Louis-Joseph. *Journal du cannonier Bricard, 1792–1802*. Paris: Delagrave, 1891.

Buchez, Philippe-Joseph-Benjamin, and Prosper-Charles Roux, eds. *Histoire parlementaire de la Révolution française, ou journal des assemblées nationales, depuis 1789 jusqu'en 1815*. Paris: Pauline, 1834–38.

Carnot, Lazare. *Correspondance générale de Carnot, publiée avec des notes historiques et biographiques, août 1792–mars 1795*. Edited by Etienne Chavaray. Paris: Imprimerie Nationale, 1892–1907.

———. *Révolution et mathématique*. Introduction and edited by Jean-Paul Charnay. Paris: L'Herne, 1985.

Caron, Pierre, ed. *Paris pendant la Terreur: Rapports des agents secrets du Ministre de l'Intérieur*. Paris: Alphonse Picard, 1910–78.

Chapelon, Jean. *Oeuvres complètes*. Edited by Annie Elsass. Saint-Etienne: Centre d'Etudes Foréziennes, 1985.

Condorcet, Marie-Jean-Antoine-Nicolas de Caritat de. *Correspondance inédite de Condorcet à Turgot*. Edited by Charles Henry. Paris: Perrin, n.d.

———*Sketch for a Historical Picture of the Progress of the Human Mind*. Translated by June Barraclough. London: Weidenfield and Nicolson, [1795], 1955.

[Conservatoire National des Arts et Métiers]. *Recueil des lois, décrets, ordonnances, arrêtés, décisions et rapports relatifs à l'origine ... du Conservatoire [National] des Arts et Métiers*. Paris: Imprimerie Nationale, 1889.

Cotty, Hermann. *Dictionnaire de l'artillerie*. Vol. 23 of *Encyclopédie méthodique*. Paris: Agasse, 1822.

———. *Mémoire sur la fabrication des armes portatives de guerre*. Paris: Magimel, 1806.

Coulomb, Charles Augustin. *Théorie des machines simples*. Paris, Bachelier, 1821.

Dartein, Charles M. S. *Traité élémentaire sur les procédés en usage dans les fonderies pour la fabrication des bouches à feu d'artillerie*. Strasbourg: Levrault, 1810.

Delambre, Jean-Baptiste-Joseph. *Grandeur et figure de la terre*. Edited by Guillaume Bigourdan. Paris: Gauthier-Villars, [1827], 1912.

Descartes, René. *Discours de la méthode*. Paris: Fayard, [1637], 1987.

Destutt de Tracy, Antoine Louis Claude. *Observations sur le système actuel d'instruction publique*. Paris: Panckoucke, year IX [1800–1801].

Dhombres, Jean, ed. *Ecole Normale de l'An III: Leçons de mathématiques*. Paris: Dunod, 1992.

Diderot, Denis. *Oeuvres complètes*. Paris: Garnier, 1876.

———. *Pensées sur l'interprétation de la nature*. Paris: Vrin, [1754], 1983.

Diderot, Denis, and Jean Le Rond d'Alembert, eds. *L'encyclopédie, ou dictionnaire raisonné des sciences, des arts et des métiers*. Paris: Briasson, 1751–72.

[Douazac]. *Dissertation sur la subordination*. 2d ed. Avignon: Dépense de la Compagnie, 1753.

Du Coudray, Philippe-Charles-Jean-Baptiste Tronson. *L'artillerie nouvelle, ou examen des changements faits dans l'artillerie française depuis 1765*. Amsterdam: n.p., 1773.

———. *L'état actuel de la querelle sur l'artillerie*. N.p., 1774.

———. *L'ordre profond et l'ordre mince, considérés par rapport aux effets de l'artillerie*. Metz: Ruault, 1776.

Du Puget, Edme Jean Antoine. *Essai sur l'usage de l'artillerie*. Amsterdam: Arkstee et Merkus, 1771.

———. *Réflexions sur la pratique raisonnée du pointement des pièces de canon*. Amsterdam: Arkstee et Merkus, 1771.

Du Teil, Jean. *De l'usage de l'artillerie nouvelle dans la guerre de campagne, connaissance nécessaire aux destinés de commander toutes les armées*. Paris: Charles-Lavauzelle, [1778], 1924.

Dumouriez, Charles-François Dupérier. *La vie et les mémoires du Général Dumouriez*. Paris: Baudoin, [1794], 1822.

Dupin, Charles. *Discours et leçons sur l'industrie*. Paris: Bachelier, 1825.

Euler, Leonard. "Recherches sur la véritable courbe que décrivent les corps jetés dans l'air." *Mémoires de l'Académie de Berlin* (1753): 321–55.

Fiosconi, Cesar, and Jordam Guserio. *Espingarda Perfeyta, or The Perfect Gun*. Translated and edited by Rainer Daehnhardt and W. Keith Neal. London: Sotheby, [1718], 1974.

Folard, Jean-Charles de. *Nouvelles découvertes sur la guerre*. 2d ed. Brussels: Foppens, 1724.

Fougeroux de Bondaroy, Auguste-Denis. "L'art du coutelier en ouvrages communs." Volume 14. *Description des arts et métiers, faites ou approuvées par Messieurs de l'Académie Royale des Sciences*. Edited by Jean Elie Bertrand. New Edition. Neuchâtel: Société Typographique, 1771–83.

Fourcroy, Antoine-François. *Sur les arts qui ont servis à la défense de la République, séance du 14 nivôse, an III.* Speech of 3 January 1795. Paris: Imprimerie Nationale, year III [1794]. (B.N. Le38 1127)

Fournial, Etienne, and Jean-Pierre Gutton, eds. *Cahiers de doléances de la province de Forez.* Saint-Etienne: Centre d'Etudes Foréziennes, 1974.

Frederick II. *Essai sur la grande guerre.* London: Compagnie, 1761.

———. *Instruction secrètte* [*sic*]. Translated by Le Prince de Ligne. Brussels: Hayez, 1787.

———. *Mémoires de Frédéric.* Paris: Plon, 1866.

Frézier, Amédée-François. *Elémens de stéréotomie à l'usage de l'architecture pour la coupe des pierres.* Paris: Jombert, 1760.

———. *La théorie et la pratique de la coupe des pierres et des bois... ou traité de stéréotomie.* Strasbourg: Doulsseker, 1737–39.

Galilei, Galileo. *Two New Sciences.* Translated by Stillman Drake. Madison: University of Wisconsin Press, 1974.

Gassendi, Jean-Jacques-Basilien. *Aide-mémoire à l'usage des officiers d'artillerie de France.* 2d–5th eds. Paris: Magimel, 1798, 1801, 1809, 1819.

———. *Aide-mémoire à l'usage des officiers du Corps-Royal de l'Artillerie.* 1st ed. Metz: Devilly, 1789.

———. *Mes Loisirs.* Dijon: Frantin, 1820.

[Gauthier, Henri]. *Instruction pour les gens de guerre.* Paris: Cognard, 1692.

Gerbaux, Fernand, and Charles Schmidt, eds. *Procès-verbaux des Comités d'Agriculture et de Commerce de la Constituant, de la Législative et de la Convention.* Paris: Imprimerie Nationale, 1906–10.

Goethe, Johann Wolfgang von. *The Campaign in France, 1792.* Translated by Robert R. Heitner. New York: Surkamp, 1987.

Gribeauval, Jean-Baptiste Vaquette de. *Tables des constructions des principaux attirails de l'artillerie proposées ou approuvées depuis 1764 jusqu'en 1789.* Edited by Jacques Charles Manson. Paris, 1792.

[Gribeauval, Jean-Baptiste de, and Joseph Florent de Vallière, *fils*], eds. *Collection des mémoires authentiques qui ont été présentés à MM. les Maréchaux de France.* Aléthopolis: Neumann, 1774. (S.H.A.T. Bib. #72359)

Grobert, Jacques François Louis. *Mémoire pour mesurer la vitesse initiale des mobiles de différens calibres.* Paris: n.p., 1804.

Guibert, Jacques-Antoine-Hippolyte de. *De la force publique considérée dans tous ses rapports.* Paris: Didot l'aîné, 1790.

———. *Défense du système de guerre moderne.* Vols. 3 and 4. In *Oeuvres militaires.* Paris: Magimel, year XII [1803].

———. *Ecrits militaires.* Paris: Copernic, 1977.

———. *Essai général de tactique.* Vols. 1 and 2. In *Oeuvres militaires.* Paris: Magimel, year XII [1803].

———. *Précis de ce qui s'est passé à mon égard à l'Assemblée du Berry, 25 mars 1789.* N.p., [1789].

Guillaume, James, ed. *Procès-verbaux du Comité d'Instruction Publique de la Convention Nationale.* Paris: Imprimerie Nationale, 1891–1907.

Guilliaud. *La fabrication d'armes de guerre à Saint-Etienne.* N.p., [1800]. (A.D.L. L835)

———. *Mémoire sur la mise en oeuvre de tous les metaux dans le Département de la Loire.* Lyon: Louis Cutty, [1795–96]. (A.D.L. L930)

Guyton-Morveau, Louis-Bernard. *Sur l'état de situation des arsenaux et l'armement des armées de terre et de mer de la République*. Speech delivered 2 February 1795. Paris: Imprimerie Nationale, year III [1795].

[Hautes Alpes]. *Procès-verbal de l'inauguration de l'Ecole Centrale*. Gap: Allier, year V [1796–97].

Hébert, Jacques-René. *Le Père Duchesne*. Paris: E.D.H.I.S., 1969.

Jalambert, M. C. *Rapport à l'Exposition de Londres en 1862: Pour l'arquebusiers*. Saint-Etienne: Robins, 1862.

Jefferson, Thomas. *The Papers of Thomas Jefferson*. Edited by Julian P. Boyd. Princeton: Princeton University Press, 1950–.

Journal des débats et des décrets. Paris: Imprimerie Nationale, 1789–1802.

Jugement du Conseil de Guerre tenu par ordre du roi, à l'Hôtel des Invalides du octobre 12, 1773. Paris: Imprimerie Royale, 1773.

La Fontainerie, François de, ed. *French Liberalism and Education in the Eighteenth Century*. New York: McGraw-Hill, 1932

La Mettrie, Julien Offray de. *L'homme machine*. Edited by Paul-Laurent Assoun. Paris: Renoel, 1981.

Laclos, Pierre-Amboise-François Choderlos de. *Les liaisons dangereuses*. Translated by P.W.K. Stone. New York: Penguin, 1961.

———. *Oeuvres complètes*. Edited by Laurent Versini. Paris, Gallimard, 1979.

Lacroix, Sylvestre-François. *Discours sur l'instruction publique, prononcé à la distribution des prix des Ecoles Centrales, le 29 thermidor, an VIII*. Speech of 17 August 1800. Paris: Duprat, year IX [1800].

[Lacuée, Jean-Gérard, comte de Cessac]. *Art militaire*. Vol. 4, supplément. In *Encyclopédie méthodique*. Paris: Panckoucke, year V [1796–97].

Le Blond, Guillaume. *L'artillerie raisonnée*. Paris: Jombert, 1761.

———. "Affût," "Artillerie," and "Canon de campagne." Vol. 1: 190–92, 603–24; vol. 2: 202–9. *Supplément à l'Encyclopédie, ou dictionnaire raisonné des sciences, des arts et des métiers*. Amsterdam: Rey, 1776.

———. *Traité de l'attaque des places*. Paris: Jombert, 1743.

Le Pelletier, Louis-August. *Une famille d'artilleurs, Mémoires de Louis-Auguste Le Pelletier, seigneur de Glatigny, Lieutenant-général des armées du roi, 1696–1769*. Paris: Hachette, 1896.

Le Roy, Jean-Baptiste, Pierre-Simon Laplace, Charles-Augustin de Coulomb, and Jean-Charles de Borda. *Rapport fait à l'Académie Royale des Sciences, le samedi 19 mars 1791, d'un Mémoire important de M. Blanc*. Paris: Moutard, 1791.

Lespinasse, Augustin. *Essai sur l'organisation de l'arme de l'artillerie*. Paris: Magimel, year VIII [1799–1800].

Lombard, Jean-Louis, ed. *Nouveaux principes d'artillerie, commentés par Léonard Euler* by Benjamin Robins. Dijon: Frantin, 1783.

———. *Tables du tir des canons et des obusiers*. Auxonne: n.p., 1787.

———. *Traité du mouvement des projectiles*. Dijon: Frantin, year V [1796–97].

Malthus, Francis. *Pratique de la guerre concernant l'usage de l'artillerie, bombes, et mortiers*. Paris: Guillemont, 1646.

Martini, Francesco di Giorgio. *Trattati di architettura ingegneria e arte militare*. Milan: Polifilo, [1475], 1967.

Mercier, Louis-Sébastien. *Tableau de Paris*. New ed. Amsterdam: n.p., 1782–88.

Mesnil-Durand, François-Jean de Graindorge d'Orgeville de. *Observations sur le canon par rapport à l'infanterie en générale*. Paris: Jombert, 1772.

———. *Projet d'un ordre français en tactique*. Paris: Boudet, 1755.

———. *Réponse à la brochure intitulée: L'ordre profond et l'ordre mince, considérés par rapports aux effets de l'artillerie*. Amsterdam: Cellot and Jombert, 1776.

Messance, Louis. *Nouvelle recherches sur la population de la France avec quelques remarques importantes sur divers objets d'administration*. Lyon, Frères Perisse, 1788.

Mirabeau, Honoré Gabriel Riquetti de. *Mémoires du ministère du duc d'Aiguillon*. 3d ed. Paris: Buisson, 1792.

Monge, Gaspard. *Description de l'art de fabriquer les canons*. Paris: Imprimerie du C.S.P., year II [1794].

———. *Géométrie descriptive, Leçons de l'an III*. Paris, year VII [1799].

———. "Géométrie descriptive." *Séances [et leçons] des Ecoles Normales, recueillies par les sténographes, et revues par les professeurs*. Paris: Reyner, year III [1795].

———. "Stéréotomie." *Journal de l'Ecole Polytechnique* 1 (year III [1795]): 1–14.

Moniteur: Réimpression de l'ancien Moniteur [universel], seule histoire authentique et inaltérée de la Révolution française depuis la réunion des Etats-Généraux jusqu'au Consulat (mai 1789–novembre 1799). Edited by A. Ray. Paris, 1858–63.

[Montalembert, Marc-René de]. *Essai sur l'intérêt des nations en général et de l'homme en particulier*. N.p., 1749.

Montesquieu, Charles de Secondat de. *Esprit des lois*. Paris: Flammarion, 1979.

Montucla, Jean Etienne. *Histoire des mathématiques*. Paris: Blanchard, [1799], 1960.

Morey, Thomas du and Chev. de Frazan. "Rapport." *Journal de physique* 16 (1780): 127–39.

Muller, John. *A Treatise of Artillery*. London: Millan, 1768.

Norton, Robert. *The Gunner, Shewing the Whole Practice of Artillerie*. [London]: Robinson, 1628.

Olivier, Théodore, ed. *Applications de la géométrie descriptive*. Paris: Carillian-Goeury, 1847.

Ozanam, Didier, and Michel Antoine, eds. *Correspondance secrète du comte de Broglie avec Louis XV*. Paris: Klincksieck, 1961.

Pascal, Blaise. *Pensées*. Edited by Louis Lafuma. Paris: Editions du Luxembourg, 1951.

Pion des Loches, Antoine-Augustin-Flavien. *Mes campagnes, 1792–1815: Notes et correspondance du colonel d'artillerie*. Edited by Maurice Chipon and Léonce Pingaud. Paris: Firmin-Didot, 1889.

Ply, Gustave. "Etude sur l'organisation du service technique dans les manufactures d'armes." *Revue d'artillerie* 32 (1888): 344–90; 33 (1888–89): 5–47, 101–42, 211–43, 297–332.

Prieur de la Côte-d'Or, Claude-Antoine. *Mémoire sur l'Ecole Centrale des Travaux Publics*. Report of 30 prairial, year III [18 June 1795]. Paris: Imprimerie de la République, messidor, year III [1795].

Proudhon, Pierre-Joseph. *Système des contradictions économiques*. Vol. 1. *Oeuvres complètes*. Geneva: Slatkine, 1982.

Puységur, Jacques-François de Chastenet de. *Art de la guerre par principes et par règles*. Paris: Jombert, 1749.

Ressons, Deschiens de. "Méthode pour tirer les bombes avec succès." *Mémoires de l'Académie des Sciences* (1716): 79–86.

Robespierre, Maximilien. *Oeuvres complètes*. Paris: Leroux, 1912.

Robins, Benjamin. *Mathematical Tracts*. Edited by James Wilson. London: Nourse, 1761.

———. *Nouveaux principes d'artillerie*. Translated by Du Puy. Grenoble: Durand, 1771.

Roland de la Platière, Jean-Marie. *Lettres écrites de Suisse, d'Italie, de Sicile et de Malthe, Par M.**, avocat en Parlement à Mlle.** à Paris en 1776, 1777 et 1778*. Amsterdam: n.p., 1780.

———. *Manufactures, arts et métiers*. 3 vols. in 5. In *Encyclopédie méthodique*. Paris: Panckoucke, 1790.

Rousseau, Jean-Jacques. *Emile, or on Education*. Translated by Allan Bloom. New York: Basic Books, 1979.

———. *Les confessions*. Paris: Gallimard, [1789], 1959.

Saint-Auban, Jacques Antoine Baratier de. *Mémoires sur les nouveaux systèmes d'artillerie*. N.p., 1775. (S.H.A.T. Bib. #72359)

Saint-Germain, Claude-Louis. *Mémoires de M. le comte de Saint-Germain*. Switzerland: Libraires Associés, 1779.

Saint-Rémy, Pierre Surirey de, ed. *Mémoires d'artillerie, recueillis*. 3d ed. Paris: Rollin, 1745.

Saxe, Maurice de. *Mes rêveries, ou mémoires sur l'art de la guerre*. Amsterdam: Arkstee et Merkus, 1757.

Scheel, Heinrich Othon von. *Mémoires d'artillerie contenant l'artillerie nouvelle, ou les changements faits dans l'artillerie française depuis 1765*. Copenhagen: Philibert, 1777.

———. *Treatise of Artillery Containing a New System or Alterations Made in the French Artillery since 1765*. Translation by Jonathan Williams. Philadelphia: Fenno, 1800.

[Servan, Joseph]. *Le citoyen soldat, ou vue patriotique sur la manière la plus avantageuse de pourvoir à la défense du royaume*. "Dans la pays de la liberté," 1780.

Smith, Adam. *An Inquiry into the Wealth of Nations*. Chicago: University of Chicago Press, 1976.

Straka, Georges, ed. *Poèmes du XVIIIe siècle en dialecte de Saint-Etienne (Loire), Edition avec commentaires philologiques et linguistiques*. Paris: Société d'Edition les Belles Lettres, 1964.

Tartaglia, Niccolò. *La nova scientia*. Venice, 1537.

———. *Three bookes of colloquies concerning the arte of shooting*. Translated by Cyprian Lucar. London: Harrison, 1588.

Tézenas du Montcel, Paul, ed. *Etude sur les assemblées provinciales: L'Assemblée du département de Saint-Etienne et sa Commission Intermédiaire, 8 octobre 1787 à 21 juillet 1790*. Paris: Champion, 1903.

Thiroux, M. *Instruction théorique et pratique d'artillerie*. 3d ed. Paris: Dumaine, 1849.

Tousard, Louis. *American Artillerist's Companion, or Elements of Artillery*. Philadelphia: Concord, 1809.

Tuetey, Alexandre, ed. *Répertoire général des sources manuscrites de l'histoire de Paris pendant la Révolution française*. Paris: Imprimerie Nouvelle, 1890–1914.

Ufano, Diego. *Artillerie, c'est à dire vraie instruction de l'artillerie*. Translated by Théodore de Brye. [Zutphen]: Aelst, 1621.

Vallière, Jean-Florent de, *père*. "Mémoire sur les charges et les portées des bouches à feu." In *Artillerie polémique de 1772*. Edited by Joseph-Florent Vallière, *fils*. Aléthopolis: Neumann, 1772.

Vallière, Joseph-Florent de, *fils*. *Mémoire touchant la supériorité des pièces d'artillerie longues et solides, lu à l'Académie des Sciences, 19 août 1775*. Paris: Imprimerie Royale, 1775. (S.H.A.T. Bib. #72227)

Vandermonde, Alexandre-Théophile. "Economie politique." Vol. 2: 233–45, 290–302, 447–63; vol. 3: 145–61, 437–45; vol. 4: 168–80, 452–71; vol. 5: 89–109. *Séances [et leçons] des Ecoles Normales, recueillies par les sténographes, et revues par les professeurs*. Paris: Reyner, year III [1795].

———. *Procédés de la fabrication des armes blanches*. Paris: Imprimerie du Département de la Guerre, year II [1793–94].

———. "Rapport fait par order du Comité de Salut Public, sur les fabriques et le commerce de Lyon, 15 brumaire, an III." Report of 5 November 1794. *Journal des arts et manufactures* 1 (year III [1794]): 1–48. (B.N. Vp5301)

Vandermonde, Alexandre-Théophile, Gaspard Monge, and Claude-Louis Berthollet. *Avis aux ouvriers en fer sur la fabrication de l'acier*. Paris: Imprimerie du Département de la Guerre, year II [1793–94].

Vauban, Jacques-Anne-Joseph Sébastien Le Prestre de. *Le triomphe de la méthode*. Edited by Nicholas Faucherre and Philippe Prost. Paris: Gallimard, 1992.

———. *Sa famille, et ses écrits: ses oisivetés et sa correspondance*. Edited by Albert de Rochas d'Aiglun. Geneva: Slatkine, 1972.

———. *Science militaire contenant l'A.B.C. d'un soldat*. La Haye: Moetjens, 1689.

———. *Traité des sièges et de l'attaque des places*. Edited by Augoyat. Paris: [1704], 1829.

Villantroys, P. L., intro. and trans. *Nouvelles expériences d'artillerie* by Charles Hutton. Paris: Magimel, year X [1801–2].

Voltaire, François-Marie Arouet. *Oeuvres complètes*. Paris: Garnier, 1877.

———. *Voltaire's Correspondence*. Edited by Theodore Besterman. Geneva: Institut et Musée Voltaire, 1953–65.

Secondary Sources

Abbott, W. *Technical Drawing*. London: Blackie, 1962.

Accampo, Elinor. *Industrialization, Family Life, and Class Relations: Saint Chamond, 1815–1914*. Berkeley: University of California Press, 1989.

Acerra, Martine. "Les constructeurs de la Marine (XVIIe–XVIIIe siècle)." *Revue historique* 273 (1985): 283–304.

Agulhon, Maurice. *Marianne into Combat: Republican Imagery and Symbolism in France, 1789 to 1880*. Translated by Janet Lloyd. Cambridge: Cambridge University Press, 1981.

Alder, Ken. "A Revolution to Measure: The Political Economy of the Metric System in France." In *The Values of Precision*. Edited by M. Norton Wise. Princeton: Princeton University Press, 1995.

———. "Forging the New Order: French Mass Production and the Language of the Machine Age, 1763–1815." Ph.D. diss., Harvard University, 1991.

Aldrich, Robert. "Late Comer or Early Starter? New Views on French Economic History." *Journal of European Economic History* 16 (1987): 89–100.

Ames, Edward, and Nathan Rosenberg. "The Enfield Armory in Theory and History." *The Economic Journal* 78 (1968): 827–42.

Aminzade, Ronald. *Ballots and Barricades: Class Formation and Republican Politics,*

1830–1871. Princeton: Princeton University Press, 1993.

Anderson, Perry. *Lineages of the Absolutist State*. London: N.L.B., 1975.

Anderton, Keith. "The Limits of Science: A Social, Political, and Moral Agenda for Epistemology in Nineteenth-Century Germany." Ph.D. diss., Harvard University, 1993.

Andrews, Richard Mowery. "Social Structures, Political Elites and Ideology in Revolutionary Paris, 1792–94: A Critical Examination of Albert Soboul's *Les sans-culottes parisiens en l'an II*." *Journal of Social History* 19 (1985): 71–112.

Appadurai, Arjun, ed. *The Social Life of Things: Commodities in Cultural Perspective*. Cambridge: Cambridge University Press, 1986.

Artz, Frederick B. *The Development of Technical Education in France 1500–1850*. Cambridge: MIT Press, 1966.

Aubry, Paul V. *Monge, le savant ami de Napoléon Bonaparte, 1746–1818*. Paris: Gauthier-Villars, 1964.

Aulard, François-Alphonse. *Etudes et leçons sur la Révolution française*. Paris: Alcan, 1893.

———. *The French Revolution: A Political History*. Translated by Bernard Maill. New York: Russell and Russell, [1910], 1965.

Auslander, Leora. "Perceptions of Beauty and the Problem of Consciousness: Parisian Furniture Makers." In *Rethinking Labor History: Essays on Discourse and Class Analysis*. Edited by Lenard R. Berlandstein. Urbana: University of Illinois Press, 1993.

———.*Taste and Power: Furnishing Modern France*. Berkeley: University of California Press, 1996.

———. "The Creation of Value and the Production of Good Taste: The Social Life of the Furniture Trade in Paris, 1860–1914." Ph.D. diss., Brown University, 1990.

Babeau, Albert. *La vie militaire sous l'ancien régime: Les officiers*. Vol. 2. Paris: Firmin-Didot, 1890.

Bachelard, Gaston. *Le nouvel esprit scientifique*. Paris: Presses Universitaires de France, 1949.

Baczko, Bronislaw. *Ending the Terror: The French Revolution after Robespierre*. Translated by Michel Petheram. Cambridge: Cambridge University Press, 1994.

Bailey, De Witt, and Douglas A. Nie. *English Gunmakers: The Birmingham and Provincial Gun Trade in the Eighteenth and Nineteenth Centuries*. New York: Arco, 1978.

Baker, Keith Michael. *Condorcet: From Natural Philosophy to Social Mathematics*. Chicago: University of Chicago Press, 1975.

———. *Inventing the French Revolution: Essays on French Political Culture in the Eighteenth Century*. Cambridge: Cambridge University Press, 1990.

Ballot, Charles. *L'introduction du machinisme dans l'industrie française*. Geneva: Slatkine Reprints, 1978.

Basset, M. A. "Essais sur l'histoire des fabrications d'armement en France jusqu'au mi-18e siècle." *Mémorial de l'artillerie* 14 (1935): 880–1280.

———. "Les grands maîtres de l'artillerie: Essai biographique." *Mémorial de l'artillerie* 11 (1932): 865ff.

Battison, Edwin A. "A New Look at the 'Whitney' Milling Machine." *Technology and Culture* 14 (1973): 592–98.

———. "Eli Whitney and the Milling Machine." *The Smithsonian Journal of History* 1 (1966): 9–34.

Baynes, Ken, and Francis Pugh. *The Art of the Engineer.* Woodstock, N.Y.: Overlook Press, 1981.

Bedini, Silvio. *Thomas Jefferson, Statesman of Science.* New York: Macmillan, 1990.

Beer, Carel de, ed. *The Art of Gunfounding: The Casting of Cannon in the Late 18th Century.* Rotherfield, U.K.: Boudriot, 1991.

Behrens, C.B.A. *Society, Government, and the Enlightenment: The Experiences of Eighteenth-Century France and Prussia.* New York: Harper and Row, 1985.

Belhoste, Bruno. "Les origines de l'Ecole Polytechnique." *Histoire de l'éducation* 42 (1989): 13–53.

Belhoste, Bruno et al. *Le moteur hydraulique en France au XIXe siècle: Concepteurs, inventeurs et constructeurs.* Nantes: Université de Nantes, 1990.

Belhoste, Bruno, Antoine Picon, and Joël Sakarovitch. "Les exercices dans les écoles d'ingénieurs sous l'ancien régime et la révolution." *Histoire de l'education* 46 (1990): 53–109.

Belofsky, Harold. "Engineering Drawing: A Universal Language in Two Dialects." *Technology and Culture* 32 (1991): 23–46.

Bennet, J. A. "The Mechanics' Philosophy and the Mechanical Philosophy." *History of Science* 24 (1986): 1–28.

Berg, Maxine. *The Age of Manufactures: Industry, Innovation, and Work in Britain, 1700–1820.* Totowa, N.J.: Barnes and Noble Books, 1985.

———. *The Machinery Question and the Making of Political Economy.* Cambridge: Cambridge University Press, 1980.

Bergeron, Louis. *France under Napoleon.* Translated by R. R. Palmer. Princeton: Princeton University Press, 1981.

Bergeron, Louis, and Guy Chaussinand-Nogaret. *Les "Masses de granit": Cent mille notables du Premier Empire.* Paris: Editions de l'Ecole des Hautes Etudes en Sciences Sociales, 1979.

Bertaud, Jean-Paul. "Napoleon's Officers." *Past and Present* 112 (1986): 91–111.

———. *The Army of the French Revolution: From Citizen-Soldiers to Instrument of Power.* Translated by R. R. Palmer. Princeton: Princeton University Press, 1988.

———. "Voies nouvelles pour l'histoire militaire de la Révolution." *Voies nouvelles pour l'histoire de la Révolution française.* Paris: B.N., 1978.

Beyler, Richard. "From Positivism to Organicism: Pascual Jordan's Interpretations of Modern Physics in Cultural Context." Ph.D. diss., Harvard University, 1994.

Bien, David. "La réaction aristocratique avant 1789: L'exemple de l'armée." *Annales: Economies, sociétés, civilisations* 29 (1974): 23–48, 505–34.

———. "Military Education in 18th-Century France: Technical and Non-Technical Determinants." In *Science, Technology and Warfare: Third Military History Symposium.* Edited by Monte Wright and L. Paszek. Washington, D.C.: General Publishing Office, 1969.

———. "The Army in the French Revolution: Reform, Reaction and Revolution." *Past and Present* 85 (1979): 68–98.

Biggs, Lindy. *Rational Production.* Baltimore: Johns Hopkins University Press, 1995.

Billaçois, François. *The Duel: Its Rise and Fall in Early Modern France.* Translated by Trita Selous. New Haven: Yale University Press, 1990.

Biot, Jean-Baptiste. *Essai sur l'histoire générale des sciences pendant la Révolution française.* Paris: Duprat, 1803.

Birembaut, Arthur. "Les écoles gratuites de dessin." In *Ecoles techniques et militaires au XVIIIe siècle*. Edited by Roger Hahn and René Taton. Paris: Hermann, 1986.

Blackmore, Howard. *British Military Firearms, 1650–1850*. London: Jenkins, 1961.

Blanchard, Anne. *Les ingénieurs du "roy" de Louis XIV à Louis XVI*. Montpellier: Dehan, 1979.

Blanning, T.C.W. *The Origins of the French Revolutionary Wars*. London: Longman, 1986.

Bodinier, Gilbert. *Les officiers de l'armée royale combattants de la Guerre d'Indépendance des Etats-Unis de Yorktown à l'an II*. Vincennes: S.H.A.T., 1983.

———. "Les officiers de l'armée royale et la Révolution." In *Le métier militaire en France aux époques des grandes transformations sociales*. Edited by André Corvisier. Vincennes: S.H.A.T., 1980.

Bonaparte, Louis-Napoléon. *Etudes sur le passé et l'avenir de l'artillerie*. Edited by Alphonse Favé. Paris: Dumain, 1846–62.

Bonnefoy, François. *Les armes de guerre portatives en France du début du règne de Louis XIV à la veille de la Révolution (1660–1789): De l'indépendance à la primauté*. Paris: Libraire de l'Inde, 1991.

———. "Maximilien Titon, directeur-général des magasins d'armes de Louis XIV, et le développement des armes portatives en France." *Histoire, Economie et Société* 5 (1980): 353–80.

Booker, Peter Geoffrey. *A History of Engineering Drawing*. London: Northgate, 1979.

Bos, H.J.M. "Mathematics and Rational Mechanics." In *The Ferment of Knowledge: Studies in the Historiography of Eighteenth-Century Science*. Edited by Roy Porter. Cambridge: Cambridge University Press, 1980.

Bosher, J. F. *French Finances, 1770–1795: From Business to Bureaucracy*. Cambridge: Cambridge University Press, 1970.

Bossenga, Gail. "La Révolution française et les corporations: Trois examples lillois." *Annales: Economies, sociétés, civilisations* 43 (1988): 405–26.

———. "Protecting Merchants: Guilds and Commercial Capitalism in Eighteenth-Century France." *French Historical Studies* 15 (1988): 693–703.

Bottet, Maurice. *Monographies de l'arme blanche (1789–1870) et de l'arme à feu portative (1718–1900) des armées françaises de terre et de mer*. Paris: Haussmann, 1959.

Bouchard, Georges. *Un organisateur de la victoire: Prieur de la Côte-d'Or, Membre du Comité de Salut Public*. Paris: Clavreuil, 1946.

Boudriot, Jean. *Armes à feu françaises: Modèles d'ordonnance*. Paris, 1961.

———. "L'évolution de la platine de l'arme d'infanterie française." *Gazette des armes* 68 (1979): 30–35.

Bouiller, Robert. "La fabrique de boutons Alcock à Roanne, au XVIIIe siècle." *La famille Alcock: 200 ans d'histoire en Roannais*. Roanne: Centre Forézien d'Ethnologie, 1985. (B.M.R.)

Bourdelais, Patrice. "Employés de la grande industrie: Les dessinateurs du Creusot, Formations et carrières (1850–1914)." *Histoire, économie et société* 8 (1989): 437–46.

Bourdieu, Pierre. *Distinction: A Social Critique of the Judgement of Taste*. Translated by Richard Nice. Cambridge: Harvard University Press, 1984.

Bourguet, Marie-Noëlle. *Déchiffrer la France: La statistique départementale à l'époque napoléonienne*. Paris: Edition des Archives Contemporaines, 1988.

Bradley, Joseph. *Guns for the Tsar: American Technology and the Small Arms Industry in Nineteenth-Century Russia.* Dekalb, Ill.: Northern Illinois University Press, 1990.

Bradley, Margaret. "Scientific Education versus Military Training: The Influence of Napoleon Bonaparte on the Ecole Polytechnique." *Annals of Science* 32 (1975): 415–49.

Braesch, Frédéric. "Essai de statistique de la population ouvrière de Paris vers 1791." *La Révolution française* 63 (1912): 289–321.

———. *La Commune du dix août, 1792: Etude sur l'histoire de Paris du 20 juin au 2 décembre 1792.* Paris: Hachette, 1911.

Breen, T. H. "Narrative of Commercial Life: Consumption, Ideology, and Community on the Eve of the American Revolution." *William and Mary Quarterly* 50 (1993), 471–501.

———. "The Meaning of Things: Interpreting the Consumer Economy in the Eighteenth Century." In *Consumption and the World of Things.* Edited by John Brewer and Roy Porter. London: Routledge, 1993.

Bret, Patrice. "Les origines de l'institutionnalisation de la recherche militaire en France (1775–1825)." In Colloque International d'histoire militaire, Actes no. 15. *L'influence de la Révolution française sur les armées en France, en Europe, et dans le monde.* Vincennes: Commission Française d'Histoire Militaire, 1991.

Brewer, John. *The Sinews of Power: War, Money, and the English State, 1688–1783.* New York: Knopf, 1989.

Brockliss, L.W.B. *French Higher Education in the Seventeenth and Eighteenth Centuries: A Cultural History.* Oxford: Clarendon Press, 1987.

Brown, Howard G. "Politics, Professionalism, and the Fate of the Army Generals after Thermidor." *French Historical Studies* 19 (1995): 133–52.

Brown, M. L. *Firearms in Colonial America: The Impact of History and Technology, 1492–1792.* Washington, D.C.: The Smithsonian Institution Press, 1980.

Bucciarelli, Louis. *Designing Engineers.* Cambridge: MIT Press, 1994.

Buckingham, Earle. *Principles of Interchangeable Manufacturing.* 2d ed. New York: The Industrial Press, 1941.

Buroway, Michael. *Manufacturing Consent: Changes in the Labor Process under Monopoly Capitalism.* Chicago: University of Chicago Press, 1979.

Burstin, Haim. "Problems of Work during the Terror." In *The Terror.* Vol. 4. *The French Revolution and the Creation of Modern Political Culture.* Edited by Keith Michael Baker. Oxford: Elsevier, 1994.

Cain, M. R. "Historique de la création de l'Atelier de Précision de l'Artillerie." *Mémorial de l'artillerie française* 28 (1954): 369–92.

Carlson, W. Bernard. *Innovation as a Social Process: Elihu Thomson and the Rise of General Electric, 1870–1900.* Cambridge: Cambridge University Press, 1991.

Carson, John. "Army Alpha, Army Brass, and the Search for Army Intelligence." *Isis* 84 (1993): 278–309.

Chagniot, Jean. *Paris et l'armée au XVIIIe siècle: Etude politique et sociale.* Paris: Economica, 1985.

Chalmin, Pierre. "La querelle des Bleus et des Rouges dans l'artillerie française à la fin du XVIIIe siècle." *Revue d'histoire économique et sociale* 46 (1968): 465- 505.

Chandler, Alfred. *The Visible Hand: The Managerial Revolution in American Business.*

Cambridge: Harvard University Press, 1977.

Chandler, David G. *The Art of War in the Age of Marlborough*. London: Batsford, 1976.

———. *The Campaigns of Napoleon*. New York: Macmillan, 1966.

Charbonnier, Prosper Jules. *Essais sur l'histoire de la ballistique*. Paris: Société d'Editions Géographiques Maritimes et Coloniales, 1928.

Charnay, Jean-Paul. "Ruptures, répétitions et contradictions dans la fortune politique et l'assise sociale de Lazare Carnot." In *Lazare Carnot, ou le savant-citoyen*. Edited by Jean-Paul Charnay. Paris: Université de Sorbonne, 1990.

Chartier, Roger. "Un recruitment scolaire au XVIIIe siècle, l'Ecole Royale du Génie à Mézières." *Revue d'histoire moderne et contemporaine* 20 (1973): 353–75.

Chaussinand-Nogaret, Guy et al. *Histoire des élites en France du XVIe au XXe siècle: L'honneur, le mérite, l'argent*. Paris: Tallandier, 1991.

Chisick, Harvey. *The Limits of Reform in the Enlightenment: Attitudes toward the Education of the Lower Classes in Eighteenth-Century France*. Princeton: Princeton University Press, 1981.

Chuquet, Arthur. *La jeunesse de Napoléon*. Paris: Colin, 1897–99.

Coase, R. H. "The Nature of the Firm," [1937]. In *The Firm, the Market and the Law*. Chicago: University of Chicago Press, 1988.

Cobb, Richard. *The People's Armies: The armées révolutionnaires, Instrument of the Terror in the Departments, April 1793 to Floréal, year II*. Translated by Marianne Elliott. New Haven: Yale University Press, 1987.

Cobban, Alfred. *The Social Interpretation of the French Revolution*. Cambridge: Cambridge University Press, 1964.

Cohen, Yves. "Inventivité organisationnelle et compétitivité: L'interchangeabilité des pièces face à la crise de la machine-outil en France autour de 1900." *Entreprises et histoire* 5 (1994): 53–72.

Colin, Jacques Lambert Alphonse. *L'infanterie au XVIIIe siècle: La tactique*. Paris: Berger-Levrault, 1907.

Collins, H. M. "The Seven Sexes: A Study in the Sociology of a Phenomenon, or the Replication of Experiments in Physics. *Sociology* 9 (1975): 205–24.

Colvin, Fred. *Gages and their Use in Inspection*. New York: McGraw-Hill, 1942.

Constant, Edward W., II. "A Model for Technological Change Applied to the Turbojet Revolution." *Technology and Culture* 14 (1973): 553–72.

Coonaert, Emile. *Les corporations en France avant 1789*. Paris: Gallimard, 1941.

Cooper, Carolyn C. "The Portsmouth System of Manufacture." *Technology and Culture* 25 (1984): 182–225.

Cooper, Carolyn C., Robert B. Gordon, Patrick M. Malone, and Michael Raber. *Model Establishment: A History of the Springfield Armory 1794–1918*. New York: Oxford University Press, forthcoming.

Corvisier, André. *L'armée française de la fin du XVIIe siècle au ministère de Choiseul*. Paris: Presses Universitaires de France, 1964.

———. "Rapport de synthèse." *Le métier militaire en France aux époques de grandes transformations sociales*. Edited by André Corvisier. Vincennes: S.H.A.T., 1980.

Corvisier, André, and Anne Blanchard et al., eds. *Histoire militaire de la France*. Paris: Presses Universitaires de France, 1992–94.

Court, W.H.B. *The Rise of the Midland Industries*. London: Oxford University Press, 1938.

Cronon, William. *Nature's Metropolis: Chicago and the Great West*. New York: Norton, 1991.

Crouzet, François. *Britain Ascendant: Comparative Studies in Franco-British Economic History*. Translated by Martin Thom. Cambridge: Cambridge University Press, 1990.

———. "Recherches sur la production d'armements en France (1815–1913)." *Revue d'histoire* 509 (1974): 45–84; 510 (1975): 409–22.

Darnton, Robert. *Mesmerism and the End of the Enlightenment in France*. Cambridge: Harvard University Press, 1968.

———. *The Literary Underground of the Old Regime*. Cambridge: Harvard University Press, 1982.

Daston, Lorraine J. "Objectivity and the Escape from Perspective." *Social Studies of Science* 22 (1992): 597–618.

———. "The Ideal and the Reality of the Republic of Letters in the Enlightenment." *Science in Context* 4 (1991): 367–86.

———. "The Physicalist Tradition in Early Nineteenth-Century French Geometry." *Studies in the History and Philosophy of Science* 17 (1986): 269–95.

Daston, Lorraine J., and Peter Galison. "The Image of Objectivity." *Representations* 40 (1992): 81–128.

Dauben, Joseph. "Marat: His Science and the French Revolution." *Archives internationales d'histoire des sciences* 22 (1969): 235–61.

Daumas, Maurice, ed. Histoire générale des techniques. Paris: Presses Universitaires de France, 1962–79.

———. *Les instruments scientifiques aux XVIIe et XVIIIe siècles*. Paris: Presses Universitaires de France, 1953.

Davis, J. C. "Science and Utopia: The History of a Dilemma." In *Nineteen Eighty-Four: Science Between Utopia and Dystopia*. Edited by Everett Mendelsohn and Helga Nowotny. Dordrecht: Reidel, 1984.

Day, Charles R. *Education for the Industrial World: The Ecoles d'Arts et Métiers and the Rise of French Industrial Engineering*. Cambridge: MIT Press, 1987.

Deforge, Yves. *Le graphisme technique: Son histoire et son enseignement*. Seyssel: Vallon, 1981.

DeJean, Joan. *Literary Fortifications: Rousseau, Laclos, Sade*. Princeton: Princeton University Press, 1984.

Descreux, Denis. *Notices biographiques stéphanoises*. Saint-Etienne: Constantin, 1868.

Devenne, Florence. "La Garde Nationale: Création et évolution, 1789–août 1792." *Annales historiques de la Révolution française* 283 (1991): 49–66.

Deyrup, Felicia Johnson. *Arms Makers of the Connecticut Valley: A Regional Study of the Economic Development of the Small Arms Industry, 1798–1870*. Northampton, Mass.: Smith College Press, 1948.

Dhombres, Jean, and Nicole Dhombres. *Naissance d'un nouveau pouvoir: Sciences et savants en France, 1793–1824*. Paris: Payot, 1989.

Dontenwill, Serge. "Roanne au dernier siècle de l'ancien régime: Aspects démographique et sociaux." *Etudes foréziennes* 4 (1971): 49–73.

Downing, Brian. *The Military Revolution and Political Change: Origins of Democracy and Autocracy in Early Modern Europe*. Princeton: Princeton University Press, 1992.

Doyle, William. *Origins of the French Revolution*. 2d ed. Oxford: Oxford University

Press, 1988.

Du Teil, Joseph. *Napoléon Bonaparte et les généraux Du Teil, 1788–1794: L'Ecole d'Artillerie d'Auxonne et le siège de Toulon.* Paris: Picard, 1897.

Dubessy, R. *Historique de la Manufacture de Saint-Etienne.* Saint-Etienne: n.p., 1900.

Duffy, Christopher. *The Fortress in the Age of Vauban and Frederick the Great, 1660–1789.* London: Routledge and Kegan Paul, 1985.

Dufour, J. E. *Dictionnaire topographique du Forez et des paroisses du Lyonnais et du Beaujolais formant le département de la Loire.* Mâcon: Protat Frères, 1946.

Dunlavy, Colleen. *Politics and Industrialization: Early Railroads in the United States and Prussia.* Princeton: Princeton University Press, 1994.

Durfee, W. F. "The First Systematic Attempt at Interchangeability in Firearms." *Cassier's Magazine* (April 1894): 469–77.

Durkheim, Emile. *Professional Ethics and Civic Morals.* Translated by Cornelia Brookfield. London: Routledge and Kegan Paul, [1898], 1957.

Duveen, Denis I., and Roger Hahn. "Laplace's Succession to Bézout's Post of Examinateur des Elèves de l'Artillerie." *Isis* 48 (1957): 416–27.

Edgerton, Samuel Y. *The Renaissance Discovery of Linear Perspective.* New York: Harper and Row, 1976.

Edmonson, James M. *From Mécanicien to Ingénieur: Technical Education and the Machine Building Industry in Nineteenth-Century France.* New York: Garland, 1987.

Egret, Jean. *The French Prerevolution.* Translated by Wesley D. Camp. Chicago: University of Chicago Press, 1977.

Ellul, Jacques. *The Technological Society.* Translated by John Wilkinson. New York: Knopf, 1964.

Etner, François. *Histoire du calcul économique en France.* Paris: Economica, 1987.

Fairchilds, Cissie. "The Production and Marketing of Populuxe Goods." In *Consumption and the World of Goods.* Edited by John Brewer and Roy Porter. London: Routledge, 1993.

Fayet, Joseph. *La Révolution française et la science, 1789–1795.* Paris: Marcel Rivière, 1960.

Ferguson, Eugene. *Engineering and the Mind's Eye.* Cambridge: MIT Press, 1992.

———. "The Measurement of the 'Man-Day.'" *Scientific American* 225 (1971): 96–103.

Ferrero, Guglielmo. *Aventure, Bonaparte en Italie, 1796–97.* Paris: Plon, 1936.

Ferrone, Vincenzo. "Les mécanismes de formation des élites de la maison de Savoie: Recrutement et sélection dans les écoles militaires du Piémont au XVIIIe siècle." *Paedagogica Historica* 30 (1994): 341–70.

Foley, Vernard et al. "Leonardo, the Wheel Lock, and the Milling Process." *Technology and Culture* 24 (1983): 399–427.

Forrest, Alan. *Conscripts and Deserters: The Army and French Society during the Revolution and Empire.* New York, Oxford University Press, 1989.

———. *The French Revolution and the Poor.* New York: St. Martin's Press, 1981.

———. *The Soldiers of the French Revolution.* Durham, N.C.: Duke University Press, 1990.

Foucault, Michel. *Discipline and Punish: The Birth of the Prison.* Translated by Alan Sheridan. New York: Random House, 1979.

Fourcy, Amboise. *Histoire de l'Ecole Polytechnique.* Edited by Jean Dhombres. Paris: Belin, 1987.

Fournial, Etienne et al. *Saint-Etienne: Histoire de la ville et de ses habitants*. Roanne: Horvath, 1976.

Fox, Allan. *Beyond Contract: Work, Power and Trust Relations*. London: Faber and Faber, 1974.

Fox, Robert. "Education for a New Age, The Conservatoire des Arts et Métiers." In *Artisan to Graduate*. Edited by D.S.L. Cardwell. Manchester: Manchester University Press, 1974.

Fox, Robert, and George Weisz, eds. *The Organization of Science and Technology in France, 1808–1914*. Cambridge: Cambridge University Press, 1980.

Fox-Genovese, Elizabeth. *The Origins of Physiocracy: Economic Revolution and Social Order in Eighteenth-Century France*. Ithaca: Cornell University Press, 1976.

Freidson, Eliot. *Professional Powers: A Study of the Institutionalization of Formal Knowledge*. Chicago: University of Chicago Press, 1986.

Fridenson, Patrick. "The Coming of the Assembly Line to Europe." In *The Dynamics of Science and Technology: Social Values, Technical Norms and Scientific Criteria in the Development of Knowledge*. Edited by Wolfgang Krohn, Edwin Layton, and Peter Weingart. Dordrecht: Reidel, 1978.

Fries, Russell I. "British Response to the American System: The Case of the Small-Arms Industry after 1850." *Technology and Culture* 16 (1975): 377–403.

Furet, François. *Penser la Révolution française*. Paris: Gallimard, 1978.

Furet, François, and Denis Richet. *La Révolution française*. Paris: Fayard, 1965–66.

Furet, François, and Mona Ozouf, eds. *A Critical Dictionary of the French Revolution*. Translated by Arthur Goldhammer. Cambridge, Mass.: Belknap, 1989.

Gaffarel, Paul. "Le Général de Gassendi." *Mémoire de la Société Bourguignonne de Géographie et d'Histoire* 19 (1903): 387–459.

Gaier, Claude. "Note sur la fabrication des 'platines identiques' et sur la Manufacture Impériale de Platines de Liège." *Le Musée d'Armes, Bulletin trimestriel des Amis du Musée d'Armes de Liège* 7 (1979): 5–9.

Galley, Jean-Baptiste. *L'élection de Saint-Etienne à la fin de l'ancien régime*. Saint-Etienne: Ménard, 1903.

———. *Saint-Etienne et son district pendant la Révolution*. Saint-Etienne: Imprimerie de la Loire Républicaine, 1903–9.

Gardette, Pierre. *Atlas linguistique et ethnographique du Lyonnais*. Paris: CNRS, 1956–76.

———. *Etudes de géographie linguistique*. Strasbourg: Société de Linguistique Romane, 1983.

Garnier, Josette. *Bourgeoisie et propriété immobilière en Forez aux XVIIe et XVIIIe siècle*. Saint-Etienne: Centre d'Etudes Foréziennes, 1982.

Gascoigne, John. "Mathematics and Meritocracy: The Emergence of the Cambridge Mathematical Tripos." *Social Studies of Science* 14 (1984): 547–84.

Gat, Azar. *The Origins of Military Thought from the Enlightenment to Clausewitz*. Oxford: Clarendon, 1989.

Gaugelin, Françoise. "Les ouvriers de la Manufacture Extraordinaire d'Armes de Paris, 23 août 1793 au 1 pluviôse, an III." Mémoire de maîtrise, Université de Paris I, 1975–76. Written under the direction of Albert Soboul. (S.H.A.T. Bib. #84889)

Gille, Bertrand. *Les origines de la grande industrie métallurgique en France*. Paris: Domat, 1947.

Gillispie, Charles C. *Science and Polity at the End of the Old Regime*. Princeton: Princeton University Press, 1980.

Gillispie, Charles C. "The *Encyclopédie* and the Jacobin Philosophy of Science: A Study of Ideas and Consequences." In *Critical Problems in the History of Science*. Edited by Marshall Clagett. Madison: University of Wisconsin Press, 1959.

———. "The Natural History of Industry." *Isis* 48 (1957): 398–407.

Gillispie, Charles C., and A. P. Youschklevitch. *Lazare Carnot, Savant: A Monograph Treating Carnot's Scientific Work*. Princeton: Princeton University Press, 1971.

Gillmor, C. Stewart. *Coulomb and the Evolution of Physics and Engineering in Eighteenth-Century France*. Princeton: Princeton University Press, 1971.

Glas, Eduard. "On the Dynamics of Mathematical Change in the Case of the Monge and the French Revolution." *Studies in the History and Philosophy of Science* 17 (1986): 249–68.

Golinski, Jan. *Science as a Public Culture: Chemistry and Enlightenment in Britain, 1760–1820*. Cambridge: Cambridge University Press, 1992.

Gonnard, Philippe. "Moeurs administrative du Premier Empire: La conscription dans la Loire." *Revue d'histoire de Lyon* 7 (1908): 81–110.

Goodman, Deena. "Governing the Republic of Letters: The Politics of Culture in the French Enlightenment." *History of European Ideas* 13 (1991): 183–99.

Goodman, John D. "The Birmingham Gun Trade." In *Birmingham and the Midland Hardware District*. Edited by Samuel Timms. London: Harwicke, 1866.

Gordon, David M. *Merchants and Capitalists: Industrialization and Provincial Politics in Mid-Nineteenth-Century France*. Tuscaloosa, Ala.: University of Alabama Press, 1985.

Gordon, Robert B. "Who Turned the Mechanical Ideal into Mechanical Reality?" *Technology and Culture* 29 (1988): 744–78.

Gordon, Robert B., and Patrick Malone. *The Texture of Industry: An Archeological View of the Industrialization of North America*. New York: Oxford University Press, 1994.

Gorman, Michael E., and W. Bernard Carlson. "Interpreting Invention as a Cognitive Process: The Case of Alexander Graham Bell, Thomas Edison, and the Telephone." *Science, Technology and Human Values* 15 (1990): 131–64.

Gough, Terrence J. "Isolation and Professionalization of the Army Officer Corps: A Post-revisionist View of *The Soldier and the State*." *Social Science Quarterly* 73 (1992): 420–36.

Gould, Stephen J. *Wonderful Life: The Burgess Shale and the Nature of History*. New York: Norton, 1989.

Gras, Louis-Joseph. *Essai sur l'histoire de la quincaillerie à Saint-Etienne*. Saint-Etienne: Théolier, 1904.

———. *Histoire de l'armurerie stéphanoise*. Saint-Etienne: Théolier, 1905.

———. *Le Forez et le Jarez navigables*. Saint-Etienne: Théolier, 1930.

Grattan-Guinness, I. "The *ingénieur-savant*, 1800–1830: A Neglected Figure in the History of French Mathematics and Science." *Science in Context* 6 (1993): 405–33.

———. "Work for the Workers: Advances in Engineering Mechanics and Instruction in France, 1800–1830." *Annals of Science* 41 (1984): 1–33.

Greenberg, John L. "Mathematical Physics in Eighteenth-Century France." *Isis* 77 (1986): 59–78.

Greer, Donald. *The Incidence of the Terror during the French Revolution: A Statistical Interpretation*. Gloucester, Mass.: Peter Smith, 1966.

Guérin, Daniel. *La lutte de classes sous la Première République: Bourgeois et "bras nues,"*

1793–97. 3d ed. Paris: Gallimard, 1946.
Guerlac, Henry. "Science and War in the Old Regime." Ph.D. diss., Harvard University, 1941.
———. "Some Aspects of Science during the French Revolution." In *Essays and Papers in the History of Modern Science*. Baltimore: Johns Hopkins University Press, 1977.
———. "Vauban: The Impact of Science on War." In *Makers of Modern Strategy from Machiavelli to the Nuclear Age*. Edited by Peter Paret. Princeton: Princeton University Press, 1986.
Guichon de Grandpont, A. "La querelle de l'artillerie (parti rouge et parti bleu) au XVIIIe siècle." *Bulletin de la Société Académique de Brest* 2d ser., 20 (1894–95): 2–103.
Guillaume, James. "Un mot légendaire: 'La république n'a pas besoin de savants.'" *La Révolution française* 38 (1900): 385–99.
Guilmartin, John Francis. *Gunpowder and Galley: Changing Technology and Mediterranean Warfare at Sea in the Sixteenth Century*. Cambridge: Cambridge University Press, 1974.
Guineau, Bernard. *La Manufacture Royale de Quincaillerie, Taillanderie, et Bijouterie de toutes sortes de métaux façon d'Angleterre de La Charité-sur-Loire, 1756–1809*. La Charité-sur-Loire: Association des Amis de La Charité-sur-Loire, 1985.
Guitton, Henri. *Industrie des rubans de soie en France*. Paris: Sirey, 1928.
Gullickson, Gay. *Spinners and Weavers of Auffay: Rural Industry and the Sexual Division of Labor in a French Village, 1750–1850*. Cambridge: Cambridge University Press, 1986.
Gutwirth, Madelyn. *The Twilight of the Goddesses: Women and Representation in the French Revolutionary Era*. New Brunswick: Rutgers University Press, 1992.
Hacker, Barton. "Greek Catapults and Catapult Technology: Science, Technology and War in the Ancient World." *Technology and Culture* 9 (1968): 34–54.
Hahn, Roger. *Anatomy of a Scientific Institution: The Paris Academy of Sciences, 1666–1803*. Berkeley: University of California Press, 1971.
———. "Fourcroy, Advocate of Lavoisier?" *Archives internationales d'histoire des sciences* 12 (1959): 285–88.
———. "L'enseignement scientifique aux écoles militaires et d'artillerie." In *Ecoles techniques et militaires au XVIIIe siècle*. Edited by Roger Hahn and René Taton. Paris: Hermann, 1986.
———. "Le rôle de Laplace à l'Ecole Polytechnique." In *La formation polytechnicienne, 1794–1994*. Edited by Bruno Belhoste, Amy Dahan Dalmedico, and Antoine Picon. Paris: Dunod, 1994.
———. "Scientific Research as an Occupation in Eighteenth-Century Paris." *Minerva* 13 (1975): 501–13.
———. "The Triumph of Scientific Activity: From Louis XIV to Napoléon." *Proceedings of the Annual Meeting of the Western Society for French History* 19 (1989): 204–11.
Hall, A. Rupert. *Ballistics in the Seventeenth Century: A Study in the Relations of Science and War with Reference Principally to England*. Cambridge: Cambridge University Press, 1952.
———. "Gunnery, Science, and the Royal Society." In *The Uses of Science in the Age of Newton*. Edited by John G. Burke. Berkeley: University of California Press, 1983.
Hanagan, Michael P. *Nascent Proletarians: Class Formation in Post-Revolutionary France*. Oxford: Basil Blackwell, 1989.

Hanagan, Michael P. *The Logic of Solidarity: Artisans and Industrial Workers in Three French Towns, 1871–1914*. Urbana: University of Illinois Press, 1980.

Harris, John. "Michael Alcock and the Transfer of Birmingham Technology to France before the Revolution." In *Essays in Industry and Technology in the Eighteenth Century*. Hampshire, U.K.: Variorum, 1992.

Headrick, Daniel. *The Tools of Empire: Technology and European Imperialism in the Nineteenth Century*. New York: Oxford University Press, 1981.

Hecht, Gabrielle. "Rebels and Pioneers: Technocratic Ideologies and Social Identities in the French Nuclear Workplace, 1955–1969." *Social Studies of Science* (August 1996), forthcoming.

Hecht, Jacqueline. "Un example de multidisciplinarité: Alexandre Vandermonde (1735–1796)." *Population* 26 (1971): 641–76.

Heilbron, J. L. "Experimental Natural Philosophy." In *The Ferment of Knowledge: Studies in the Historiography of Eighteenth-Century Science*. Edited by Roy Porter. Cambridge: Cambridge University Press, 1980.

———. *Weighing Imponderables and Other Quantitative Science around 1800*. Historical Studies in the Physical and Biological Sciences, Supplement to vol. 24. Berkeley: University of California Press, 1993.

Henderson, W. O. *Industrial Britain Under the Regency: The Diaries of Escher, Bodmer, May, and de Gallois, 1814–1818*. London: Cass, 1968.

Hennebert, Eugene. *Gribeauval, lt.-général des armées du roy*. Paris: Berger-Levrault, 1896.

Hessen, Boris. *The Social and Economic Roots of Newton's "Principia."* New York: Howard Fertig, 1971.

Higonnet, Patrice. "Cultural Upheaval and Class Formation During the French Revolution." In *The French Revolution and the Birth of Modernity*. Edited by Ferenc Fehér. Berkeley: University of California Press, 1990.

———. "The Politics of Linguistic Terrorism and Grammatical Hegemony during the French Revolution." *Social History* 5 (1980): 41–69.

Hilaire-Pérez, Liliane. "Invention and the State in Eighteenth-Century France." *Technology and Culture* 32 (1991): 911–31.

Hobsbawm, E. J. *The Age of Revolution, 1789–1848*. Cleveland: World Publishing, 1962.

Hoke, Donald R. *Ingenious Yankees: The Rise of the American System in the Private Sector*. New York: Columbia University Press, 1990.

Hounshell, David. *From the American System to Mass Production, 1800–1932: The Development of Manufacturing Technology in the United States*. Baltimore: Johns Hopkins University Press, 1983.

Hublot, Emmanuel. *Valmy, ou la défense de la nation par les armes*. Paris: Fondation pour les Etudes de la Défense Nationale, 1987.

Hughes, Basil Perronet. *Firepower: Weapons Effectiveness on the Battlefield, 1630–1850*. New York: Scribner's, 1974.

Hughes, Thomas P. *American Genesis: A Century of Invention and Technological Enthusiasm, 1870–1970*. New York: Penguin, 1989.

———. *Networks of Power: Electrification in Western Society, 1880–1930*. Baltimore: Johns Hopkins University Press, 1983.

———. "The Electrification of America: The Systems Builders." *Technology and Culture* 20 (1979): 124–61.
———. "The Evolution of Large Technological Systems." In *The Social Construction of Technological Systems: New Directions in the Sociology and History of Technology*. Edited by Wiebe Bijker, Thomas P. Hughes, and Trevor Pinch. Cambridge: MIT Press, 1987.
Hull, David. "In Defense of Presentism." *History and Theory* 18 (1979): 1–15.
Hunt, Lynn. "Committees and Communes: Local Politics and National Revolution in 1789." *Comparative Studies in Society and History* 18 (1976): 321–46.
———. *Politics, Culture, and Class in the French Revolution*. Berkeley: University of California Press, 1984.
———. *Revolution and Urban Politics in Provincial France: Troyes and Reims, 1786–1790*. Stanford: Stanford University Press, 1978.
Huntington, Samuel. *The Soldier and the State: The Theory and Politics of Civil-Military Relations*. Cambridge, Mass.: Belknap Press, 1957.
Israel, Paul. *From Machine Shop to Industrial Laboratory: Telegraphy and the Changing Context of American Invention, 1830–1920*. Baltimore: Johns Hopkins University Press, 1992.
Ivins, W. M. *On the Rationalization of Sight*. New York: Metropolitan Museum, 1938.
Jacob, Margaret. *Living the Enlightenment: Free Masonry and Politics in Eighteenth-Century Europe*. New York: Oxford University Press, 1991.
———. *The Radical Enlightenment: Pantheists, Freemasons and Republicans*. London: Allen and Unwin, 1981.
Jarlier, Pierre. *Répertoire d'arquebusiers et de fourbisseurs français avec des noms de directeurs et contrôleurs des manufactures d'armes françaises*. Saint-Julian-du-Sault: Clobies, 1976.
Jaurès, Jean. *Histoire socialiste de la Révolution française*. Paris: Editions Sociales, 1952.
Johnson, William A., ed. and trans. *Christopher Polhem: The Father of Swedish Technology*. Hartford: Trinity College Press, 1963.
Jones, Colin. "Bourgeois Revolution Revivified: 1789 and Social Change." In *Rewriting the French Revolution*. Edited by Colin Lucas. Oxford: Clarendon Press, 1991.
———."The Great Chain of Buying: Medical Advertisement, the Bourgeois Public Sphere, and the Origins of the French Revolution." *American Historical Review* 101 (1996): 13–40.
———. "The Military Revolution and the Professionalisation of the French Army under the Ancien Régime." In *The Military Revolution and the State, 1500–1800*. Edited by Michael Duffy. Exeter, U.K.: Exeter Studies in History, 1980.
Joravsky, David. *The Lysenko Affair*. Cambridge: Harvard University Press, 1970.
Julia, Dominique. *Les trois couleurs du tableau noir: La Révolution*. Paris: Belin, 1981.
Julia, Dominique et al. *Atlas de la Révolution française*. Paris: Editions de l'Ecole des Hautes Etudes en Sciences Sociales, 1988.
Kaiser, David. *Politics and War: European Conflict from Philip II to Hitler*. Cambridge: Harvard University Press, 1990.
Kaplan, Steven L. *Provisioning Paris: Merchants and Millers in the Grain and Flour Trade during the Eighteenth Century*. Ithaca: Cornell University Press, 1984.
———. "Réflexions sur la police du monde du travail, 1700–1815." *Revue historique*

261 (1979): 17–77.

———. "Social Classification and Representation in the Corporate World of Eighteenth-Century France: Turgot's 'Carnival.'" *Work in France: Representations, Meaning, Organization, and Practice.* Edited by Steven L. Kaplan and Cynthia J. Koepp. Ithaca: Cornell University Press, 1986.

Kasson, John. *Civilizing the Machine: Technology and Republican Values in America, 1776–1900.* New York: Penguin, 1976.

Keller, Alexander. "Mathematics, Mechanics and the Origins of the Culture of Mechanical Invention." *Minerva* 23 (1985): 348–61.

Kelly, George Armstrong. *Victims, Authority, and Terror: The Parallel Deaths of d'Orléans, Custine, Bailly, and Malesherbes.* Chapel Hill: University of North Carolina Press, 1982.

Kennard, Arthur Norris. *French Pistols and Sporting Guns.* Hamlyn, N.Y.: Country Life Collectors' Guides, 1972.

Kennedy, Paul. *Rise and Fall of the Great Powers: Economic Change and Military Conflict from 1500–2000.* New York: Random House, 1987.

Kennett, Lee. *The French Armies in the Seven Years' War: A Study in Military Organization and Administration.* Durham, N.C.: Duke University Press, 1967.

Kline, Ronald. "Constructing 'Technology' as 'Applied Science': Public Rhetoric of Scientists and Engineers in the U.S., 1880–1945." *Isis* 86 (1995): 194–221.

Koepp, Cynthia J. "The Alphabetic Order: Work in Diderot's *Encyclopédie.*" In *Work in France: Representations, Meaning, Organization, and Practice.* Edited by Steven L. Kaplan and Cynthia J. Koepp. Ithaca: Cornell University Press, 1986.

Koyré, Alexandre. "La dynamique de Niccolò Tartaglia." *La science au seizième siècle, Colloque international de Royaumont.* Paris: Hermann, 1960.

Kranakis, Eda. "The Social Determinants of Engineering Practice: A Comparative View of France and America in the Nineteenth Century." *Social Studies of Science* 19 (1989): 5–70.

Kriedte, Peter, Hans Medick, and Jürgen Schlumbohn. *Industrialization before Industrialization: Rural Industry in the Genesis of Capitalism.* Cambridge: Cambridge University Press, 1981.

Lacombe, Christiane. "Introduction du machinisme dans les fabrications d'armement en France au XIXe siècle." *Revue internationale d'histoire militaire* 41 (1979): 37–49.

Landes, David. "Religion and Enterprise: The Case of the French Textile Industry." In *Enterprise and Entrepreneurs in Nineteenth- and Twentieth-Century France.* Edited by Edward C. Carter II, Robert Foster, and John N. Moody. Baltimore: Johns Hopkins University Press, 1976.

———. *Revolution in Time: Clocks and the Making of the Modern World.* Cambridge: Harvard University Press, 1983.

———. *The Unbound Prometheus: Technological Change and Industrial Development in Western Europe from 1750 to the Present.* Cambridge: Cambridge University Press, 1969.

———. "What Do Bosses Really Do?" *The Journal of Economic History* 46 (1986): 585–623.

Langins, Janis. *La République avait besoin des savants; les débuts de l'Ecole Polytechnique: L'Ecole Centrale des Travaux Publics et les cours Révolutionnaires de l'an III.* Paris: Belin, 1987.

———. "Sur l'enseignement et les examens à l'Ecole Polytechnique sous le Directoire: A propos d'une lettre inédite de Laplace." *Revue d'histoire des sciences* 40 (1987): 145–77.

———. "The Ecole Polytechnique (1794–1804): From Encyclopedic School to Military Institution." Ph.D. diss., University of Toronto, 1979.

———. "Words and Institutions during the French Revolution: The Case of 'Revolutionary' Scientific and Technical Education." In *The Social History of Language*. Edited by Peter Burke and Roy Porter. Cambridge: Cambridge University Press, 1987.

Larousse, Pierre. *Le grand dictionnaire universel du XIXe siècle*. Geneva: Slatkine, [1867], 1982.

Lasseray, André. *Les français sous les treize étoiles, 1775–1783*. Mâcon: Protat Frères, 1935.

Latour, Bruno. "Drawing Things Together." In *Representations in Scientific Practice*. Edited by Michael Lynch and Steve Woolgar. Cambridge: MIT Press, 1990.

———. *Science in Action: How to Follow Scientists and Engineers Through Society*. Cambridge: Harvard University Press, 1987.

———. *The Pasteurization of France*. Translated by Alan Sheridan and John Law. Cambridge: Harvard University Press, 1988.

Lauerma, Matti. *Jacques-Antoine-Hippolyte de Guibert, 1743–1790*. Annales Academiae Scientiarum Fennicae. Series B, vol. 229. Helsinki: Suomalainen Tiedeakatemia, 1989.

———. *L'artillerie de campagne française pendant les guerres de la Révolution: Evolution de l'organisation et de la tactique*. Annales Academiae Scientiarum Fennicae. Series B, vol. 96. Helsinki, 1956.

Laurent, Gustave. "Un mémoire historique du chimiste Hassenfratz." *Annales historiques de la Révolution française* 1 (1924): 163–64.

Law, John. "Technology and Heterogeneous Engineering: The Case of Portuguese Expansion." In *The Social Construction of Technological Systems: New Directions in the Sociology and History of Technology*. Edited by Wiebe E. Bijker, Thomas P. Hughes, and Trevor Pinch. Cambridge: MIT Press, 1987.

Layton, Edwin. "Escape from the Jail of Shape: Dimensionality and Engineering Science." In *Technological Development and Science in the Industrial Age*. Edited by Peter Kroes and Martijn Bakker. Dordrecht: Kluwer, 1992.

———. "Millwrights and Engineers: Science, Social Roles, and the Evolution of the Turbine in America." In *The Dynamics of Science and Technology: Social Values, Technical Norms and Scientific Criteria in the Development of Knowledge*. Edited by Wolfgang Krohn, Edwin Layton, and Peter Weingart. Dordrecht: Reidel, 1978.

———. "Mirror-Image Twins: The Communities of Science and Technology in Nineteenth-Century America." *Technology and Culture* 12 (1971): 562–80.

———. *Revolt of the Engineers: Social Responsibility and the American Engineering Profession*. Baltimore: Johns Hopkins University Press, 1986.

———. "Veblen and the Engineers." *American Quarterly* 14 (1960): 64–72.

Le Puillon de Boblaye, Théodore. *Esquisse historique sur les Ecoles d'Artillerie pour servir à l'histoire de l'Ecole d'Application de l'Artillerie et du Génie*. Metz: Rousseau-Pallez, [1858].

Lefebvre, Georges. *The French Revolution: From 1793 to 1799*. Translated by J. H. Stewart and J. Friguglietti. Vol. 2. New York: Columbia University Press, 1964.

———. "L'affaire du donjon de Vincennes." *Annales historiques de la Révolution française* 22 (1950): 263–65.

Lehning, James R. *The Peasants of Marles: Economic Development and Family Organization in Nineteenth-Century France*. Chapel Hill: University of North Carolina Press, 1980.

Lenk, Torsten. *The Flintlock: Its Origin and Development*. Translated by G. A. Urquatt. London: Holland Press, 1965.

Léon, Antoine. *La Révolution française et l'education technique*. Paris: Société des Etudes Robespierristes, 1968.

Lequin, Yves. *Ouvriers de la région lyonnaise*. Lyon: Presse Universitaire de Lyon, 1977.

Leseure, E. *Historique des mines de houille du département de la Loire*. Saint-Etienne: Théolier, 1901.

Leslie, Stuart W. *The Cold War and American Science: The Military-Industrial-Academic Complex at MIT and Stanford*. New York: Columbia University Press, 1993.

Lespinasse, René de. *Les métiers et corporations de la ville de Paris*. Paris: Imprimerie Nationale, 1892.

Lindqvist, Svante. "Labs in the Woods: The Quantification of Technology During the Late Enlightenment." In *The Quantifying Spirit in the Eighteenth Century*. Edited by Tore Frängsmyr, J. L. Heilbron, and R. Rider. Berkeley: University of California Press, 1990.

Linebaugh, Peter. *The London Hanged: Crime and Civil Society in the Eighteenth Century*. Cambridge: Cambridge University Press, 1992.

Liu, Tessie P. *The Weaver's Knot: The Contradictions of Class Struggle and Family Solidarity in Western France, 1750–1914*. Ithaca: Cornell University Press, 1994.

Lombarès, Michel de. *Histoire de l'artillerie française*. Paris: Charles-Lavauzelle, 1984.

Long, Pamela O. "The Openness of Knowledge: An Ideal and Its Context in Sixteenth-Century Writings on Mining and Metallurgy." *Technology and Culture* 32 (1991): 318–55.

Longino, Helen E. *Science as Social Knowledge: Values and Objectivity in Scientific Inquiry*. Princeton: Princeton University Press, 1990.

Lubar, Steve. "Representation and Power." *Technology and Culture* 36 (1995): S54-S81.

Lucas, Colin. "Nobles, Bourgeois and the Origins of the French Revolution." *Past and Present* 60 (1973): 84–126.

———."Revolutionary Violence, the People, and the Terror." In *The Terror*. Vol. 4. *The French Revolution and the Creation of Modern Political Culture*. Edited by Keith Michael Baker. Oxford: Elsevier, 1994.

———. *The Structure of the Terror: The Example of Javogues and the Loire*. Oxford: Oxford University Press, 1973.

Lundgreen, Peter. "Engineering Education in Europe and the U.S.A., 1750–1930: The Rise to Dominance of School Culture and the Engineering Profession." *Annals of Science* 47 (1990): 33–75.

Lynch, Michael. "Discipline and the Material Form of Images: An Analysis of Scientific Visibility." *Social Studies of Science* 15 (1985): 37–65.

Lynn, John A. "En avant! The Origins of the Revolutionary Attack." In *Tools of War: Instruments, Ideas and Institutions of Warfare, 1445–1871*. Edited by John A. Lynn. Urbana: University of Illinois Press, 1990.

———. "French Opinion and the Military Resurrection of the Pike, 1792–1794." *Mil-*

itary Affairs 41 (1977): 1–7.
———. "Recalculating French Army Growth during the Grand Siècle, 1610–1715." *French Historical Studies* 18 (1994): 881–906.
———. *The Bayonets of the Republic: Motivation and Tactics in the Army of Revolutionary France, 1791–94*. Urbana: University of Illinois Press, 1984.
———. "The *trace italienne* and the Growth of Armies: The French Case." *Journal of Military History* 55 (1991): 297–330.
MacKenzie, Donald. *Inventing Accuracy: An Historical Sociology of Nuclear Missile Guidance*. Cambridge: MIT Press, 1990.
MacKenzie, Donald, and Graham Spinardi. "Tacit Knowledge, Weapons Design and the Uninvention of Nuclear Weapons." *American Journal of Sociology* 101 (1995): 44–99.
Maguin, Paul. *Les armes de Saint-Etienne*. Saint-Etienne: Maguin, 1990.
Marglin, Stephen. "What Do Bosses Do?: Part I." *The Review of Radical Political Economy* 6 (1974): 60–112.
Marx, Leo. *The Machine in the Garden: Technology and the Pastoral Ideal in America*. New York: Oxford University Press, 1964.
Mathias, Peter. "Skills and the Diffusion of Innovations from Britain in the Eighteenth Century. " In *The Transformation of England*. New York: Columbia University Press, 1979.
Mathiez, Albert. "Compte Rendue." Review of *Le Comité de Salut Public et les fabrications de guerre sous la Terreur* by Camille Richard. *Annales Révolutionnaires* 14 (1922): 166–68.
———. *La victoire en l'an II: Esquisses sur la défense nationale*. Paris: Félix Alcan, 1916.
———. "Les divisions dans les comités de gouvernement à la veille du 9 thermidor d'après quelques documents inédits." *Revue historique* 118 (1915): 70–87.
———. *The French Revolution*. Translated by C. A. Phillips. New York: Russell and Russell, [1922–27], 1962.
Maza, Sarah. "Luxury, Morality, and Social Change: Why There was No Middle Class Consciousness in Prerevolutionary France." Unpublished typescript, 1995.
———. "Politics, Culture, and the Origins of the French Revolution." *Journal of Modern History* 61 (1989): 704–23.
———. *Private Lives and Public Affairs: The Causes Célèbres of Prerevolutionary France*. Berkeley: University of California Press, 1994.
McConnell, David. *British Smooth-Bore Artillery: A Technological Study*. Ottawa: Minister of the Environment, 1988.
McKendrick, Neil. "Commercialization and the Economy." In *The Birth of a Consumer Society: The Commercialization of Eighteenth-Century England*. Edited by Neil McKendrick, John Brewer, and J. H. Plumb. Bloomington: Indiana University Press, 1982.
McNeill, William H. *The Pursuit of Power: Technology, Armed Force, and Society since A.D. 1000*. Chicago: University of Chicago Press, 1982.
Mention, Léon. *Le comte de Saint-Germain et ses réformes*. Paris: Clavel, 1884.
Merley, Jean. "La situation de bassin houiller stéphanois en 1783." *Bulletin du Centre d'Histoire Régionale de Saint-Etienne* 2 (1977): 55–69.
———. "Réflexions sur le cours des grains à Saint-Etienne et au Puy aux XVIIe et XVIIIe siècles (1660–1793)." *Bulletin du Centre d'Histoire Régionale, Université de Saint-Etienne* 1 (1981): 37–70.

Mérot, Catherine. "Le recruitment des Ecoles Centrales sous la Révolution." *Revue historique* 274 (1985): 357–85.

Merton, Robert K. *Science, Technology and Society in Seventeenth-Century England*. New York: Howard Fertig, [1938], 1970.

Mervaud, Christiane. *Voltaire et Frédéric II: Une dramaturgie des lumières, 1736–1778*. Volume 234 in *Studies on Voltaire and the Eighteenth Century*. Oxford: The Voltaire Foundation, 1985.

Meyer, Stephen. *The Five Dollar Day: Labor Management and Social Control in the Ford Motor Company, 1908–1921*. Albany: State University of New York Press, 1981.

Meyssonnier, Simone. *La balance et l'horloge: La genèse de la pensée libérale en France au XVIIIe siècle*. Montreuil: Editions de la Passion, 1989.

Michelet, Jules. *Histoire de la Révolution française*. Paris: Chamerot, 1847–53.

Misa, Thomas. "Military Needs, Commercial Realities, and the Development of the Transistor, 1948–1958." In *Military Enterprise and Technological Change: Perspectives on the American Experience*. Edited by Merritt Roe Smith. Cambridge: MIT Press, 1985.

Mizroeff, Nicolas. "Pictorial Sign and Social Order: L'Académie Royale de Peinture et Sculpture, 1638–1752." Ph.D. diss., University of Warwick, 1990.

Mokyr, Joel. "Technological Inertia in Economic History." *Journal of Economic History* 52 (1992): 325–38.

———. *The Lever of Riches: Technological Creativity and Economic Progress*. New York: Oxford University Press, 1990.

Molloy, Peter Michael. "Technical Education and the Young Republic: West Point as America's Ecole Polytechnique, 1802–1833." Ph.D. diss., Brown University, 1975.

Morison, Elting E. "Gunfire at Sea: A Case Study of Innovation." *Men, Machines, and Modern Times*. Cambridge: MIT Press, 1966.

Mornet, Daniel. *Les origines intellectuelles de la Révolution française*. Paris: Colin, 1933.

Motley, Mark. *Becoming a French Aristocrat: The Education of the Court Nobility, 1580–1715*. Princeton: Princeton University Press, 1990.

Mousnier, Roland. *Les hiérarchies sociales de 1450 à nos jours*. Paris: Presses Universitaires de France, 1969.

Moutet, Aimée. "Introduction de la production à la chaine en France du début du XXe siècle à la grande crise de 1930." *Histoire, économie et société* 6 (1983): 63–82.

Mumford, Lewis. *Technics and Civilization*. New York: Harcourt, 1934.

Musson, Albert Edward, and Eric Robinson. *Science and Technology in the Industrial Revolution*. Manchester: Manchester University Press, 1969.

Nardin, Pierre. *Gribeauval, Lieutenant-général des armées du roi, 1715–1789*. Paris: La Fondation pour les Etudes de Défense Nationale, 1982.

———. "Le Comité de l'Artillerie et la réalisation des matériels d'armement dès origines à 1870." *Memorial de l'artillerie française* 50 (1976): 471–504.

Nef, John U. *War and Human Progress: An Essay on the Rise of Industrial Civilization*. Cambridge: Harvard University Press, 1950.

Noble, David. *America by Design: Science, Technology, and the Rise of Corporate Capitalism*. Oxford: Oxford University Press, 1977.

———. "Command Performance: A Perspective on the Social and Economic Consequences of Military Enterprise." In *Military Enterprise and Technological Change: Perspectives on the American Experience*. Edited by Merritt Roe Smith. Cambridge: MIT Press, 1985.

———. *Forces of Production: A Social History of Industrial Automation.* New York: Knopf, 1984.

Nosworthy, Brent. *The Anatomy of Victory: Battle Tactics, 1689–1763.* New York: Hippocrene, 1990.

O'Brien, Patrick, and Caglar Keyder. *Economic Growth in Britain and France, 1780–1940: Two Paths to the Twentieth Century.* London: Allen and Unwin, 1978.

Osborne, Michael. "Applied Natural History and Utilitarian Ideals: 'Jacobin Science' at the Muséum d'Histoire Naturelle, 1789–1870." In *Re-creating Authority in Revolutionary France.* Edited by Bryant T. Ragan and Elizabeth Williams. New Brunswick, N.J.: Rutgers University Press, 1992.

Outram, Dorinda. "The Ordeal of Vocation: The Paris Academy of Sciences and the Terror, 1793–95." *History of Science* 21 (1983): 251–73.

Ozouf, Mona. *Festivals and the French Revolution, 1789–1799.* Translated by Alan Sheridan. Cambridge: Harvard University Press, 1988.

———. "The Terror after the Terror? An Immediate History." In *The Terror.* Vol. 4. *The French Revolution and the Creation of Modern Political Culture.* Edited by Keith Michael Baker. Oxford: Elsevier, 1994.

Palmer, R. R. "Frederick II, Guibert, Bülow: From Dynastic to National War." *Makers of Modern Strategy from Machiavelli to the Nuclear Age.* Edited by Peter Paret. Princeton: Princeton University Press, 1986.

———. "The National Idea in France before the Revolution." *Journal of the History of Ideas* 1 (1940): 95–111.

Parent, Michel, and Jacques Verroust. *Vauban.* Paris: Fréal, 1971.

Paret, Peter. "Napoleon." In *Makers of Modern Strategy from Machiavelli to the Nuclear Age.* Edited by Peter Paret. Princeton: Princeton University Press, 1986.

Parker, Geoffrey. *The Military Revolution: Military Innovation and the Rise of the West, 1500–1800.* Cambridge: Cambridge University Press, 1988.

Parker, Harold T. *The Bureau of Commerce in 1781 and Its Policies with Respect to French Industry.* Durham, N.C.: Carolina Academic Press, 1979.

Passac, Philibert-Jérome Gaucher. "Précis sur M. de Gribeauval," [1816]. In *Revue d'artillerie* 34 (1889): 98–109.

Patrick, Alison. *The Men of the First French Republic: Political Alignments in the National Convention.* Baltimore: Johns Hopkins University Press, 1972.

Paulin-Desormeaux, A. *Manuel de l'armurier, du fourbisseur et de l'arquebusier.* Paris: Roret, 1832.

Pauze, André. "L'émeute des ouvriers canonniers en 1831." *Bulletin des Amis du Vieux Saint-Etienne* (1991): 59–62.

Payen, Jacques. *Capital et machine à vapeur au XVIIIe siècle: Les frères Périer et l'introduction en France de la machine à vapeur de Watt.* Paris: Mouton, 1969.

Pemberton, David McKinney. "Industrialization and the Bourgeoisie in Nineteenth-Century France: The Experience of Saint-Etienne, 1820–1872." Ph.D. diss., Rutgers State University, 1984.

Perkin, Harold. *The Origins of Modern English Society.* London: Ark, 1969.

———.*The Third Revolution: Professional Elites in the Modern World.* London: Routledge, 1996.

Perrin, Noel. *Giving Up the Gun: Japan's Reversion to the Sword, 1543–1879.* Boston: Godine, 1979.

Peyrière, Monique. "Recherches sur la machine à coudre en France, 1830–1889."

Mémoire de maîtrise, Ecole des Hautes Etudes en Sciences Sociales, 1990.

Picard, Ernest, and Louis Jouan. *Artillerie française au XVIIIe siècle*. Paris: Berger-Levrault, 1906.

Pickering, Andrew, ed. *Science as Practice and Culture*. Chicago: University of Chicago Press, 1991.

Picon, Antoine. *Architectes et ingénieurs au siècle des lumières*. Marseilles: Parenthèses, 1988.

———. "Gestes ouvriers, opération et processus technique: La vision de travail des encyclopédistes." *Recherches sur Diderot et sur l'Encyclopédie* 13 (1992): 131–47.

———. *L'invention de l'ingénieur moderne: L'Ecole des Ponts et Chaussées, 1747–1851*. Paris: Presses de l'Ecole Nationale des Ponts et Chaussées, 1992.

Pinch, Trevor J., and Wiebe Bijker. "The Social Construction of Technological Facts and Artifacts: Or How the Sociology of Science and the Sociology of Technology Might Benefit Each Other." In *The Social Construction of Technological Systems: New Directions in the Sociology and History of Technology*. Edited by Wiebe Bijker, Thomas P. Hughes, and Trevor Pinch. Cambridge: MIT Press, 1987.

Piore, M. J. "Dualism as a Response to Flux and Uncertainty" and "The Technological Foundations of Dualism and Discontinuity." In *Dualism and Discontinuity in Industrial Societies*. Edited by S. Berger and M. J. Piore. Cambridge: Cambridge University Press, 1980.

Poisson, Georges. *Choderlos de Laclos, ou l'obstination*. Paris: Grasset, 1985.

Polanyi, Michael. *Personal Knowledge*. London: Routledge and Kegan Paul, 1958.

———. *The Tacit Dimension*. London: Routledge and Kegan Paul, 1967.

Porter, Theodore. *Trust in Numbers: Objectivity in Science and Public Life*. Princeton: Princeton University Press, 1995.

Pothier, Francisque. *Roanne pendant la Révolution, 1789–1796*. Roanne: Durand, 1868.

Potofsky, Allan. "Work and Citizenship: Crafting Images of Revolutionary Builders, 1789–1791." In *The French Revolution and the Meaning of Citizenship*. Edited by Renée Waldinger, Philip Dawson, and Isser Woloch. Westport, Conn.: Greenwood Press, 1993.

Pouchet, Georges. *Les sciences pendant la Terreur d'après les documents du temps et les pièces des Archives Nationales*. Edited by James Guillaume. Paris: Société de l'Histoire de la Révolution Française, [1870], 1896.

Pritchard, James. *Louis XV's Navy, 1748–1762: A Study of Organization and Administration*. Kingston, Ont.: McGill-Queen's University Press, 1987.

Proctor, Robert. *Value-Free Science: Purity and Power in Modern Knowledge*. Cambridge: Harvard University Press, 1991.

Quimby, Robert S. *The Background of Napoleonic Warfare: The Theory of Military Tactics in Eighteenth-Century France*. New York: Columbia University Press, 1957.

Rabinbach, Anson. *The Human Motor: Energy, Fatigue and the Origins of Modernity*. New York: Basic Books, 1990.

Ragan, Bryant T., and Elizabeth Williams, eds. *Re-creating Authority in Revolutionary France*. New Brunswick, N.J.: Rutgers University Press, 1992.

Randall, Adrian. "Work and Resistance in the West of England." In *Regions and Industries: A Perspective on the Industrial Revolution in Britain*. Edited by Pat Hudson. Cambridge: Cambridge University Press, 1989.

Reddy, William. "The Moral Sense of Farce: The Patois Literature of Lille Factory

Laborers, 1848–70." In *Work in France: Representations, Meaning, Organization, and Practice*. Edited by Steven L. Kaplan and Cynthia J. Koepp. Ithaca: Cornell University Press, 1986.

——— *The Rise of Market Culture, The Textile Trade and French Society, 1750–1900*. Cambridge: Cambridge University Press, 1984.

Reinhard, Marcel. *Le Grand Carnot, De l'ingénieur au conventionnel, 1753–1792*. Vol. 1. Paris: Hachette, 1950.

———. *Le Grand Carnot, L'organisateur de la victoire, 1792–1823*. Vol. 2. Paris: Hachette, 1952.

Reynolds, Terry. *Stronger than a Thousand Men: A History of the Vertical Water Wheel*. Baltimore: Johns Hopkins University Press, 1983.

Richard, Camille. *Le Comité de Salut Public et les fabrications de guerre sous la Terreur*. Paris: Rieder, 1921.

Richards, W. A. "The Importance of Firearms in West Africa in the Eighteenth Century." *Journal of African History* 21 (1980): 43–59.

Rider, Robin. "Measure of Ideas, Rules of Language: Mathematics and Language in the 18th Century." In *The Quantifying Spirit in the Eighteenth Century*. Edited by Tore Frängsmyr, J. L. Heilbron, and R. Rider. Berkeley: University of California Press, 1990.

Riley, James C. *The Seven Years' War and the Old Regime in France: The Economic and Financial Toll*. Princeton: Princeton University Press, 1986.

Riskin, Jessica. "The Quarrel over Method in Natural Science and Politics during the Late Enlightenment." Ph.D. diss., University of California at Berkeley, 1995.

Roche, Daniel. *Le siècle des lumières en province: Académies et académiciens provinciaux, 1680–1789*. Paris: Mouton, 1978.

———.*The Culture of Clothing: Dress and Fashion in the "Ancien Régime."* Translated by Jean Birrell. Cambridge: Cambridge University Press, 1993.

Roe, J. W. *English and American Tool Builders*. New Haven: Yale University Press, 1916.

Roehl, Richard. "French Industrialization: A Reconsideration." *Explorations in Economic History* 13 (1976): 233–81.

Rolt, L.T.C. *A Short History of Machine Tools*. Cambridge: MIT Press, 1965.

Rose, R. B. *The Enragés: Socialists of the French Revolution?* Sydney: Sydney University Press, 1965.

———. *The Making of the Sans-Culottes: Democratic Ideas and Institutions in Paris, 1789–1792*. Manchester: Manchester University Press, 1983.

Rosen, Howard. "The Système Gribeauval: A Study of Technological Development and Institutional Change in Eighteenth-Century France." Ph.D. diss., University of Chicago, 1981.

Rosenberg, Nathan, ed. and intro. *Great Britain and the American System of Manufacturing*. Edinburgh: Edinburgh University Press, 1969.

Ross, Steven T. *From Flintlock to Rifle: Infantry Tactics, 1740–1866*. London: Associated University Press, 1979.

———. "The Development of the Combat Division in Eighteenth-Century French Armies." *French Historical Studies* 4 (1965–66): 84–94.

Rossi, Paolo. *Philosophy, Technology and the Arts in the Early Modern Era*. Translated by Salvator Attanasio. New York: Harper and Row, 1970.

Rousseau, Jean. "La vie quotidienne dans les manufacture d'armes à l'époque de la

Révolution et de l'Empire." Mémoire de maîtrise, IVe Section de l'Ecole Pratique des Hautes Etudes, 1969. (S.H.A.T. Bib. #8402)

Rowlands, Marie B. *Masters and Men in the West Midland Metalware Trades before the Industrial Revolution*. Manchester: Manchester University Press, 1975.

Rudé, George. *The Crowd in the French Revolution*. Oxford: Clarendon Press, 1959.

Rule, John. "The Property of Skill in the Period of Manufactures." In *The Historical Meanings of Work*. Edited by Patrick Joyce. Cambridge: Cambridge University Press, 1987.

Sabel, Charles. *Work and Politics: The Division of Labor in Industry*. Cambridge: Cambridge University Press, 1982.

Sabel, Charles, and Jonathan Zeitlin. "Historical Alternatives to Mass Production: Politics, Markets and Technology in Nineteenth-Century Industrialization." *Past and Present* 108 (1986): 133–76.

———. "Stories, Strategies, Structures: Rethinking Alternatives to Mass Production." Introduction to *Worlds of Possibility: Flexibility and Mass Production in Western Industrialization*, forthcoming. Typescript published by New York: Center for Law and Economic Studies at Columbia School of Law, 1993.

Safford, Frank. *The Ideal of the Practical: Columbia's Struggle to Form a Technical Elite*. Austin: University of Texas Press, 1976.

Sakarovitch, Joël. "La géométrie descriptive, Une reine déchue." In *La formation polytechnicienne, 1794–1994*. Edited by Bruno Belhoste, Amy Dahan Dalmedico, and Antoine Picon. Paris: Dunod, 1994.

Sawyer, John E. "The Entrepreneur and the Social Order: France and the United States." *Men In Business*. Edited by A. H. Cole and W. Miller. Cambridge: Harvard University Press, 1952.

———. "The Social Basis of the American System of Manufacturing." *Journal of Economic History* 14 (1954): 361–79.

Schaffer, Simon. "Enlightenment and the Automaton." In *The Sciences in Enlightened Europe*. Edited by William Clark, Jan Golinski, and Simon Schaffer. Chicago: Chicago University Press, forthcoming

———. "Natural Philosophy." In *The Ferment of Knowledge: Studies in the Historiography of Eighteenth-Century Science*. Edited by Roy Porter. Cambridge: Cambridge University Press, 1980.

Schneider, Mark. "Gerard Desargues, The Architectural and Perspective Geometry: A Study in the Rationalization of Figure." Ph.D. diss., Virginia Polytechnic Institute, 1984.

Scott, Melissa. "The Victory of the Ancients: Tactics, Technology, and the Use of Classical Precedent." Ph.D. diss., Brandeis University, 1992.

Scott, Samuel. "The French Revolution and Professionalization of the French Officer Corps: 1789–93." In *On Military Ideology*. Edited by Morris Janowitz and Jacques van Doorn. Rotterdam: Rotterdam University Press, 1971.

———. "The Regeneration of the Line Army during the French Revolution." *Journal of Modern History* 42 (1970): 307–30.

———. *The Response of the Royal Army to the French Revolution*. Oxford: Oxford University Press, 1978.

Scranton, Philip. "Diversity in Diversity: Flexible Production and American Industrialization, 1880–1930." *Business History Review* 65 (1991): 27–90.

Séris, Jean-Pierre. *Machine et communication: Du théâtre des machines à la mécanique industrielle*. Paris: Vrin, 1987.

Sewell, William H., Jr. *A Rhetoric of Bourgeois Revolution: The Abbé Sieyes and "What is the Third Estate?"* Durham, N.C.: Duke University Press, 1994.

———. "Etats, Corps, and Ordre: Some Notes on the Social Vocabulary of the French Old Regime." In *Sozialgeschichte Heute: Festschrift für Hans Rosenberg zum 70. Geburtstag*. Edited by Hans-Ulrich Wehler. Göttingen: Vandenhoeck und Ruprecht, 1974.

———. "Ideologies and Social Revolutions: Reflections on the French Case." *Journal of Modern History* 57 (1985): 57–85.

———. "The Sans-culotte Rhetoric of Subsistence." In *The Terror*. Vol. 4. *The French Revolution and the Creation of Modern Political Culture*. Edited by Keith Michael Baker. Oxford: Elsevier, 1994.

———. "Visions of Labor: Illustrations of the Mechanical Arts before, in, and after Diderot's *Encyclopédie*." In *Work in France: Representations, Meaning, Organization, and Practice*. Edited by Steven L. Kaplan and Cynthia Koepp. Ithaca: Cornell University Press, 1986.

———. *Work and Revolution in France: The Language of Labor from the Old Regime to 1848*. Cambridge: Cambridge University Press, 1980.

Shapin, Steven. *A Social History of Truth: Civility and Science in Seventeenth-Century England*. Chicago: University of Chicago Press, 1994.

———. "The House of Experiment in Seventeenth-Century England." *Isis* 79 (1988): 373–404.

Shapin, Steven, and Simon Schaffer. *Leviathan and the Air-Pump: Hobbes, Boyle, and the Experimental Life*. Princeton: Princeton University Press, 1985.

Shelby, Lon. "The Geometrical Knowledge of the Medieval Master Masons." *Speculum* 47 (1972): 395–421.

Sherry, Michael. *In the Shadow of War: The United States Since the 1930s*. New Haven: Yale University Press, 1995.

Shinn, Terry. *L'Ecole Polytechnique, 1794–1914*. Paris: Presses de la Fondation Nationale des Sciences Politiques, 1980.

Silberner, Edmond. *La guerre dans la pensée économique du XVIe au XVIIIe siècle*. Paris: Sirey, 1939.

Simon, Herbert A. *The Sciences of the Artificial*. Cambridge: MIT Press, 1969.

Skinner, Quentin. *Reason and Rhetoric in the Philosophy of Hobbes*. Cambridge: Cambridge University Press, 1996.

Skocpol, Theda. *States and Social Revolution: A Comparative Analysis of France, Russia, and China*. Cambridge: Cambridge University Press, 1979.

Skocpol, Theda, and Meyer Kestnbaum. "Mars Unshackled: The French Revolution in World-Historical Perspective." In *The French Revolution and the Birth of Modernity*. Edited by Ferenc Fehér. Berkeley: University of California Press, 1990.

Smeaton, W. A. *Fourcroy, Chemist and Revolutionary, 1755–1809*. Cambridge, U.K.: Heffer, 1962.

Smith, Jay Michael. *The Culture of Merit: Nobility, Royal Service, and the Making of Absolute Monarchy in France, 1600–1789*. Ann Arbor: University of Michigan Press, 1996.

Smith, Merritt Roe. "Army Ordnance and the 'American System' of Manufacturing,

1815–1861." In *Military Enterprise and Technological Change: Perspectives on the American Experience*. Edited by Merritt Roe Smith. Cambridge: MIT Press, 1985.

———. *Harpers Ferry Armory and the New Technology: The Challenge of Change*. Ithaca: Cornell University Press, 1979.

Soboul, Albert. *Les sans-culottes parisiens en l'an II, Histoire politique et sociale des sections de Paris, 2 juin 1793–9 thermidor an II*. La Roches-sur-Yon: Potier, 1958.

Sombart, Werner. *Krieg und Kapitalismus*. New York: Arno Press, [1913], 1975.

Sonenscher, Michael. *The Hatters of Eighteenth-Century France*. Berkeley: University of California Press, 1987.

———. "The Sans-culottes of the Year II: Rethinking the Language of Labor in Revolutionary France." *Social History* 9 (1984): 301–28.

———. *Work and Wages: Natural Law, Politics and the Eighteenth-Century French Trades*. Cambridge: Cambridge University Press, 1989.

Sorel, Albert. *L'Europe et la Révolution française*. Paris, 1891.

Speier, Hans. "Militarism in the Eighteenth Century." *Social Research* 3 (1936): 304–36.

Stafford, Barbara. *Body Criticism: Imaging the Unseen in Enlightenment Art and Medicine*. Cambridge: MIT Press, 1991.

Starobinsky, Jean. *1789: The Emblems of Reason*. Translated by Barbara Bray. Charlottesville: University Press of Virginia, 1982.

Staum, Martin. "Science and Government in the French Revolution." In *Science, Technology and Culture in Historical Perspective*. Edited by Louis A. Knafla et al. Calgary: University of Calgary Press, 1976.

Steele, Brett D. "Muskets and Pendulums: Benjamin Robins, Leonard Euler, and the Ballistics Revolution." *Technology and Culture* 35 (1994): 348–82.

———. "The Ballistics Revolution: Military and Scientific Change from Robins to Napoléon." Ph.D. diss., University of Minnesota, 1994.

Stone, Bailey. *The Genesis of the French Revolution: A Global-Historical Interpretation*. Cambridge: Cambridge University Press, 1994.

Sturgill, Claude C. "The French Army's Budget in the Eighteenth Century: A Retreat from Loyalty." In *The French Revolution in Culture and Society*. Edited by David G. Troyansky, Alfred Cismaru, and Norwood Andrews, Jr. New York: Greenwood Press, 1991.

Styles, John. "Manufacture, Consumption, and Design in Eighteenth-Century England." In *Consumption and the World of Goods*. Edited by John Brewer and Roy Porter. London: Routledge, 1993.

Suleiman, Ezra N. *Elites in French Society: The Politics of Survival*. Princeton: Princeton University Press, 1978.

Taine, Hippolyte-Adolphe. *La Révolution*. Paris: Hachette, 1878–85.

Taton, René. *L'oeuvre scientifique de Gaspard Monge*. Paris: Presses Universitaires de France, 1951.

Taylor, George. "Non-capitalist Wealth and the Origins of the French Revolution." *American Historical Review* 72 (1967): 469–96.

Thomas, Selma. "'La plus grande économie et la précision la plus exacte,' L'oeuvre d'Honoré Blanc." *Le Musée d'Armes, Bulletin trimestriel des Amis du Musée d'Armes de Liège* 7 (1979): 1–4.

Thompson, E. P. *The Making of the English Working Class*. New York: Vintage Books, 1963.

———. "Time, Work-Discipline, and Industrial Capitalism." *Past and Present* 38 (1967): 56–97.

Tilly, Charles. *Coercion, Capital, and European States, A.D. 990–1992*. Rev. ed. Cambridge, Mass.: Blackwell, 1992.

———. "War-Making and State-Making as Organized Crime." In *Bringing the State Back In*. Edited by Peter B. Evans, Dietrich Rueschemeyer, and Theda Skocpol. Cambridge: Cambridge University Press, 1985.

Tocqueville, Alexis de. *The Old Régime and the French Revolution*. Translated by Stuart Gilbert. Garden City, N.Y.: Doubleday, [1856], 1955.

Tomas, F. "Problèmes de démographie historique: Le Forez au XVIIIe siècle." *Cahiers d'histoire* 13 (1968): 381–99.

Tomlinson, H. C. *Guns and Government: The Ordnance Department under the Later Stuarts*. London: Royal Historical Society, 1979.

Tønnesson, Kåre D. *La défaite des sans-culottes: Mouvement populaire et réaction bourgeoise de l'an III*. Drammen, Norway: Presses Universitaires d'Oslo, 1959.

Trebilcock, Clive. "'Spin-Off' in British Economic History: Armaments and Industry, 1760–1914." *Economic History Review* 2d ser., 22 (1969): 474–90.

Tresse, René. "Les dessinateurs du Comité de Salut Public." *Techniques et civilisations* 5 (1956): 1–10.

———. "Un atelier pilote de la Révolution française: L'Atelier de Perfectionnement des Armes Portatives, mai–décembre 1794." *Techniques et civilizasions* 5 (1956): 54–64.

Truant, Cynthia. "Independent and Insolent: Journeymen and Their 'Rites' in the Old Regime Workplace." In *Work in France: Representations, Meaning, Organization, and Practice*. Edited by Steven L. Kaplan and Cynthia J. Koepp. Ithaca: Cornell University Press, 1986.

Truesdell, Clifford. *Essays in the History of Mechanics*. Berlin: Springer-Verlag, 1968.

van Creveld, Martin. *Technology and War: From 2000 B.C. to the Present*. New York: The Free Press, 1989.

Veblen, Thorstein. *The Engineers and the Price System*. New York: Huebsch, 1921.

Vérin, Hélène. *La gloire des ingénieurs: L'intelligence technique du XVIe au XVIIIe siècle*. Paris: Albin, 1993.

Versini, Laurent. "Un maître de forges au service des Lumières et de la Révolution américaine: François Ignace de Wendel, 1741–1795." *Studies on Voltaire and the Eighteenth Century* 155 (1976): 2137–2206

Vincenti, Walter G. "Engineering Knowledge, Type of Design, and Level of Hierarchy: Further Thoughts about *What Engineers Know*." In *Technological Development and Science in the Industrial Age*. Edited by Peter Kroes and Martijn Bakker. Dordrecht: Kluwer, 1992.

———. *What Engineers Know and How They Know It: Analytical Studies from Aeronautical History*. Baltimore: Johns Hopkins University Press, 1990.

Wallace, Anthony F. C. *The Social Context of Innovation: Bureaucrats, Families, and Heroes in the Early Industrial Revolution, as Foreseen in Bacon's "New Atlantis."* Princeton: Princeton University Press, 1982.

Weber, Max. *Economy and Society: An Outline of Interpretive Sociology*. Berkeley: University of California Press, 1968.

———. *From Max Weber*. Edited and translated by H. H. Gerth and C. Wright Mills. New York: Oxford University Press, 1946.

Weiss, John Hubbel. "Bridges and Barriers: Narrowing Access and Changing Structure in the French Engineering Profession, 1800–1850." In *Professions and the French State*. Edited by Gerald Geison. Philadelphia: University of Pennsylvania Press, 1984.

———. *The Making of Technological Man: The Social Origins of French Engineering Education*. Cambridge: MIT Press, 1982.

Werner, Stephen. *Blueprint: A Study of Diderot and the Encyclopédie Plates*. Birmingham, Ala.: Summa, 1993.

Westfall, Robert. "Science and Technology during the Scientific Revolution: An Empirical Approach." *Renaissance and Revolution: Humanists, Scholars, Craftsmen, and Natural Philosophers in Early Modern Europe*. Edited by J. V. Field and Frank A.J.L. James. Cambridge: Cambridge University Press, 1993.

Wetzler, Peter. *War and Subsistence: The Sambre and Meuse Army in 1794*. New York: Lang, 1985.

Williams, L. Pierce. "The Politics of Science in the French Revolution." In *Critical Problems in the History of Science*. Edited by Marshall Clagett. Madison: University of Wisconsin Press, 1959.

Williams, Rosalind. "Cultural Origins and Environmental Implications of Large Technological Systems." *Science in Context* 6 (1993): 377–403.

Willmouth, Frances. "Mathematical Sciences and Military Technology: The Ordnance Office in the Reign of Charles II." In *Renaissance and Revolution: Humanists, Scholars, Craftsmen, and Natural Philosophers in Early Modern Europe*. Edited by J. V. Field and Frank A.J.L. James. Cambridge: Cambridge University Press, 1993.

Winner, Langdon. "Do Artifacts Have Politics?" In *The Whale and the Reactor*. Chicago: University of Chicago Press, 1986.

Wise, M. Norton, ed. *Values of Precision*. Princeton: Princeton University Press, 1995.

———. "Meditations: Enlightenment Balancing Acts, or the Technologies of Rationalism." In *World Changes: Thomas Kuhn and the Nature of Science*. Edited by Paul Horwich. Cambridge: MIT Press, 1993.

Woloch, Isser. "Napoleonic Conscription: State Power and Civil Society." *Past and Present* 111 (1986): 101–29.

———. "The Right to Primary Education in the French Revolution: From Theory to Practice." In *The French Revolution and the Meaning of Citizenship*. Edited by Renée Waldinger, Philip Dawson, and Isser Woloch. Westport, Conn.: Greenwood Press, 1993.

Woodbury, Robert S. *Studies in the History of Machine Tools*. Cambridge: MIT Press, 1972.

———. "The Legend of Eli Whitney and Interchangeable Parts." *Technology and Culture* 1 (1959): 235–53.

Woronoff, Denis. *L'industrie sidérurgique en France pendant la Révolution et l'Empire*. Paris: Editions des Hautes Etudes en Science Sociales, 1984.

Wright, John. "Sieges and Customs of War at the Opening of the Eighteenth Century." *American Historical Review* 39 (1934): 629–44.

Zeitlin, Jonathan. "Flexibility and Mass Production at War: Aircraft Manufacture in Britain, the United States, and Germany, 1939–1945." *Technology and Culture* 36 (1995): 46–79.

Zilsel, Edgar. "The Sociological Roots of Science." *The American Journal of Sociology* 47 (1942): 544–62.

INDEX